# Techniques of Radar Reflectivity Measurement

# Techniques of Radar Reflectivity Measurement

edited by
Nicholas C. Currie

ARTECH HOUSE, INC.
610 Washington St.
Dedham, MA 02026



International Standard Book Number: 0-89006-131-9
Library of Congress Catalog Card Number: 83-072777

## TABLE OF CONTENTS

## PREFACE

There have been a number of books written in recent years on various aspects of radar, but there have been only a very few written on the subject of radar reflectivity measurement techniques, even though a great deal of work has been going on in this area for many years. This book attempts to help fill this void. The material was originally prepared as a text for a continuing education course in radar reflectivity measurement techniques taught at Georgia Tech, and thus is intended to serve as both a tutorial for the relative novice and as a reference for the experienced practitioner. Each of the contributing authors has been involved in radar reflectivity experiments for many years, and thus is uniquely qualified to present the material from a fully practical point of view.

The book is organized such that the first six chapters present a tutorial of radar measurement techniques including: experimental procedures, calibration, recording techniques, and data analysis procedures. The remaining nine chapters present special topics such as bistatic measurements, far field measurement ranges, compact ranges, scale model measurements, experiment planning, ground truth measurements, radar angle measurements, and even radiometric (passive RCS) measurements. Useful formulas for RCS measurement calculations are sprinkled liberally throughout the book, and references are included for other papers and texts which can provide more detailed information on a particular subject.

It is hoped that this book will be useful in assisting both newcomers to the field and experienced hands in furthering the very important science of radar reflectivity measurement.

*Nicholas C. Currie*
*August, 1983*

# CHAPTER 1
# INTRODUCTION TO RADAR REFLECTIVITY TECHNIQUES

**Nicholas C. Currie**
**Engineering Experiment Station**
**Georgia Institute of Technology**



## 1.1 OVERVIEW

The properties of the interaction of electromagnetic energy with matter have been of considerable interest to scientists and engineers ever since Hertz first transmitted and received an electromagnetic signal in his laboratory in 1886. Since World War II, applications of radar (*ra*dio *d*etection *a*nd *r*anging) have proliferated to the point where thousands of radar systems are in current use for both civilian and military applications. Today's radar systems are very sophisticated, utilizing time domain and frequency processing techniques to extract the characteristics of targets of interest from noise and clutter. However, a radar (no matter how simple or complex) can only process the information which is imposed on a transmitted waveform when it interacts with a physical body which scatters energy back to the radar receiver. For this reason, it is extremely important to know the scattering properties of both desired targets and targets of no interest (called clutter) to assist in designing radar systems. This explains the large amount of effort which has been put into the characterization of radar reflectivity over the last 40 years.

The basic problem of radar reflectivity measurement is illustrated in Figure 1-1. The desired goal is to accurately describe the interaction mechanism between a target and the electromagnetic energy that may impinge upon it. This is a complex problem, even if other factors which affect such a measurement are ignored, since the target's interaction with incident electromagnetic energy will depend upon the number, size (in terms of wavelengths), and distances (in wavelengths) between scatterers on the target as well as the

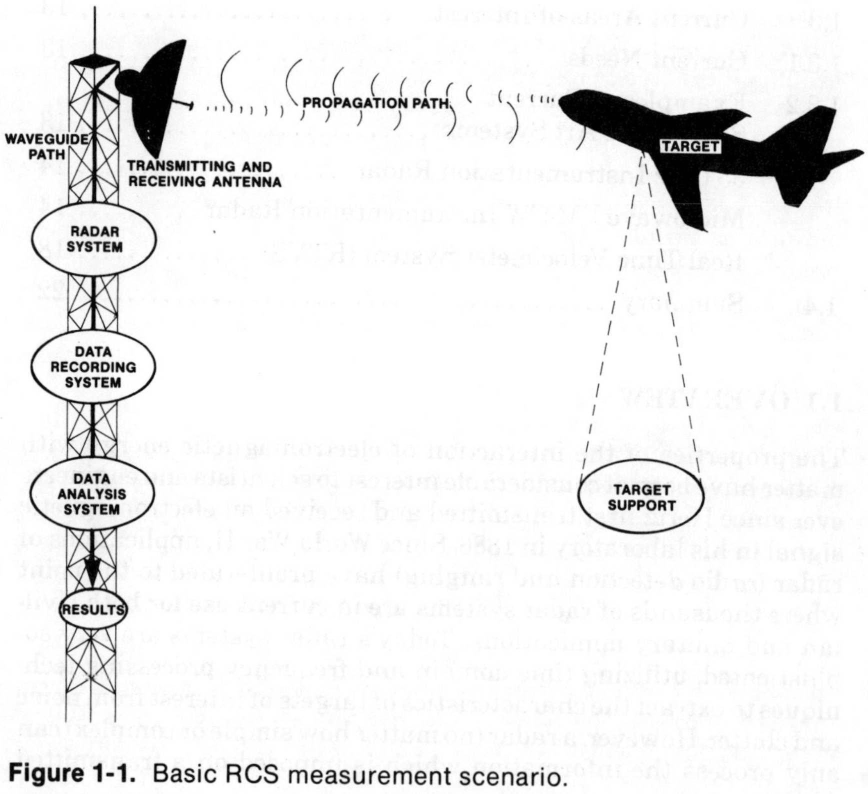


**Figure 1-1.** Basic RCS measurement scenario.

polarization of the wave relative to the orientation of each scatterer. If the other physical factors which affect such a measurement are also considered, such as the target support interactions with the target, the propagation path effects (such as multipath, ducting, and diffraction), the antenna effects, the radar system parameters, the distortion due to data recording, and the approximations made during data analysis, then the measurement problem can become extremely complex. Thus, the procedure has traditionally been to determine some relatively simple descriptors that can be related to the true target interaction mechanism, and thus give useful results about a radar's ability to detect a target. Such descriptors that are commonly used include: average or median radar cross section, standard deviations of radar cross section, and amplitude distribution type and width. For these types of measures it is possible to define the effects of the scenario on the measurement results so as to establish a high degree of confidence in the measured data.

There are two views on the usefulness of reflectivity data which are often expressed. Non-experimentalists, who may be unaware of the problems involved collecting reflectivity data, often assume that when a data set is published it is absolutely correct to a fraction of a dB and can be plugged into a radar system model which will forevermore predict the system performance accurately. Others, who have tried to reconcile the typically large discrepancies in data sets on similar targets under different but supposedly similar experimental conditions, may come to believe that all data are worthless and that measurements are a waste of time. In actuality, neither of these views is correct. Although specific measurements can be made very accurately, if care is taken, usually only a small subset of the physical parameters that affect reflectivity can be measured. Thus, two sets of measurements which appear to be made under identical conditions may actually differ because of changes in some key parameter, which affects the reflectivity, and which has not been identified, resulting in widely varying results. For this reason, reflectivity must, in general, be dealt with on a statistical basis, and physical parameters as well as electromagnetic scattering properties must be measured to the extent practical in order to assist in relating various data sets and extrapolating data measurements.

In this book there are three primary goals which will be addressed. First, the current techniques for measuring radar reflectivity will be discussed along with problems encountered and solutions from a practical point of view. Secondly, the characteristics of reflectivity data will be discussed so as to provide insight into the interpretation of data and the assessment of the validity of data for a given application. Finally, nearly all of the formulas and equations necessary for reflectivity calculations are given so that this material may serve as a convenient reference for the knowledgeable experimentalist.

## 1.2 A BRIEF HISTORY OF RADAR REFLECTIVITY

### 1.2.1 Early Years*

Although a great deal of experimental radar work has been carried out in recent years, the work has consisted of extending data bases which were essentially established by the end of World War II. Table 1-1 summarizes some of the major early events in radar reflectivity characterization.

**TABLE 1-1**
**Some Important Early Events in Radar Scattering Development**

| YEAR | EVENT | PERSON/ORGANIZATION |
|---|---|---|
| 1898 | Reflection of Radio Waves from Metallic Objects | Heinrich Hertz |
| 1903 | Reflection of Radio Waves from Ships | Hulsmeyer |
| 1925 | Reflection of Radio Waves from the Ionosphere Using FM-CW Radar Techniques | Appleton and Barnett |
| 1925 | Reflection of Radio Waves From the Ionosphere Using Pulsed Radar Techniques | Tuve and Breit |
| 1935 | Reflection of Radio Waves From Flying Aircraft | Watson-Watt/British Air Ministry |
| 1940-1945 | Various Experiments on Targets, Clutter, and Models | MIT Radiation Laboratory |
| 1946 | Definition of Radar Cross Section per Unit Area ($\sigma^o$) for Area Targets | Naval Research Laboratories/ Herbert Goldstein |
| 1946 | Definition of the Detailed Radar Range Equation | Army Signal Corps/ Omberg and Norton |

*Portions of this section are adapted from Blacksmith, *et al.*, [1];© 1965 IEEE.

Qualitative observations that radio waves are scattered and reflected by almost everything have been reported since the first experiments with radio waves by Hertz.[1] Attempts to reduce extraneous reflections with lossy materials were reported in the late 1890s. The design and construction of broadcast antennas led to a painful realization of the importance of reflection and scattering from the supporting towers and surrounding structures. There was also an early awareness of the potential utility of such reflections. Ridenour[2] notes that as early as 1903, experiments to detect radio waves reflected from ships were carried out by Hulsmeyer, who received a patent for an "obstacle detector" the following year.[3] Despite the pressing need for a device to help vessels "see through fog," practical forms of such ideas were slow in coming, and the idea was again suggested in 1922 by Marconi[1] on the occasion of his being awarded the Medal of Honor by the Institute of Radio Engineers.

The detection method suggested by Marconi in 1922 is similar in concept to the unmodulated, or CW, method which is still used for making scattering measurements. However, present-day backscatter systems use a hybrid tee to "screen the receiver from the local tramsmitter." By 1925, the remaining two techniques that are basic to radar, and widely used for scattering measurements, had been tried. Both were first used for studies of the reflection and scattering of radio waves from the ionosphere. In England, Appleton and Barnett[5] introduced frequency modulation (FM). In the United States, Tuve and Breit[6] of the Carnegie Institution of Washington introduced the pulse method for estimating ionospheric heights.

Due to the imminence of war in Europe, the British development of radar preceded that of the United States. Measurements were made of the reflections from a Heyford bomber flying at 10,000 feet under the auspices of Sir Watson-Watt in 1935 leading to the development of the British Home Chain of early warning radar stations.[2]

Despite these early beginnings, there were no quantitative radar cross section measurements until the development of radar created a demand for the data. It is probably true that most operators of the earliest radar systems obtained rough information on the scattering area of their targets by observation. The MIT Radiation Laboratory, which was established in the United States in 1940, was very active in the development of radar, and was engaged in the study of radar return from aircraft, ships, sea, and ground using operational radars.[7] This type of measurement, involving both operational radars and targets, has come to be known as dynamic radar cross section measurement, in contrast to another common practice today, where operational targets are replaced by models.

The first known measurements of radar cross sections where models were employed are those described in a 1942 MIT Radiation Laboratory report.[8] These measurements were made in a room, the walls of which had been "made black to the 10-cm radiation employed." Scattering measurements were made on a model aircraft with a 55-cm wing span. Others at the same organization proposed using optical methods to determine radar cross section.[9] Since coherent light sources were not available, optical methods did not provide a complete simulation of radar return, but optical techniques were used for a study of some highly polished metal models of aircraft. (*See* Chapter 10 for a detailed discussion of the development of model measurements.)

Countless measurements of radar propagation and clutter backscatter characteristics were performed at MIT Radiation Laboratory during the period from 1941 to 1946, which led to the development of two important concepts: (1) a detailed radar range equation including the propagation factor; and (2) the development of the concept of $\sigma^\circ$ and $\eta$ for area and volume scatterers. The detailed radar range equation is first discussed in a report by Omberg and Norton for the U.S. Army in 1943.[10] The concept of $\sigma^\circ$ is included in Goldstein's discussion of sea echo in reference [7] published in 1946. Thus, by the end of World War II, the basics of radar propagation and scattering were well understood, and data existed for targets and clutter up through X-band (10 GHz).

Undoubtedly, substantial RCS measurement programs were carried out in England and Germany during the war. Unfortunately, information on these is not readily available, but smatterings of information surface from time to time. It is known, for example, that the use of rubber substances loaded with iron filings was extensively investigated in Germany as a possible radar cross section reduction material for periscopes and snorkels. (The author would appreciate learning about any of these earlier RCS related programs.)

### 1.2.2 Recent History

Since World War II, work in radar reflectivity has concentrated on extending the quality of the measurements obtained in terms of the number of measured parameters, as well as increasing the frequencies of interest up through the millimeter wave bands. Table 1-2 lists some of the important measurement programs that have been conducted in the last 40 years.

It is believed that the first comprehensive radar cross section measurement program using models was that of Ohio State University (OSU).[11] The objective of their program was to obtain radar cross section data of aircraft for use in the design of decoys as radar confusing devices.

Within a few years after World War II, CW radar cross section measurements were also being made by the Naval Research Laboratory, the U.S. Army Evans Signal Laboratory,[12] and McGill University.[13] The first known use of a pulsed system on a model cross section range was by Sichak,[14] who used this technique to determine the nose-on cross section of V-2 rockets. In order to be able to measure the cross section characteristics of larger targets, Ohio State University also added pulsed equipment to their facility.[15] In this same time period, Fails and Fubini[16] made a significant contribution with their study of "Methods of Measuring Radar Cross Sections."

A number of organizations were working on characterizing the land and sea surface during the 1950s. Some of the more significant measurement programs included: the measurement of low angle sea return at 6 GHz and 35 GHz, by the Georgia Institute of Technology, in which a 110-foot tower was constructed to raise and lower the measurement radar system so that the radar depression angle could be changed while illuminating the same patch of sea.[17] Also, the Georgia Institute of Technology undertook measuring the reflectivity properties of land clutter at X-band as a function of polarization, terrain type, and weather conditions.[18] At about the same time, measurements on land clutter were also performed by the University of Texas at X-band, $K_u$-band, and $K_a$-band.[19] Measurements were made from bridges throughout Texas, Louisianna, and Georgia to obtain a variety of depression angles and terrain types.

In the early 1960s, a comprehensive measurement program was carried out by Ohio State University to characterize bare ground, grass, snow, and pavement as a function of depression angle, frequency, and polarization.[20] The data from this program became the standard for the reflectivity of terrain, although a repeat of the measurements by Ohio State personnel 10 years later yielded higher values, thereby illustrating the difficulties in obtaining accurate absolute measurements.[21]

**TABLE 1-2**
**Some Major Reflectivity Programs Since World War II**

| YEAR | EVENT | FREQUENCY | ORGANIZATION |
|---|---|---|---|
| 1945-1952 | RCS Measurement of Model Aircraft | 10 GHz | Ohio State University |
| 1952 | Low Angle Sea Clutter Measurements | 6.8, 35 GHz | Georgia Institute of Technology |
| 1957 | Measurement of Polarization Properties of Land Clutter | 9.3 GHz | Georgia Institute of Technology |
| 1957 | Measurements of Land Clutter | 9.4, 24, 35 GHz | University of Texas |
| 1958 | MMW Laser Modeling Measurements | 35 GHz (10 GHz - 2 THz*) | EMI Electronics |
| 1959 | Low Angle Land Clutter, Desert Clutter | 10 GHz | Goodyear Aircraft Corporation |
| 1960 | Measurements of Land Clutter as Functions of Surface Roughness, Polarization, Incidence Angle, Dielectric Constant, and Frequency | 10, 16, 35 GHz | Ohio State University |
| 1964 | RATSCAT Target Facility Begins Operation | 30 MHz - 24 GHz (30 MHz-95 GHz*) | Holloman Air Force Base, New Mexico |
| 1967 | Measurement of Tree Spectrum | 10 GHz | U.S. Army ERADCOM |
| 1968 | Airborne Measurements of Land Clutter | 0.4, 1, 5, 10 GHz | Naval Research Laboratory |
| 1968 | Airborne Sea Clutter Measurements | 0.4, 1, 5, 10 GHz | Naval Research Laboratory |
| 1970 | MMW Low Angle Sea Return | 10, 95 GHz | Georgia Institute of Technology/ Naval Air Development Center |
| 1974 | Radar Backscatter from Crops as a Function of Soil Moisture | 4 - 8 GHz (4 - 35 GHZ*) | University of Kansas Center for Research, Inc. |

| | | | |
|---|---|---|---|
| 1974 | Millimeter Backscatter Measurements of Rain | 10, 35, 70, 95 GHz | U.S. Army Ballistic Research Laboratory/Georgia Institute of Technology |
| 1975 | Low Angle Land Clutter Measurements | 10, 16, 35, 95 GHz | Georgia Institute of Technology |
| 1976 | Diurnal Snow Backscatter Measurements | 35 GHz | Georgia Institute of Technology |
| 1977 | Diurnal Snow Backscatter Measurements | 6, 8, 10, 16, 35 GHz | University of Kansas, Center for Research, Inc. |
| 1978 | Diurnal Snow Backscatter Measurements | 35, 95, 140 GHz | Rome Air Development Center |
| 1978 | Millimeter Wave Rain Attenuation Measurements | 94 GHz | TNO Physics Laboratory, The Netherlands |
| 1979 | Low Angle Millimeter Wave Sea Clutter Measurements | 10, 16, 35, 95 GHz | Georgia Institute of Technology/ Naval Surface Weapons Center |
| 1980 | Airborne Measurements of Military Targets and Clutter (HOWLS) | 16, 35 GHz | M.I.T. Lincoln Laboratories |
| 1981 | NATO Sea Backscatter Program | 10, 16, 35, 95 GHz | NSWC, USA; TNO Physics Laboratory, The Netherlands; RSRE, England; FGAN, Germany |
| 1982 | RCS and Attenuation Measurements of Falling Snow | 35, 95, 140, 220 GHz | U.S. Army (Snow 1-A) |
| 1982-1983 | Low Angle Measurements of Arctic Tundra | 0.165, 0.43, 1.25, 3.4, 9.1 GHz | MIT Lincoln Laboratories |

*Current Capabilities

Another important measurement program was conducted about the same time by the Goodyear Aircraft Corporation at X-band.[22] The measurements included both low and high angle data for irrigated farmland and desert terrain.

A modeling RCS range was established by EMI Electronics, Ltd. in Wells, Somerset, England in 1958 with the capability of performing RCS measurements on scale models at 35 GHz.[23] Since that time the frequency coverage has been extended over an order of magnitude to 2 THz (2000 GHz). Also, coherent operation is currently available through the use of lasers at the high frequencies which provide for target cross range imaging capabilities.

The 1960s were a very productive period for radar reflectivity work. The U.S. Air Force's RATSCAT facility opened in 1964 with capabilities of continuous frequency coverage from 30 MHz to 24 GHz, and the ability to perform measurements on full scale targets up to 50,000 lbs in weight.[24] Current RATSCAT capabilities include frequency coverage of selected bands up through 95 GHz and the ability to record four frequencies simultaneously, using their integrated radar measurements system (IRMS).

The 1960s also saw a number of important clutter measurement programs. Measurements performed by the U.S. Army ERADCOM on tree return spectra in 1967 established that clutter spectra were not necessarily Gaussian in nature, but rather, seemed to have a roll-off characteristic closer to the form $1/(1 + (f/f_c)^n)$ where $f_c$ is the 3 dB cutoff frequency and $n$ is equal to three.[25] A measurement program conducted by the Georgia Institute of Technology six years later verified these results, and determined that $n$ is a function of the transmitted frequency.[26]

In 1968 the Naval Research Laboratory assembled its now famous four-frequency airborne measurement radar system, and performed a large number of measurements on land and sea backscatter at P, L, S, and X-bands.[27] The systems were capable of transmitting *HH* and *VV* alternately, and data were obtained for various types of terrain and the sea for 5-degree to 70-degree depression angles.

The 1970s saw increased interest in obtaining radar reflectivity data at the higher millimeter wave frequencies. In 1970 a low angle sea return experiment was conducted at 10 GHz and 95 GHz by the Georgia Institute of Technology and the Naval Air Development Center.[28] One of the key conclusions from this experiment was that the polarization characteristics of the backscatter that have been observed at microwave frequencies are not seen at millimeter waves.

For example, at microwave frequencies for low depression angles, horizontal polarization almost always is lower in the average value of the reflectivity than that of vertical polarization, while it was observed that at 95 GHz the reverse is true.

In the mid-1970s, a number of important measurements were made on terrain, rain, and snow at microwave frequencies. In 1974 the University of Kansas conducted measurements over a 4-8 GHz range using their FM-CW spectrometer (*see* Chapter 3).[29] Their investigations concentrated on determining the effects of soil moisture on microwave backscatter from fields and crops. Later on, the frequency coverage of the Kansas system was extended up through $K_a$-band, and extensive measurements were performed on a variety of terrain types including trees,[30] grass, pavement,[31] and snow-covered ground [32] primarily for high depression angles.

In 1974 a comprehensive, multifrequency rain backscatter experiment was conducted jointly by the U.S. Army Ballistic Research Laboratory and the Georgia Institute of Technology.[33] The radar backscatter from thunderstorms was measured at 9, 35, 70, and 95 GHz, along with measurements of the rain rate and drop size distributions. Spectral distributions were measured on the return and were found to have a power function roll-off characteristic similar to that observed earlier by Fishbein for tree clutter.[24]

The next year, multifrequency backscatter measurements were conducted by the Georgia Institute of Technology to fill a data void for low angles at frequencies above 16 GHz.[26] Data were collected for frequencies of 9, 16, 35, and 95 GHz over depression angles of 2 degrees up to 15 degrees. Once again the spectral characteristic shapes of the tree returns were found to be power functions of the form $1/(1 + (f/f_c)^n)$ where $n$ decreases with increasing frequency.

During the next several years, three important measurement programs were conducted to characterize snow-covered ground at millimeter wave frequencies. In 1976 the Georgia Institute of Technology measured the reflectivity of snow-covered ground at 35 GHz at a site near Steamboat Springs, Colorado.[34] It was discovered that the snow had at least two distinct states from a reflectivity point of view, and that snow-covered ground could vary 10 dB or more in reflectivity during a daily diurnal cycle. This effect was linked to the presence of so-called "metamorphic snow" which is defined as snow that has been modified by successive freeze and melt cycles after falling to the earth.

The University of Kansas conducted more extensive tests over a frequency range of 6-35 GHz[31] at the same site in 1977. The results of this experiment showed that the diurnal reflectivity variations observed the previous year for 35 GHz were also present (although to a lesser extent) at lower frequencies.

In 1978 a series of measurements were performed by the U.S. Army Rome Air Development Center (RADC) on the backscatter of snow-covered ground at 35, 95, and 140 GHz.[35] This experiment showed that the diurnal backscatter effect was present for frequencies above 35 GHz as well as for lower frequencies (although once again the effect at higher frequencies was less than at 35 GHz.)

Also in 1978, an experiment was reported by the TNO Physics Laboratory in the Netherlands involving the measurement of the attenuation through rain at 94 GHz.[36] This experiment related actual data on attenuation through rain to theoretical predictions for the first time at 95 GHz, thus verifying the theoretical models for rain attenuation at millimeter waves.

In 1979 a millimeter wave low angle sea backscatter measurement program was conducted jointly by the Georgia Institute of Technology and the Naval Surface Weapons Center.[37] Measurements were made at 9.5, 16, 35, and 95 GHz for depression angles less than 10° as a function of polarization, wave height and direction, and wind speed and direction. This program was followed two years later by a NATO experiment to measure sea return and sea targets over frequencies of 10 to 95 GHz. Representative organizations from the United States, Great Britain, Germany, and the Netherlands participated in the experiment, which was conducted in Great Britain and Stadt, Norway.

In the last three years, numerous measurement programs have been carried out, including the HOWLS airborne radar measurements by the MIT Lincoln Laboratories at 16 and 35 GHz, the SNOW ONE and SNOW ONE-A measurement programs sponsored by the U.S. Army's Cold Regions Laboratory in which falling snow and snow-covered ground have been characterized from 35 GHz through 220 GHz, and low angle measurements of arctic tundra at frequency bands of UHF through microwave frequencies by the MIT Lincoln Laboratories. Thus, although detailed RCS measurement programs have been conducted for over 40 years, new measurement programs are on the increase as new frequencies and more detailed RCS descriptions become necessary to support modern radar technology system needs.

## 1.3 CURRENT AREAS OF INTEREST

### 1.3.1 Current Needs

The major difference between reflectivity measurement programs of today and those of World War II vintage is due to the advances in sensor and data processing technology.

The development of sensors, which can produce wide bandwidth coherent waveforms, and digital processors, which can perform real-time analysis on data, allows measurements to be performed today which were envisioned in the 1940s but were beyond the state of the art at that time.

The radar needs of today include: the ability to detect desired targets in adverse weather and in the presence of clutter; to classify targets by type; and even in some cases to recognize specific targets. Since many radars are to be used on airborne or mobile platforms, the physical size and weight of the radar sensor has a definite limit. The state of the art in computer processing is such that if a unique target or clutter characteristic (or signature) can be established, then real-time data processing should be capable of exploiting that signature in an operational situation. Thus, the current RCS measurement problem is to determine all of the significant information that can be measured on the reflectivity of a target or clutter so that appropriate processing will be able to exploit the target information in a smart sensor. It can be shown that all of the reflectivity information about a simple scatterer is contained in the polarization scattering matrix. (*See* Chapter 2 for a detailed definition of the polarization scattering matrix.) Thus, all of the RCS information can be measured for a complex target if the polarization scattering matrix can be determined for each of the major scatterers which together make up the RCS of the complex target. Thus, the primary thrust of present-day measurement programs is to determine the polarization scattering matrix of targets and clutter with a high enough resolution capability so as to resolve the individual scatterers of a complex target. Of course, these measurements must be performed for varying environmental, geometric, and target configuration conditions to be useful for developing smart sensors.

### 1.3.2 Examples of Current State of the Art Systems

Three recently developed RCS measurement systems will be described here to give an idea of the current state of the art in RCS measurement technology. The first system is a $K_a$ band pulsed instrumentation radar capable of transmitting 128 coherent frequencies over a 600 MHz bandwidth, thus allowing a compressed range resolution of approximately 1 foot while gathering the full polarization matrix. The second system is an FM-CW system recently developed by Scientific Atlanta, Inc. which provides a 5-cm range resolution while providing the ability to measure any linear polarization. The third system is a CW instrumentation tracking radar, recently developed for the U.S. Army White Sands Missile Range, which uses a multitone technique to measure velocity and range accurately.

*35 GHz Instrumentation Radar*

Recently, a 35 GHz instrumentation radar was developed by the Georgia Institute of Technology for the purpose of measuring the full polarization scattering matrix of a complex target with suitable resolution to resolve major target scatterers. Figure 1-2 gives a block diagram of the system while Table 1-3 summarizes the parameters. (A picture of this system is reproduced as Figure 3-30 in Chapter 3.) A frequency synthesizer is used to generate the transmitted and local oscillator frequencies. The 240 MHz IF difference frequency is established between the transmit and local oscillator signals by use of a single sideband modulator which multiplies a 240 MHz stable local oscillator signal with the transmit RF signal. A series of pin modulators and power amplifiers generate the 3 W, 100 *ns* transmit pulse. Both the transmit frequency and the polarization are under computer control. Thus, for a fixed frequency, the in-phase and quadrature components of *VV, VH, HH,* and *HV* or *RR, RL, LL,* and *LR* polarizations can be recorded. Then the frequency is changed and the process repeated at a 5 kHz burst rate. A total of 128 5-MHz steps can thus be recorded on digital tape. Fourier processing of the coherent frequency steps yields a high resolution range profile of the target being examined. In practice, a resolution of 1 foot has been achieved with the system. (For example, Figure 6-9 in Chapter 6 was recorded with this system.) Since the system is fully coherent, for a rotating target, inverse synthetic aperture radar (ISAR) techniques can be used to obtain high cross range resolution as well.

*Microwave FM-CW Instrumentation Radar*

A new FM-CW instrumentation radar has recently been introduced by Scientific-Atlanta, Inc. called the Model 2084 radar cross section analyzer.[38] This system performs a linear frequency sweep over a range of 4 GHz within the 8 to 18 GHz frequency band in 0.5 seconds, resulting in a processed minimum range resolution of 5 *cm*. Table 1-4

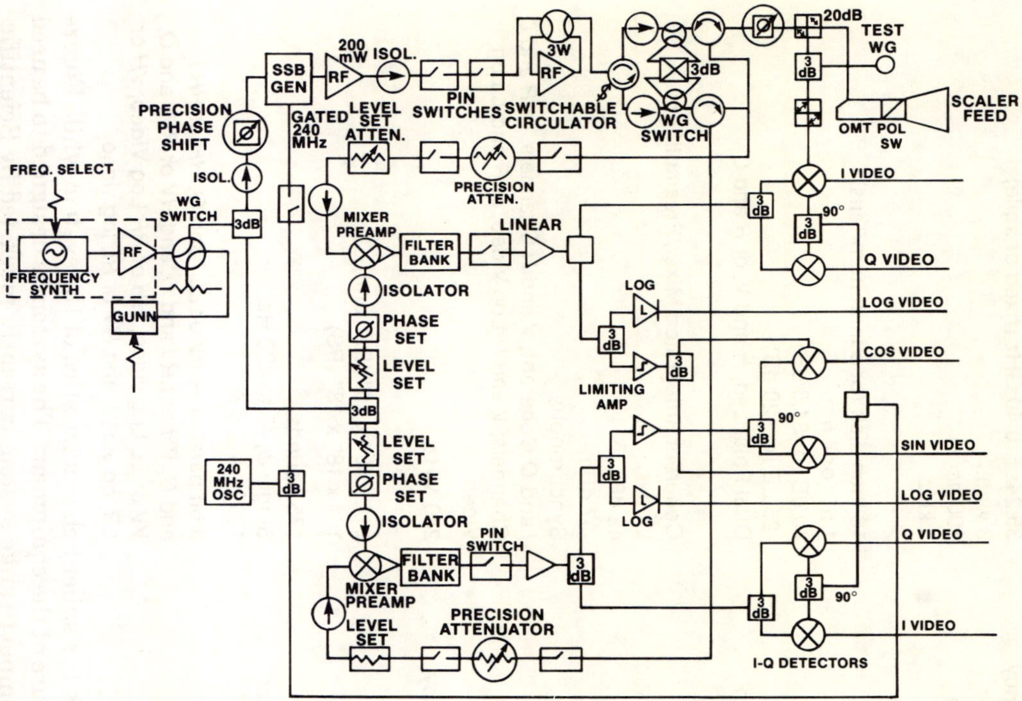


**Figure 1-2.** Simplified block diagram of a millimeter measurement radar system incorporating the capability to record both coherent data and polarimetric data (courtesy of the Georgia Institute of Technology).

**TABLE 1-3**
**Operating Parameters of Georgia Tech 35 GHz Instrumentation Radar**

| PARAMETER | CHARACTERISTIC |
|---|---|
| *Transmitter:* | |
| RF Frequency | 35.24 ± 0.40 GHz (fixed or agile) |
| Peak Power | 3 watts |
| Pulsewidth | 100 *ns* |
| *prf* | 5 kHz |
| *Antennas:* | |
| Type | 18 inch dish (or 5 inch horn/lens) |
| Beamwidth | 1.5° (or 4.5°) |
| Gain | 40 dB (or 34.5 dB) |
| Sidelobes | -25 dB (or -20 dB) |
| Polarization | Dual Polarized, *H* and *V*, or *R* and *L* |
| *Receiver:* | |
| Type | Coherent, Integrated Mixer/Preamplifier |
| Characteristic | Linear |
| Bandwidth | 42 dB |
| Sensitivity | -97 dBm |
| Detection | Synchronous |
| Output | *I* and *Q* Coherent, *V* and *H* Channels, or *R* and *L* Channels; *V* and *H* Log Video, or *R* and *L* Log Video |
| IF Frequency | 240 MHz |
| *Physical:* | |
| Size | 21″ x 40″ x 16″ (R/T unit)<br>15″ x 18″ x 13″ (PS) |
| Weight | 175 pounds |
| Prime Power | Single $\phi$, 47 to 420 Hz<br>120 V, AC, 5 A |
| *Data Output:* | 8 bit parallel — *VV* or *LL I* and *Q*, *HH* or *RR I* and *Q*, *VH* or *LR I* and *Q*, and *HV* or *RL I* and *Q*; *VV* or *LL* Log video, *HH* or *RR* Log Video, *VH* or *LR* log video, and *HV* or *RL* log video |

summarizes the salient characteristics of the system while Figure 1-3 is a picture of the equipment. The system is designed to be used with the compact range system currently marketed by Scientific-Atlanta to provide an indoor high resolution RCS measurement capability for targets up to $6m$ by $6m$ in extent. The system includes a real time Fast Fourier transform processor and a Hewlett Packard

**Figure 1-3.** Model 2084 radar cross section analyzer (courtesy of Scientific-Atlanta, Inc.).

minicomputer to provide detailed plots of RCS in various formats, including isometric or contour plots of RCS *versus* range and cross range, as illustrated in Figure 1-4. The cross range resolution is obtained by calculating the spectrum of a rotating target for a fixed frequency and separating the frequency components of target scatterers rotating at different velocities relative to the radar. The minicomputer allows operator control of the data acquisition and analysis process, and provides for storage of data on either a cassette, floppy disk, or Winchester disk.

This system is typical of much of the modern RCS work in which an old concept (FM-CW radar was demonstrated by Appleton and Barnett in 1922) was coupled with new advances in microwave components and digital technology to produce significant enhancements in RCS measurement capabilities.

**TABLE 1-4**
**Primary Characteristic of the Scientific-Atlanta, Model 2084 Radar Cross Section Analyzer**
*(from ref. [34];© 1983 Scientific-Atlanta, Inc.)*

| | |
|---|---|
| Frequency Coverage | 8-18 GHz |
| Transmitted Power | 10 mW, CW radiated |
| Receiver Noise Figure | 6 dB |
| Polarization | Any linear independently rotatable transmit and receive antenna |
| Minimum Discernable RCS Level | -60 dBsm at maximum range (20 *m*) |
| Dynamic Range | 60 dB |
| Output | CRT displays or hard copy of RCS *vs.* range for fixed aspect, RCS *vs.* frequency for fixed aspect, RCS *vs.* aspect for any frequency, RCS *vs.* range and cross range for any fixed angle<br>Plots of RCS *vs.* range, RCS *vs.* aspect, RCS *vs.* frequency, RCS *vs.* down range and cross range in isometric or contour format |
| Target Size | Up to 6 meters |
| Frequency Granularity | 30 MHz |
| Two-Dimensional Resolution (128 x 128 pixels) | 5 *cm* range x 10 *cm* cross range (6 *m* x 6 *m* target) |
| Data Rate | 2 Hz |
| Data Recording | Cassette, floppy disk, or Winchester disk |

*Real-Time Velocimeter System* [39]

Recently, an instrumentation tracking radar was developed by Datron Systems, Inc., with support from the Georgia Institute of Technology and the New Mexico State Physical Sciences Laboratory, for the U.S Army White Sands Missile Range (WSMR). The mission of the radar is to determine the velocity and range histories of projectiles launched from artillery, and to determine the impact points on the ground. The system approach involves transmitting three simultaneous CW, X-band signals and recovering the Doppler-shifted return signals. These signals are then processed through a

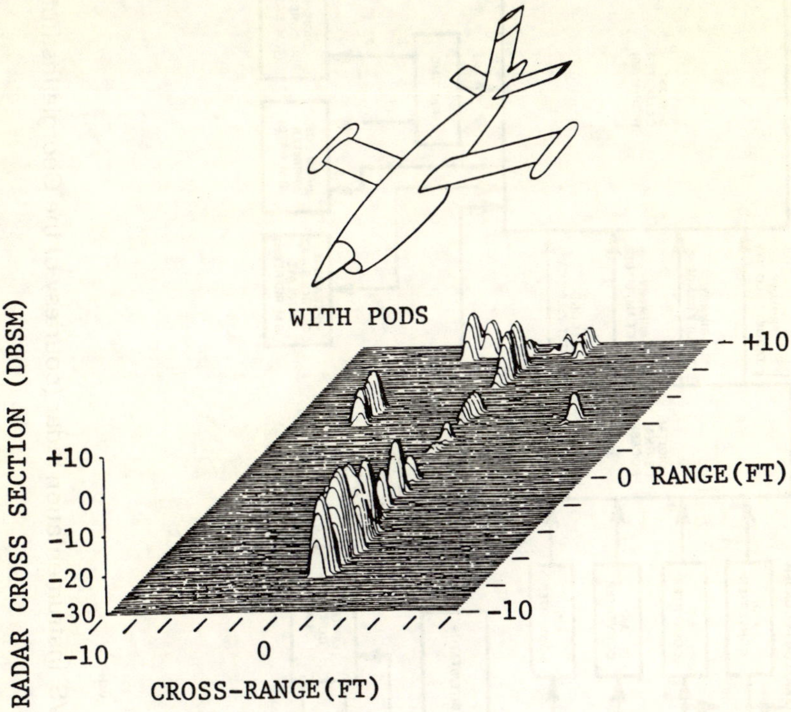


**Figure 1-4.** Isometric plot of RCS *versus* range and cross range for a simulated aircraft (reprinted from Mensa, ref. [40]; ©1981 Artech House, Inc.).

high speed, complex-to-complex *fast Fourier transform (FET),* and target detection and selection are made on the basis of amplitude and frequency. The range to the target is then obtained by computing the phase difference between two pairings of the three Doppler-shifted return signals. The three signals allow for unambiguous range measurement to 50 km, with a range measurement accuracy of greater than 10 meters, and an unambiguous Doppler measurement for targets moving at speeds up to 3750 *m/s* with an accuracy of better than 0.3 *m/s*. The quality of the information from the radar signals is strongly dependent upon the performance of the radar system hardware. Figure 1-5 is a simplified block diagram of the RTVS mechanization, while Figure 1-6 gives a view of the separate low-sidelobe transmitting and receiving antennas. The transmitter generates three X-band CW frequencies between 10.0 and 10.26 GHz. The frequencies are coherent and are related as frequency $F_1$, fre-

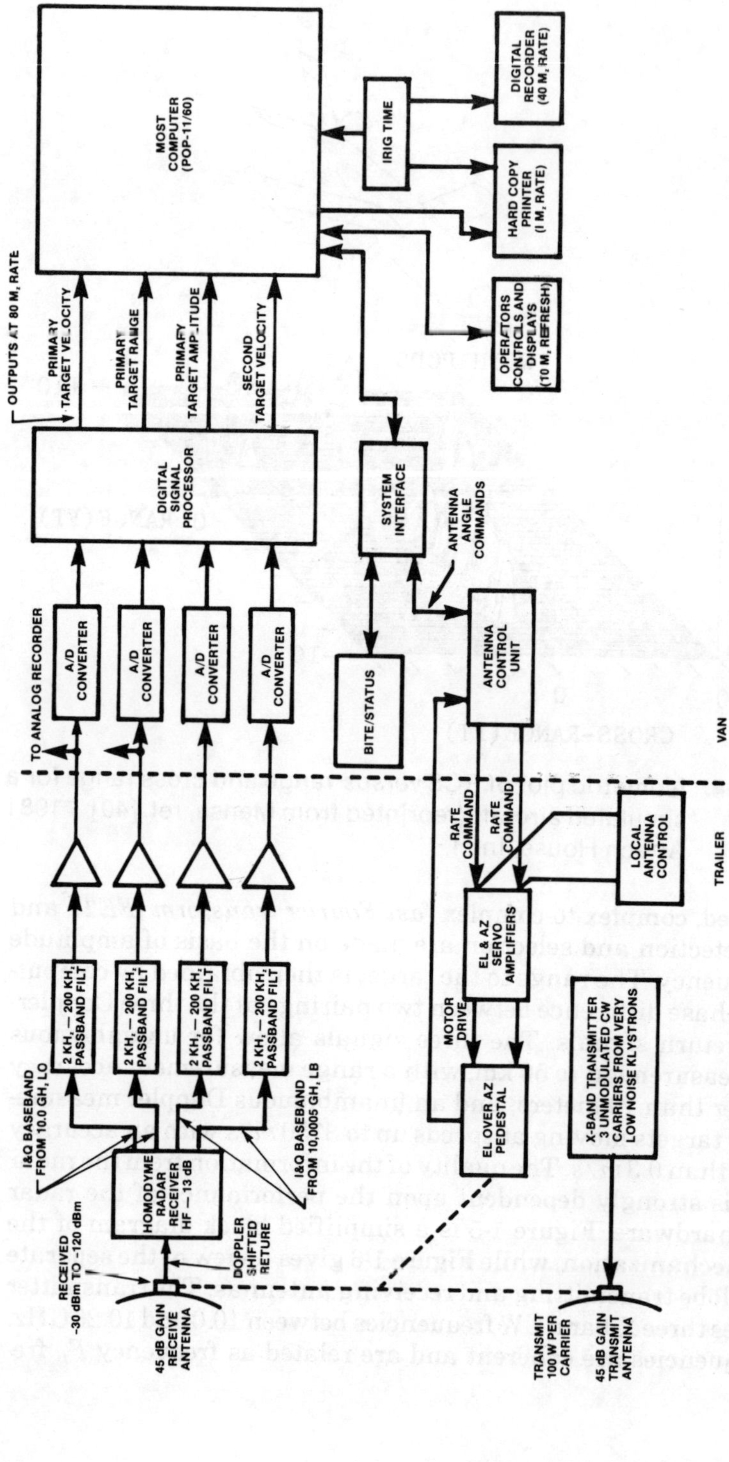


**Figure 1-5.** Simplified block diagram of the RTVS instrumentation radar (courtesy of the Georgia Institute of Technology).

quency $F_2 = F_1 + 500$ kHz, and frequency $F_3 = F_1 + 499$ kHz, or $F_1 + 492$ kHz, or $F_1 + 468$ kHz. The 1 kHz offset between $F_1$ and $F_2$ allows unambiguous ranging over a 150 km interval. The 500 kHz frequency offset between $F_1 F_2$ allows unambiguous ranging over a 300-meter range interval. By using the separate frequencies, a coarse range and fine range interval resolution can be developed to resolve target range to an accuracy better than 10 meters. The frequencies are generated by using a low-noise, cavity-stabilized klystron oscillator and up-converter on three separate channels by using UHF signals generated by an ultra-low noise synthesizer.

The FFT programmable digital signal processor is a MARS 232 special purpose digital computer that combines the features of both an array processor and a hard-wired signal processor to achieve "throughput" 50 to 100 times faster than a minicomputer. It permits real-time processing of signal and data streams at several million words per second, without the intervention of the host computer.

The application of FFT techniques and the use of a high-speed

**Figure 1-6.** Real time velocimeter system (RTVS) trailer, pedestal, and antenna (courtesy of Brian H. Hudson).

programmable array processor have greatly enhanced the versatility, accuracy, and speed of Doppler, angle, and range measurements in the CW multifrequency radar system. The processing not only makes possible the real-time analysis of the Doppler spectrum, and implementation of the threshold, detection, and tracking algorithms that complete the functional radar requirements, but also provides an adaptive tool for extraction of target reflectivity characteristics. Once again an old idea has been implemented with modern technology to achieve surprising results.

## 1.4 SUMMARY

In this chapter a brief history of radar reflectivity measurements has been given. It was seen that radar reflectivity research was relatively sparse until World War II when a considerable amount of work was done in just a few years. Since that time, work has concentrated on filling in "holes" in the data and taking advantage of new measurement techniques made possible by technological advances. However, despite these technological advances one need only look at the large variations in the measured quantities among experimental data sets to see that the basic problems of making accurate reflectivity measurements still plague us.

Thus, it should always be kept in mind that the basic physics of radar reflectivity measurement never change. Even though high frequencies, wide bandwidths, or sophisticated digital processing techniques may be used in modern measurements systems, the requirements for carefully defining system calibration, measurement system stability, and environmental characterization never cease. The following chapters in this book discuss methods of achieving these fundamental requirements.

## REFERENCES

1. Blacksmith, Jr., P., R. E. Hiatt, and R. B. Meek, "Introduction to Radar Cross Section Measurements," *Proc. IEEE,* Vol. 53, August 1965, pp. 902-903.

2. Ridenour, L. N., *Radar System Engineering,* Vol. 1 of the M.I.T. Radiation Laboratory Series, McGraw-Hill Book Company, 1947.

3. Hulsmeyer, G., "Hertzian Wave Projection and Receiving Aparatus Adapted to Indicate and Give Warning of the Presence of a Metallic Body Such as a Ship or a Train in the Line of Projecting Such Waves," British Patent 13170, September 22, 1904.

4. Marconi, G., "Radio Telegraphy," *Proc. IRE,* Vol. 10, 1922, pp. 218-238.

5. Appleton, E. V., and M. A. F. Barnett, "Local Reflections of Wireless Waves From the Upper Atmosphere," *Nature,* Vol. 115, March 1925, pp. 333-334.

6. Tuve, M. A., and G. Breit, "Note on a Radio Method of Estimating the Height of the Conducting Layer," *Terrestrial Magnetism and Atmospheric Electricity,* Vol. XXX, March 1925, pp. 15-6.

7. Kerr, D. E., *Propagation of Short Radio Waves,* Vol. 13 of the M.I.T. Radiation Laboratory Series, McGraw-Hill Book Company 1947.

8. Raymond, R. C., "Scattering of 10 *cm* Radiation by a Model Airplane," M.I.T. Radiation Laboratory Report No. 156, Cambridge, Massachusetts, May 1942.

9. Goudsmit, S. A., and P. R. Weiss, "Possible Measurement of Radar Echoes by the Use of Model Targets," MIT Radiation Laboratory Report No. 196, Cambridge, Massachusetts, March 1943.

10. Omberg, A. C., and K. A. Norton, "The Maximum Range of a Radar Set," *Proc. IRE,* Vol. 35, January 1947, pp. 4-24.

11. Yates, Y. P., "A Continous-Wave Method of Measuring Radar Cross Sections and Reflection Patterns by Means of Models," The Ohio State University Research Foundation, Columbus, Ohio, Report No. 759-33, October 31, 1945.

12. Gorr, B. B., "Air Force Cambridge Research Center Radar Cross Section Measurement Equipment and Range," Electromagnetic Radiation Lab., U.S. Air Force Cambridge Research Center, L. G. Hanscom Field, Bedford, Massachusetts, Report No. AFCRC-TR-59-137, 1959.

13. Woonton, G. A., D. R. Hay, and D. C. Hogg, "Radar Cross Sections of Aircraft," McGill University, Montreal, Quebec, Canada, Report No. 1 on Project DRBX 27, October 1951.

14. Sichak, W., "Missile Detection," Federal Telecommunications Lab., Inc., Final Report on Contract W36-039SC 32119, August 1948.

15. "Final Report, Modification of a Radar System for Automatic Recording of Backscatter Radiation Patterns," Vol. 1, The

Ohio State University Research Foundation, Columbus, Ohio, Report No. 406-1, July 1951 (ATI 117-588); Vol. 2, Report No. 406-2, May 1952 (ATI 161-644).

16. Fails, W. A., and E. G. Fubini, "Methods of Measuring Radar Cross Sections," Airborne Instruments Lab., Inc., Mineola, New York, Report No. 380-1, January 1949.

17. Flynt, E. R., *et al.*, "Sea Return Study," Final Report on Contract NObsr-49063, Georgia Institute of Technology, Atlanta, Georgia, August 1957.

18. Hayes, R. D., C. H. Currie, and M. W. Long, "An X-Band Polarization Measurements Program," *Record of the Third Annual Radar Symposium,* University of Michigan, February 1957.

19. Grant, C. R., and B. S. Yaplee, "Back Scattering from Water and Land at Centimeter and Millimeter Wavelengths," *Proc. IRE,* Vol. 45, July 1957, pp. 976-82.

20. Cosgriff, R. L., W. H. Peake, and R. C. Taylor, "Terrain Scattering Properties for Sensor System Design" *(Terrain Handbook II),* Engineering Experiment Station Bulletin, Vol. 29, No. 3, Ohio State University, Columbus, Ohio, May 1960.

21. Peake, W. H., and T. L. Oliver, "The Response of Terrestrial Surfaces at Microwave Frequencies," Technical Report AFAL-TR-70-301, Ohio State University, Columbus, Ohio, May 1971.

22. Goodyear Aircraft Corporation, "Radar Terrain Return Study, Final Report: Measurements of Terrain Back-Scattering Coefficients with an Airborne X-Band Radar," Final Report on Contract NOas-59-6186-C, Report No. GERA-463, September 20, 1959.

23. Cram, L. A., and S. C. Woolcock, "Review of Two Decades of Experience Between 30 GHz and 900 GHz in the Development of Model Radar Systems," *AGARD Conference Proceedings No. 245 on Millimeter and Submillimeter Wave Propagation and Circuits,* September 1978, pp. 6-1 to 6-15.

24. Marlowe, H. C., *et al.*, "The RATSCAT Cross Section Facility," *Proc. IEEE,* Vol. 53, August 1965, pp. 946-954.

25. Fishbein, W., S. W. Graveline, and O. E. Rittenback, "Clutter Attenuation Analysis," U. S. Army Electronics Command, Technical Report No. ECOM-2808, March 1967.

26. Currie, N. C., F. B. Dyer, and R. D. Hayes, "Radar Land Clutter Measurements at Frequencies of 9.5, 16, 35, and 95 GHz," Technical Report No. 3 on Contract DAAA-25-73-C-0256, Georgia Institute of Technology, Atlanta, Georgia, April 1975.

27. Daley, J. C., *et al.*, "NRL Terrain Clutter Study, Phase II," Naval Research Laboratory Report 6749, October 21, 1968.

28. Rivers, W. K., "Low Angle Sea Return at 3 mm Wavelengths," Final Report on Contract N62269-70-C-0489, Georgia Institute of Technology, November 1970.

29. Ulaby, F. T., "Radar Measurement of Soil Moisture Content," *IEEE Transactions on Antennas and Propagation,* Vol. AP-22, January 1975, pp. 36-45.

30. Bush, T., *et al.*, "Seasonal Variations of the Microwave Scattering Properties of Deciduous Trees as Measured in the 1-18 GHz Spectral Range," NASA Report No. NASE9-10261, The University of Kansas Center for Research, June 1976.

31. Stiles, W. H., F. T. Ulaby, and E. Wilson, "Backscatter Response of Roads and Roadside Surfaces," Report No. SAND78-7069 on Contract DE-AC04-76DP00789, The University of Kansas Center for Research, March 1979.

32. Stiles, W. H., and F. T. Ulaby, "Microwave Remote Sensing of Snowpacks," NASA Report No. 3263, The University of Kansas Center for Research, June 1980.

33. Currie, N. C., F. B. Dyer, and R. D. Hayes, "Analysis of Radar Rain Return at Frequencies of 9.375, 35, 70, and 95 GHz," Technical Report No. 2 on Contract DAAA 25-73-C-0256, Georgia Institute of Technology, Atlanta, Georgia, February 1975.

34. Currie, N. C., F. B. Dyer, and G. W. Ewell, "Radar Millimeter Backscatter Measurements from Snow," Technical Report No. AFATL-TR-77-4 on Contract F08635-77-C-0221, Georgia Institute of Technology, Atlanta, Georgia, January 1977.

35. Hayes, D. J., *et al.*, "Millimeter Wave Scattering From Snow Backgrounds," *Digest of the 1979 IEEE APS International Symposium,* No. 79CH1456-3 AP, pp. 499-502.

36. Keizer, W. P. M. N., "Rain Attenuation Measurements at 94 GHz: Comparison of Theory and Experiment," *AGARD Conference Proceedings No. 245 on Millimeter and Submillimeter*

*Wave Propagation and Circuits,* September 1978, pp. 44-1 to 44-9.

37. Trebits, R. N., N. C. Currie, and B. Perry, "Multifrequency Radar Sea Clutter Measurements," Final Technical Report on Contract N6921-78-C-A179, Georgia Institute of Technology, Atlanta, Georgia, April 1980.

38. "Model 2084 Radar Cross Section Analyzer," Scientific-Atlanta, Inc., Sales Brochure, June 1983.

39. Bomar, L. C., *et al.*, "CW Multitone Radar Ranging Using DFT Techniques," *Proceedings of the Thirteenth European Microwave Conference,* September 1983.

40. Mensa, D. L. *High Resolution Radar Imaging*, Artech House, Dedham, Massachusetts, 1981.

# CHAPTER 2

# RADAR CROSS SECTION

**Robert N. Trebits**

**Engineering Experiment Station**
**Georgia Institute of Technology**



## 2.1 DEFINITION OF RADAR CROSS SECTION

The monostatic radar cross section of a target can be defined in terms of the portion of incident power on a target which is re-radiated back to the radar. The power density at a target at range $R$ from a directive antenna of gain $G_t$ is

$$\frac{P_t G_t}{4\pi R^2} \tag{2-1}$$

where $P_t$ is the transmitted power. The power density at the receiving antenna of the signal re-radiated by the target is

$$\left[\frac{P_t G_t}{4\pi R^2}\right] \left[\frac{\sigma}{4\pi R^2}\right] \tag{2-2}$$

where $\sigma$ represents the radar cross section of the target.
The received power $P_r$ captured by the receiving antenna of effective aperture $A_e$ is

$$P_r = \left[\frac{P_t G_t}{4\pi R^2}\right] \left[\frac{\sigma}{4\pi R^2}\right] A_e \tag{2-3}$$

The gain $G_r$ of the receiving antenna can be expressed in terms of the effective aperture by means of the relationship

$$G_r = \frac{4\pi A_e}{\lambda^2} \tag{2-4}$$

where λ is the radar wavelength. The received power can then be expressed

$$P_r = \left[\frac{P_t G_t}{4\pi R^2}\right] \left[\frac{\sigma}{4\pi R^2}\right] \left[\frac{G_r \lambda^2}{4\pi}\right] \tag{2-5}$$

or

$$P_r = \frac{P_t G^2 \sigma \lambda^2}{(4\pi)^3 R^4} \tag{2-6}$$

for monostatic radar operation ($G_t = G_r = G$). Thus, the radar cross section parameter $\sigma$ quantitively relates the transmitted power to the received power re-radiated by a target. Equation (2-6) represents the simplest form of what is conventionally called the "radar range equation."

An implication of this physical definition is that a plane wave is incident on the target, since radar cross section is range independent and is strictly a characteristic of the target itself, its orientation with respect to the transmitter/receiver, and the incident radar signal. This plane wave requirement defines the "far-field" condition: the target is physically removed from the source of transmitted energy such that the wavefronts incident at the target are essentially planar in practical applications.

An equivalent theoretical definition of radar cross section also implies an incident plane wave at the target by assuming an infinite range:

$$\sigma = \lim_{R\to\infty} 4\pi R^2 \left|\frac{E_r^2}{E_i^2}\right| \tag{2-7}$$

where

$E_r$ = electric field magnitude at the receiver
$E_i$ = electric field magnitude incident at the target

An alternate definition for radar cross section can also be made using analogous magnetic field parameters:

$$\sigma = \lim_{R\to\infty} 4\pi R^2 \left|\frac{H_r^2}{H_i^2}\right| \tag{2-8}$$

Qualitatively the radar cross section of an object is a measure of its "size" as seen at a particular radar wavelength and polarization. Radar cross section (RCS) has units of square meters ( $m^2$ ) and is often expressed in decibels ( $dBm^2$ ) relative to a square meter, i.e.,

$$\sigma\,(\mathrm{dBm}^2) = 10 \log_{10} [\,\sigma\,(m^2)\,] \tag{2-9}$$

Note that a multiplicative factor of 10 is used to compute the decibel relationship since $\sigma$ is proportional to power rather than voltage.

To account for the polarization dependency of radar cross section, consider the relationship between the transmitted and received electric fields in terms of orthogonal, linear polarization components $E_H$ and $E_V$, where the $H$ subscript denotes horizontal and $V$ denotes vertical polarization, respectively. Then $E_H$ and $E_V$ can be expressed in terms of the reflectivity of a target illuminated by both polarizations via proportionality constant $a_{ij}$, where the first subscript denotes the transmitted polarization and the second subscript denotes the received polarization:[1]

$$\begin{aligned} E_H^r &= a_{HH} E_H^t + a_{VH} E_V^t \\ E_V^r &= a_{HV} E_H^t + a_{VV} E_V^t \end{aligned} \tag{2-10}$$

or in matrix format:

$$\begin{bmatrix} E_H^r \\ E_V^r \end{bmatrix} = \begin{bmatrix} a_{HH} & a_{VH} \\ a_{HV} & a_{VV} \end{bmatrix} \begin{bmatrix} E_H^t \\ E_V^t \end{bmatrix} \tag{2-11}$$

The constants $a_{ij}$ may be considered in this context to be range independent, but complex valued, since the phase relationship between electric field components must be accommodated. For monostatic radar operation $a_{HV}$ and $a_{VH}$ are equal, from the consideration of the reciprocity relationship between transmitting and receiving antennas. The matrix in Equation (2-11) defines the relationship in amplitude and in phase between components of the transmitted and received electric field and is, therefore, called the target's scattering matrix.

A matrix expression analogous to Equation (2-11) may be defined for orthogonal polarization components designated right and left circular polarizations. Using a subscript $R$ for right circular and $L$ for left circular polarization, this matrix expression is

$$\begin{bmatrix} E_R^r \\ E_L^r \end{bmatrix} = \begin{bmatrix} c_{RR} & c_{LR} \\ c_{RL} & c_{LL} \end{bmatrix} \begin{bmatrix} E_R^t \\ E_L^t \end{bmatrix} \tag{2-12}$$

As in the linear polarization case, the off-diagonal elements $c_{RL}$ and $c_{LR}$ are equal for monostatic operation.

The relationship between the constants $a_{ij}$ and $c_{ij}$ can be derived by expressing the circular polarization electric field components in terms of linear ones. The resulting set of relationships is

$$
\begin{aligned}
|c_{RR}| &= |1/2\,(a_{HH} - a_{VV}) + j\,a_{HV}| \\
|c_{LL}| &= |1/2\,(a_{HH} - a_{VV}) - j\,a_{HV}| \\
|c_{RL}| &= |1/2\,(a_{HH} + a_{VV})|
\end{aligned}
\qquad (2\text{-}13)
$$

The relative phase between the horizontal and vertical polarization components of the received electric field is called the polarimetric phase Φ. This parameter has been used in recent years for target and clutter discrimination. The polarimetric phase can be obtained by performing the dot product vector operation between the horizontal and vertical polarization electric field components.

$$\vec{E}_H^r \cdot \vec{E}_V^r = |E_H^r|\,|E_V^r| \cos\Phi \qquad (2\text{-}14)$$

or

$$\cos\Phi = \frac{a_{HH}\,a_{HV} + a_{VH}\,a_{VV}}{\sqrt{a_{HH}^2 + a_{VH}^2}\,\sqrt{a_{HV}^2 + a_{VV}^2}} \qquad (2\text{-}15)$$

Matrix expressions (2-11) and (2-12) justify the definition of the radar cross section scattering matrix

$$\sigma = \begin{bmatrix} \sigma_{11} & \sigma_{21} \\ \sigma_{12} & \sigma_{22} \end{bmatrix} \qquad (2\text{-}16)$$

where each $\sigma_{ij}$ is proportional to the square of the associated $a_{ij}$ or $c_{ij}$, i.e., $\sigma_{ij} \propto |a_{ij}|^2$ or $\sigma_{ij} \propto |c_{ij}|.^2$ Thus, $\sigma_{11}$ relates the reflectivity of a target illuminated by polarization #1 and "seen" by the same polarization, and $\sigma_{21}$ relates the target's reflectivity illuminated by polarization #2 and "seen" by polarization #1. The radar cross section polarization matrix, therefore, contains *all* the reflectivity information available from that target.

The target scattering matrix defined in Equation (2-11) may, therefore, be expressed in terms of radar cross section components. For example, for linear polarizations

$$\begin{bmatrix} a_{HH} & a_{VH} \\ a_{HV} & a_{VV} \end{bmatrix} = \begin{bmatrix} \sqrt{\sigma_{HH}}\, e^{j\phi HH} & \sqrt{\sigma_{VH}}\, e^{j\phi VH} \\ \sqrt{\sigma_{HV}}\, e^{j\phi HV} & \sqrt{\sigma_{VV}}\, e^{j\phi VV} \end{bmatrix} \qquad (2\text{-}17)$$

In Equation (2-12) $\sqrt{\sigma_{ij}}$ represents the magnitude of each matrix component and $\phi_{ij}$ represents the phase of the associated element.

For the case of circular polarization transmission and linear polarization reception the components of the radar cross section polarization matrix are

$$\sigma_{LH} \propto 1/2 \mid a_{HH} + j\, a_{HV} \mid$$
$$\sigma_{LV} \propto 1/2 \mid a_{HV} + j\, a_{VV} \mid$$
$$\sigma_{RH} \propto 1/2 \mid a_{HH} - j\, a_{HV} \mid$$
$$\sigma_{RV} \propto 1/2 \mid a_{HV} - j\, a_{VV} \mid \qquad (2\text{-}18)$$

## 2.2 ANALYTICAL METHODS OF RADAR CROSS SECTION PREDICTION

The analytical calculation of the radar cross section of an object is essentially the solution to a boundary value problem defined by Maxwell's equations, the object, and the incident field. While this description is succinct, the number of shapes for which closed form solutions can be determined (using such mathematical techniques as separation of variables) is very limited. These particular shapes include spheroids, toruses, and semi-infinite surfaces such as cones and paraboloids, which do not represent most real-world objects particularly well. Thus, while such exact solutions represent mathematical fulfillment, the calculation of the radar cross section of an object with otherwise less tractable geometry must be performed using approximation techniques.

The most familiar approximation technique is *geometric optics,* in which radar "rays" are treated according to the physical laws of reflection and refraction. This technique traces individual rays by reflection off object surfaces and refraction through transmissive surfaces. While such treatment is esthetically pleasing, geometric optics cannot take into account either polarization effects or effects due to the wave nature of electromagnetic energy, e.g., diffraction.

An approximation technique which utilizes electromagnetic properties is *physical optics.* The surface of an object is divided into incremental units, and current densities are calculated for each unit, assuming each unit to be a portion of an infinite plane. Furthermore, geometric shadowed regions of the object are assumed to contribute no current at those locations. The re-radiated fields from each illuminated unit are then calculated and summed to yield the orientation-dependent radar cross section of the object. Physical optics also cannot account for polarization-dependent characteristics of an object's radar cross section.

The *geometrical theory of edge diffraction* (GTD) was developed in the mid-1850s by Joseph B. Keller, who postulated that the edge-diffracted rays lie along the generators of a forward cone centered on the point of diffraction. The cone half-angle is equal to the angle subtended by the edge and the incident ray, and the cone axis coincides with the edge. The polarization of the diffracted field is determined by the components of both the incident electric and magnetic fields along the edge, and is therefore felt to be more realistic than either geometric optics or physical optics in this regard. The far diffracted field is given by the product of four quantities: the incident field strength, a divergence factor accounting for the spreading of rays away from the point of diffraction, a diffraction coefficient that depends on the incident polarization and the angles of incidence and diffraction, and a phase factor that accounts for the electrical distance between the point of diffraction and a field point along a diffracted ray. The theory is quite good for field points well removed from reflection and shadow boundaries because the diffraction coefficients become singular in those directions. However, several asymptotic formulations have been proposed since Keller's original work, and these seem to be effective in the troublesome transition regions. For more detailed information on radar cross section prediction techniques *see* reference [2].

## 2.3 RADAR CROSS SECTION OF SIMPLE SHAPES

As might be expected, the perfectly conducting sphere is the simplest geometric shape whose radar cross section can be determined exactly. With its three-dimensional symmetry, the sphere's radar cross section is aspect independent, which makes this object an obvious calibration target for measurements. Kerr has shown that the radar cross section of a sphere of radius $a$, normalized to its "optical" cross section $\pi a^2$, is given by the expression

$$\frac{\sigma}{\pi a^2} = \frac{1}{(ka^2)} \left| \sum_{n=1}^{\infty} (-1)^n (2n + 1)(a_n^s - b_n^s) \right|^2 \qquad (2\text{-}19)$$

where

$k = 2\pi/\lambda$

$a_n^s$ and $b_n^s$ = coefficients listed on page 448 in Kerr [3]

Figure 2-1 depicts the plot of the normalized radar cross section of a shpere as a function of radius normalized to wavelength. Note that three distinct regions are apparent in Figure 2-1. (While the curve in Figure 2-1 is for a conducting sphere, the three regions shown are characteristic of all objects.) In the region where $2\pi a/\lambda < 1$, which defines the Rayleigh region, the radar cross section can be seen to vary inversely as the fourth power of the radar wavelength, since

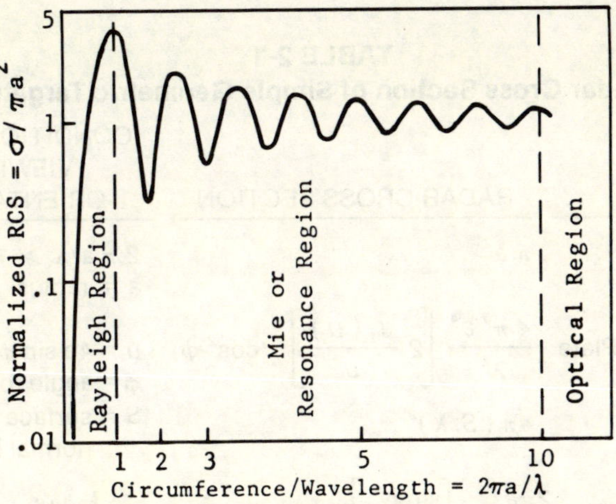


**Figure 2-1.** Radar cross section of a conducting sphere showing three regions.

**the slope of the curve within this wavelength region is +4.**

**In the wavelength region where $1 < 2\ \pi 2/\lambda\ v\ 10$, the radar cross section is oscillatory, defining the Mie region. In this resonance type of region, a so-called *creeping wave* travels around the sphere and back toward the receiver, interfering constructively or destructively with the specular backscatter, depending on the sphere radius.**

**In the wavelength region where $2\pi a/\lambda > 10$, the radar cross section of a sphere approaches its geometric projected area $\pi a^2$, which defines the optical region. In general, for very small wavelengths compared to the dimensions of an object, the object's radar cross section will essentially equal its geometric cross section.**

**Table 2-1 lists the radar cross section of several simple geometric shapes.**

**Figure 2-2 depicts the measured radar cross section of a conducting, square, flat plate, oriented as shown, as a function of the angle (Θ) incident to the surface. Near normal incidence to the plate, the radar cross section may be approximately described by the expression**

$$\sigma(\Theta) = \frac{4\pi A^2}{\lambda^2}\left[\frac{\sin(kb\sin\Theta)}{kb\sin\Theta}\right]^2\cos^2\Theta \qquad (2\text{-}20)$$

**where**

**$A$ = area of the square plate**
**$b$ = dimension of plate side**
**$\Theta$ = angle from normal to the plate**

**TABLE 2-1**
**Radar Cross Section of Simple Geometric Targets**

| SHAPE | RADAR CROSS SECTION | CONDITIONS AND VIEWING ORIENTATION |
|---|---|---|
| Sphere | $\pi a^2$ | $2\pi a/\lambda > 10$<br>$a$ = radius |
| Flat Ciroular Plate | $\frac{4\pi^3 a^4}{\lambda^2}\left[2\frac{J_1(u)}{u}\right]^2 \cos^2\phi$ | $u = 4a\sin\phi/\lambda$<br>$\phi$ = angle to normal |
| Flat Plate | $4\pi(S/\lambda)^2$ | $S$ = surface area, normal incidence |
| Cylinder | $\frac{2\pi a \ell^2}{\lambda}\left[\frac{\sin N}{N}\right]^2 \cos\phi$ | $a$ = radius<br>$\ell$ = length<br>$N = 2\pi\ell\sin\phi/\lambda$<br>$\phi$ = angle to side normal |
| Cone | $\pi a^2 \tan^2\alpha$ | $\alpha$ = half angle<br>$a$ = base radius, nose on incidence |

The pattern of the curve of Figure 2-2 represents the measured backscatter pattern of a square plate as the plate is rotated at various angles to the receiver (observer).[4] In a manner analogous to that of an antenna pattern, the plate's backscatter pattern may be described in terms of a mainlobe width (at the -3dB points) and sidelobe levels (relative to the mainlobe peak). For purposes of radar backscatter calibration, the flat plate backscatter pattern shown in Figure 2-2 is seen to have high sidelobes and mainlobe width which decreases with increasing plate size, for a fixed radar wavelength.

Figure 2-3 depicts the measured radar cross section of this same plate rotated 45 degrees about its center and line-of-sight to the radar as a function of incidence angle $\Theta$ to its surface.[4] Note that the magnitudes of the sidelobes are depressed relative to those shown in Figure 2-2. This effect can be explained by regarding the plate reflectivity pattern in Figure 2-2 as the result of a uniform illumination of a surface, and that of Figure 2-3 as the result of a linearly tapered illumination. The square plate oriented as shown in Figure 2-3 is a desirable calibration target from the standpoint of multipath interference rejection (low sidelobes), but its relatively narrow mainlobe width may make this flate plate reflector a difficult target to align to the radar line-of-sight.

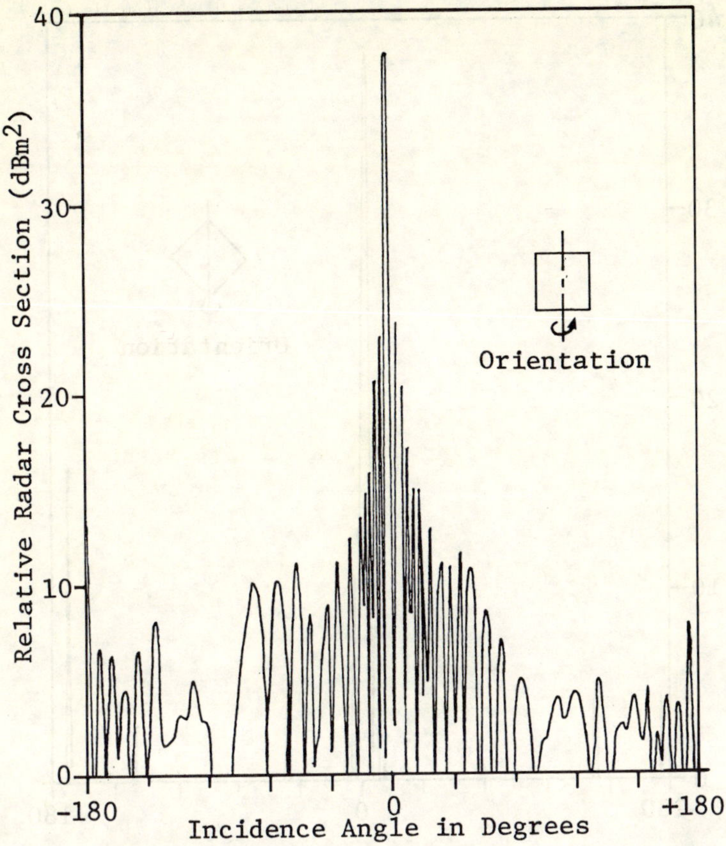


**Figure 2-2.** Measured radar cross section of square, conducting, flat plate as a function of incidence angle to the surface (reprinted from Trebits, ref. [4]).

Several flat conduting plates may be combined, normal to each other, to create targets for radar calibration with particularly advantageous backscatter patterns. A combination of two plates at right angles defines a diplane or dihedral corner reflector, while a combination of three mutually perpendicular flat plates defines a trihedral corner reflector. Both reflectors have the general property of completely reversing the direction of an incident ray by means of multiple bounces of the rays on the reflector's sides. This reflective property gives these types of reflectors a wide degree of latitude in alignment to a radar, and makes them extremely useful for various radar calibration procedures.

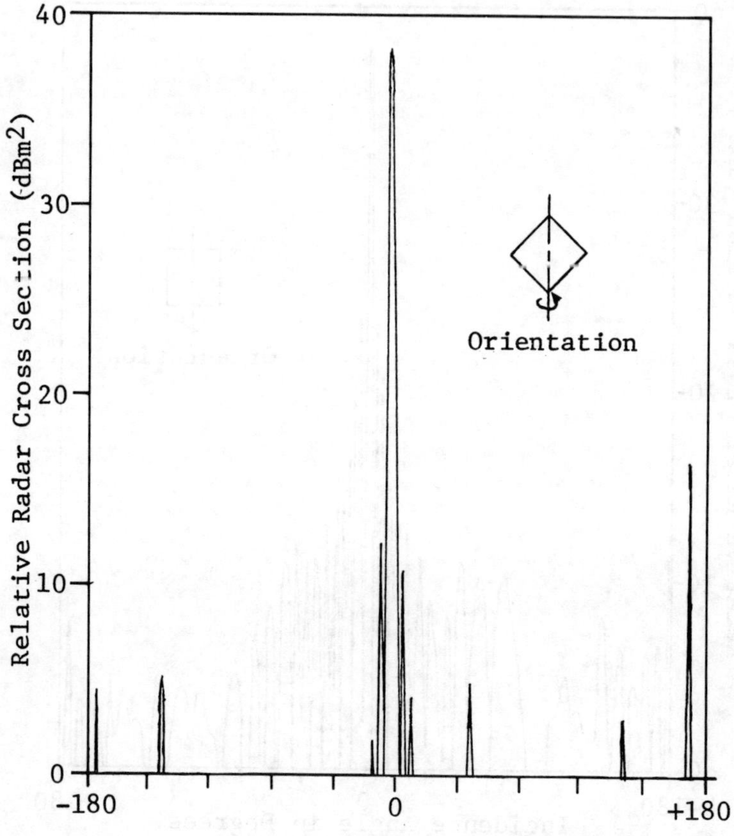


**Figure 2-3.** Measured radar cross section of a square, conducting, flat plate rotated 45° to the incident polarization as a function of the incidence angle to the surface (reprinted from Trebits, ref. [4]).

The trihedral reflector is the most commonly used corner reflector and is designed with a variety of shaped sides, including triangles, circular sectors, and squares, as shown in Figure 2-4. Table 2-2 summarizes the peak radar cross section of each of these trihedral corner reflector configurations and the azimuthal mainlobe width of each, when all reflectors are oriented with one crease vertical.

In Table 2-2 the dimension $a$ represents the crease length, and the mainlobe width is defined between the –3 dB points of the reflector backscatter pattern.

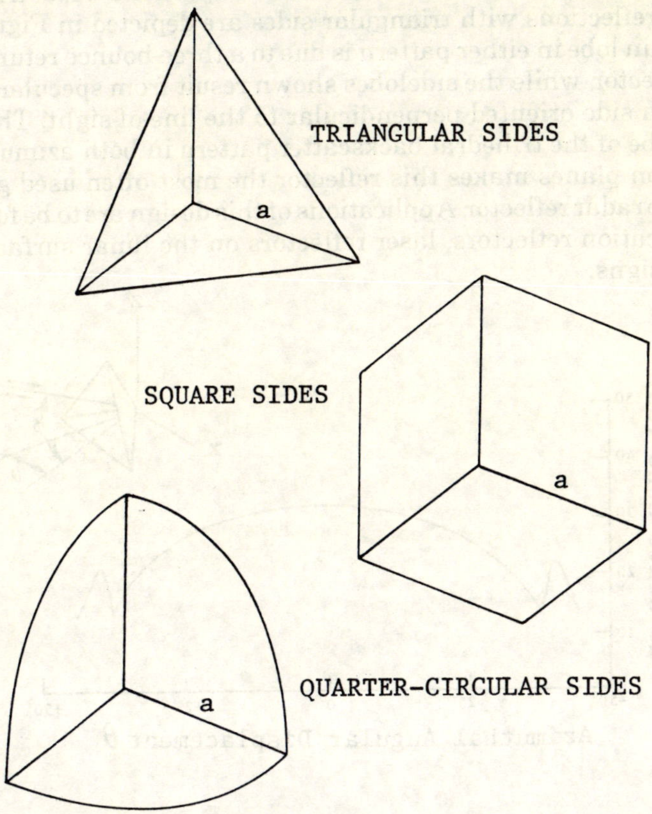


**Figure 2-4.** Trihedral corner reflectors with various shaped sides.

**TABLE 2-2**
**Trihedral Radar Cross Section**

| | $\sigma$ ( *max* ) | MAINLOBE WIDTH |
|---|---|---|
| Square Side Reflector | $12\ \pi a^4/\lambda^2$ | 23° |
| Quarter Circle Side Reflector | $.507\, \pi^3 a^4/\lambda^2$ | 30° |
| Triangle Side Reflector | $4\, \pi a^4/3\lambda^2$ | 40° |

The azimuth and elevation backscatter patterns of a trihedral corner reflections with triangular sides are depicted in Figure 2-5. The main lobe in either pattern is due to a three-bounce return from the reflector, while the sidelobes shown result from specular reflecion off a side oriented perpendicular to the line-of-sight. The wide mainlobe of the trihedral backscatter pattern in both azimuth and elevation planes makes this reflector the most often used general purpose radar reflector. Applications of this design are to be found in boat location reflectors, laser reflectors on the lunar surface, and traffic signs.

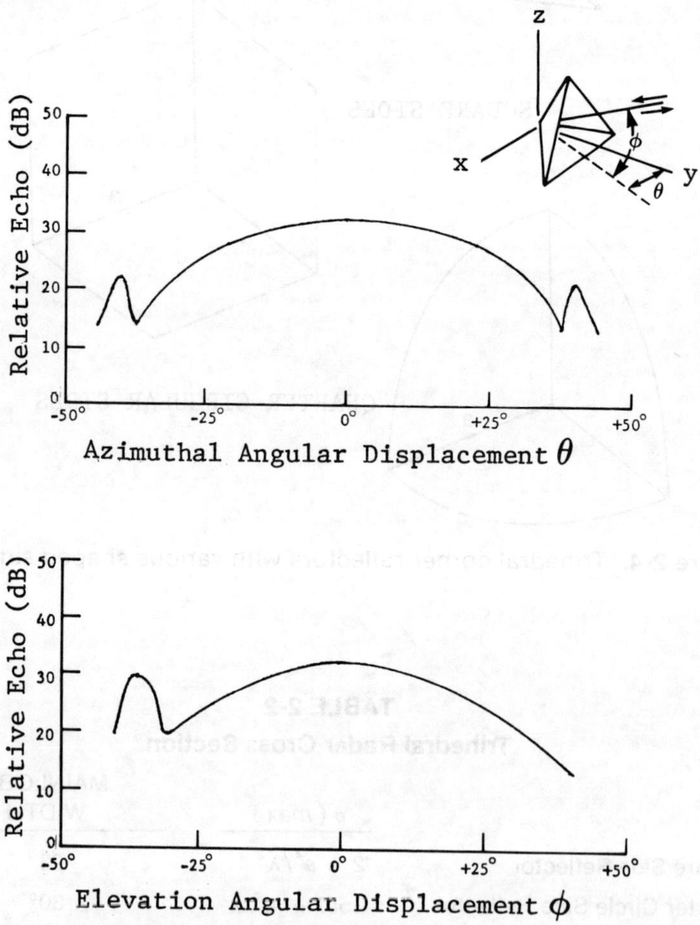


**Figure 2-5.** Backscatter pattern for a trihedral corner reflector with triangular sides.

The trihedral corner reflector, as a retroreflector, is an odd-bounce object, with its three-bounce area formed by a hexagon defined by cutting off the corners of the projection of the reflector viewed on its symmetry axis, as shown in Figure 2-6. Linearly polarized incident radiation is backscattered in a specular manner: the reflected polarization is the same as the incident. However, the reflection of circularly polarized incident radiation is of the opposite sense. That is, right circularly polarized incident radiation will be reflected as left circular radiation, an effect which may make the trihedral corner reflector useless as a target for circularly polarized signals. However, grids or dielectric coatings have been used to permit the use of corner reflectors with circularly polarized signals.

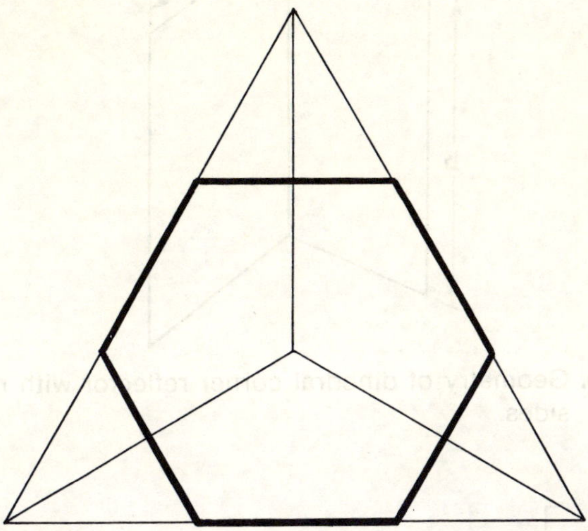

**Figure 2-6.** Hexagon three-bounce area of a trihedral corner reflector with triangular sides viewed at maximum cross section.

The dihedral corner reflector is made up of two mutually perpendicular flat metal plates, each usually rectangular in shape, as shown in Figure 2-7(a). If the dimensions of the rectangular sides are *a* and *b*, the maximum radar cross section for the dihedral corner reflector is:

$$\sigma(max) = \frac{8\pi a^2 b^2}{\lambda^2} \tag{2-21}$$

The backscatter pattern for the dihedral corner reflector in the azimuth plane (rotation axis at the reflector seam) is shown in

Figure 2-7(b). The two sidelobes symmetrically located at ± 45° from the maximum value are the result of specular reflection from each rectangular side. The analogous pattern in the elevation plane is identical to that of a rectangular flat plate of height $b$ and width $a\sqrt{2}$ as described in Equation (2-20). Thus, the dihedral backscatter pattern is wide in the azimuth plane and typically narrow in the elevation plane, depending on the height of the reflector (in wavelengths).

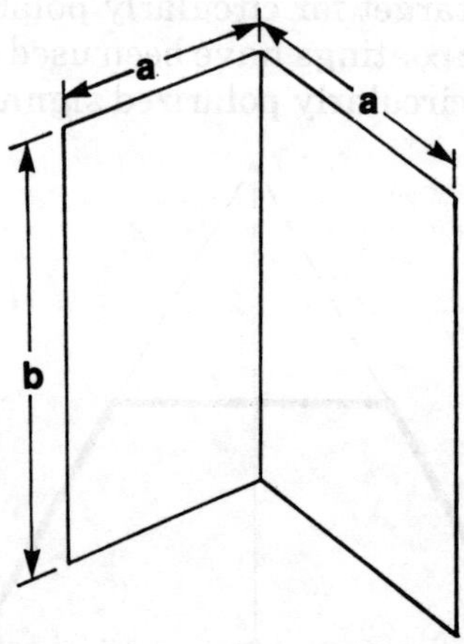


**Figure 2-7a.** Geometry of dihedral corner reflector with rectangular sides.

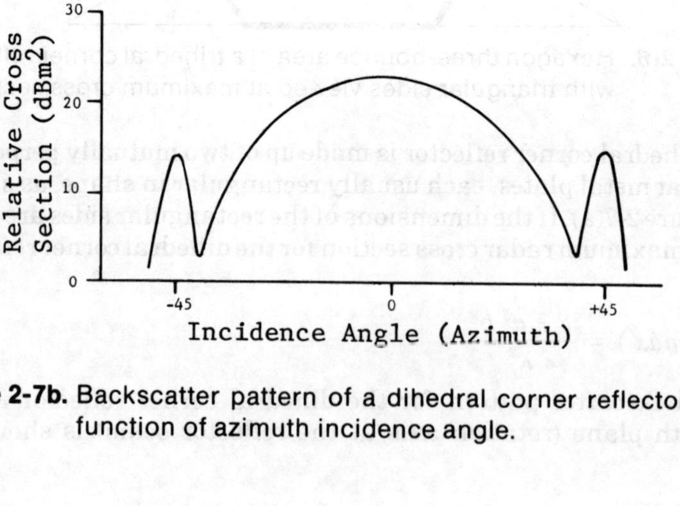


**Figure 2-7b.** Backscatter pattern of a dihedral corner reflector as a function of azimuth incidence angle.

Linearly polarized radiation incident on the dihedral corner reflector with an electric field vector either parallel or perpendicular to the seam will be reflected with the same polarization. Circularly polarized incident radiation will be reflected also with the same sense circular polarization, a characteristic of an even-bounce object.

An unusual feature of the polarization sensitivity of the dihedral corner reflector is its effect on incident linearly polarized radiation striking it at some arbitrary angle with respect to the seam, as shown in Figure 2-8. The reflected radiation polarization will be linear, but will be oriented at an equal, but opposite, angle to the dihedral seam. If the dihedral reflector is rotated about an axis parallel to the radar line of sight, the reflected linear polarization will rotate in the opposite direction.

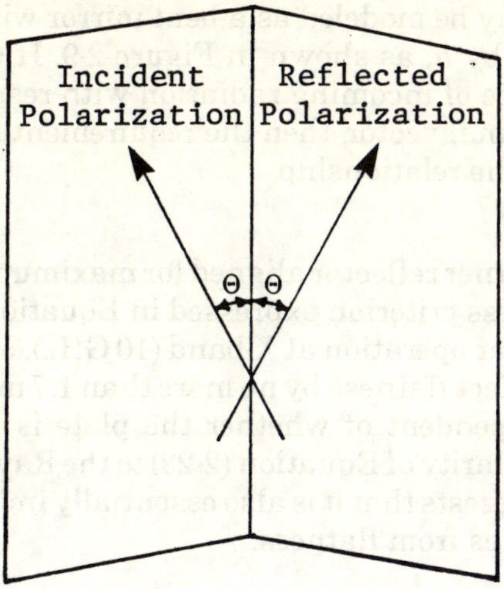


**Figure 2-8.** Polarization sensitivity of dihedral corner reflector.

The polarization sensitivity of the dihedral corner reflector can be exploited in the calibration of the cross-polarization receiver of a radar transmitting linearly polarized energy (*see* Chapter 4). Recall that a trihedral corner reflector or a dihedral corner reflector, with its seam parallel to the incident linear polarization, can be used to calibrate the amplitude transfer function of the co-polarized radar receiver. If the dihedral corner reflector is rotated 45° about the radar line-of-sight, the seam will be oriented at that same angle to the incident radiation polarization direction. The backscatter radiation polarization will be rotated 45° with respect to the seam in the opposite direction, and therefore will be rotated 90° with respect to the incident polarization. Thus, all reflected energy will appear in the cross-polarized receiver of the radar. Analogously, rotation of the dihedral by 22.5° from the incident radiation polarization will result in equal power in both polarization receiver channels of the radar.

The experimental accuracy of using a corner reflector for quantitative radar measurements may be compromised by inaccuracies of the reflector construction, which may take two forms. The first involves the degree of flatness of each plate, while the second involves the accuracy with which the plates are attached orthogonally to each other.

Reflector sides can become "non-flat" through bowing of too thin metal, inaccurate machining, or surface irregularities in the metal itself. A plate may be modeled as a bent mirror with an amount of concavity given by $h$, as shown in Figure 2-9. If $\Theta$ represents the angle of incidence of incoming radiation with respect to the undeformed plate normal vector, then the requirement for flatness may be described by the relationship

$$h \cos \Theta \leq \lambda/8 \tag{2-22}$$

For a dihedral corner reflector aligned for maximum cross section, $\Theta$ is 45°. The flatness criterion expressed in Equation (2-22) then implies that for radar operation at X-band (10 GHz), each plate should deviate from perfect flatness by no more than 1.7 mm. This flatness criterion is independent of whether the plate is bent concave or convex. The similarity of Equation (2-22) to the Rayleigh roughness criterion also suggests that it is also essentially independent of how each plate deviates from flatness.

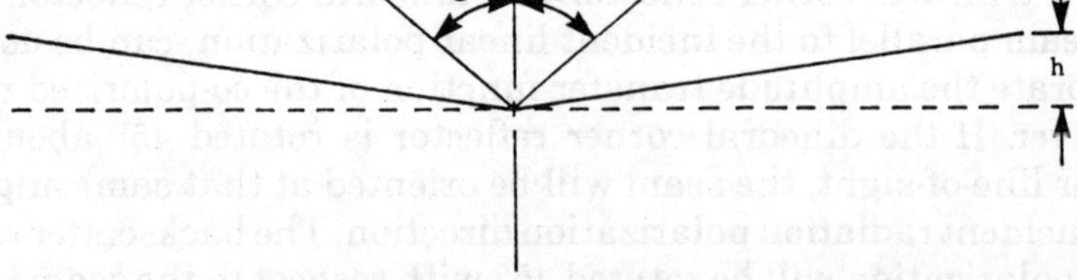


**Figure 2-9.** Bent mirror model of a flat plate (reprinted from ref. [6]).

Angular errors degrade corner reflector performance by causing the retroreflected energy to be split into two beams for each non-perpendicular angle. Thus, for a trihedral corner reflector, when one angle differs from 90°, the triply reflected energy is split into two beams. When two angles differ from 90°, four divergent beams result, and when all three angles differ from 90°, six divergent beams result. The net effect of this beam splitting is to decrease the radar cross section of the corner reflector.

A pertinent parameter to quantify the radar cross section degradation due to improper angle alignment is denoted by Δ, the linear error at the extremity of an edge, which is required to reduce the cross section by half (–3 dB). For a dihedral corner reflector with square sides, the amount of edge deviation is predicted by the expression

$$\Delta = 0.31\,\lambda \tag{2-23}$$

For a trihedral corner reflector with triangular sides, the amount of single edge deviation required to lower the reflector's cross section by 3 dB is given by the expression

$$\Delta = 0.70\,\lambda \tag{2-24}$$

If all three angles are misaligned equally, the analogous expression becomes

$$\Delta = 0.35\,\lambda \tag{2-25}$$

Table 2-3 summarizes several combinations of angular error for various corner reflector cases.[6]

**TABLE 2-3**
**Edge Deviations in Reflector Sides**
(*Adapted from ref.* [6])

| Corner Reflector | No. Errors | Side | EDGE DEVIATIONS (-1dB) | (-3dB) | (-10dB) |
|---|---|---|---|---|---|
| Dihedral | 1 | Square | .17 λ | .31 λ | .61 λ |
| Trihedral | 1 | Square | .24 λ | .40 λ | .66 λ |
| Trihedral | 1 | Triangular | .42 λ | .70 λ | 1.21 λ |
| Trihedral | 3 | Square | .14 λ | .24 λ | .44 λ |
| Trihedral | 3 | Triangular | .20 λ | .35 λ | .62 λ |

## 2.4 RADAR CROSS SECTION OF COMPLEX OBJECTS

The backscatter patterns from real world objects tend to exhibit complexity in terms of variation in cross section amplitude with relatively small changes in target aspect with respect to the radar line-of-sight. Targets such as aircraft and vehicles can be seen to be made up of many discrete scattering centers. The energy reflected from these scattering centers will add constructively or destructively as a function of target aspect due to the spatial arrangement, orientation, and individual reflectivity of these centers. Thus, there will be a rapid fluctuation of radar cross section with target aspect angle, an observation which characterizes large targets.

Figure 2-10 shows the radar cross section of a B-26 (World War II bomber) as a function of azimuth aspect angle.[7] Reflectivity data are presented in polar format, which highlights the symmetry of this aircraft's backscatter pattern with the physical symmetry of the aircraft itself. In general, this type of target exhibits higher radar cross sections at the so-called principal aspects: head-on, tail-on, and broadside. If the same aircraft were scanned for azimuth reflectivity with the radar elevated with respect to the horizontal plane of the aircraft, the resulting backscatter pattern would perhaps bear a gross resemblance to Figure 2-10, but would differ greatly in the details of null locations and relative peak levels.

The complexity of the backscatter pattern of this aircraft requires

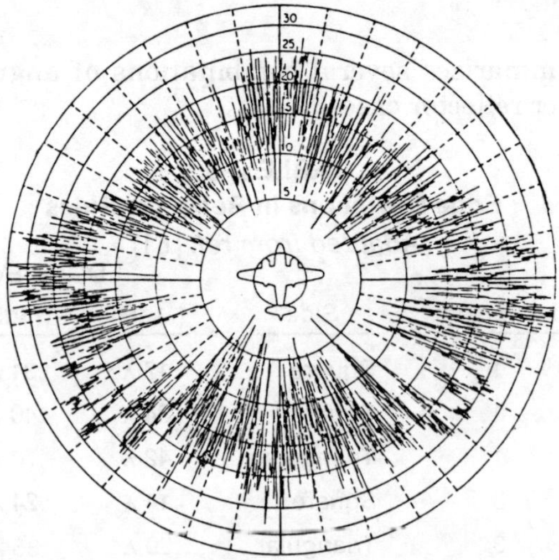


**Figure 2-10.** Backscatter of a B-26, two engine bomber as a function of azimuth angle (reprinted from Ridenour, ref. [7]; ©1947 McGraw-Hill Book Company).

that care be taken in defining the radar cross section of such a target for radar system performance evaluation, for instance. Several different means of defining a target RCS may be appropriate depending on the target and its involvement with the radar. A median RCS value is often used, calculated over a range of target aspect angles, where the median value represents the fifty percent point of the amplitude cumulative probability distribution function of RCS values. For a radar landing system the range of target aspects might reasonably be limited to ±10 degrees from head-on orientation. For a surveillance radar application, it may be more reasonable to select the ten percentile point for the aircraft RCS, such that 90 percent of target aspects have RCS values which exceed the 10 percentile value. Such reasoning purposely underestimates the reflectivity of the aircraft and increases the design burden on the radar system required to detect and/or track this target.

Figure 2-11 shows the full 360 degree backscatter patterns for an automobile at 70 GHz.[8] The upper pattern corresponds to $\sigma_{HH}$, and the lower pattern corresponds to $\sigma_{HV}$. The parallel-polarized radar cross section for this automobile assumes its highest value at broadside orientation (0 degrees), with other peaks at head-on and tail-on orientations. The cross-polarized radar cross section pattern shows similar, but smaller, amplitude aspect dependency.

## 2.5 RADAR CROSS SECTION OF DISTRIBUTED OBJECTS

Quantification of the radar reflectivity of land/sea clutter and precipitation clutter requires additional definition of terminology. In the case of land or sea backscatter the relevant backscatter parameter is the radar reflectivity per unit area illuminated by the radar, or

$$\sigma^\circ = \sigma / A \qquad (2\text{-}26)$$

where

$\sigma^\circ$ = radar reflectivity per unit area (called *sigma zero*)

$A$ = illuminated ground/sea patch

$\sigma$ = radar cross section of illuminated area

The units of $\sigma^\circ$ are obviously $m^2/m^2$ and are usually specified in decibels, e.g., -15 dB.

In the literature authors will also use the terms "normalized cross section" and "scattering coefficient" for $\sigma^\circ$. Care must be exercised, however, since these same terms are occasionally used in reference to the parameter $\gamma$, where

$$\sigma^\circ = \gamma \sin \Theta \qquad (2\text{-}27)$$

and $\Theta$ is the grazing angle to the local surface.

It is implicit in the definition of $\sigma^\circ$ that the illuminated surface contains a large number of scattering elements whose individual

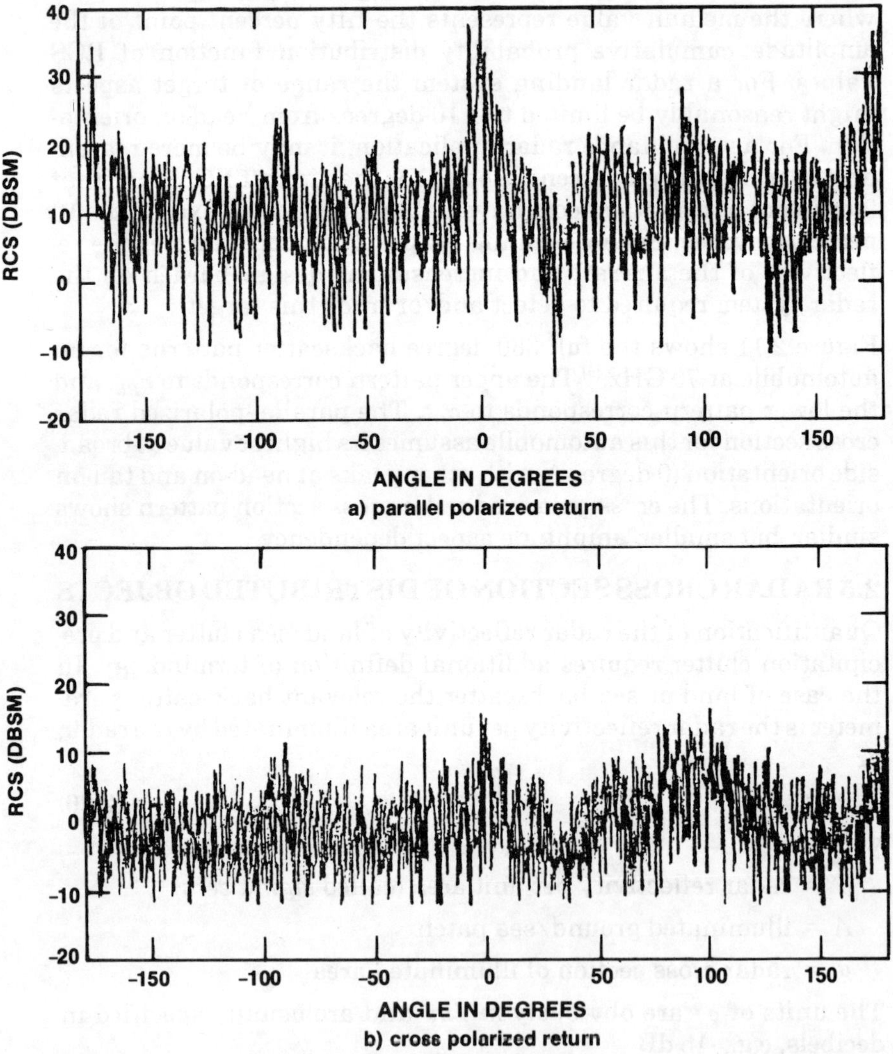


**Figure 2-11.** Radar cross section of an Oldsmobile Cutlass at 69.8 GHz for horizontally polarized transmission: a) parallel polaried return; b) cross polarized return (reprinted from Martin, ref. [8]).

phases are statistically independent. This condition permits the superposition of power in computation of average reflected signal levels. The size of the illuminated surface that permits this assumption is dependent on radar wavelength and geometry, since size must be interpreted in terms of wavelength units on the surface.

The parameter $\sigma^\circ$ is a function of radar frequency and polarizaton, surface content and roughness, and geometry. Ground reflectivity depends on the grazing angle, foliage condition, snow cover, and wetness, for example. Sea reflectivity depends on the sea state, wind/wave direction, and wind speed, for example.

The analogous reflectivity parameter for atmospheric backscatter is the radar reflectivity per unit volume illuminated by the radar, or

$$\eta = \sigma / V \tag{2-28}$$

where

$\eta$ = radar reflectivity per unit volume (sometimes also referred to as $\sigma_v$)

$V$ = volume of resolution cell

$\sigma$ = radar cross section of resolution cell

The units of $\eta$ are $m^2/m^3$ or $m^{-1}$, and $\eta$ is also often specified in decibels per meter$^{-1}$.

The parameter $\eta$ is also a complex function of radar frequency and polarization, volume content, and geometry. Rain backscatter, for example, depends on rainfall rate, drop size distribution, and look angle, all of which can vary from location to location and from one time of year to another. Frozen hydrometeor backscatter, furthermore, depends on the mixture of rain, sleet, snow, and hail prevailing within the illuminated volume, and the backscatter of dust depends on the size distribution and dielectric properties of the particles.

Although $\sigma^\circ$ and $\eta$ themselves are target parameters independent of the radar, that quantity which one actually measures is the radar cross section of the illuminated resolution cell. The reflectivity parameters $\sigma^\circ$ and $\eta$ are then derived from the measured radar cross section $\sigma$ using either Equation (2-26) or (2-28), as appropriate. This derivation requires calculation of either the illuminated area $A$ on the ground/sea surface or the volume $V$ of the resolution cell.

The antenna beam shape is one factor in the determination of either $A$ or $V$, and one-way antenna patterns (3 dB beamwidths, azimuth and elevation) have been used in many reported results. Note, however, that some use of two-way, 3 dB beamwidths has also been documented. Therefore, care must be exercised when comparing reflectivity data from different investigators and different measurements exercises.

In the determination of surface reflectivity, the calculation of the illuminated area $A$ depends on the geometry. Two different geometries are defined, as determined by the elevation beamwidth $\phi_{el}$, the slant range $R$, and the radar pulse length $\tau$. The pulse length limited case is defined when the radar range resolution projected onto the surface is smaller than the range extent of the area illuminated by the antenna beam, as shown in Figure 2-12. The area of the largest resolution cell for this case is approximately rectangular and can be calculated by multiplying the projected range resolution $(c\tau/2)\sec\Theta$ by the width of the illuminated ellipse $2R\tan(\phi_{az}/2)$. For the pulse length limited case then, the area $A$ is given by

$$A = 2R\,(c\tau/2)\tan(\phi_{az}/2)\sec\Theta \tag{2-29}$$

where

$$\tan\Theta < \frac{2R\tan(\phi_{el}/2)}{c\tau/2} \tag{2-30}$$

and uniform mainlobe illumination is assumed. A beam shape factor of 1.33 may be included to account for non-uniform illumination.

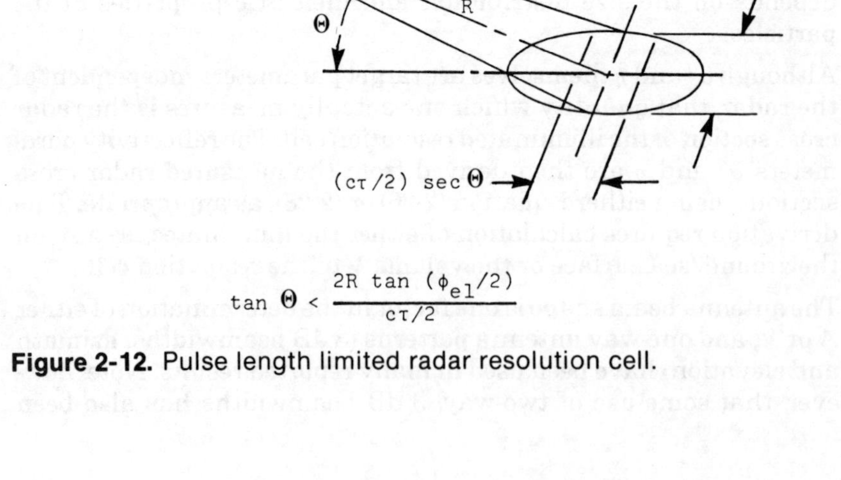


**Figure 2-12.** Pulse length limited radar resolution cell.

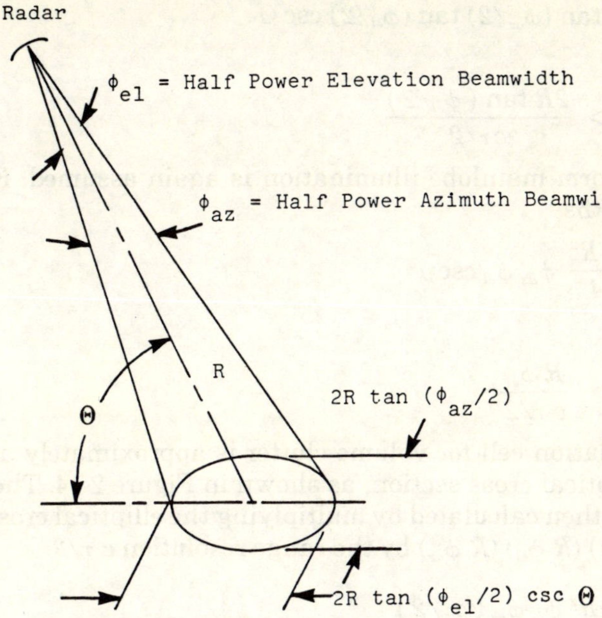


$$\tan \Theta > \frac{2R \tan (\phi_{el}/2)}{c\tau/2}$$

**Figure 2-13.** Beamwidth limited radar resolution cell.

For beamwidth values less than approximately 10°, the small angle approximation may be used:

$$A = R\ (c\tau/2)\ \phi_{az} \sec \Theta \tag{2-31}$$

where

$$\tan \Theta < \frac{R\,\phi_{el}}{c\,\tau/2} \tag{2-32}$$

The beamwidth limited case is defined when the radar range resolution projected onto the ground/sea surface is larger than the range extent of the area illuminated by the antenna beam, as shown in Figure 2-13. The resolution cell on a flat earth is an ellipse, with a range axis diameter of approximately $2R \tan(\phi_{el}/2) \csc \Theta$ and an azimuth axis diameter of approximately $2R \tan(\phi_{az}/2)$. The area $A$ of the resolution cell for the beamwidth limited case is then given by:

$$A = R^2 \tan(\phi_{az}/2) \tan(\phi_{el}/2) \csc \Theta \tag{2-33}$$

where

$$\tan \Theta > \frac{2R \tan(\phi_{el}/2)}{c\tau/2} \tag{2-34}$$

and uniform mainlobe illumination is again assumed. For small beamwidths

$$A - \frac{\pi R^2}{4} \phi_{az} \psi_{el} \csc\Theta \tag{2-35}$$

where

$$\tan \Theta > \frac{R \phi_{el}}{c\, \tau/2} \tag{2-36}$$

The resolution cell for volume clutter is approximately a cylinder with elliptical cross section, as shown in Figure 2-14. The cell volume $V$ is then calculated by multiplying the elliptical cross section area $(\pi/4)(R\ \phi_{el})(R\ \phi_{az})$ by the range resolution $c\ \tau/2$:

$$V = \frac{\pi}{4} R^2 \phi_{el} \phi_{az} (c\tau/2) \tag{2-37}$$

The value of $V$ calculated from Equation (2-37) is often reduced to account for the two-way illumination of the volume of the resolution cell, calculated from one-way 3 dB beamwidths.

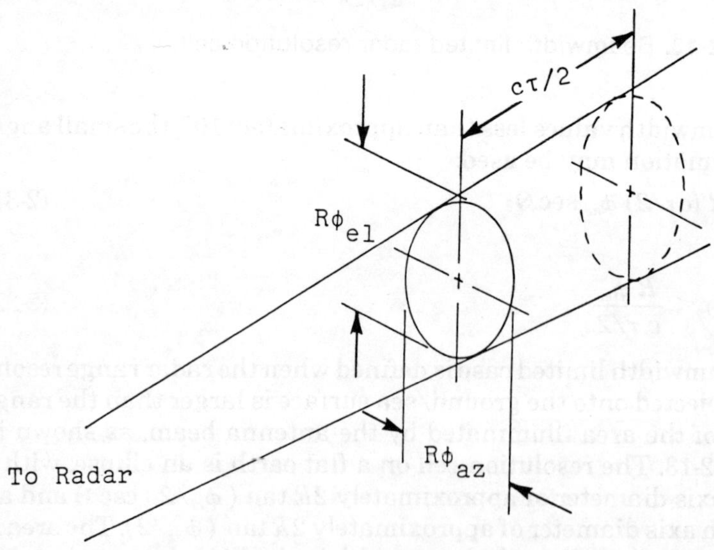


**Figure 2-14.** Volumetric radar resolution cell.

Unlike a stationary object, a distributed target such as land, sea, or precipitation will cause the backscatter radar signal to fluctuate over a wide range of amplitudes, due to the interference effects resulting from signals reflected from moving scattering centers. Therefore, $\sigma^\circ$ and $\eta$ will vary in time, and temporal averaging is necessary to describe "the reflectivity." In addition, mean values and median values of $\sigma^\circ$ and $\eta$ are often confused with each other when comparing different data sources. Differences between mean and median reflectivity values depend on the clutter source, geometry, and radar parameters themselves, and can differ by many dB, making comparisons all the more complicated.

In describing the general characteristics of $\sigma^\circ$, one may define three distinct regions of incidence angle, as shown in Figure 2-15. In the incidence angle region near grazing incidence, the ground or sea surface may appear "smooth," using Rayleigh's surface roughness criterion:

$$\Delta h \sin \Theta \leq /8 \tag{2-38}$$

where

$\Delta h$ = root-mean-square height of surface variation
$\Theta$ = incidence to the "plane" surface
$\lambda$ = radar wavelength

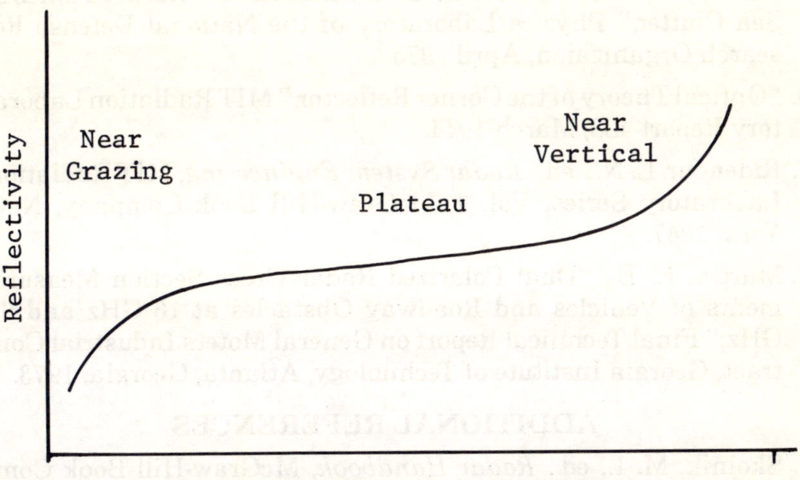


**Figure 2-15.** Three regions of incidence angle dependence for surface clutter backscatter.

In general, $\sigma^\circ$ increases rapidly in this region with increasing incidence angle. In the middle, or plateau, region the surface appears rough and $\sigma^\circ$ changes only slowly with increasing incidence angle. In the third, near vertical incidence region, specular scattering becomes dominant and $\sigma^\circ$ increases rapidly with incidence angle. Any surface reflectivity data which do not follow this general rule should be held highly suspect until the cause of the deviation is determined.

## REFERENCES

1. Long, M. W., *Radar Reflectivity of Land and Sea,* Artech House, Dedham, Massachusetts, 1975.
2. Ruck, G. T., ed., *Radar Cross-Section Handbook,* Vols. I and II, Plenum Press, New York and London, 1970.
3. Kerr, D. E., *Propagation of Short Radio Waves,* Radiation Laboratory Series, Vol. 13, McGraw-Hill Book Company, New York, 1951.
4. Trebits, R. N., "Sea Backscatter Test Plan," Technical Report on Contract N60921-77-C-A168, Georgia Institute of Technology, Atlanta Georgia, August 1978.
5. Sittrop, H., "X and $K_u$-band Radar Backscatter Characteristics of Sea Clutter," Physics Laboratory of the National Defense Research Organizaion, April 1975.
6. "Optical Theory of the Corner Reflector," MIT Radiation Laboratory Report 433, March 1944.
7. Ridenour, L. N., ed., *Radar System Engineering,* MIT Radiation Laboratory Series, Vol. 1, McGraw-Hill Book Company, New York, 1947.
8. Martin, E. E., "Dual Polarized Radar Cross Section Measurements of Vehicles and Roadway Obstacles at 16 GHz and 69 GHz," Final Technical Report on General Moters Industrial Contract, Georgia Institute of Technology, Atlanta, Georgia, 1973.

## ADDITIONAL REFERENCES

1. Skolnik, M. I., ed., *Radar Handbook,* McGraw-Hill Book Company, New York, 1979.
2. Skolnik, M. I., *Introduction to Radar Systems,* McGraw-Hill Book Company, New York, 1980.

3. "Radar Cross Section of Simple Shapes," Lockheed Missile and Space Company Report LMSC-D051213.
4. Meeks, M. L., *et al.*, "Reflection Properties of Radar Targets," Georgia Institute of Technology, Atlanta, Georgia, 1954.
5. Bonkowski, R. R., *et al.*, "Studies in Radar Cross Section VI," Project MIRO, Willow Run Research Center, Michigan, October 1953.
6. Jasik, Henry, ed., *Antenna Engineering Handbook,* McGraw-Hill Book Company, New York, 1961.
7. Trebits, R. N., *et al.*, "MM-Wave Reflectivity of Land and Sea," *Microwave Journal*, August 1978.
8. Nathanson, F. E., *Radar Design Principles,* McGraw-Hill Book Company, New York, 1969.

# CHAPTER 3

# BASIC RCS MEASUREMENT CONCEPTS AND SYSTEMS

**George W. Ewell**

**Engineering Experiment Station**
**Georgia Institute of Technology**

## 3.1 FUNDAMENTAL CONCEPTS

The objective of many measurement programs is measurement of the radar cross section (RCS) of targets (or clutter). The radar cross section is defined as

$$\sigma = \lim_{R\to\infty} 4\pi R^2 \left| \frac{E_s}{E_i} \right|^2 \tag{3-1}$$

where

$E_i$ = magnitude of incident electric field

$E_s$ = magnitude of scatered electric field in the direction of the observer

$R$ = distance from target to observer

and it is assumed that the incident and scattered polarizations are the same.

The value of radar cross section is rarely determined by direct measurement as suggested in Equation (3-1) above, but rather by measurement of received power reflected from the target. Received power, radar parameters, range and RCS are related by the *radar equation* which for free-space propagation may be expressed as

$$P_r = \frac{P_t G_t G_r \lambda^2 \sigma}{(4\pi)^3 R_1^{\,2} R_2^{\,2}} \tag{3-2}$$

where

$P_r$ = received power

$P_t$ = transmitted power

$G_t, G_r$ = transmitting and receiving antenna gains

$\lambda$ = operating wavelength

$R_1, R_2$ = ranges from transmitter and receiver to target, respectively.

If the transmitting and receiving site are collocated then Equation (3-2) assumes a more familiar form

$$P_r = \frac{P_t G_t G_r \lambda^2 \sigma}{(4\pi)^3 R^4} \tag{3-3}$$

where in this case

$R_1 = R_2 = R$ = range from radar-to-target

In many cases free-space conditions will not be satisfied and energy propagating by other than the direct path between the target and the radar may influence the received signals. The effects of such rays may be accounted for by inclusion of the propagation factor $F$, and the radar equation becomes

$$P_r = \frac{P_t G_t G_r \lambda^2 \sigma F_t^2 F_r^2}{(4\pi)^3 R^4} \tag{3-4}$$

In addition, there are applications for which system losses and atmospheric propagation losses must be separately accounted for in the radar equation, normally as loss factors (greater than one) which appear in the denominator.

The definition of RCS given by Equation (3-1) is completely general, but is a function of target, aspect, polarization, and frequency. Polarization dependence of a target is often expressed in terms of the polarization scattering matrix, which may be expressed

$$S = \begin{bmatrix} \sqrt{\sigma_{11}}\, e^{j\phi_{11}} & \sqrt{\sigma_{12}}\, e^{j\phi_{12}} \\ \sqrt{\sigma_{21}}\, e^{j\phi_{21}} & \sqrt{\sigma_{22}}\, e^{j\phi_{22}} \end{bmatrix}$$

where the $\sigma$'s are for various combinations of (1) transmit and (2) receive polarizations, and the $\phi$'s define the reflection phase.

In this formulation any two orthogonal polarizations may be selected, and often horizontal and vertical polarization are chosen, for which case

$$S = \begin{bmatrix} \sqrt{\sigma_{HH}}\, e^{j\phi_{HH}} & \sqrt{\sigma_{HV}}\, e^{j\phi_{HV}} \\ \sqrt{\sigma_{VH}}\, e^{j\phi_{VH}} & \sqrt{\sigma_{VV}}\, e^{j\phi_{VV}} \end{bmatrix}$$

Note that using this notation, for example, $\sigma_{HH}$ is the RCS for horizontally polarized transmission and reception and $\sigma_{HV}$ is the cross-polarized RCS. In the case of linear, reciprocal conditions, $\sigma_{HV} = \sigma_{VH}$.

Another widely used variation of the radar equation arises when descriptors of extended radar clutter are desired. Since the RCS of extended clutter is a function of the amount of clutter illuminated, it is often desirable to describe clutter in terms of some normalized parameter. In the case of surface clutter

$$\sigma_c = \sigma^o \cdot A \tag{3-5}$$

where

$\sigma_c$ = clutter cross section (RCS)

$\sigma^\circ$ = RCS per unit area

$A$ = illuminated area

Often at low angles, the area may be approximated by

$$A \sim kR\Theta \frac{c\tau}{2} \sec \psi \tag{3-6}$$

where

$\Theta$ = 3 dB antenna beamwidth

$c$ = velocity of light

$\tau$ = transmitted pulse length

$\psi$ = depression angle

$k$ = a constant whose value depends upon beam shape, is normally close to unity, and is often assumed equal to one

More precise values or higher angles must include beam shape, and are available in the literature.[1] For volumetric clutter,

$$\sigma_c = \eta V \tag{3-7}$$

and a normal approximation for illuminated volume is

$$V = k \frac{\Omega R^2 c\tau}{2} \tag{3-8}$$

where

$\Omega$ = solid angle defined by the antenna 3 dB beamwidths

$k$ = a constant dependent upon beam shape, often assumed equal to one

In almost all cases, an RCS measurement program is intended to provide data for subsequent analysis of system performance. Questions concerning the validity of RCS data are usually minimized if the principal radar parameters of the measurement system and the system to be analyzed are similar; significant differences in frequency, polarization or resolution should be closely examined to assess their effect on the utility of any resulting RCS data. In addition, the measurement system must be accurately calibrated so that meaningful measurements may be performed, and the targets must be accurately described so that it may be determined when the various data should be utilized. Recorded data must be of such a nature that they can be utilized for system analysis, and such data must accurately describe the behavior of the targets of interest over all applicable geometries and target conditions. Often targets have amplitudes and/or phases which vary with time and space, so that in addition to a measure of average or median value, some measure of the shape of the amplitude distribution, the temporal and spatial correlation functions, or the temporal and spatial frequency spectra and will be desired.

The particular type of system utilized to perform measurements may strongly influence the characteristics of the data. In particular, the decision to utilize a CW or pulsed system, the overall system resolution, and the frequency of operation may often impact the requirements for the measurement system.

In addition to the selection of general radar measurement system type, a number of other factors are important in planning and carrying out a program of reflectivity measurement, including:

Adequate Sensitivity
Dynamic Range
System Stability and Interference Levels
System Resolution
Range Sampling
System Calibration
Environment/Target Description
Basic Data Objective/Data Quantity

These factors are discussed in the following sections, first in terms of a conventional pulsed radar measurement system, but with extensions to other types of systems as appropriate. The remainder of the chapter is divided into several sections including characterization of the measurement system and discussion of the data requirements for a measurement program. The chapter concludes with sections discussing some types of measurement systems and a summary of some representative measurement procedures.

## 3.2 SYSTEM CHARACTERIZATION AND CALIBRATION

Perhaps the most common radar cross section measurement system used to gather data in the field is a pulsed, single frequency, moderately short pulse (0.1 to 1.0$\mu s$) radar system, and will be used to illustrate many of the factors involved in characterization of a measurement system. Such a pulsed system may function as shown in the simplified block diagram of Figure 3-1. In such a system, the transmitter generates a pulse of the desired duration, which is sent to the antenna through the duplexer and radiated into space. Upon reception, the signal is routed to the receiver, mixed with the local oscillator, amplified at an intermediate frequency and detected to form the video signal. This video signal is sampled at one or more desired points, and the resulting information is recorded on a permanent storage medium. The behavior of the sampling circuit and its approximation to "ideal" performance may be important, particularly for shorter pulses. In many cases, recorded data will consist of an analog or digital record of the received signal present in one or more range gates. The recorded signals may then be analyzed in order to extract the pertinent features off-line.

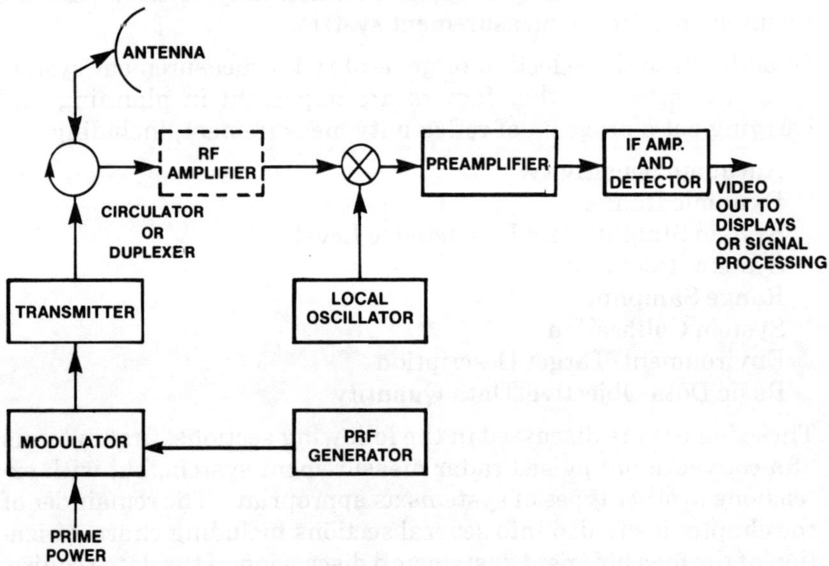


**Figure 3-1.** Simplified pulsed radar block diagram.

In order for a radar measurement program to provide meaningful results, the radar used for the measurement must be both characterized and calibrated. Characterization involves the measurement of parameters which define performance of the radar, such as bandwidth, power, dynamic range, and minimum detectable signal. Calibration, on the other hand, is used to establish the connection between measured quantities (such as video voltage) and values of radar cross section. In addition, if phase measurements are to be performed, the characterization of the radar must include those parameters which are necessary if meaningful coherent spectral and correlation values are to be obtained.

In general, a radar measurement program may be carried out with either a coherent or an incoherent radar system. However, even if a coherent system is utilized it is normal practice to operate in an incoherent mode or an amplitude-only mode when acquiring many radar cross section measurements. If coherent target and clutter spectra are of interest, often the output of the phase channels are separately recorded in order to obtain such data. Thus, there are two basically different types of measurements which are performed with radar systems: amplitude and phase, and different characterization, calibration, and validation procedures are utilized for the two system types.

### 3.2.1 Amplitude Calibration

There are a number of different methods for amplitude calibration of a radar system. While there is no universally accepted convention, the four most common and straightforward may be classed as:

Absolute Calibration

Relative Calibration

Mixed Calibration

Indirect Calibration

Each of these calibration procedures has its own advantages and disadvantages as discussed in the following paragraphs.

*Absolute Calibration*

When an absolute calibration is performed, some feature of the target return is referenced or compared with the return from a standard target of known cross section at the same range. Such standard targets may include spheres, corner reflectors, or Luneburg lenses. A true absolute measurement should use a reference target at the same range and having the same RCS as the target being measured. However, standard targets of convenient RCS are often utilized at different ranges and a range correction factor applied. If the standard target is at a different range, additional factors such as multipath returns, TR tube recovery, sensitivity time control, automatic gain control, and possible ac coupling all need to be taken into account in any such range correction factor.

*Relative Calibration*

A relative calibration serves to establish the transfer function of the receiver; i.e., to define a relationship between power input and voltage output. A more complete characterization is obtained if the relative calibration also includes the properties of the data acquisition system; that is, establishing a relationship between input power and data stored on the permanent storage medium. Such a calibration may be accomplished by varying an input signal in known increments throughout the receiver and data acquisition system dynamic range, while recording the resulting signals with the data acquisition system.

The direct relationship between input power and received data values is in general known at only a few specific points and often some means of interpolation is required to determine intermediate values. Piecewise linear, least mean squares linear, and least mean squares cubic fits are often employed for such interpolation. Figure 3-2 shows such fits for a calibration of a logarithmic receiver and a six-bit data recording system. There is some merit in the use of the cubic fit [Figure 3-2 (c)] due to its realistic behavior for both large and small signal values.

While a relative calibration by itself does not permit determination of radar cross section, it will permit calculation of fluctuation statistics, and may be combined with other techniques for calculation of radar cross section of targets or clutter.

When conducting a relative calibration, the presence of automatic gain control, possible ac coupling of the receiver, and saturation of stages within the receiver may be factors which should be taken into account.

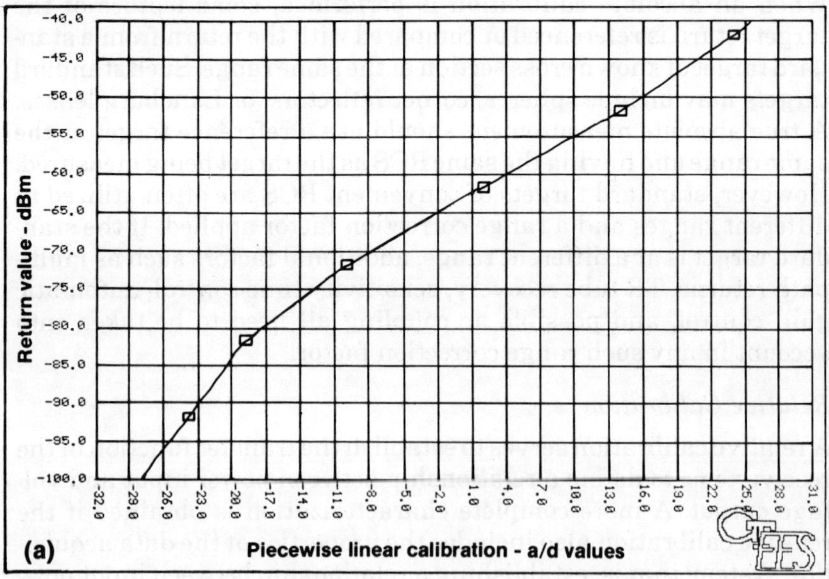


(a) Piecewise linear calibration - a/d values

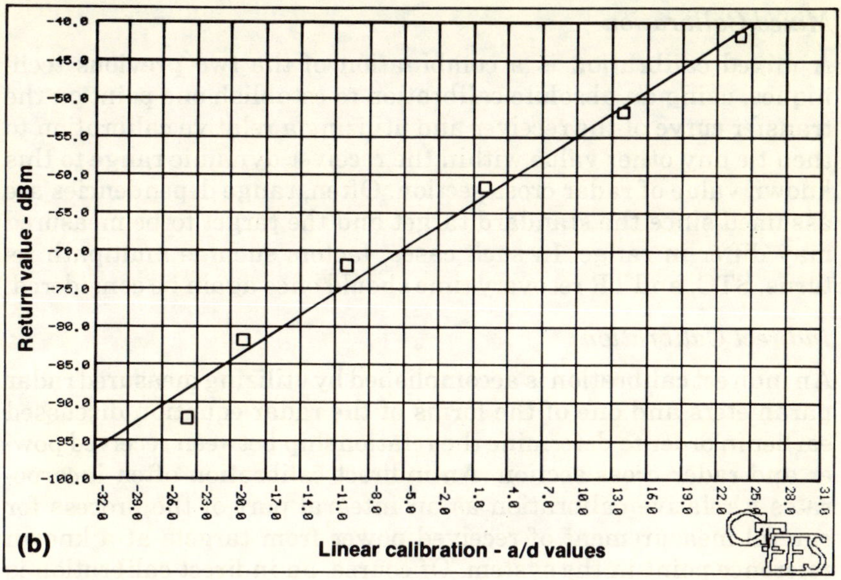


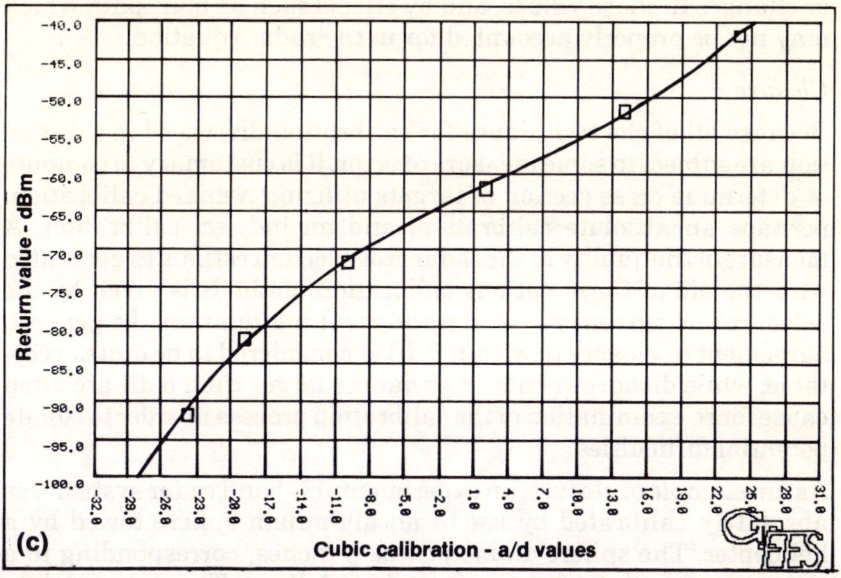


**Figure 3-2.** Calibration curves for a high resolution radar showing injected signal levels as squares along with fitted curves using: (a) piecewise linear interpolation, (b) least mean square linear fit, and (c) least mean square cubic fit.

*Mixed Calibration*

A mixed calibration is a combination of the two previous techniques, using an absolute calibration to establish one point on the transfer curve of the receiver and utilizing a relative calibration to then tie any other value within the receiver dynamic range to this known value of radar cross section. Often, range dependencies are assumed since the standard target and the target to be measured may differ in range. In such cases, factors such as multipath returns, STC, and TR recovery time should once again be considered.

*Indirect Calibration*

An indirect calibration is accomplished by utilizing measured radar parameters and one of the forms of the radar equation discussed earlier in order to determine the relationship between received power and radar cross section. An indirect calibration often incorporates a relative calibration as an integral part of the process for careful measurement of received power from targets at a known reference point in the system. Of course, an indirect calibration is only as good as the measured values, and may be affected by errors or changes in those values, and by effects such as multipath which may not be properly accounted for in the radar equation.

*Closure*

Because all of the techniques for calibration discussed in this section are subject to some measure of error, it is customary to compute or determine cross section of targets utilizing a mixed calibration, perhaps an absolute calibration, and an indirect calibration. A measure of the quality of the radar cross section estimates generated with the aid of these various calibration methods is given by the "closure" or agreement among the resulting answers. In general, agreement or closure to within 3 dB is considered to be quite acceptable, while disagreements or variances larger than 6 dB are often cause for re-examination of the calibration process in order to isolate potential difficulties.

As an example of closure, an experimental X-band radar system was absolutely calibrated by use of an aluminum sphere towed by a helicopter. The sphere diameter was 8 inches, corresponding to a radar cross section of approximately -15 dBsm. The recorded data were processed using the radar equation and measured radar parameters to generate estimates of radar cross section. The radar was scanning and the target elevation angle varied with range. To compensate for this, the elevation of the antenna was varied to ensure that the peak of the antenna beam swept over the target during the measurement session. Thus, one would properly assume that alignment of the target and the antenna beam corresponded to the maximum signal received during the measurement. Figure 3-3 is the calculated amplitude distribution of specific RCS for one such calibration run, indicating closure to within 2 dB (-17 measured *versus* -15 expected).

When closure is first attempted, good agreement is rarely obtained. Likely sources of error include mathematical error, water in the waveguide, multipath returns, misalignment of the antennas, unaccounted losses, difference in receiver response to injected and actual signals, to name only a few, and considerable effort may be required to obtain satisfactory closure. In the previous example, additional refinement resulted in an additional 2.2 dB adjustment which produced closure to approximately 0.2 dB, an agreement certainly less than experimental error.

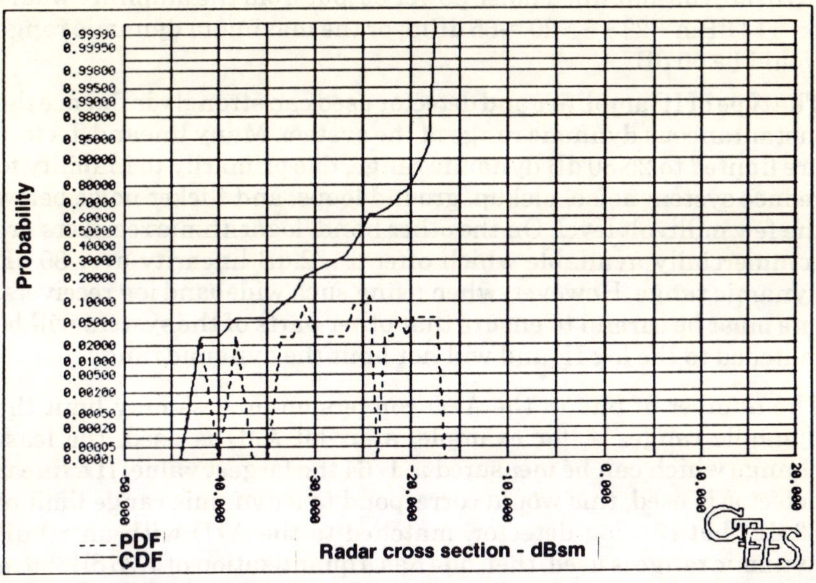


**Figure 3-3.** Measured distribution of radar cross section for an 8 inch diameter sphere towed from a helicopter. Associating the largest value of return with alignment of the sphere and the radar beam yields closure with 2 dB.

### 3.2.2 Dynamic Range

Target RCS is rarely a fixed value but may vary over some substantial range; in addition, it is often the case that both large and small targets must be accomodated simultaneously. The ability of the system to encompass such a range of amplitude values is characterized by its dynamic range, and may be limited by signal-to-noise ratio, number of A/D bits, receiver gains, IF/detector type, system noise and pick-up, or delay stability.

The magnitude of the received signal relative to thermal noise in the receiver may place a limit on achievable dynamic range. For example, if the peak signal-to-noise ratio for a target is 30 dB, then (unless special signal processing is used) the dynamic range of signal measurement would be less than 30 dB.

Whenever an amplifier is included in the receiver chain, it represents a potential limit on achievable receiver dynamic range. If, for example, one included an RF preamplifier with a gain of 20 dB, noise figure of 6 dB, bandwidth of 200 MHz, and saturated power output of -15 dBm, the amplified noise power output from the amplifier would be -114 dBm + 23 + 6 + 20 = -65 dBm, or the maximum dynamic range would be 50 dB.

The type of IF amplifier and detector used can strongly influence the instantaneous dynamic range of the system. Many linear detectors are limited to 25-30 dB dynamic range, due primarily to inability to reduce system noise, pickup, ground loops, and flicker noise below the few millivolt level. On the other hand, logarithmic receivers are commercially available which offer $\pm 1/2$ dB linearity over 80 dB dynamic range. However, when using such wideband log receivers, one must be careful to ensure that other parts of the system will be matched to the log IF and will not limit the dynamic range.

The number of bits in the A/D conversion process may limit the dynamic range. If, for example, a six-bit A/D is used, the least change which can be measured is 1/64 the largest value. If a linear detector is used, this would correspond to a dynamic range limit of 36 dB, but if a log detector, matched to the A/D with an 80 dB dynamic range is used, then one has a quantization of 1.25 dB/bit, a constraint of a different type.

Uncertainty in the point in time at which the A/D converter actually performs the conversion may also affect the system dynamic range, such uncertainties acting to reduce the effective number of A/D bits. This uncertainty in conversion may be related to the slew rate of the voltage input or the frequency of the input signal, and an equivalent error voltage. Consider the situation in Figure 3-4 where the error in voltage, $\Delta V$, corresponding to an error in sample time of $\Delta t$ is given by

$$\Delta V = \frac{dV}{dt} \Delta t \tag{3-9}$$

Similar relationships may be derived for sinusoidal signals, leading to the relationship between sample rate, aperture uncertainty, and number of A/D bits shown in Figure 3-5.

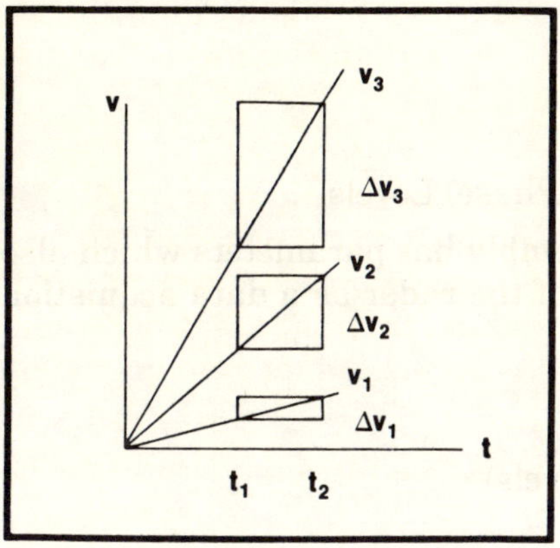


Time Error translated to signal conversion error $\Delta v$ for different rates dv/dt.

$$\Delta v_1 = \frac{dv_1}{dt}(t_2 - t_1)$$

$$\Delta v_2 = \frac{dv_2}{dt}(t_2 - t_1)$$

$$\Delta v_3 = \frac{dv_3}{dt}(t_2 - t_1)$$

It can be seen from the above equations, that the larger the slew rate (dv/dt) for a given aperture uncertainty ($t_2 - t_1$), the greater is the error introduced into signal conversion ($\Delta V$).

**Figure 3-4.** Effect of uncertainty in sample time on the accuracy of voltage conversion in an A/D converter.

### 3.2.3 Subsystem Characterization

If a meaningful measurement program is to be carried out, the radar system must be characterized both to define the parameters required for indirect calibration of the system, and also to ensure that the radar system has properties which are suitable for the measurements to be carried out. The four main portions of the system which require characterization are the receiver, transmitter, antenna and waveguide path, and the data acquisition system.

A number of parameters are required to characterize the performance of a radar receiver, including:

Bandwidth
TR Recovery

IF/Preamplifier Recovery
Gain Variations with Range and Gain Setting
Dynamic Range and Transfer Function
Level of Intermodulation Products
Pulse Rise Time and Ripple
Minimum Detectable Signal
Signal Coupling

A radar transmitter must also be characterized, and significant features include:

Peak Power
Frequency
Pulsewidth
*prf*
RF Spectra
Stability (both Amplitude and Phase) Levels

The antenna and waveguide assembly has parameters which also strongly influence performance of the radar as a data acquistion system, including:

Gain
Sidelobe Level
Beamwidth (1, 3, and 10 dB Levels)
Losses
Polarization and Cross-Polarization Levels and Isolation

Finally, the data acquisition system has characteristics which can strongly affect the recorded data, including:

Amplitude Resolution
Stability
Time Aperture Width
Slew Characteristics
Time Jitter

## 3.3 DATA

Once the system has been calibrated and closure is complete, measured values of received power are easily converted into values of target radar cross section. For the case of a discrete target, this usually suffices; however, for extended targets such as sea, land, or atmospheric clutter some normalized measure of clutter strength, such as radar cross section per unit area or unit volume, is normally desired. As described in section 3.1, land and sea clutter RCS, $\sigma$, may be converted to radar cross section per unit area, $\sigma°$, by dividing this

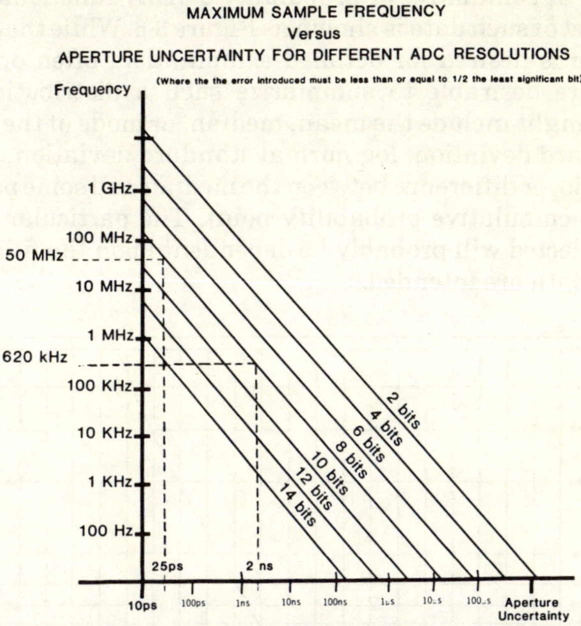


**Figure 3-5.** Relationship between input signal frequency, sample time uncertainty, and effective A/D converter resolution (reprinted from ref. [2]; © 1980 Lecroy Research Systems).

value by the effective illuminated area $A$; i.e.,

$$\sigma^\circ = \sigma_c/A \tag{3-10}$$

Similarly, for volumetric clutter, the radar cross section per unit volume, $\eta$,may be computed by dividing the radar cross section by an equivalent volume, $V$, described by Equation (3-8); i.e.,

$$\eta = \sigma_c/V \tag{3-11}$$

One problem which one should be aware of when measuring volumetric clutter, such as clouds, is that the intervening path attenuation may affect the measured values of cross section. This is particularly true for higher frequency systems, where atmospheric losses may become appreciable.

### 3.3.1 Presentation Format

Once data from such a pulsed system have been recorded on a permanent storage medium and related to useful units such as RCS, then they must be presented in a meaningful form. Often targets are fluctuating, and the complete definition of target scattering requires calculation of cumulative or probability density functions; a representative set of such data is shown as Figure 3-6. While the complete distribution is desired for detailed examination, often one or two numbers are desirable to summarize such a distribution. Such measures might include the mean, median, or mode of the distribution, standard deviation, log normal standard deviation, mean-to-median ratio, or difference between the median and some point such as the 95% cumulative probability point. The particular measure which is selected will probably be dependent upon the final use for which the data are intended.

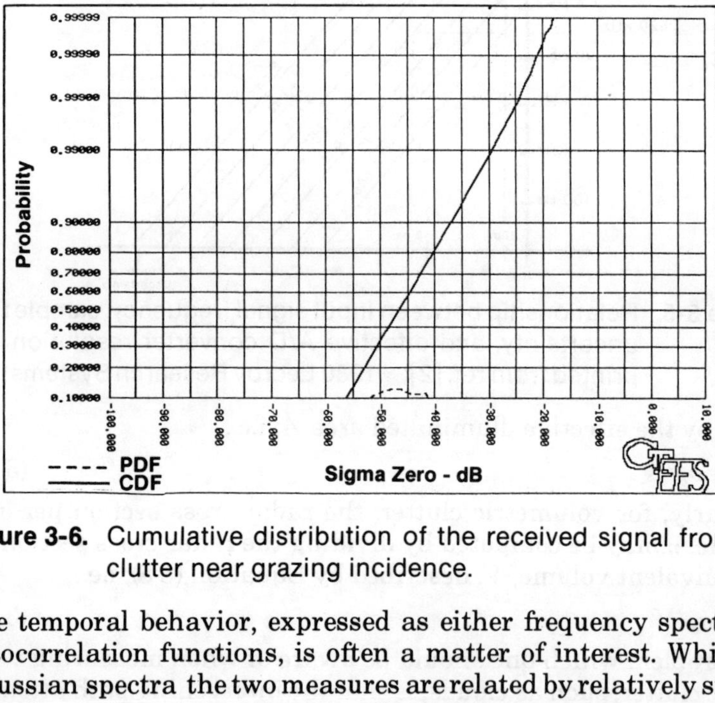


**Figure 3-6.** Cumulative distribution of the received signal from sea clutter near grazing incidence.

The temporal behavior, expressed as either frequency spectra or autocorrelation functions, is often a matter of interest. While for Gaussian spectra the two measures are related by relatively simple transformations, in many cases of interest, the spectra are considerably broader than Gaussian. For example, Figure 3-7 shows a comparison of measured data, Gaussian spectra, and various power law fits to experimental data. As can be seen, many spectra may be decidedly non-Gaussian. Often one may choose to use FFT procedures to calculate the appropriate spectra as opposed to direct transformation of the autocovariance function to obtain power spectral density. In many cases where such transforms are used, suitable weighting of the data is an integral part of the data analysis proce-

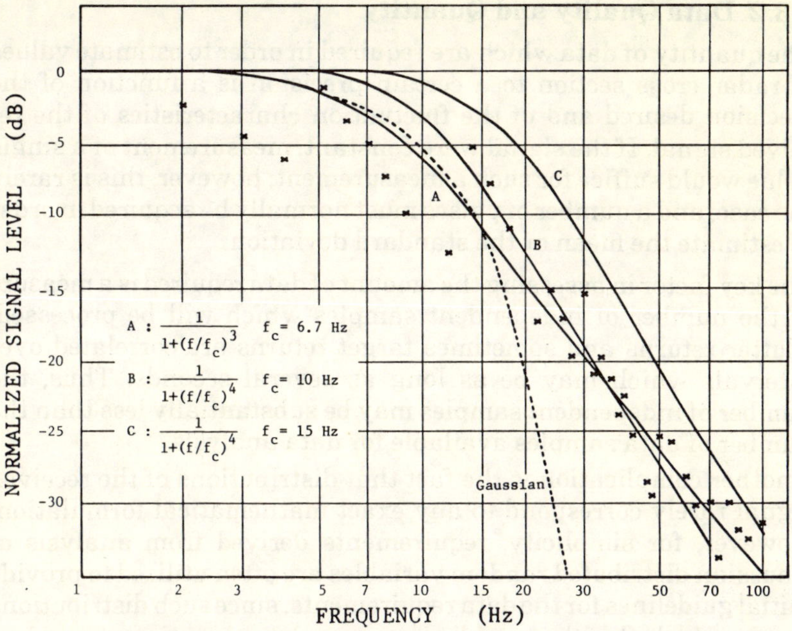


**Figure 3-7.** Spectral shapes from land clutter return, illustrating departure from Gaussian spectra. The x's are data points, the solid lines are polynominal approximations to the data, and the dotted line is a Guassian spectral shape (reprinted from Fishbein *et al.*, ref. [22]).

dure. There are a number of different weighting functions which may be used for such a purpose, the more common of which are summarized in reference [3].

Finally, a complete physical description of the target being measured is important both in an attempt to determine under which conditions the data should be applied, and perhaps in an attempt to extract the dependencies of the radar return on various physical or environmental parameters. For hard targets, a description of the size, altitude, orientation, and surface parameters will often suffice. For clutter, various additional environmental measures may also be important, such as wind speed and direction, atmospheric moisture, moisture content of leaves or snow, or wave height and direction, to name only a few. (*See* Chapter 12 for further descriptions of ground truth parameter measurements.) Careful characterization of the target and environment being observed is a vital, and often neglected, aspect of any radar measurement program.

### 3.3.2 Data Quality and Quantity

The quantity of data which are required in order to estimate values of radar cross section to a certain precision is a function of the precision desired and of the fluctuation characteristics of the received signal. If the signal were constant, measurement of a single value would suffice for such a measurement; however, this is rarely the case, and a number of pulses must normally be acquired in order to estimate the mean or the standard deviation.

The key factor in assessing the amount of data required is a measure of the number of independent samples which will be processed. Clutter returns and sometimes target returns are correlated over intervals which may be as long as several seconds. Thus, the number of independent samples may be substantially less than the number of data samples available for data analysis.

Another complication is the fact that distributions of the received signal rarely correspond to any exact mathematical formulation. However, for simplicity, requirements derived from analysis of Gaussian distributed random variables are often utilized to provide initial guidelines for the data requirements, since such distributions are easy to deal with, providing an easy determination of required data sets for specified error bounds and confidence limits.

To illustrate the procedure, if it is desired to estimate the mean of a fluctuating process to within 1 dB at the 95% confidence level, what are the required number of samples? In order to address this problem one must first have some knowledge of the shape of the distribution, its variance, and its decorrelation time. One approach might assume that the process is lognormal, and that a logarithmic receiver is used; thus, the video signal would be normally distributed, and the widely known properties of the Gaussian distribution may be used. Then the 95% confidence interval is given by:[4]

$$[\bar{X} + 1.96\,\sigma\, n^{-1/2},\ \bar{X} - 1.96\,\sigma\, n^{-1/2}] \qquad (3\text{-}12)$$

where

$\bar{X}$ = median value of the process

$\sigma$ = standard deviation of the process

$n$ = number of independent samples.

Thus, for our situation,

$$1\ dB = 1.96\,\sigma\, n^{-1/2} \qquad (3\text{-}13)$$

and if one assumes $\sigma = 6$ dB, then $n \simeq 138$ independent samples. If the process decorrelates in 0.1 seconds, then approximately 14 seconds of data would be required to obtain the desired confidence limits.

The approach outlined above for estimation of the mean requires a knowledge of standard deviation and decorrelation time of the process. Often a "wild guess" must be used to initially size the experiment, which must be refined on the basis of initial experimental data. As an example of such an approximation, one might bound the standard deviation by guessing the maximum range of fluctuation and dividing by a factor of from four to six. Such procedures are justified by the observation that almost all occurrences of a normal process occur within two to three standard deviations from the mean.

If it is desired to estimate the standard deviation of a process, one may resort to relationships such as those given in Figure 3-8 to estimate data required in order to estimate the standard deviation. However, if tails of the amplitude distribution are of interest, even larger data sets may be required; for example, if it is desired to look at the $10^{-6}$ point on the amplitude distribution, millions of independent samples will be required.

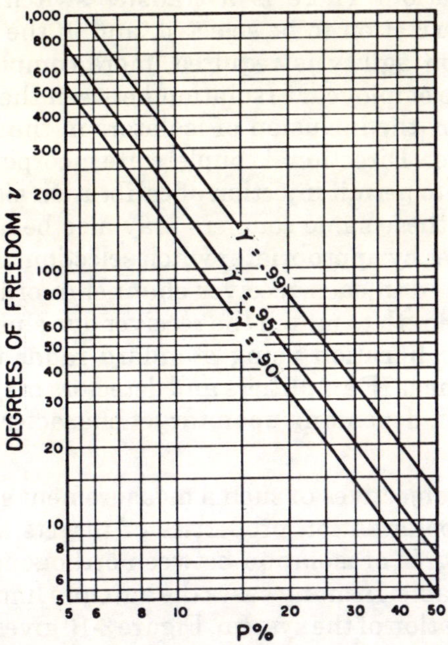


**Figure 3-8.** Number of independent samples requires to estimate the standard deviation within P% of its true value with precision γ (reprinted from Natrella, ref. [4]).

Different data size requirements are encountered when measurements of frequency spectra are desired. In this case the length of the data set determines the frequency resolution, i.e., a 0.1 Hz resolution requires a 10-second data sample. The highest frequency which may be analyzed is given by half the sample rate. Often, a number of spectra are averaged together, and the number of spectra which must be averaged together to achieve a desired precision may be determined using the same procedures as were used for estimation of confidence levels of the median of a fluctuating process.

## 3.4 MEASUREMENT SYSTEMS

### 3.4.1 Conventional Pulsed Radar

As was discussed earlier in this chapter, many of the RCS measurement radars are relatively simple pulsed radar systems. However, measurement radars often must have a number of features not found in operational systems. Figure 3-9 is a simplified block diagram of a dual-polarized measurement radar system.

In Figure 3-9 note that there are two receiver channels, one for each received polarization. There is a transfer switch to permit the transmitted polarization to be selected, and in the event pulse-to-pulse polarization agility is required, more complex phase shift networks may be employed.[5] Isolation between the receiver channels is assured by the inclusion of isolators in the local oscillator path to each mixer. Directional couplers are incorporated into each receiver channel to permit injection of calibration signals into each receiver. One of these same couplers may also be used to monitor transmitted power by appropriate switch selection. Precision attenuators are included in each receiver channel in order to adjust received signal strength to match the receiver dynamic range as well as to facilitate calibration using standard radar targets. Either logarithmic or linear IF amplifiers and detectors may be used in the receiver channels, depending upon target characteristics and data requirements.

One of the main objectives of such a measurement system is measurement of the polarization properties of targets and clutter, for which a high degree of isolation between orthogonal polarization channels is desirable. In many cases the antenna limits the ultimate polarization isolation of the system. Figure 3-10 gives a set of antenna patterns for a dual-polarized front-fed paraboloidal dish antenna. While excellent polarization isolation is achievable on-axis, for targets off-axis or for clutter which extends over an extended area,

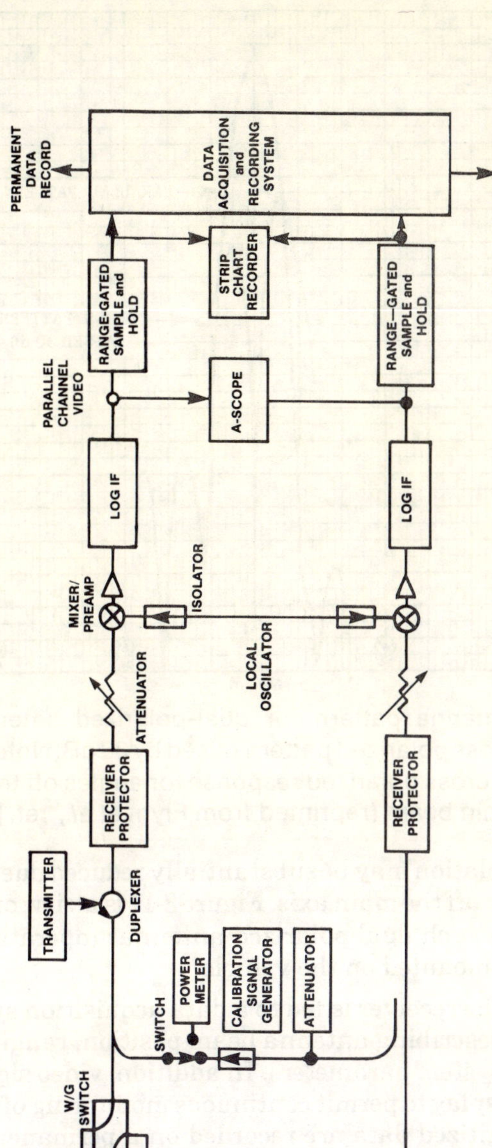


**Figure 3-9.** Simplified block diagram of a dual-polarized measurement radar. Note the provisions for injection of calibration signals in both parallel and cross polarization receiver channels.

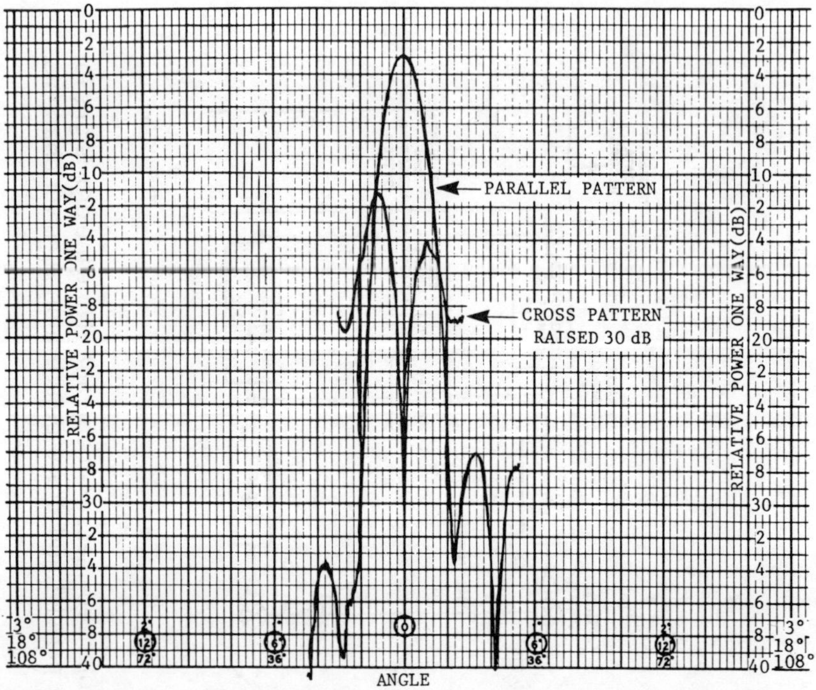


**Figure 3-10.** Antenna pattern for dual-polarized antenna with the cross polarized pattern raised by 30 dB. Note the increase in cross polarized response for angles off the peak of the main beam (reprinted from Flynt *et al.,* ref. [6]).

the effective isolation may be substantially reduced due to the reduction in isolation off the main axis. Figure 3-11 is a view of a radar test van with three such dual-polarized antennas (operating at X, $K_u$, and $K_a$ bands) mounted on the vehicle.

The output of the receiver is fed to a data acquisition system along with the data describing antenna beam position, range, and status of certain key system parameters. In addition, video signals are fed to an analog display to permit continuous monitoring of the received signal. The digitized data are recorded on a permanent recording medium such as hard disk or magnetic tape. In addition, certain data are presented on a display for comparison with input data and to verify proper functioning of the data acquisition system. A valuable feature is to have the capability to generate hard copies of selected data to permit additional on-line review and data evaluation.

**Figure 3-11.** Photograph of a radar measurement van with three dual-polarized paraboloidal dishes (X-, $K_u$, and $K_a$ bands) mounted on the roof (courtesy of the Georgia Institute of Technology).

While many of the features desirable in a measurement system are illustrated in the system described above, factors such as extremely high range resolution requirements or recording of Doppler information may require specialized systems.

### 3.4.2 High Resolution Measurement Systems

Often, the desire for a high resolution radar system generates the requirement for values of target and clutter RCS taken with a short pulse measurement system. In addition, extremely short pulse measurement systems are often a useful diagnostic tool for guiding the theoretical development of radar cross section models, and for indicating dominant scatterers on complex targets whose RCS may be reduced using radar cross section reduction techniques.

Problems associated with operation of short pulse measurement systems include the relatively poor noise performance associated with the wide bandwidth of such systems, the difficulty of implementing a data acquisition system for relatively short pulses, and acquisition of suitable calibration targets. In addition, the concept of uniformly distributed clutter often breaks down when high resolution radar systems are utilized, and the resulting highly varying signals often require wide dynamic range receivers which are difficult to implement for short pulses. Relative calibration of such wideband systems may be made more difficult by the fact that the response of some receivers is different for wide and narrow pulses, thus making it difficult to utilize an easily generated long pulse for relative calibration purposes.

There are a number of means of implementing short pulse radar measurement systems, including diode gating of longer pulses, the incorporation of grid modulated TWTs, use of magnetrons and pedestal modulators, Hertzian generators, and pulse compression systems.

A system which is based on the generation of the short RF pulse by a pulsed oscillator is shown in Figure 3-12, which shows a millimeter wave system employing either a pulsed magnetron or extended interaction oscillator for generation of the approximately 50 *ns* pulse which is used by the system. Except for the required receiver bandwidths, much of the remainder of the system is quite conventional.

Figure 3-13 shows a gated oscillator followed by a higher power amplifier in order to obtain the required pulse length, approximately 0.6 *ns* for this example. An interesting feature of such a system is the fact that in addition to the short pulse, noise surrounding the short pulse is also amplified. Particularly for an amplifier having a relatively high noise figure, such amplified noise may limit the dynamic range of the receiver. An approach which minimizes this limitation of dynamic range is one in which the TWT grid is pulsed directly. Such systems have been used extensively for measurement of radar cross section of models, and can generate pulses as short as a few *ns*.

Pulse compression systems may also be used for such applications, and the block diagram of such a system (the AN/APS-116) is shown in Figure 3-14. The use of linear amplifiers in such a system limits the system dynamic range considerably. Values of from 30 to 40 dB dynamic range have been measured for such a system which has an effective compressed pulsewidth of approximately 3 *ns*. This basic system has been modified for a number of measurement programs,

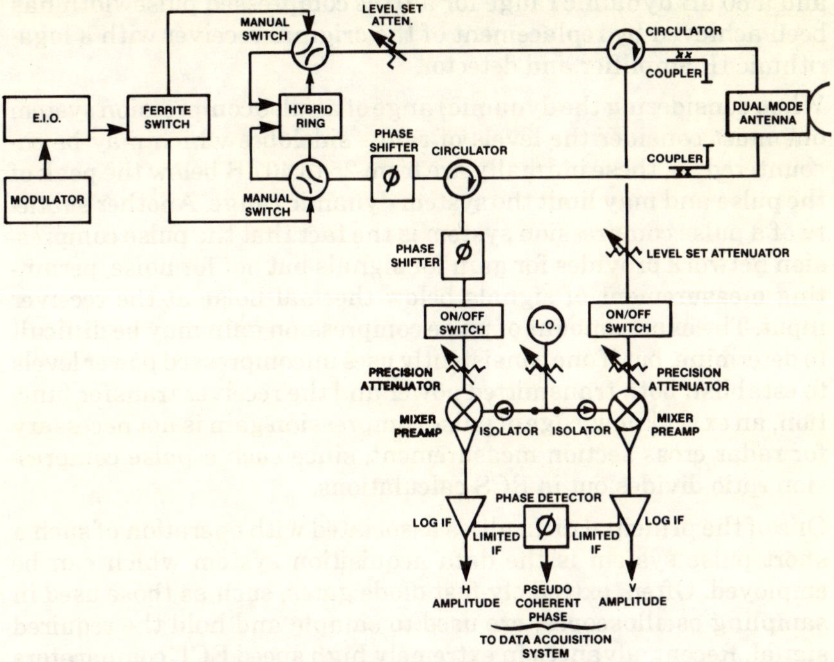


**Figure 3-12.** Simplified block diagram of a 95 GHz short pulse, pulsed oscillator (1 kW EIO) (courtesy of the Georgia Institute of Technology).

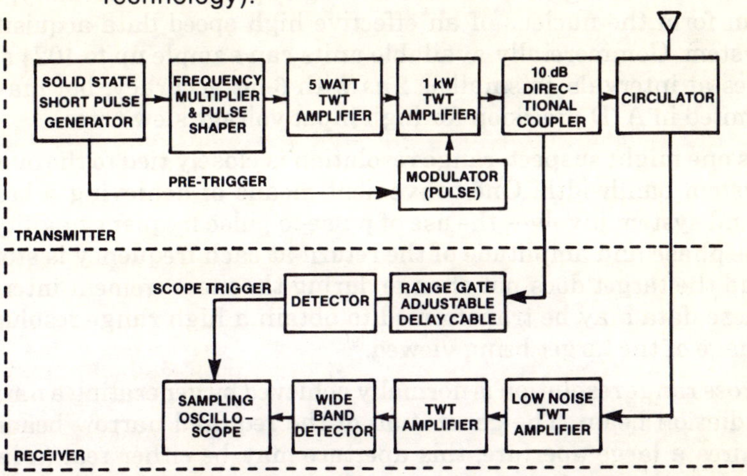


**Figure 3-13.** Simplified block diagram of a short pulse (0.6 ns) 1.8 kW peak power measurement radar (reprinted from Bachman, ref. [7]; © 1975 IEEE).

and a 60 dB dynamic range for a 10 *ns* compressed pulsewidth has been achieved by replacement of the original receiver with a logarithmic IF amplifier and detector.

When considering the dynamic range of a pulse compression system one must consider the levels of range sidelobes which may be encountered — these normally are from 25 to 40 dB below the peak of the pulse and may limit the system dynamic range. Another subtlety of a pulse compression system is the fact that the pulse compression network provides for gain for signals but not for noise, permitting measurement of signals below thermal noise at the receiver input. The exact amount of pulse compression gain may be difficult to determine, but if one consistently uses uncompressed power levels to establish both transmitted power and the receiver transfer function, an exact knowledge of pulse compression gain is not necessary for radar cross section measurement, since such a pulse compression ratio divides out in RCS calculations.

One of the principal difficulties associated with operation of such a short pulse system is the data acquisition system which can be employed. Often, extremely fast diode gates, such as those used in sampling oscilloscopes, are used to sample and hold the required signal. Recent advances in extremely high speed ECL comparators have permitted the generation of up to 6-bit, 2 *ns* A/D converters. There are commercially available units which incorporate such converters along with an internal high speed buffer memory, and can form the nucleus of an effective high speed data acquisition system. Commercially available units can sample up to 1024 samples at intervals as small as 2 *ns* with 6-bit accuracy, but may be limited in A/D precision for high input voltage slew rates.

As one might suspect, range resolution is closely tied to the overall system bandwidth. One convenient means of achieving a broadband system involves the use of pulse-to-pulse frequency agility. If the phase and amplitude of the return at each frequency is stored, and the target does not change during the measurement interval, these data may be transformed to obtain a high range resolution image of the target being viewed.[9]

Cross range resolution is normally achieved by generating a narrow radiation beam. The generation of the required narrow beam requires a large aperture; this aperture may be either real (a large antenna) or synthetic (as in a synthetic aperture radar).[10] In recent years it has come to be more widely realized that synthetic aperture techniques are only one form of cross range imaging which is possi-

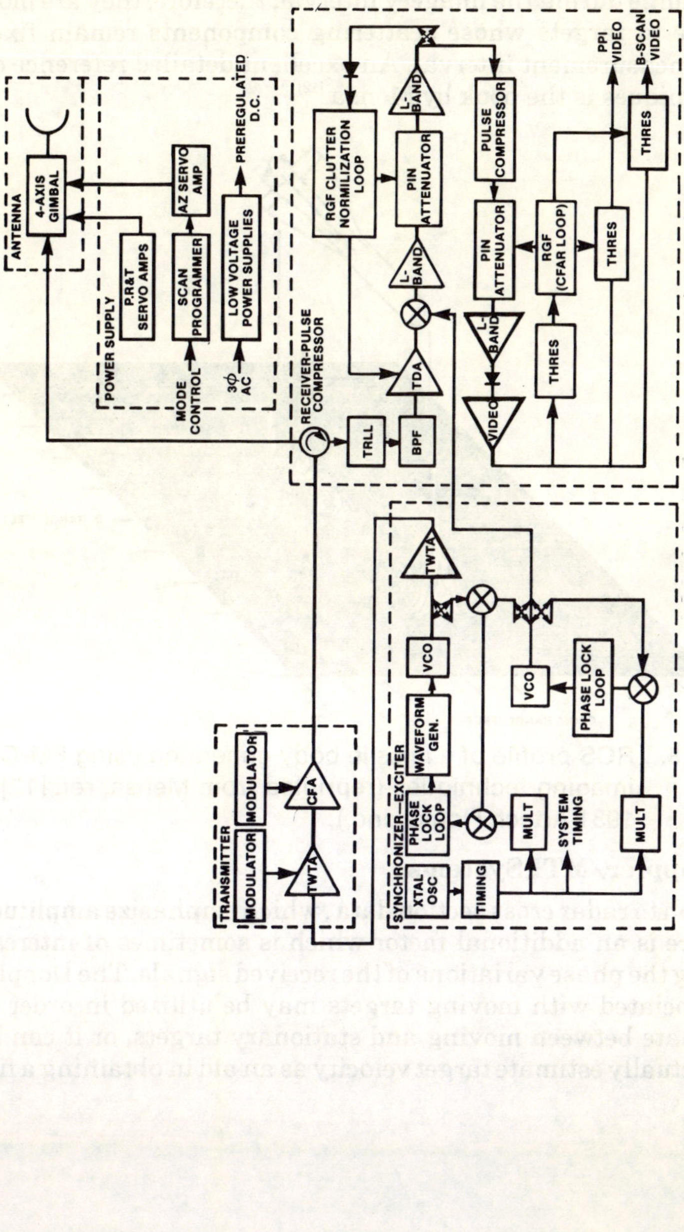


**Figure 3-14.** Block diagram of AN/APS-116 pulse compression radar system. For measurement applications the AFC loops may be disabled and the diode detector replaced with a receiver of wider dynamic range (reprinted from Smith and Logan, ref. [8]; © 1980 IEEE).

ble when either the observer moves past the target or the target rotates in a precisely known manner.[11] On RCS measurement ranges conditions for high resolution imaging are often satisfied, and appropriate transformations of recorded data can yield high quality imagery of complex targets as is shown in Figure 3-15. All of these imaging techniques which require extended time periods to image the target assume that the target scattering characteristics do not change during the imagery interval; therefore, they are most applicable to targets whose scattering components remain fixed over the measurement interval. An excellent detailed reference on such techniques is the book by Mensa.[12]

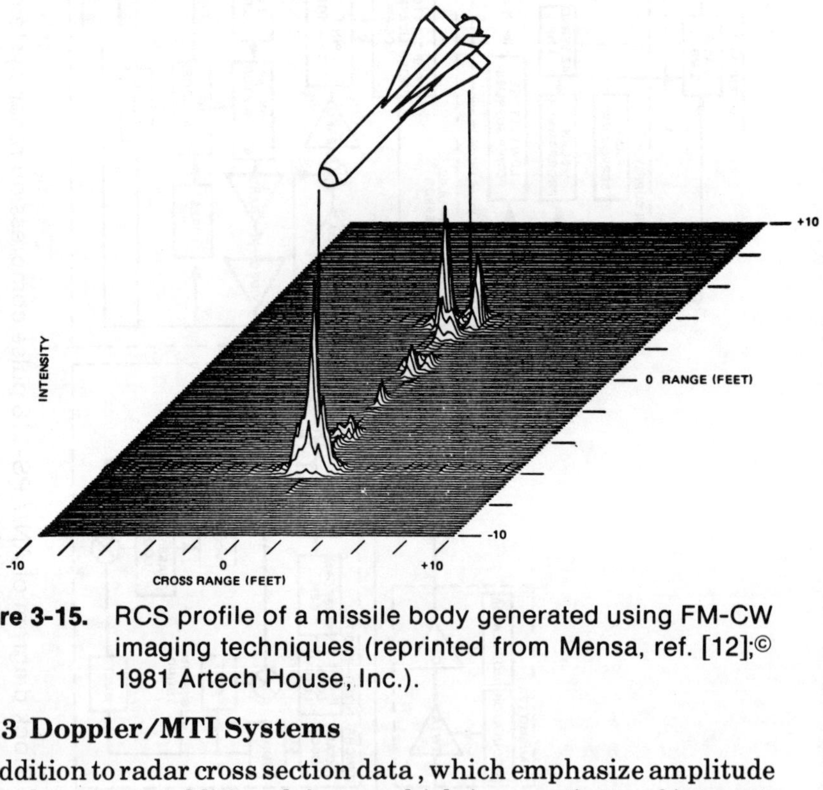


**Figure 3-15.** RCS profile of a missile body generated using FM-CW imaging techniques (reprinted from Mensa, ref. [12];© 1981 Artech House, Inc.).

### 3.4.3 Doppler/MTI Systems

In addition to radar cross section data, which emphasize amplitude only, there is an additional factor which is sometimes of interest, that being the phase variations of the received signals. The Doppler shift associated with moving targets may be utilized in order to discriminate between moving and stationary targets, or it can be used to actually estimate target velocity as an aid in obtaining a fire control solution. There are two general types of coherent systems: the MTI (moving target indication) systems, which have unambiguous range, but ambiguous velocity; and the pulse Doppler systems, which have unambiguous velocity, but often have ambiguous range. The end use to which the data are to be placed is an important consideration in designing the measurement system used to acquire such data.

If the phase of the received signal is to be accurately measured, frequency stability in the radar system of a high order is required, since phase variations must be kept to a minimum. There are two different configurations which are often used for coherent measurements, the *master-oscillator power-amplifier* (MOPA) system, shown in Figure 3-16, or the coherent on receive system, shown in Figure 3-17. The principal difference between the two systems is in the method of obtaining the phase reference, the MOPA requiring a stable oscillator and amplifier while the coherent on receive system can employ a much less expensive and complex high power oscillator such as a magnetron, but requires a lockable *coherent local oscillator* (COHO) in order to remember transmitted phase.

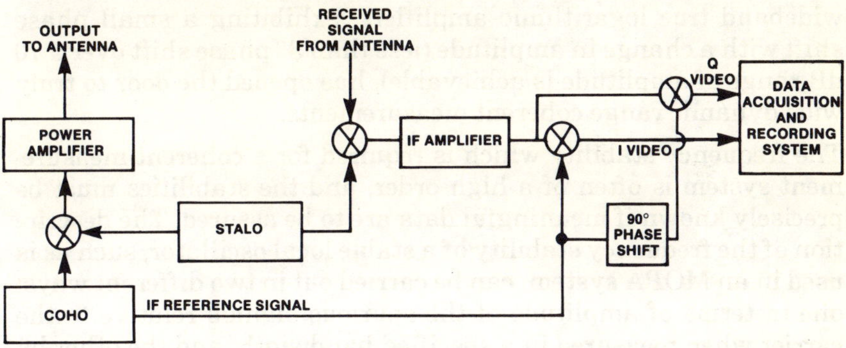


**Figure 3-16.** Simplified block diagram of a MOPA coherent radar which records both in-phase (I) and quadrature (Q) components of the received signal.

Note that two receiver channels, both in-phase and quadrature signals (*I* and *Q*), are required for a full description of any target, and limiting may or may not be desirable in the receiver channels, depending upon the final application of these measured data.

Until recently, linear detectors were often used for measurements of *I* and *Q* channel amplitudes. However, recent developments in

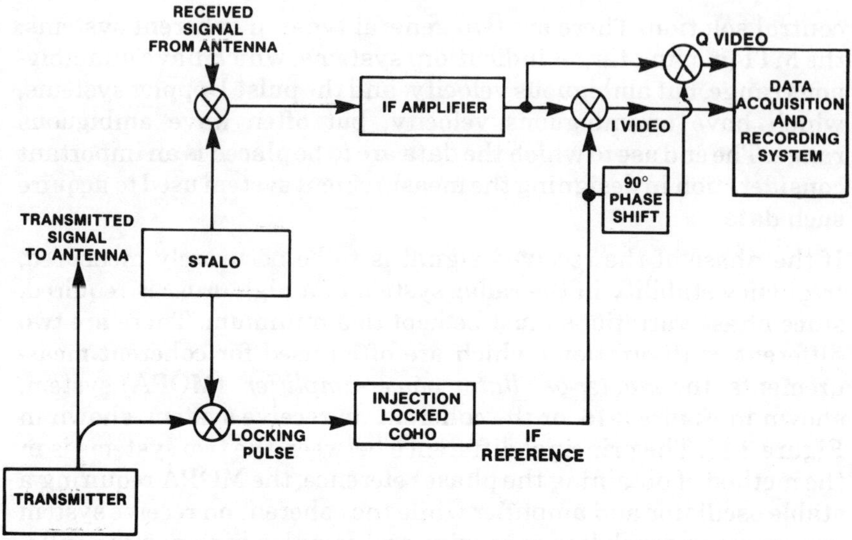


**Figure 3-17.** Simplified block diagram of a coherent-on-receive coherent radar which uses a sample of the transmitted pulse to phase lock the reference coherent oscillator (COHO).

wideband true logarithmic amplifiers, exhibiting a small phase shift with a change in amplitude (less than 3° phase shift over a 70 dB range of amplitude is achievable), has opened the door to truly wide dynamic range coherent measurements.

The frequency stability which is required for a coherent measurement system is often of a high order, and the stabilities must be precisely known if meaningful data are to be assured. The description of the frequency stability of a stable local oscillator, such as is used in an MOPA system, can be carried out in two different ways: one in terms of amplitude of the spurious signals relative to the carrier when measured in a specified bandwidth, and the other by the frequency deviation in a specified bandwidth of spurious signals at some frequency removed from the carrier frequency. For small frequency deviations such as those which would ordinarily be found in a high quality stable local oscillator, the two are related by[13]

$$\frac{P_s}{P_c} = -20 \log \left( \frac{2^{1/2} F_m}{\Delta f} \right) \tag{3-14}$$

where

$P_s$ = noise power in specified band $F_m$ Hz from carrier
$P_c$ = carrier signal power

$F_m$ = modulation rate or separation from carrier in Hz
$\Delta f$ = *rms* frequency deviation in specified bandwidth

The ability of the measurement system to measure relatively narrow spectral bandwidths, either those associated with spreading of a specific spectral line associated with a moving target, or associated with the spreading of the zero frequency line associated with approximately stationary clutter, may be limited by the achievable stability of the stable local oscillator and transmitter. For example, a high quality MTI system may have a power spectral distribution from stationary targets at video as shown in Figure 3-18 (for a more detailed discussion of the relationship between transmitted spectra and spectra at video or IF, which is range dependent, *see* reference[14]). Also shown in Figure 3-18 are curves of some representative Gaussian and power law spectral distributions for stationary ground clutter at L-band frequencies. Note that the range of spectral shapes from clutter which can be meaningfully reported may be limited by the stability or spectral width of the measurement system.

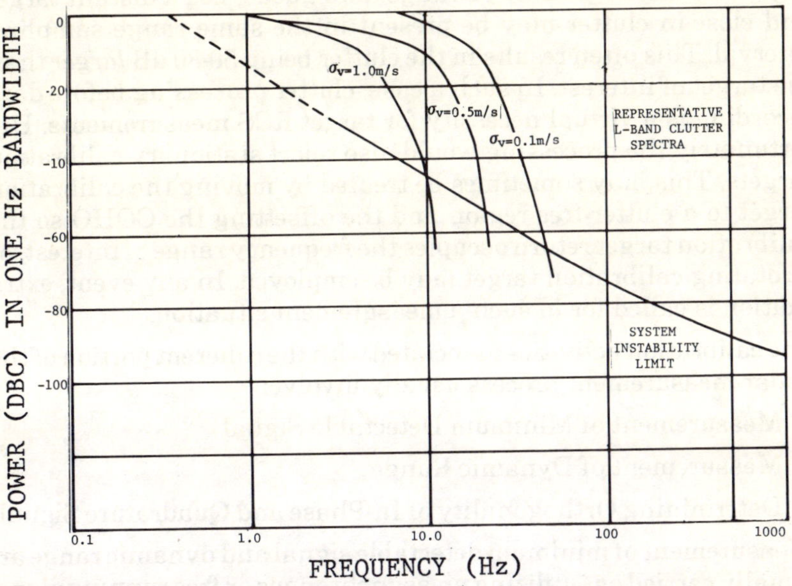


**Figure 3-18.** Spectral shapes of L-Band clutter returns and basic system instabilities at IF, illustrating that system stability may limit the spectral measurement capability of the system.

Calibration of coherent data is made somewhat more complex by the fact that the receiver output signals are related not only to the amplitude but to the phase of the received signal, and that much of the normal signal processing rejects stationary targets. Normal signal generators can be used to inject known signals, but they are in general not phase locked with the system frequency standard, and will thus exhibit "beating," or will be modulated by the difference frequency. If both the $I$ and $Q$ signals are directly accessible, the amplitude may be computed as the square root of the sum of the squares of the $I$ and $Q$ components; however, this is often inconvenient. In the event a stationary external standard target reference is used, the magnitude of the output of a given receiver channel depends upon the relative phase of the signal at the receiver. In addition, the coherent outputs of many radars are limited or "clipped" at some level; spectral data required for system design usually must reflect the spectral spreading associated with such limiting if they are to be useful in system analysis and design.

In many pulse Doppler radar systems, the *prf* may be sufficiently high that the target may be range-ambiguous, i.e., a distant target and close-in clutter may be present in the same range sampling interval. This often results in the clutter being 50-80 dB *larger* than the target of interest. In such a case, clutter processing before data recording is a virtual necessity for target RCS measurements. Unfortunately the processing would also reject stationary calibration targets. This may sometimes be treated by moving the calibration target to a clutter-free region and the offsetting the COHO so the calibration target return occupies the frequency range of interest, or a rotating calibration target may be employed. In any event, extra caution is called for in such a measurement situation.

The calibration activities associated with the coherent portion of the radar measurement process usually involve:

Measurement of Minimum Detectable Signal

Measurement of Dynamic Range

Determining Orthogonality of In-Phase and Quadrature Signals

Measurement of minimum detectable signal and dynamic range are usually carried out utilizing an asynchronous or free running signal generator, following the same procedure as used for a relative calibration. Of course, the output signal will be a sampled sine wave whose amplitude is related to the strength of the received signal, and whose frequency is given by the difference between the injected test signal and the actual transmitted signal frequency.

Orthogonality of the in-phase ($I$) and quadrature ($Q$) signals may be verified by injecting a test signal as before, and computing the square of the $I$ and $Q$ components. For large signal-to-noise ratios, the resultant magnitude should be constant, with the phase varying. For lower signal-to-noise ratios, the amplitude may vary but should be bounded, and the angle should be uniformly distributed through 360°.

Due to the difficulty of accurate amplitude calibration of a coherent receiver, often a coherent measurement system includes an additional amplitude channel, and the output of this amplitude channel is used to obtain amplitude information, while the phase channels are reserved for aquisition of data describing frequency spectra. In that event, only limiting in the channels, and orthogonality of the $I$ and $Q$ channels are of major interest. In particular, orthogonality of $I$ and $Q$ channels may be verified by displaying the outputs of the two receiver channels in a scatter scope format (where one signal drives the x-axis, and the other signal the y-axis), and the resultant appearance should then be approximately circularly distributed for receiver noise or for an injected incoherent test signal.

Coherent data are often analyzed for frequency content, and the data normalized to zero frequency or some standard reference frequency component. The amount of data and the spacing of the data samples is intimately connected with the ability to extract the desired information, the length of the data set determining the resolution, and the spacing or sampling rate determining the maximum unambiguous frequency which can be analyzed. In general, the resolution of the data is determined by the length of the set used to describe the target; for example, a one-second data set gives a resolution of approximately 1 Hz and a two-second data set provides approximately 1/2 Hz resolution. The spacing or sampling rate determines the unambiguous frequency, or the frequency at which aliasing will occur. For example, sampling at 1000 pulses per second will permit analysis of frequency components up to 500 Hz; in general, the sampling rate should be chosen sufficiently high so that no significant aliasing of the data occurs. As an alternative, the data can be filtered with a low pass filter to eliminate components which might result in aliasing if the sample rate is not high enough to record all components.

Analysis of the data may actually be carried out by performing an FFT on the data set, and squaring the magnitude of the resulting signal to generate the power spectrum of the signal; or the data set's autocorrelation may be computed, and then Fourier transformed to produce a power spectral density. In many cases the processes are ergodic so that the power spectrum and power spectral density are identical, but the data should be carefully examined to ensure this is true before any such assumption is made.

In order to minimize difficulties associated with truncation of the data sets, "tapering" or "weighting" of the data are often employed, and there are a number of so-called windows or weights which are utilized for such a purpose. Some of the more common are Gaussian, Kaiser-Bessel, Tchebycheff, or Hamming weights. An excellent review of a number of weight functions has been published,[3] and it is a fortunate circumstance that in many cases the particular choice of weight functions is not critical to meaningful analysis of the data.

When computing spectra, often a number of spectra from a given target are computed, perhaps being overlapped by approximately one-half the data interval, and the composite set of frequency components averaged together in order to provide a composite return. Such an approach has the advantage and/or disadvantage of suppressing any anomalous behavior which may have occured during one particular data set, and providing a more consistent and uniform set of results.

### 3.4.4 CW Radars

*Low Power Short Range*

It is possible to utilize a continuous wave radar system for radar cross section measurements, rather than the pulsed systems described in the previous section, and such systems have been utilized extensively in anechoic chambers and compact reflectivity ranges (*see* Chapter 9). A simplified block diagram of such a CW radar system, employing a single antenna for transmission and reception, is given in Figure 3-19. Such a CW radar system is relatively inexpensive, but has a number of limitations which are primarily associated with the fact that both leakage from the transmitter and extraneous reflections must be nulled at the receiver. Such a nulling is often difficult to achieve, requiring a good match of microwave components, freedom from physical vibration, and a high order of frequency stability in the transmitter.

In practice, RF tuners are used to null out received energy in the receiver in the absence of the target. The degree of cancellation that is required is a function of the accuracy of the measurement which is desired, the size of the target, and the transmitted power. Figure 3-20

**is a plot of the error in radar cross section as a function of the uncancelled energy incident on the receiver. The fact that the relative phase of the target and this leakage energy can be a random**

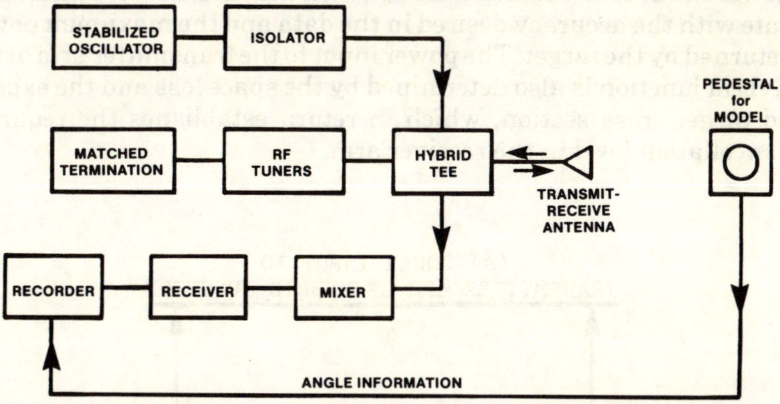


**Figure 3-19.** Simplified block diagram of a CW radar suitable for model measurements (reprinted from Blacksmith *et al.*, ref. [15];© IEEE).

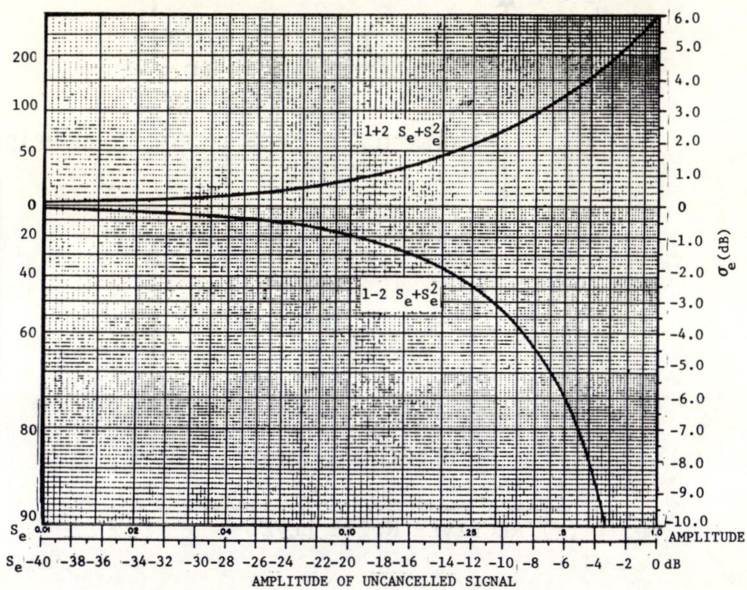


**Figure 3-20.** Bounds on measurement error as a function of uncancelled signal in a CW measurement system (reprinted from Blacksmith *et al.*, ref. [15];© 1965 IEEE).

function can produce variations in measured RCS. Figure 3-21 gives a graphical representation of the amount of cancellation which is required by the system. As can be seen, the residual uncancelled power in the receiver arm is fixed by the minimum level commensurate with the accuracy desired in the data and the maximum power returned by the target. The power input to the transmitter arm of the hybrid junction is also determined by the space loss and the expected target cross section, which in return establishes the required cancellation level in the receiver arm.

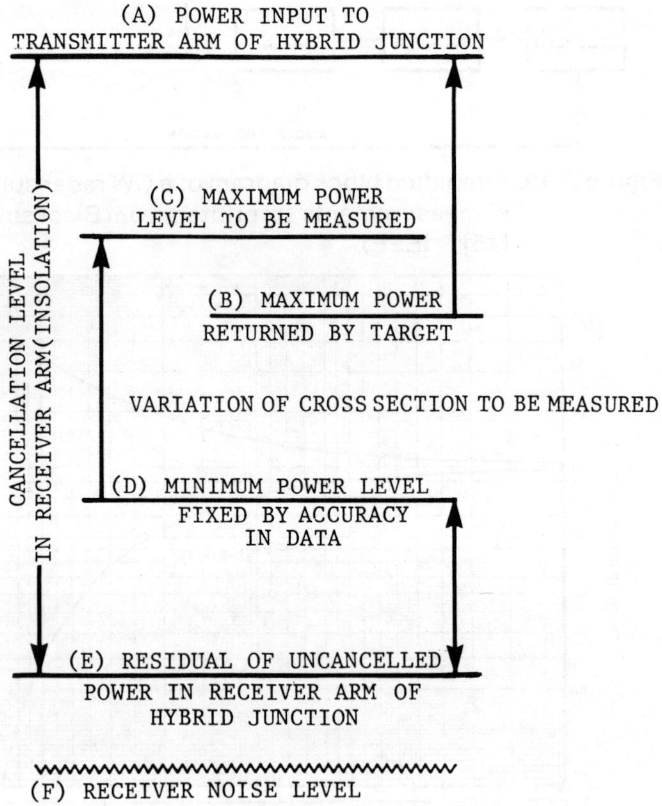


**Figure 3-21.** Cancellation requirements in a CW measurement system (reprinted from Blacksmith *et al.*, ref. [15];© 1965 IEEE).

Since the degree of cancellation which can be achieved is dependent upon the frequency stability of the source, often extremely stable frequency oscillators are used in the system. Typically, at X-band a stability of one part in $10^5$ would be required for 85 dB of isolation, and 5 parts in $10^6$ would be required for an isolation of 100 dB. By using a Pound-stabilized oscillator with invar waveguide components at X-band frequencies, isolations in excess of 120 dB have been achieved for periods of up to 30 minutes.[15]

*High Power CW Radar*

Often, in attempts to achieve increased isolation between the transmitter and the receiver, separate antennas for transmission and reception are incorporated as shown in Figure 3-22, which also includes a feed-through nulling circuit designed to null any existing leakage between the transmitter and receiver. Use of such techniques permits sufficiently high CW powers to be employed so that *in situ* measurements of targets and clutter can be performed with the CW system. Unfortunately, such a procedure has a number of difficulties associated with it. The primary difficulty in CW clutter

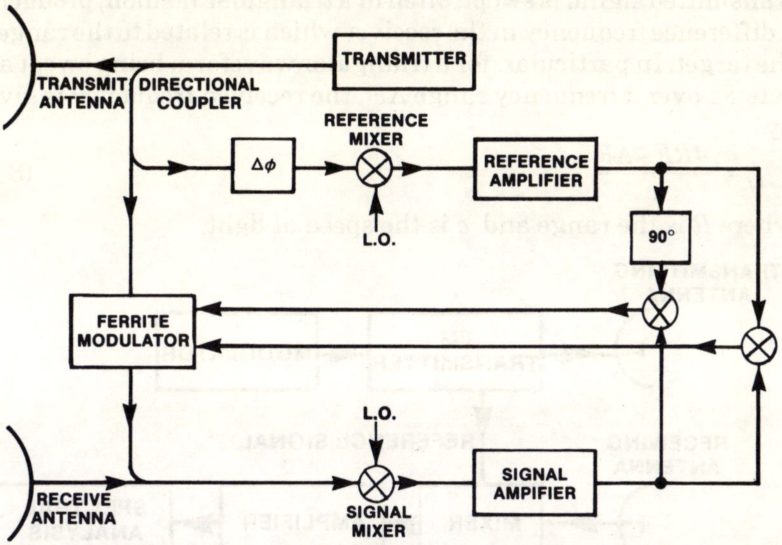


**Figure 3-22.** Simplified block diagram of a feed-thru nulling CW receiver. The system uses a sample of the transmitted signal to cancel leakage and close-in clutter.

measurements (particularly at angles near grazing) is identification of the area which makes the major contribution to received clutter power. This, coupled with signal decorrelation as a function of range, makes meaningful interpretation of such CW measurements often difficult, particularly when it is desired to generate an engineering description of clutter such as $\sigma°$. For that reason, such measurements are often made with systems having parameters close to those of operational systems, and data are interpreted strictly in terms of signal-to-noise ratios at various points in the receiver.

### *FM-CW Radar*

It is possible to build a continuous wave radar which employs frequency modulation in order to circumvent many of the difficulties associated with a conventional CW radar system described earlier. Such systems have the potential advantage of providing extremely high range resolution, which is tied to the overall system frequency deviation, without the requirements for large bandwidth in the signal processing section. A simplified block diagram of such an FM-CW radar system is shown in Figure 3-23. In such a system, the transmitted signal is mixed with the received signal, while the transmitted signal is swept, often in a triangular fashion, producing a difference frequency in the receiver which is related to the range to the target. In particular, for a triangular waveform being swept at a rate $F_m$ over a frequency range $\Delta F$, the received frequency is given by

$$f_f = \frac{4RF_m \, \Delta F}{c} \tag{3-15}$$

where $R$ is the range and $c$ is the speed of light.

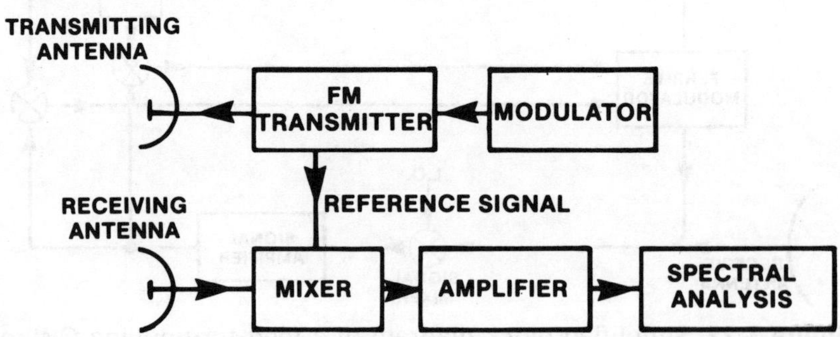


**Figure 3-23.** Simplified block diagram of a FM-CW radar. The FM modulation is usually triangular, but other waveforms may be employed (adapted from Blacksmith, *et al.*, ref. [15]; © 1965 IEEE).

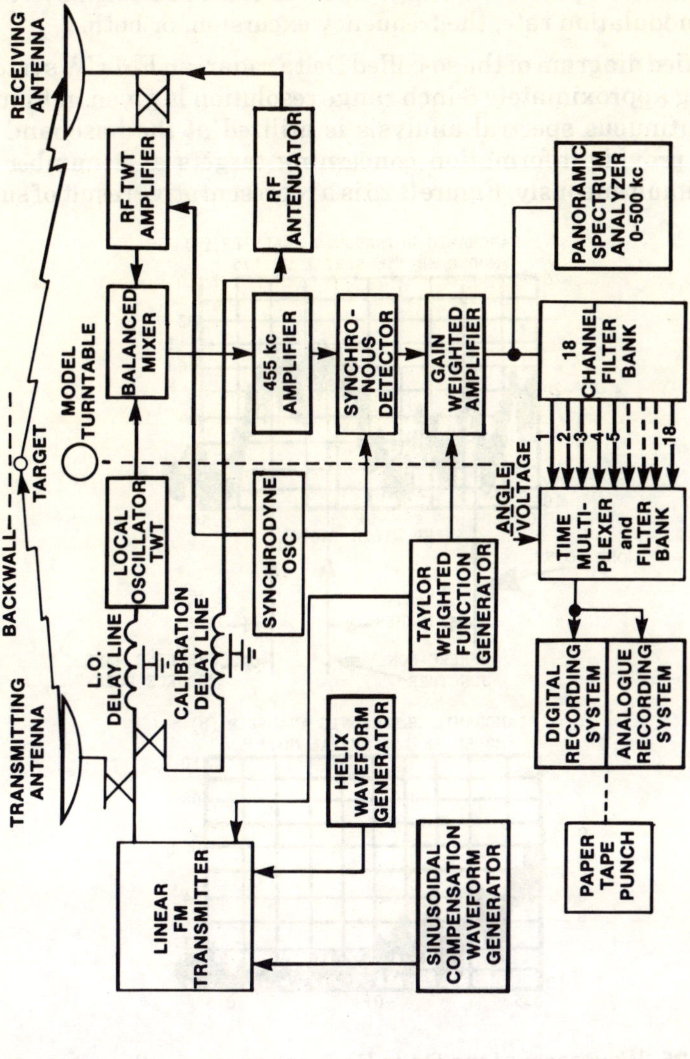


**Figure 3-24.** Block diagram of the Delta Radar, a high resolution FM- CW system (reprinted from Alongi, *et al.*, ref. [17]; © 1965 IEEE).

Thus, through frequency analysis of the received signal, frequency can be directly related to range. Analysis by a contiguous filter bank or its equivalent can provide simultaneous information from a number of ranges. Similarly, if only a single frequency is analyzed in the receiver, a particular range interval can be selected by varying the modulation rate, the frequency excursion, or both.

A simplified diagram of the so-called Delta radar, an FM-CW system providing approximately 6-inch range resolution is given in Figure 3-24. Continuous spectral analysis is utilized at the baseband in order to provide information concerning targets at a number of ranges simultaneously. Figure 3-25 is a representative result of such

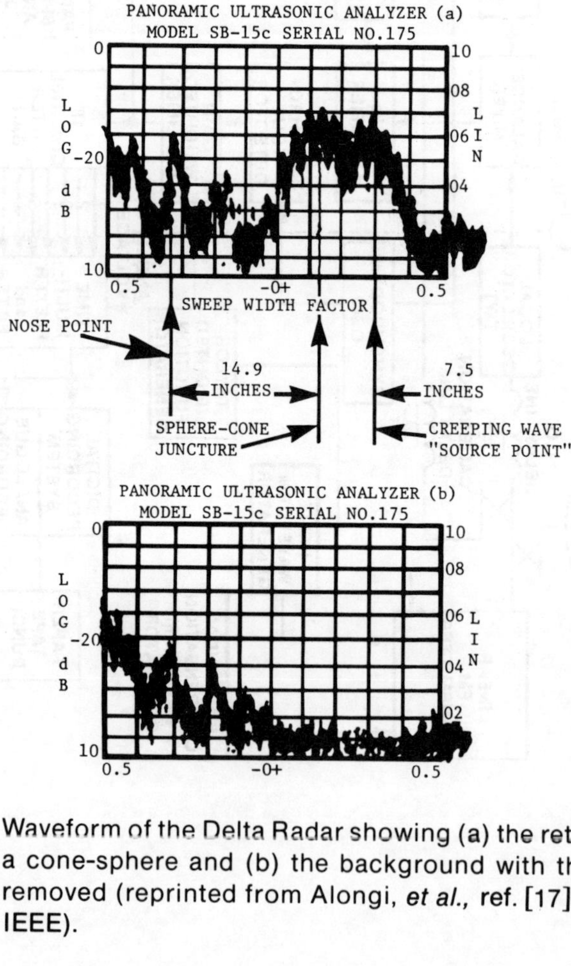


**Figure 3-25.** Waveform of the Delta Radar showing (a) the return from a cone-sphere and (b) the background with the target removed (reprinted from Alongi, *et al.*, ref. [17]; © 1965 IEEE).

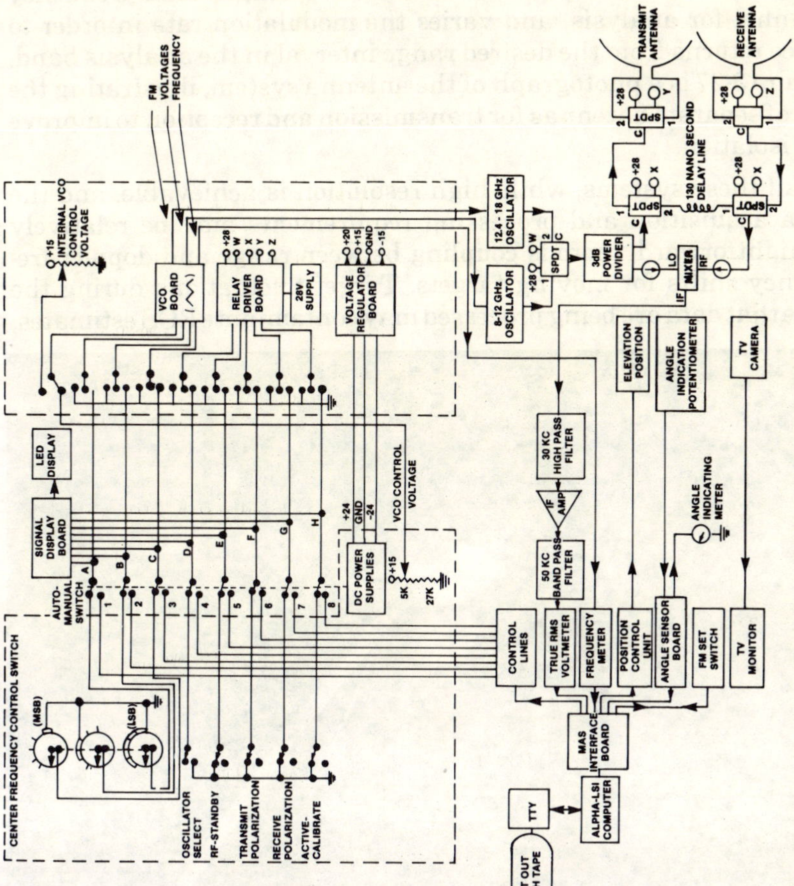


**Figure 3-26.** FM-CW radar used by the University of Kansas for earth resources reflectivity measurements (reprinted from Stiles *et al.*, ref. [18]).

spectral analysis for a sphere-type target, showing the resolution of the direct and creeping wave contributions to the received signal.

Figure 3-26 is a block diagram of an FM-CW system utilized by the University of Kansas for a number of reflectivity programs. This system is interesting in that it utilizes a single fixed-frequency channel for analysis, and varies the modulation rate in order to place returns from the desired range interval in the analysis band. Figure 3-27 is a photograph of the antenna system, illustrating the use of separate antennas for transmission and reception to improve the isolation.

In all these systems, while high resolution is achievable, and the data acquisition and processing requirements may be relatively straightforward, there is coupling between range and doppler frequency shifts for moving targets. Target fluctuations during the time that data are being processed may contaminate RCS estimates,

**Figure 3-27.** View of the transmitting and receiving antennas of the FM-CW radar of Figure 3-26 mounted on an extendable mount during a series of snow reflectivity measurements.

and relative calibration of the system may be difficult. Thus, the calibration of such systems is often accomplished by the use of standard targets of known radar cross section, and their applicability is often limited to stationary or slowly moving targets which exhibit low levels of high-frequency RCS variations.

### 3.4.5 Millimeter Radar

Millimeter radar systems, i.e., those which operate at frequencies above 30 GHz, are similar to lower frequency systems in many ways. However, the equipment limitations associated with operation at millimeter frequencies and the inherently high resolution of many of these systems often pose particular measurement problems. The powers available for millimeter wave radar systems generally decrease with increasing frequency, while noise figures of such systems increase with increasing frequency. Recently, several relatively complete discussions of such factors have become available, and should be consulted for specific details.[19,20]

Another problem associated with operation at millimeter wavelengths involves the accuracy requirements for standard targets. Since accuracy of a standard target is related to the ratio of physical dimensions to wavelength, as the wavelength gets smaller, the accuracy requirements of such standard targets may become significant. In addition, for targets such as flat plates, for a given size target the backscattering lobe pattern becomes narrower with increasing frequency, making aiming of targets much more difficult. When operating in a multipath environment, the spacing of the multipath induced vertical lobe structure may become smaller, and in fact for sufficiently large calibration targets, the physical extent of the calibration target may span an appreciable portion of such lobes, thus reflecting a signal which represents the integrated average of the incident power on the reflector as shown in Figure 3-28.

Since identification and target discrimination are often important factors in the design of many millimeter wave radar systems, the careful description of both targets and clutter becomes even more important in such cases. Careful knowledge of range, azimuth, and target or clutter type is thus an integral portion of any such millimeter measurement program.

In order to perform meaningful systems analyses at millimeter wavelengths, it is necessary to have available signature and cross section data from targets and clutter of interest. In order to acquire such data, a number of different instrumentation and/or measurement radar systems have been built. An example of a relatively high

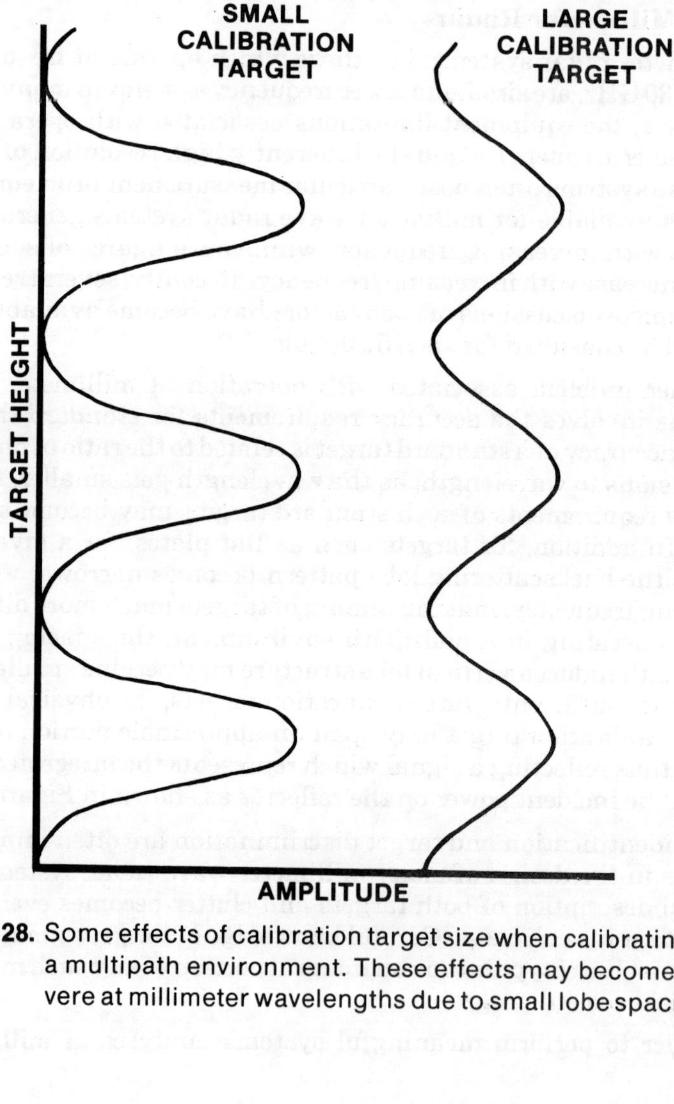


**Figure 3-28.** Some effects of calibration target size when calibrating in a multipath environment. These effects may become severe at millimeter wavelengths due to small lobe spacing.

power system utilized for such purposes is a 95 GHz system employing a pulsed magnetron or EIO oscillator, which has been used for data collection from a number of different target and clutter types. A list of principal radar characteristics is given in Table 3-1, and a block diagram of the system was shown earlier in Figure 3-12. Note in particular the incorporation of a variable attenuator in the receiver front end, thus permitting calibration over the system's dynamic range utilizing a single fixed external corner reflector, and the capability for interfacing a number of different data acquisition systems with the receiver output, depending upon the specific data requirements. The relatively large bandwidths, and correspondingly high IF center frequencies are not unusual for many millimeter measurement systems. The IF frequencies involved are usually selected to achieve the desired resolution, accommodate system frequency uncertainities, and suppress unwanted intermodulation products. While the use of larger than matched bandwidths can reduce the system performance when operating in a thermal noise limited situation, no other significant impacts are generated by these somewhat unconventional parameters.

There are a number of applications where more sophistication in signal parameters is required than provided above. An example of such a system is a lower power, coherent, frequency and polarization agile 35 GHz solid-state radar[20] using a synthesized RF source which is shown in block diagram form in Figure 3-29; Figure 3-30 gives a view of the system, and the characteristics of this system are also summarized in Table 3-1.

The two earlier experimental systems discussed above are configured such that they can be operated simultaneously with radars in other frequency bands, in many cases permitting truly simultaneous multifrequency pulse-by-pulse measurements to be made from targets and clutter. Another example of a system which provides multiple frequency band measurements capability is the RATSCAT radar cross section measurement facility at Holloman AFB. This facility has the capability for making millimeter wave reflectivity measurements at 35, 52, 70, and 94 GHz utilizing antennas which yield measurement beamwidths on the order of one degree.

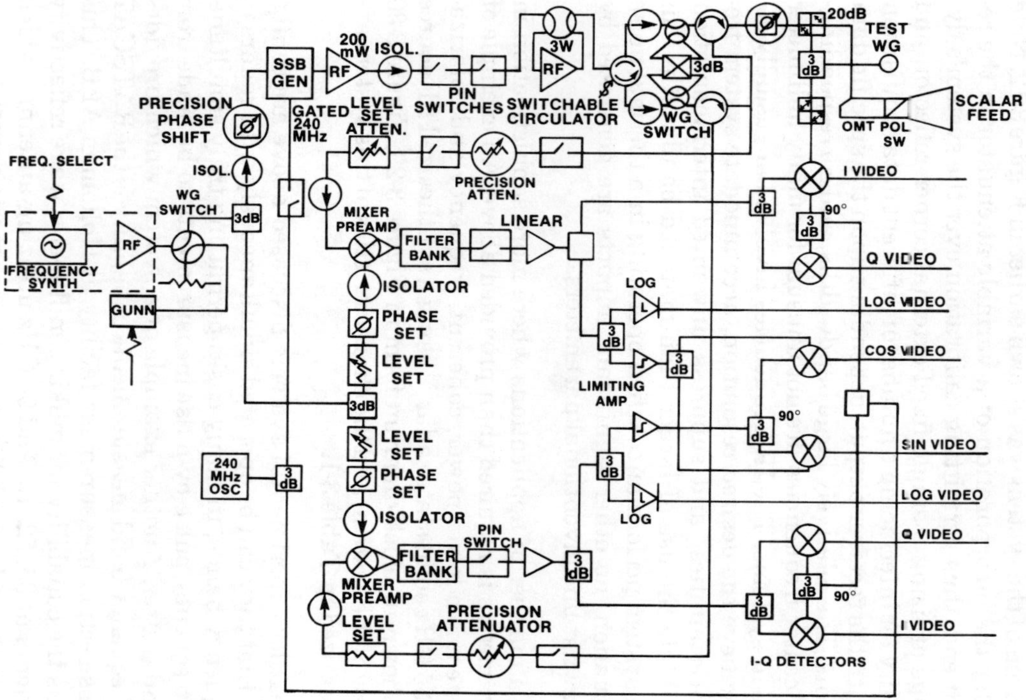


**Figure 3-29.** Simplified block diagram of a millimeter measurement radar system incorporating the capability to record both coherent data and polarimetric phase between the two received polarizations (courtesy of the Georgia Institute of Technology).

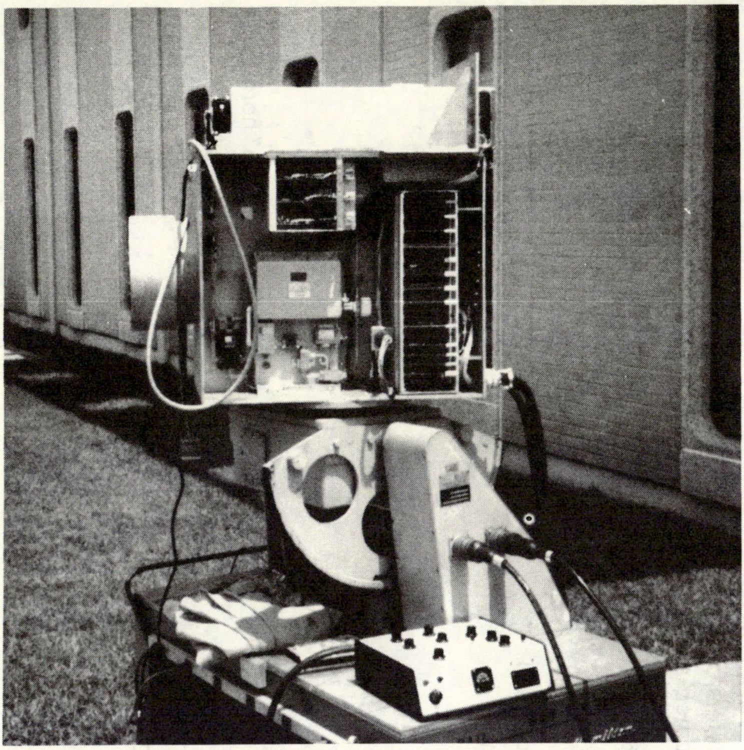

**Figure 3-30.** Side view of the millimeter measurement radar of Figure 3-29 (courtesy of the Georgia Institute of Technology).

One of the uses of a millimeter reflectivity range such as RATSCAT is for frequency-scaled measurements of model targets; that is, predicting the low frequency radar cross section behavior of a target by viewing a scaled-down version of the object at a higher operating frequency. An example of a facility dedicated to such measurements is the U.K. National Radio Modeling Facility, run by EMI Electronics, Ltd., in England.[21] This facility has radars operating as high as 2000 GHz consisting of over thirty different specialized systems. Both coherent and incoherent systems are utilized, and both amplitude and angle tracking measurements have been performed at his facility. A plan view of the modeling facilities is given in Figure 3-31, showing provisions for measurements of radar cross section of suspended targets and of reflections from artificially generated water waves.

**TABLE 3-1**
**Operating Parameters of Pulsed 35 and 95 GHz Instrumentation Radars**

| | 35 GHz | | 95 GHz | |
|---|---|---|---|---|
| TRANSMITTER | | | | |
| RF Frequency: | 35.24 ± 0.40 GHz (Synthesized fixed or agile) | | 95 GHz (nominal) | |
| Peak Power: | 200 *mW* (10 *ns* to CW)<br>3 watts (10-500 *ns*) | | 1 kW | |
| Pulsewidth: | 10 *ns* to CW (2 *ns* compressed) | | 20 *ns* | |
| *prf:* | variable to 40 kHz | | 0-4000 Hz | |
| ANTENNAS | | | | |
| Type: | 18″ Dish | 5″ horn/lens | 12″ Cassegrain | 3″ horn/lens |
| Beamwidth: | 1.25° | 4.5° | 0.7° | 3° |
| Gain: | 43 dB | 31 dB | 47 dB | 35 dB |
| Sidelobes: | -20 dB | -20 dB | -20 dB | -20 dB |
| Polarization: | Dual Polarized | | Dual Polarized | |
| Received: | H,V | | H,V | |
| Transmitted: | H,V,R,C,LC (agile or fixed) | | H,V,R,C,LC (agile or fixed) | |

| RECEIVER | | | |
|---|---|---|---|
| Type: | Coherent, Integrated Mixers/Preamps<br>Linear | Logarithmic | Logarithmic |
| Bandwidths: | 5 or 500 MHz | 160 MHz | 100 MHz |
| Dynamic Range: | 42 dB (5 MHz)<br>30 dB (500 MHz) | 60 dB | 70 dB |
| Sensitity: | -92 dBm (5 MHz)<br>-80 dBM | -85 dBm | -82 dBm |
| Detection: | Square Law Amplitude<br>Coherent (*I* and *Q*)<br>Pseudo-Coherent (*I* and *Q*) | | Pseudo-Coherent (phase) |
| IF: | 240 MHz | 240 MHz | 160 MHz |

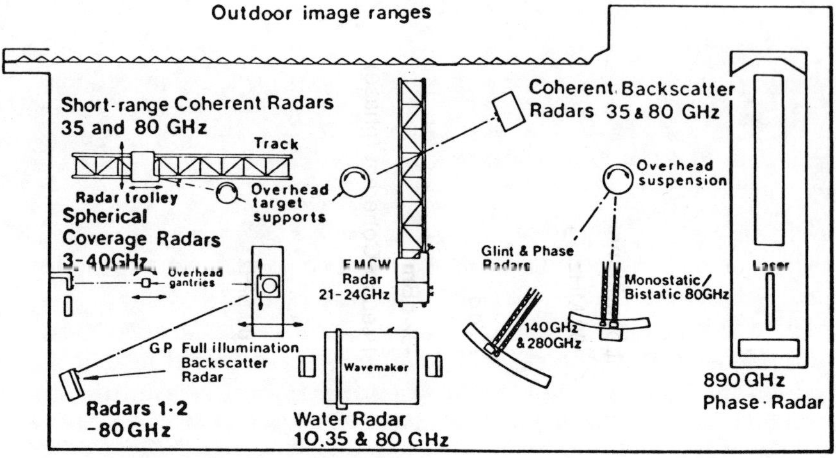


**Figure 3-31.** Plan view of the U.K. National Radio Model Facility for model measurements at frequencies as high as 2,000 GHz (reprinted from Cram and Woolcock, ref. [21]).

## 3.5 MEASUREMENT PROCEDURES

While each RCS measurement program is unique, there are a number of features and requirements which are common to any measurement program. In this section an attempt is made to highlight some of these common features.

Strange as it may sound, many measurement programs fail to fulfill expectations because measurements are taken without a clear understanding of how the data will eventually be used, how much data are required, or how the data will eventually be processed. While all of the details of the RCS dependencies of the target are rarely known before the experiment starts, an initial test plan should be prepared which would include a definition of desired data quality and quantity, and which would outline preliminary data analysis requirements. This initial test plan should provide adequate opportunities for review and restructuring of the program based on data acquired and operating experience. With such an initial experiment plan as a guide, one is ready to begin the acutal measurement program which may be broken into three principal parts: (1) pre-test, (2) the measurement themselves, and (3) data analysis and evaluation.

A number of tasks need to be accomplished before the measurement section of the program begins. These tasks include:

Take data to characterize the radar system, i.e., measure the radar parameters

Validate the data aquisition and analysis routines using known signals

Exercise the radar against simple targets having known characteristics

Once these initial steps have been carried out and one has developed reasonable confidence in the radar and data acquisition system, the measurement of target characteristics is ready to begin. A number of procedures must be followed in order to ensure data from targets of interest are both representative and accurately calibrated. A representative sequence of measurement steps might be first, to verify key radar parameters, then to perform a relative calibration, followed by viewing a standard target. At this time the system might be considered ready to collect data from targets of interest, and substantially more confidence may be placed on system stability if this data collection is immediately followed by another relative calibration and verification of key radar parameters. In addition, it may be desirable to continuously monitor certain key radar parameters, such as transmitted power, throughout the course of the data collection.

Thus, a representative set of measurement steps might include:

Warm-up Equipment

Preliminary Performance Check — set knobs, etc., record all set-up in data log

Measure Transmitted Power

Relative Calibration Using Injected Signal (check minimum detectable signal)

Record Absolute Data from Calibration Target

Check Calibration

Relative Calibration

Record Data

Measure Transmitted Power

Relative Calibration

Record System Control Settings

Absolute Calibration

Turn Off Radar

Examine Data for Anomalies

One may notice that an inordinate amount of time may appear to be consumed with calibration and data verification. A considerable amount of experience has shown that virtually every set of RCS data has some questionable feature. Numerous calibrations along with careful recording of the positions of all system controls will often either explain or substantiate the data set, and the careful and methodical recording of such pertinent auxillary data is an integral part of any RCS measurement program.

Once data have been recorded, a number of evaluations of these data need to be conducted off-line, with some results almost immediately available to highlight possible difficulties in either equipment or procedures. These off-line tests normally include tests for closure of calibrations, examination of selected frequency spectra for evidence of unwanted or peculiar signals (such as 60 or 400 Hz pickup), and evaluation of amplitude distributions for such factors as dropped or "stuck" bits in the conversion process. The use of modern portable microcomputers in the field greatly facilitates such data evaluations.

Once data have received this initial examination, then the detailed data analysis can begin and should include checking of closure of the calibrations, calibration of recorded data in terms of meaningful engineering units, and statistical analysis of these calibrated data in order to yield results such as amplitude distributions or frequency spectra which would be useful to the system designer. The final step in any such program is the interpretation of results, and attempts to establish correlations and dependencies on various environmental and physical factors.

## 3.6 SUMMARY

In this chapter, an attempt has been made to provide a broad overview of problems, equipment, and procedures in RCS measurement programs. Many of the topics, which are outlined here, are discussed in greater detail in other chapters. Chapter 4 gives a more comprehensive description of calibration procedures including the characteristics of calibration targets and sources of error. Chapter 5 discusses the fundamentals of data sampling and recording including error sources and commonly encountered problems. Chapter 6 discusses data analysis procedures, and Chapter 11 covers experiment planning in greater detail.

## REFERENCES

1. Barton, D.K., *Radar System Analysis,* p.96, Artech House, Dedham, Massachusetts, 1976.
2. "Aperture Uncertainity," from Le Croy Application Note, Le Croy Research Systems of California, Palo Alto, California, November 1980.
3. Harris, F. J., "On the Use of Windows for Harmonic Analysis with the Discrete Fourier Transform," *Proc. IEEE*, Vol.66, No. 1, January 1978, pp. 51-83.
4. Natrella, M. G., *Experimental Statistics*, NBS Handbook 91, Government Printing Office, Washington, D.C., August 1963.
5. Ewell, G.W., R. P., Zimmer, and M. E., Wallace, "Antenna Achieves Polarization Agility," *Microwave Journal*, Vol. 9, No. 6, June 1970, pp. 56-58.
6. Flynt, E. R.,*et al.*,"Appendices to A-725 Final Report," Appendix to Final Technical Report on Contract NObsr-91024, Georgia Institute of Technology, Atlanta, Georgia, December 1967.
7. Bachman, C. G., "Some Recent Developments in RCS Measurement Techniques," *Proc. IEEE*, Vol. 53, No. 8, August 1975, pp. 962-972.
8. Smith, J. M., and R. H. Logan, "AN/APS-116 Periscope Detecting Radar," *IEEE Transactions on Aerospace and Electronic Systems*, Vol. AES 16, No. 1, January 1980, pp. 66-73.
9. Ruttenberg, K., and L. Chonzit, "High Range Resolution By Means of Pulse to Pulse Frequency Shifting," *IEEE EASCON 68 Record*, pp. 47-51.
10. Walker, J. L., "Range-Doppler Imaging of Rotating Objects," *IEEE Transactions on Aerospace and Electronic Systems*, Vol. AES-16, No. 1, January 1980, pp. 23-52.
11. Mensa, D. L., S. Halevy, and G. Wade, "Coherent Doppler Tomography for Microwave Imaging," *Proc. IEEE*, Vol. 47, No. 2, February 1983, pp. 254-261.
12. Mensa, D. L., *High Resolution Radar Imaging*, Artech House, Dedham, Massachusetts, 1982.
13. Johnson, S. L., *et al.*, "Noise Spectrum Characteristics of Low-Noise Microwave Tubes and Solid-State Devices," *Proc. IEEE*, Vol. 54, No. 2, February 1966, pp. 258-265.

14. Raven, R. S., "Requirements for Master Oscillators for Coherent Radar," *Proc. IEEE*, Vol. 54, No. 2, February 1966, pp. 237-243.

15. Blacksmith, P. Jr., *et al.*, "Introduction to Radar Cross Section Measurements," *Proc. IEEE*, Vol. 53, No 8, August 1965, pp. 901-920.

16. O'Hara, F. J., and G. M. Moore, "A High Performance CW Receiver Using Feed-thru Nulling," *Microwave Journal*, Vol.6, No. 9, September 1963, pp. 63-71.

17. Alongi, A. V., *et al.*, "A High Resolution X-Band FM/CW Radar for RCS Measurements," *Proc. IEEE*, Vol. 53, No. 8, August 1965, pp. 1072-1076.

18. Stiles, W. H., B. C. Hanson, and F. T. Ulaby, "Microwave Remote Sensing of the Snow: Experiment Description and Preliminary Results," University of Kansas Center for Research, Remote Sensing Laboratory, Report RSL 340-1, June 1977.

19. Reedy, E. K., and G. W. Ewell, "Millimeter Radar," Chapter 2 in Volume 4 of *Infrared and Millimeter Waves*, Academic Press, 1981.

20. Currie, N. C., *et al.*, "MM-Wave Instrumentation Radar Systems," *Microwave Journal*, Vol. 17, No. 8, August 1978, pp. 35-43.

21. Cram, L. A., and S. C. Woolcock, "Review of Two Decades of Experience Between 30 GHz and 900 GHz in the Development of Model Radar Systems," in Agard Conference Proceedings, No. 295, *Millimeter and Submillimeter Wave Propagation and Circuits*, September 1978.

22. Fishbein, W., *et al.*, "Clutter Attenuation Analysis," Technical Report No. ECOM-2808, USAECOM, March 1967, AD665352.

# CHAPTER 4

# RADAR REFLECTIVITY CALIBRATION PROCEDURES

**James A. Scheer**
**Engineering Experiment Station**
**Georgia Institute of Technology**

## 4.1 OVERVIEW

The need for the collection of radar reflectivity data through radar reflectivity measurements has already been established. The usefulness of the data, once collected, is directly related to how well the data collection system is calibrated. Calibration of the radar system is required to provide accurate quantitative results from the data collected. The entire measurement system, including the sensor, the data acquistion equipment, and the playback equipment must be calibrated to provide this accurate quantitative description of the measured parameters. This chapter describes the basic methods of system calibration, describes the procedures associated with those methods, and finally provides an analysis of the errors which are encountered in calibrating the system. An estimate of the measurement accuracy achievable, both from a relative point of view and an absolute point of view, is presented as a function of RF frequency.

## 4.2 CALIBRATION METHODS

### 4.2.1 Amplitude

Two radar reflectivity amplitude calibration methods are discussed here. The first makes use of known calibration targets, and the second involves signal generator signals which are injected into the radar system front-end. The use of calibration targets in calibrating the radar system involves boresighting the radar antenna on a known size calibration target and recording the backscatter characteristics of this known target. The radar system response to this known target, or a variety of targets, is recorded and used as a reference for determining the characteristics of unknown targets or clutter cells. The basic result from an amplitude calibration using calibrated targets is a curve of receiver response *versus* target cross section for various target sizes achieved by either using various sized calibration targets, which is called an absolute calibration, or by using a single target and inserting attenuation in the radar receiver channel (which is known as a relative calibration). Phase calibration is achieved by inserting known amounts of phase shift in the receiver channel. Advantages of this type of radar calibration include the fact that the whole system, including RF losses, antenna gain, and atmospheric and propagation losses, are all accounted for in the calibration. Sources of error associated with this calibration procedure include non-precise positioning of the radar antenna on the calibration target, errors in calibration target orientation, the absolute *versus* predicted cross section of the calibration target, and the unknown effects of multipath, which could contaminate the calibration precision.

The second type of radar calibration technique commonly utilized is the injected signal generator calibration, which is called an indirect calibration. In this method, an external signal is injected into the radar receiver and varied in amplitude (or phase) to provide a transfer function for the receiver which relates voltage out of the receiver to RF power at the injection point. The various amplitudes can be achieved by either varying the output of the signal generator, or by varying a precision attenuator in the receiver channel. The voltage out of the receiver can then be related to radar cross section through use of the radar range equation along with knowledge of the radar parameters. In a similar manner, phase calibration can be achieved by varying a phase shifter which is either in the injection path or in in the receiver channel. The signal generator calibration can use either an external precision signal generator or a built-in generator which can be used repeatedly and periodically to provide the calibration signal. An advantage of the signal generator calibration is that one need not be concerned about the pointing accuracy of the radar or the pointing accuracy of the calibration corner reflector in providing the calibration curve. Some sources of error associated with the signal generator calibration include the unknown propagation losses which must be considered as a part of the calibration of the data, errors in the determination of radar parameters such as transmitted power, and the fact that the spectrum of the signal generator pulse may not be the same as the spectrum of the actual radar transmitted pulse, in which case the receiver response for the two would be different. The following section describes, in more detail, the procedures associated with these two types of calibration methods as well as the use of both methods simultaneously to obtain a measure of the calibration accuracy which is known as "calibration closure."

### 4.2.2 Coherent Tests

Two methods are also used for measuring the integrity of the $I$ and $Q$ phase detector system in a coherent radar. One method involves the injection of a test signal, and the other method involves the use of a reference reflector. As with the amplitude calibration, the phase calibrations of the two techniques should agree with each other.

The coherence of a system is a measure of the degree to which the system can determine phase changes on a pulse-to-pulse basis. Such phase changes indicate target motion. The phase stability of the oscillators must be such that no apparent target motion exists for stationary targets. The test for coherence can be made using a stable injection signal or by looking at a stationary reference target.

### 4.2.3 Polarization

Several types of polarization-related features are employed in instrumentation radars. Multiple transmit polarizations, polarization agility, and dual polarization receivers are among them. Several tests are required to confirm the integrity of these features. Polarization isolation, aspect ratio, dual mode coupler performance, and calibration of the polarimetric phase are among them. Measurement of these parameters will be discussed below.

## 4.3 AMPLITUDE CALIBRATION PROCEDURES

### 4.3.1 Calibration Targets

One of the most often used methods for calibrating radars involves the use of calibration reference targets. These targets typically are flat plates, diplanes, trihedrals, spheres, or more sophisticated calibration targets such as Luneberg and constant-K lenses.

Figure 4-1 shows several of the commonly used calibration reference targets and gives formulas for the calculated maximum cross section of each.[1]

*General Procedure*

The general procedure for calibrating the amplitude of a radar system using the relative calibration method involves pointing the radar antenna at the calibration target and inserting known values of receiver attenuation from 0 dB up to maximum attenuation in known increments (typically 5 dB), and recording the receiver output for each value of inserted attenuation. Figure 4-2 shows the block diagram of an instrumentation radar providing precision var-

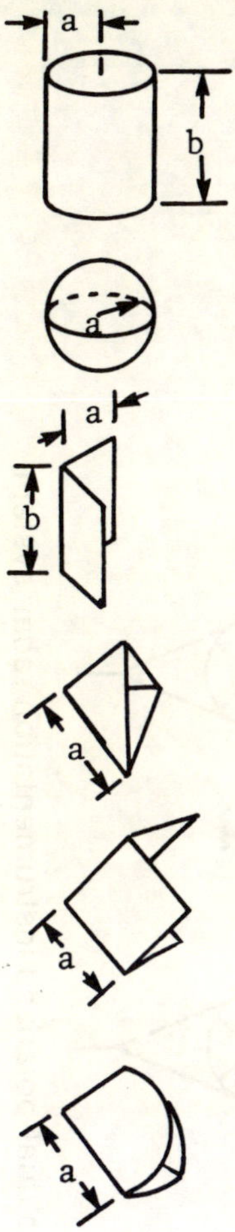


| Target | Maximum Cross Section |
|---|---|
| Cylinder | $\sigma = \dfrac{2ab^2}{\lambda}$ |
| Sphere | $\sigma = \pi a^2$ (optical region) |
| Diplane | $\sigma = \dfrac{8\pi a^2 b^2}{\lambda^2}$ |
| Triangular Trihedral | $\sigma = \dfrac{4\pi a^4}{3\lambda^2}$ |
| Square Trihedral | $\sigma = \dfrac{12\pi a^4}{\lambda^2}$ |
| Circular Trihedral | $\sigma = \dfrac{15.7a^4}{\lambda^2}$ |

**Figure 4-1.** Various types of calibration targets (adapted from Jasik, ref. [1]; ©1961 McGraw-Hill Book Company).

iable attenuators in the receiver paths (one for each channel). The known RCS value of the calibration reference target in square meters is noted, as is the range to that target. During the analysis of the recorded data, a calibration curve is generated, relating radar cross section to receiver output voltage for the calibration run. Data taken on an unknown target are then calibrated by comparing their output voltage value with the calibration curve to determine the radar cross section, assuming the target is at the same slant range as the reference target or that the effect of different slant ranges between

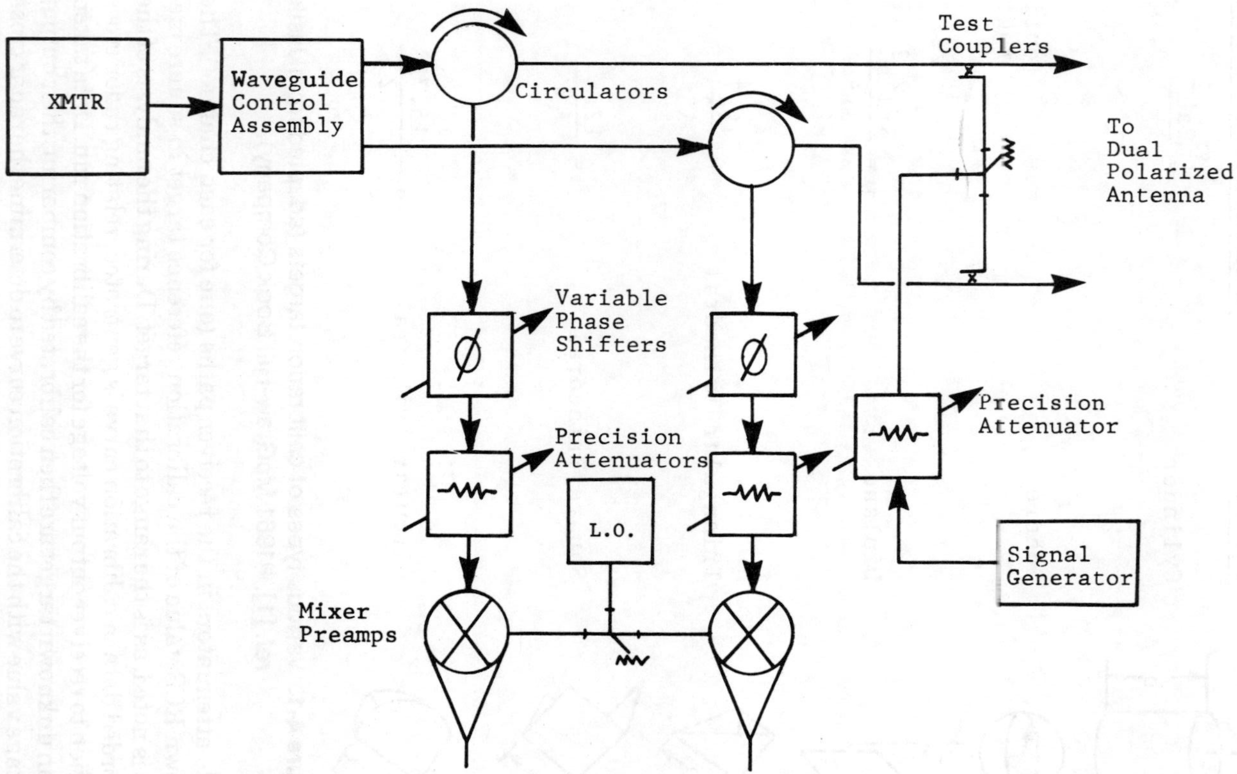


**Figure 4-2.** Block diagram of a dual-polarized instrumentation radar system.

the reference and unknown target is accounted for. The true target cross section ($\sigma_T$) is determined from the reference by use of a modified form of the radar range equation.

$$\sigma_T = \sigma_{ref} \left( \frac{V_T}{V_{ref}} \right) \left( \frac{R_T}{R_{ref}} \right)^4 \qquad (4\text{-}1)$$

where

$\sigma_{ref}$ = the RCS of the reference
$V_T$ = the receiver voltage level from the target
$V_{ref}$ = the voltage level of the reference target
$R_T, R_{ref}$ = the ranges to the target and references reflector, respectively

There are various methods for constructing the curve of receiver output voltage *versus* signal strength given the discrete calibration points. In some cases, a linear least squares fit to the data points is constructed. In other cases where a system is not particularly linear or not truly logarithmic, then a more complex curve is drawn to minimize the error at the collected points. In these cases, typically, a third or higher order curve is constructed which provides the best fit to the calibration points. In general, the curve should provide less than 0.5 dB error at each of the measured points (*see* Chapter 6 for more detailed discussion of calibration curve fitting).

In addition to calibrating the amplitude response of the system, calibration of the phase response is sometimes desired. This is true for coherent systems where absolute phase is a measured parameter, and for polarimetric systems where the phase between two channels is measured. The phase response of the radar is calibrated in much the same way as is the amplitude response, except that a calibrated precision phase shifter, as seen in Figure 4-2, is varied while recording the receiver response instead of a precision attenuator. This phase shifter can be either in the receiver path or in the local oscillator path to the mixer. Here again, as with the amplitude calibration, a best fit curve is drawn through the calibration curve. More typically, a higher order function is required with the best being a cosine function whose amplitude and end points are varied to provide the best fit through the data points.

It is most advisable to insert the attenuation and the phase shift at the radio frequency end of the radar. In some cases, this is not convenient and an alternate method involves insertion of the calibration attenuation and phase shift at the intermediate frequency

(IF) section. When this is done, it is required that the radar system be linear ahead of the attenuator; that is to say, the strongest signal encountered will not saturate the mixer preamp and the smallest signal encountered will be well above the noise level, so that the mixer preamp response is linear. The IF calibration procedure is equivalent to the RF calibration only if mixer preamp linearity exists.

An alternative method for varying the amplitude of the calibration signal is to use a variety of calibration targets of varying sizes rather than just one target and inserting attenuation (absolute calibration). The range of calibration target sizes and the number of different sizes between the smallest and the largest must be consistent with the dynamic range of targets and clutter cells expected to be encountered and must provide enough independent points so that a calibration curve can be constructed through the points. This implies that on the order of six or more different corner reflector sizes with a total dynamic range of, at least, 50 dB is desired. A problem typically encountered in this approach is the fact that the smaller sized targets will tend to be contaminated with ground clutter return and thereby will not provide a clean calibration signal. This is avoided by using a single large calibration target, because as attenuation is inserted, the ground clutter is attenuated just as much as the target. The absolute calibration approach is most often used in airborne test programs where calibration arrays of different sizes are placed along the flight path.

*Sources of Error*

Three sources of error will be considered here. The first involves the accuracy with which the radar antenna can be positioned and the accuracy with which the calibration target can be positioned. The second involves the accuracy with which the calibration target's true radar cross section is known. The third error source is the multipath effect which can create large errors in the calibration amplitude.

The radar antenna positioning is highly critical to ensure accurate calibration. For example, if the radar has a one degree, 3 dB beamwidth, then a one-half degree error in positioning would cause a 6 dB round trip error in measurement. To keep the error due to antenna pointing angle well within 1 dB, the positioning error would have to be a small percentage of the 3 dB beamwidth.

Although much less critical than the radar antenna pointing angle, the pointing angle of the calibration target must also be considered. These targets, particularly in the case of a trihedral, tend to have a much broader illumination pattern. The diplane is specular in one direction, that is, in the direction of the seam, and the flat plate is specular in all directions. The sphere has theoretically no positioning sensitivity. Calibration reflectors like the Luneberg lens and constant-K lens have broad calibration beamwidths. Figure 4-3 shows the beam patterns for several types of calibration targets.[2]

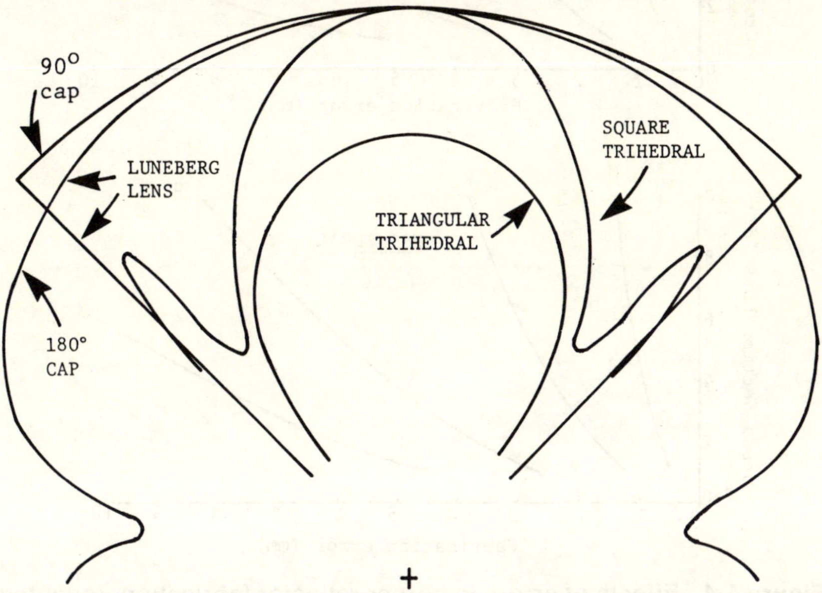


**Figure 4-3.** Calibration target radiation patterns (adapted from Jasik, ref. [2]; © 1961 McGraw-Hill Book Company).

To use a calibration target for calibrating the radar system, the true radar cross section of the target must be known. Machining tolerances inherent in fabricating the corner reflector must be considered in determining its cross section. This is more critical at shorter wavelengths, particularly at millimeter wavelengths which have received a lot of attention in the recent past in measurements programs. Figure 4-4 shows the errors in the cross section for a calibration target as a function of errors in machining the surfaces, expressed in millimeters.[3]

A very important source of potential error in calibrating the system and making target measurements involves the multipath environment for the system.[4] This is particularly true for some scenarios

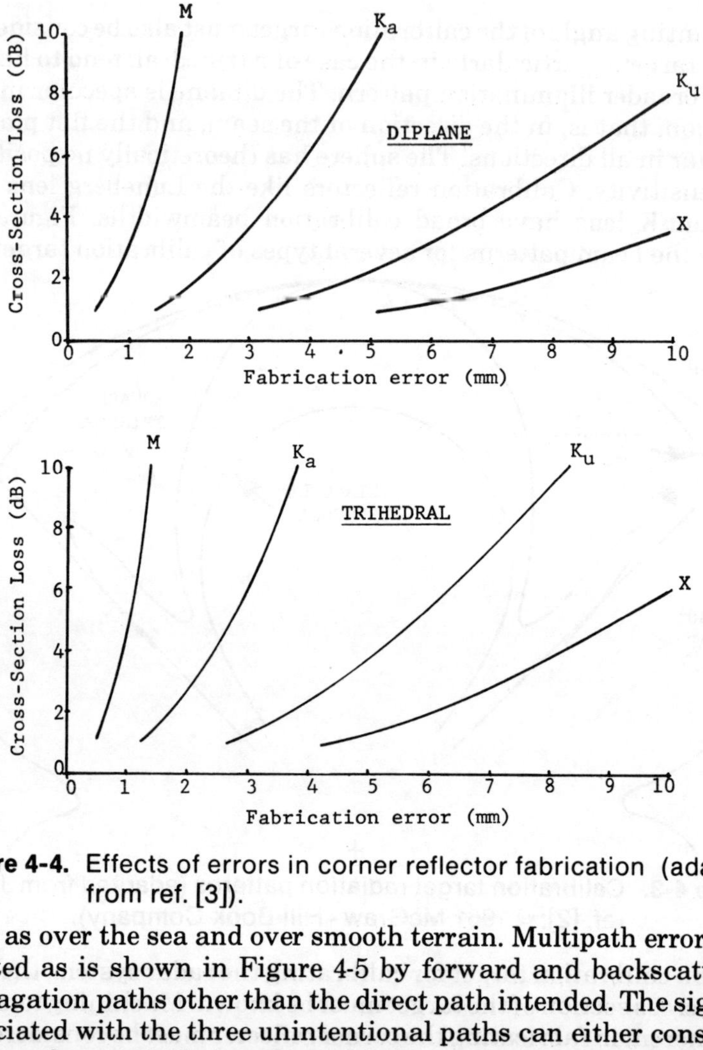


**Figure 4-4.** Effects of errors in corner reflector fabrication (adapted from ref. [3]).

such as over the sea and over smooth terrain. Multipath errors are created as is shown in Figure 4-5 by forward and backscattered propagation paths other than the direct path intended. The signals associated with the three unintentional paths can either constructively or destructively interfere with that of the intended direct path to create up to a 12 dB enhancement of the target or can create a null, depending on the geometry of the scenario. In those cases where the multipath problem has a potential for contaminating the data, then a probe of the multipath field must be made. This involves raising the calibration target in the vertical direction to determine what lobing is present in the environment. If multipath lobing is apparent, then the calibration corner must be raised to such a height that the lobing subsides to a value acceptable for the purpose of the tests,

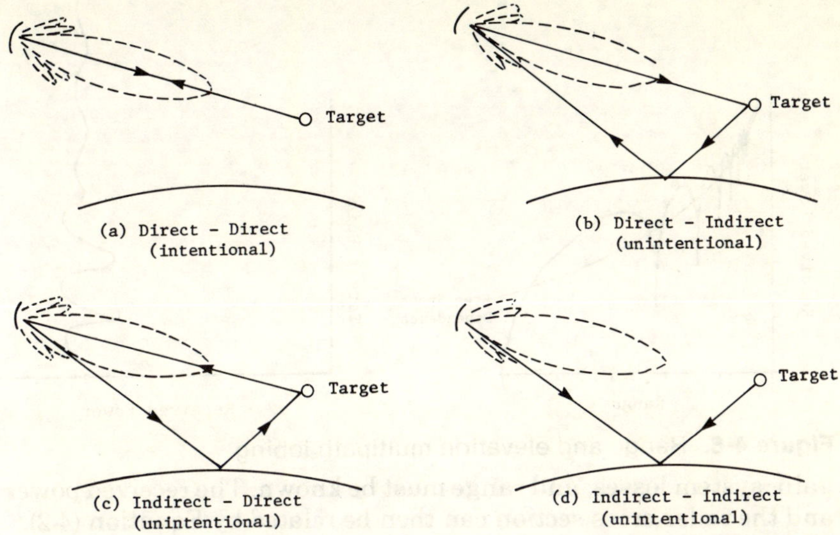


**Figure 4-5.** The four transmission paths between a radar antenna and a target near the earth which leads to multipath effects on the received signal.

(this occurs because the reflected ray path falls outside the main antenna beam) or the multipath field must be calculated. Figure 4-6 shows a typical range and elevation multipath lobe structure.

## 4.3.2 Signal Generator Calibration

*General Procedure*

An alternate method for calibrating the radar system, more commonly used as a check on the absolute calibration, is the signal generator or indirect calibration. This procedure involves injecting a pulsed signal from a signal generator into the test coupling of the radar receiver as shown in Figure 4-2 and varying its amplitude in increments over the full dynamic range of the radar receiver from saturation to receiver noise level. The result of this type of calibration is a curve of receiver output voltage as a function of input power level. This curve is much the same as the one generated in the calibration target procedure, except that the ordinate axis is received power rather than target radar cross section. The normal procedure is to use the radar range equation to calculate the radar cross section from the received power. In order to use the radar equation the values of transmitted power, wavelength, antenna

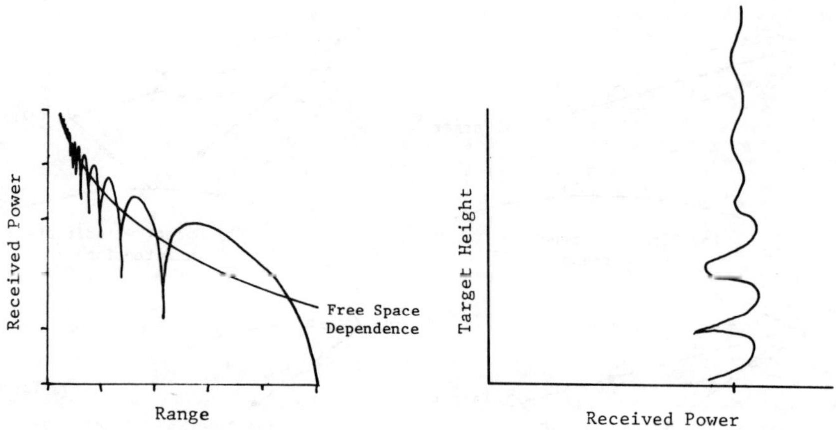


**Figure 4-6.** Range and elevation multipath lobing.

gain, system losses, and range must be known. The received power and the radar cross section can then be related by Equation (4-2).

$$\sigma_T = (4\pi)^3 \frac{P_R R^4}{L_T G_T G_R P_T \lambda^2 L_R} \tag{4-2}$$

where

$\sigma_T$ = the target radar cross section
$P_R$ = the received power
$L_T, L_R$ = the transmit and receive losses (values less than one)
$R$ = the range
$G_T, G_R$ = the transmit and receive antenna gains
$P_T$ = the transmitted power
$\lambda$ = the wavelength

*Sources of Error*

There are many elements which can contribute to an inability to obtain accurate measurements of RCS using the indirect method. These include unknown path losses, unknown or poorly estimated antenna gain, errors in determining the transmitted power (particularly since the peak transmitted power is normally calculated from average power measurements), and the inability to provide a signal generator spectrum that is equivalent to the spectrum of the transmitted pulse. This effect is most pronounced if the receiver bandwidth is matched to the spectrum of the transmitted pulse. Obviously, if the receiver bandwidth is very wide, compared to the transmitted pulse spectrum then it will respond in a similar fashion to various transmitted spectral responses so long as they are contained within the receiver bandwidth. However, with the more typically encountered narrow receiver bandwidth, if the signal generator spectrum is either different in center frequency or in shape from the spectrum of the transmitter pulse, then the receiver will respond differently.

Because of the greater likelihood of errors occuring in the indirect calibration, it is normally used only as a check on the absolute calibration except where the absolute calibration is difficult such as over the sea. In many cases, where a sea environment is not involved, the signal generator calibration is performed only occasionally during the program to determine the integrity of the system with the absolute calibration technique being employed as the primary calibration method.

A variation of the signal generator calibration is sometimes used. This involves the use of a built-in signal generator to periodically calibrate the radar system. This technique is employed in cases where the radar is not easily accessible for calibration procedures, or in a dynamic environment where it is difficult to boresight the antenna to a calibration target. In this case the built-in signal generator is periodically, either automatically or manually, switched into the receiver to provide at least a one point calibration during data acquisition. The added requirement to this calibration technique is that the built-in signal generator needs to be stabilized because it may be subject to frequency drift, and thus could be less stable than the radar itself. Here again, the important thing is to understand the potential causes of error in calibrating the radar system and data acquisition system and to understand all of the elements which can contribute to the errors.

*Closing the Calibrations*

Each of the two calibration procedures described above yields a curve relating the receiver output voltage to the input signal. In one case, the input is an RF signal from a signal generator. In the other case, the input is the signal received from a calibration target. The received signal power $P_r$ is related to the target cross section $\sigma_T$ by Equation (4-2). If a known calibration target is illuminated, and the radar cross section is calculated from Equation (4-2), then the difference between the known RCS value and the calculated value is the level of "closure" between the absolute and indirect calibration methods.

Agreement between the two methods is equivalent to satisfactorily verifying Equation (4-2), using the radar. Each of the terms of Equation (4-2) is determined by the following method:

- $P_r$ = measured radar response
- $P_T$ = measured (normally the peak power is calculated from the average power and duty factor, ($prf$ · pulsewidth)
- $G$ = measured or calculated
- $\lambda$ = computed (frequency is measured)
- $\sigma$ = calculated from dimensions or measured on a high precision RCS range
- $L_R$, $L_R$ = radar losses measured, propagation losses estimated

The two terms subject to the most error are $\sigma$, at least at the millimeter wave frequencies, $L_T$ and $L_R$, which, as far as the atmospheric and propagation elements are concerned, is only an estimated value (multipath errors and non-equivalent spectrum erors are included in these terms), and the transmitted peak power is normally calculated from average power measurements.

## 4.4 COHERENT SYSTEM CALIBRATION PROCEDURES

The significant feature of a coherent radar system is its ability to measure the phase of the received signal (as well as its amplitude) relative to the phase of the signal which was transmitted. Normally, the absolute phase is not important since that quantity is ambiguous for every half-wavelength of distance between the radar and the target of interest. The change in phase on a pulse-to-pulse basis is, however, important, so that a signal which is stable for the time period required for the signal to propagate to and from the target must be used as a reference from which the received signal phase can be compared.

The coherent detection process results in a vector quantity, as opposed to an amplitude signal for a non-coherent system. The vector quantity can be expressed in one of two ways — in a $\rho$, $\Theta$ (circular) coordinate system, or in an x,y (rectangular) coordinate system. The more common technique is the latter. The vector is expressed as two bipolar quantities — the component which is in-phase ($I$) with the reference signal and the component which is in quadrature ($Q$) with the reference. The resultant vector can be converted to the circular coordinate system ($\rho$, $\Theta$) by using the transformation:

$$\rho = \sqrt{I^2 + Q^2} \tag{4-3}$$

and

$$\Theta = \tan^{-1} \frac{Q}{I} \tag{4-4}$$

Figure 4-7 shows the block diagram of the related components of a coherent radar system. The transmitted signal is derived from mixing a *stable local oscillator* (STALO) and an IF *coherent oscillator* (COHO). When the received signal is mixed with the STALO signal, the resulting IF signal is the same frequency as that of the COHO (except for any Doppler offset which may be imparted by the target), and can therefore be compared to the COHO to determine the relative phase. The *I* and *Q* components are generated by applying the received signal to each of two phase sensitive detectors (synchronous detectors), the references for which are derived from the COHO signal which has been equally divided with one path shifted by 90° relative to the other.

Calibration of the coherent system is done to insure that the "quadrature" detection process indeed represents a 90° phase separation, and that the *I* and *Q* amplitudes are balanced. This confirmation should be done at a variety of input signal amplitudes to be sure that the quadrature detection is not amplitude sensitive.

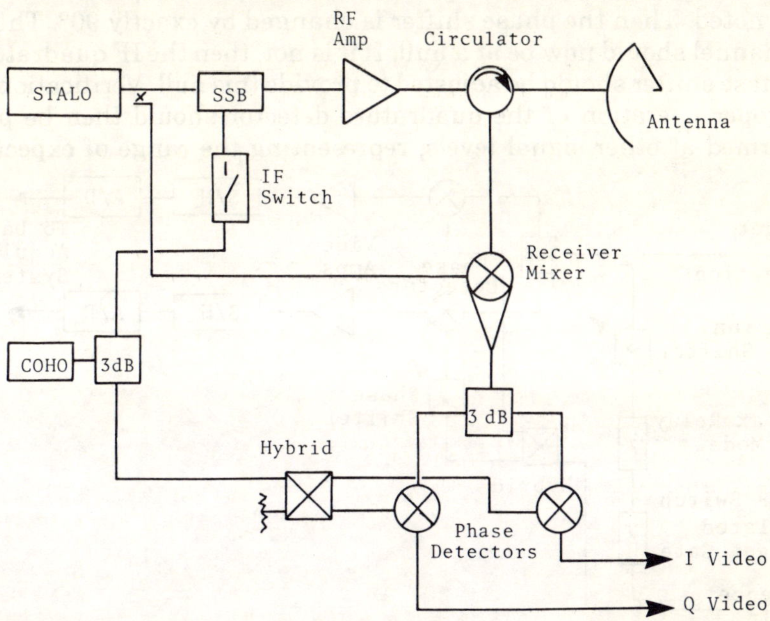


**Figure 4-7.** Simplified block diagram of a coherent radar system.

### 4.4.1 Signal Injection

The most straightforward method for setting up the coherent detector involves injecting a coherent signal into the detector at either RF or IF. RF injection provides a more thorough check of the system. However, it is usually more complex and costly than IF injection. For RF injection a sample of the transmit signal is coupled into the receiver test port. For IF injection a sample of the COHO is coupled into the IF portion of the receiver. In either case a precision phase shifter must be inserted into the injected signal path such that the signal can be accurately adjusted in phase. Typically, phase setting accuracy on the order of one degree (1°) is required.

Figure 4-8 shows the block diagram of a typical coherent (*I* and *Q*) detector with associated data acquisition circuits (i.e., sampler and analog-to-digital converter). A method for injecting a coherent IF signal into the detector is shown in the figure. The alignment and calibration procedure involves allowing the simulated target to be injected into the detector by energizing the IF test relay and activating the target simulator switch. The ouputs of the *I* and *Q* data acquisition circuits are monitored. The target phase shifter is adjusted so that the *I* channel provides zero output. The phase reading is noted. Then the phase shifter is changed by exactly 90°. The *Q* channel should now be at a null. If it is not, then the IF quadrature phase shifter should be adjusted to provide this null. Verification of proper operation of the quadrature detector should then be performed at other signal levels, representing the range of expected

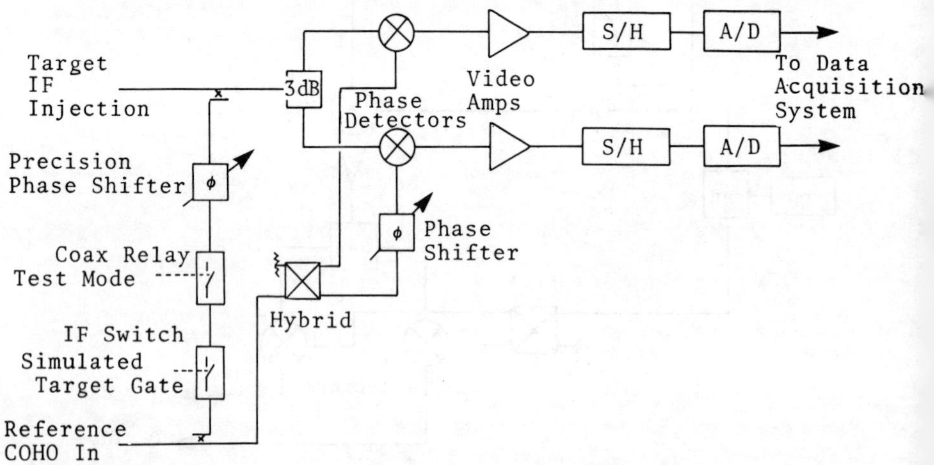


**Figure 4-8.** Block diagram of coherent radar detector and data acquisition circuits for coherent measurements.

signals into the detector. The same procedure is used for calibration of a coherent system with an RF injected signal. The precision phase shifter, electronic target simulation switch, and test mode relay functions would instead be implemented at RF using a sample of the transmitted signal. An example of this implementation is shown in Figure 4-9.

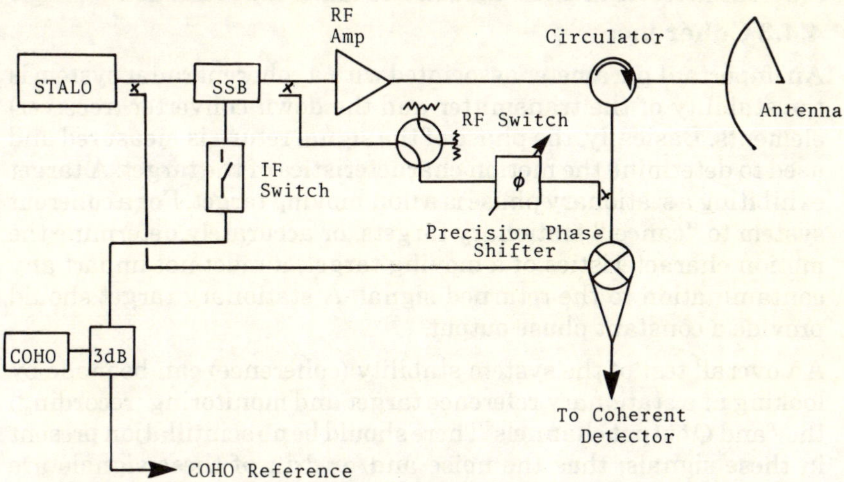


**Figure 4-9.** Block diagram of coherent instrumentation radar with provisions for RF signal injection.

### 4.4.2 Reference Target Calibration

The most accurate and complete method for calibrating the coherent detector involves the use of an injected signal as described above. Sometimes, however, a radar is not instrumented to allow for injecting such a signal, or even if it is, it is desirable to employ a simplified technique for quick checking purposes when in the field during a measurement program. Such a simplified test can be made by looking at a reference reflector, in a manner similar to the amplitude calibration described in section 4.3. The procedure requires that a precision phase shifter be installed in the local oscillator path to the mixers.

When looking at a reference target which is nominally stationary, care must be taken to ensure that there is no relative motion between the radar and the target. A test for 1° integrity of a coherent detector, for example, would require that during the test, motion between the reference reflector and the radar not exceed $\pm$ 40 $\mu m$ (.04mm) when operating at X-band. This means that both the radar and the target must be firmly mounted. The test is performed by adjusting the phase shifter to produce a null out of the *I* channel, noting the phase, and readjusting the phase shifter to achieve a null out of the *Q* channel. The difference between the two phase settings should be 90°. If not, then the IF quadrature phase shifter in the coherent detector must be adjusted to make the correction.

### 4.4.3 Coherence

An important parameter associated with a coherent radar system is the stability of the transmitter and the down converter (receiver) elements. Basically, the phase of the signal return is measured and used to determine the motion characteristics of the target. A target exhibiting a stationary phase is a non-moving target. For a coherent system to "cancel" stationary targets, or accurately determine the motion characteristics of a moving target, it must not impart any contamination to the returned signal. A stationary target should provide a constant phase output.

An overall test of the system stability (coherence) can be made by looking at a stationary reference target and monitoring (recording) the *I* and *Q* output channels. There should be no scintillation present in these signals; thus the noise and/or drift of these signals are measures of the system short term and long term stability (or instability as the case may be). The bandwidth of the noise and the rate of drift of the outputs represent the Doppler resolution and minimum detectable Doppler of the system. Spectral analysis of the output signals derived from a stationary target exposes and quantifies any system instabilities.

## 4.5 POLARIZATION CALIBRATION PROCEDURES

Many of the modern instrumentation systems employ a variety of types of polarization agility (sometimes called polarization diversity), and dual polarization receivers. When performing research to develop an understanding of the polarization related properties of targets and clutter, it is important to understand the polarization characteristics and limitations of the measurement systems themselves. Various features which need to be measured are polarization isolation, aspect ratio, and polarimetric phase.

### 4.5.1 Measuring Polarization Isolation

Polarization isolation is a measure of separation between two channels in a dual polarized radar. For example, it is a measure of how much a vertically polarized signal contaminates the horizontally polarized receiver channel. Polarization contamination can occur in the transmit section, in the antenna, and/or in the receiver section. A test of polarization isolation indicates the degree to which polarization effects can be studied. For example, to determine the polarization ratio of a target, the system isolation must be greater than the polarization ratio to be measured.

An overall system test can be performed by illuminating a large (however, non-saturating) standard reference target with a signal of one polarization (say, vertical) and measuring how much energy is received in the orthogonal (horizontal) channel relative to the copolarized (vertical) channel. Contamination of the horizontal channel (in this example) would occur if 1) the transmitted waveform were not purely vertically polarized, 2) the reference reflector contaminated the return polarization due to its own imperfections or the existence of clutter in the cell, 3) the dual mode coupler (orthomode transducer) in the antenna system did not perfectly separate the horizontal component from the vertical component, or 4) there existed a leakage path between the receiver channels, such as at waveguide flange joints or through the local oscillator divider network.

The effects of the reference reflector can be separated by using a variety of targets. Although a flat plate is generally the most difficult to position properly, it is generally the easiest to make with precision, relative to a diplane, a trihedral, or a sphere.

The receiver related effects can be determined by injecting a signal into one of the receiver channels and measuring the output which may result in the other channel. Once these effects have been eliminated, any remaining contamination is attributable to the transmitter switching components and/or the antenna/feed components. The transmitter RF assembly effects can be determined by temporarily deleting all of the switching and splitting elements and exciting only one polarization with as much simplicity as possible. The remaining effects are caused primarily by the antenna system (maybe a radome could be causing depolarization), which is usually the major contributor to such contamination.

A test can be performed to determine the isolation between any two orthogonal polarizations (e.g., *V* and *H*, RHC and LHC, +45° and -45° keeping in mind the polarization properties of the reference reflectors being used. Table 4-1 shows the response of various reference targets to various illumination polarizations. The most useful target for testing linear polarization isolation is the diplane, while a diplane and corner reflector is required to test circular polarization.

**TABLE 4-1**
**Reflected Polarization for Various Reflectors**

| REFLECTOR | ILLUMINATED POLARIZATION | | | | | |
|---|---|---|---|---|---|---|
| | *V* | *H* | RHC | LHC | +45° | −45° |
| Flat Plato | *V* | *H* | LHC | RHC | +45° | −45° |
| Sphere | *V* | *H* | LHC | RHC | +45° | −45° |
| Trihedral | *V* | *H* | LHC | RHC | +45° | −45° |
| Diplane ( *V* ) | *V* | *H* | RHC | LHC | −45° | +45° |
| Diplane ( 45° ) | *H* | *V* | RHC | LHC | +45° | −45° |
| Diplane ( 22.5° ) | +45° | −45° | RHC | LHC | *V* | *H* |

The test is carried out by transmitting one polarization and using the calibration target to reflect the orthogonal polarization. The resulting signal in the parallel channel is a measure of the polarization isolation when compared to a target of the same RC which returns a parallel polarization.

### 4.5.2 Aspect Ratio

The polarization aspect ratio is the difference in amplitude between the horizontally polarized signal and the vertically polarized signal. Separating the transmitted component from the received component requires the use of separate test equipment. Measuring the actual transmitted signal requires a separate receiver placed in the far field of the radar system. This receiver must have a dual polarized antenna, or an antenna which is rotatable so that *H* and *V* can both be received. The transmitter for the radar under test is used to illuminate the test receiver in the two polarization modes. The balance between *H* and *V* can then be measured. If a measure of circular polarization is desired than a dual polarized test receiver is required. The amplitude match and relative phase can be measured with such a system. The integrity of the test is limited by the polarization parameters of the test receiver, e.g., its polarization isolation, polarization aspect ratio, and polarimetric phase accuracy.

### 4.5.3 Using a Dual Mode Coupler (Ortho-Mode Transducer) (OMT)

In many instrumentation radar systems, polarization adaptability is required. For this reason, many of these systems use a dual mode coupler at the antenna feed to produce a variety of polarizations. One can generate *H, V,* +45°, -45°, RHC, and LHC polarizations by exciting one or both ports simultaneously with the appropriate amplitude and phase relationship. Figure 4-10 shows a block diagram of such a system.

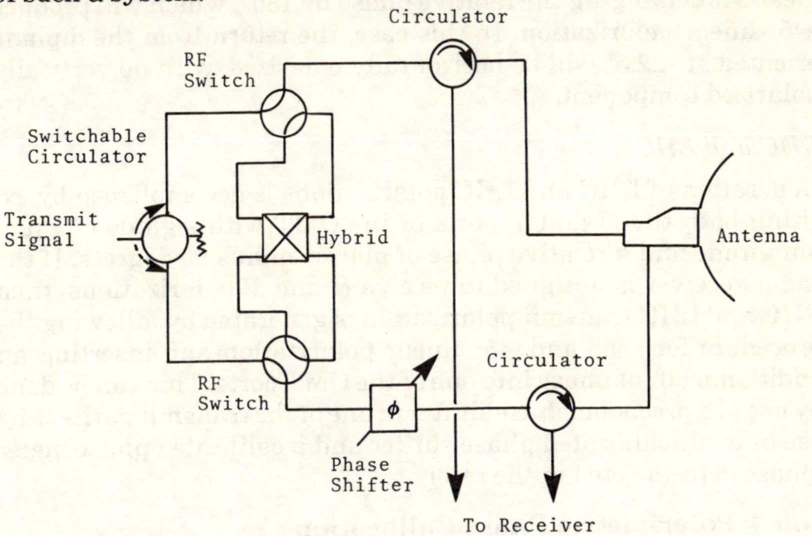


**Figure 4-10.** Block diagram of a polarization agile instrumentation radar system.

*H and V*

The trivial case for generating a signal using an OMT is horizontal or vertical polarization which is generated by exciting only the appropriate ports. A test of the polarization isolation, as described later in this section, will provide a measure of the integrity of these signals.

*+45°* and *-45°*

Forty-five degree linear polarization is generated by exciting both the *H* and *V* ports simultaneously with equal amplitude signals, either in phase or 180° out of phase. A phase shifter is required in one arm of the coupler to insure that the proper relative phase can be achieved. The following test can be made to optimize the system for 45° linear operation.

A diplane should be installed within line-of-sight of the system, such that the angle of its seam is at +22.5° relative to the vertical. If this reflector is illuminated with +45° linear polarization, the return signal will be vertically polarized and there will be no horizontally polarized component. The phase shifter should be adjusted so that this situation exists (i.e., a null exists in the horizontally polarized receive channel). The system is now set up for +45° linear polarization. If the system is polarization agile, then it most likely has the means for changing the relative phase by 180°, which will produce -45° linear polarization. In this case, the return from the diplane oriented at +22.5° will be horizontally polarized, with no vertically polarized component.

*RHC and LHC*

Generation of RHC and LHC polarizations is accomplished by exciting both the *H* and *V* ports of the OMT with signals of equal amplitude and a relative phase of plus or minus 90 degrees. If the radar receiver is designed to receive *H* and *V* polarizations, then RHC and LHC transmit polarization is generated by following the procedure for +45° and -45° linear polarization and inserting an additional 90° of phase into one of the OMT ports. This can be done by use of a precision phase shifter in one of the transmit paths or by use of an uncalibrated phase shifter and a calibrated phase meter (phase detector, etc.) in the receiver.

### 4.5.4 Polarimetric Phase Calibration

Many instrumentation radar systems are dual polarized, meaning that there are actually two receiver channels — one receiving the polarization which is transmitted (co-polarized receiver), and one receiving the orthogonal polarization (cross polarized receiver). In some systems, the two received polarizations are horizontal (*H*) and vertical (*V*), regardless of what is transmitted. Recently, much research has been done to determine the polarization related characteristics of targets and clutter. One aspect of this research involves the study of the polarimetric phase — the phase of one of the polarized returns relative to the orthogonal return. This phase is generated by applying the two received signals (at IF) to a phase detector; the resulting output is proportional to the sine or the cosine of the relative phase if the signals are limited in amplitude. The use of a single phase detector leaves an ambiguity in the phase signal because over a complete circle (360°), the sine or cosine function can take on a given value for two different angles. This ambiguity is resolved by use of two detectors, a sine and a cosine detector. The function is implemented as shown in Figure 4-11.

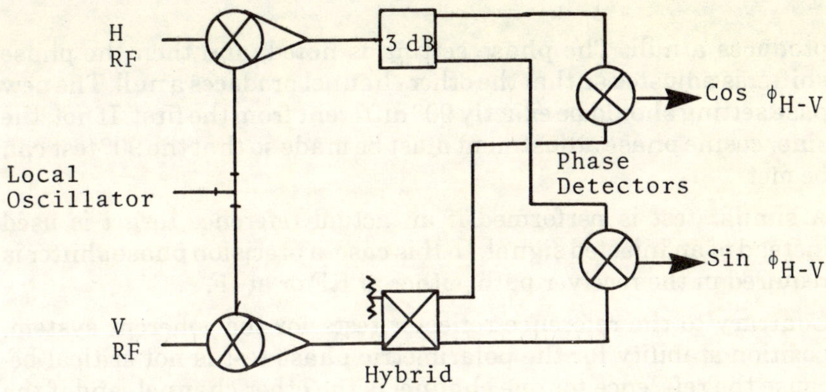


**Figure 4-11.** Implementation of in-phase (*I*) and quadrature (*Q*) detectors.

In calibrating a radar providing the polarimetric phase signal, it is important to insure the integrity of the synchronous (phase) detectors. This is done by one of two methods. Either a test target can be injected into the radar at RF or IF, or a reference target can be used.

The setup for calibrating the phase detector by signal injection is illustrated in Figure 4-12. Whether the synthetic signal is injected at RF or IF, the signal must be injected into both receiver channels simultaneously, and a precision phase shifter must be employed in at least one of the injection ports. The procedure involves adjusting the phase shifter so that one of the phase channels (sine or cosine)

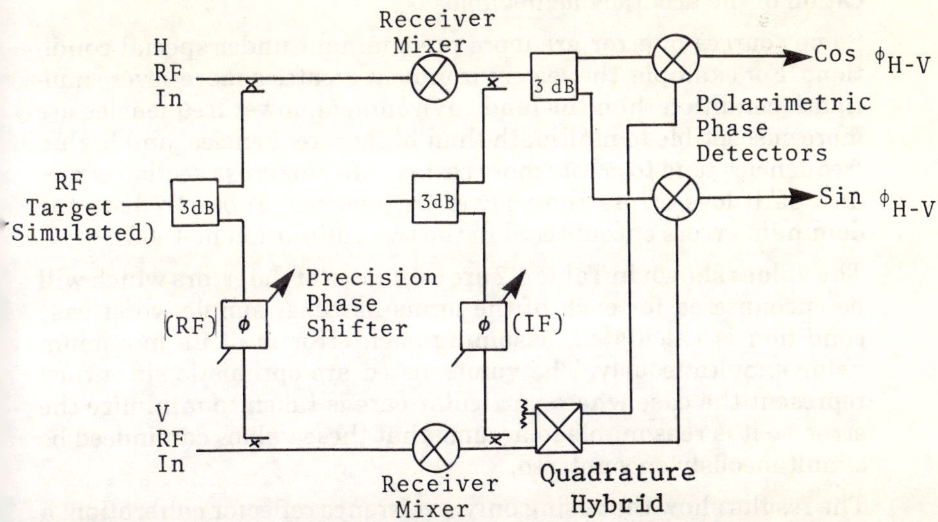


**Figure 4-12.** Receiver phase calibration using signal injection.

produces a null. The phase setting is noted, and then the phase shifter is adjusted so that the other channel produces a null. The new pase setting should be exactly 90° different from the first. If not, the sine/cosine phase adjustment must be made so that the 90° test can be met.

A similar test is performed if an actual reference target is used instead of an injected signal. In this case, a precision phase shifter is required in the receiver path, either at RF or at IF.

Contrary to the reference reflector tests for the coherent system, position stability for the polarimetric phase test is not critical because the reference for one channel is the other channel, and if the phase in one changes due to relative motion, then the phase in the other will change by the same amount. As with the amplitude calibration it is desirable to perform both the reference reflector test and the injection test to build confidence in the system integrity, which results when the two tests "close" within one or two degrees.

## 4.6 ACCURACY ACHIEVABLE IN RADAR MEASUREMENTS

There is no prescribed method to determine the accuracy with which one can determine the true radar cross section of the targets and clutter, but a review of the basic sources of error and a tabulation of the estimated magnitude of each of these will help provide an indication of the accuracy achievable.

Some sources of error are more predominant under special conditions. For example, the sea environment creates a more severe multipath condition than the land environment, lower frequencies are more susceptible to multipath than higher frequencies, and higher frequencies tend to create more corner reflector cross section errors and path loss errors than lower frequencies. Table 4-2 lists the dominant errors encountered in the two calibration methods.

The values shown in Table 4-2 are estimates of the errors which will be encountered for each of the terms listed. A simple worst case condition is calculated, assuming each error is at its maximum value simultaneously. The values listed are optimistic since they represent the case where particular care is taken to minimize the error, so it is reasonable to assume that these values can indeed be simultaneously encountered.

The results show that using only a reference reflector calibration, a 1 dB bias error is expected, with a variation from ±2.5 dB to ±3.5 dB depending on the scenario. Using a signal generator calibration, the bias error is eliminated, giving a variation of ±3 dB to ±4 dB depending on the scenario.

These results assume that special care has been taken to keep the errors near to what the state of the art permits. Often time and resources do not permit the luxury of the most sophisticated technology, increasing the errors to higher values. It is suggested that in typical measurement programs an additional 3 dB of error is introduced by lack of precision corner reflectors, lack of complete characterization of the multipath field, and radar and data acquisition system drift.

**TABLE 4-2**
**Summary of Errors**

| | μ WAVE | | mm WAVE | |
|---|---|---|---|---|
| | Land (dB) | Sea (dB) | Land (dB) | Sea (dB) |
| REFERENCE REFLECTOR CALIBRATION | | | | |
| Radar Positioning | -0.5 | -0.5 | -0.5 | -0.5 |
| Corner Reflector Positioning | -0.3 | -0.3 | -0.3 | -0.3 |
| Corner Reflector Accuracy | -0.2 | -0.2 | -0.2 | -0.2 |
| Multipath | ±1 | ±1.5 | ±0.5 | ±1 |
| Drift, Sampling, Recording | ±1 | ±1 | ±1 | ±1 |
| Calibration Curve Matching | ±1 | ±1 | ±1 | ±1 |
| Accumulated Error | -1±3 | -1±3.5 | -1±2.5 | -1±3 |
| SIGNAL GENERATOR CALIBRATION | | | | |
| Path Loss Uncertainty | ±0.5 | ±1 | ±1 | ±1.5 |
| Spectrum Equivalence | ±0.5 | ±0.5 | ±0.5 | ±0.5 |
| Drift, Sampling, Recording | ±1 | ±1 | ±1 | ±1 |
| Calibration Curve Matching | ±1 | ±1 | ±1 | ±1 |
| Accumulated Error | ±3 | ±3.5 | ±3.5 | ±4 |

As noted earlier, "closing" the calibration (i.e., obtaining agreement between the absolute calibration and the indirect calibration) yields more accuracy in the measurements by building confidence in the calibration. The improvement factor would result in the new error bounds being equal to the *rms* of the errors due to the two independent calibration methods.

It should be noted that what might appear to be a lack of repeatability of results of a particular experiment, is actually a lack of ability to precisely respect the *conditions* of the experiment; namely, the propagation parameters and physical characteristics of the target or clutter cell. This is particularly true at millimeter wave frequencies where humidity and suspended particles affect the propagation losses severely, and where clutter moisture content affects the reflectivity (backscatter). Wet and dry foliage, wet and dry soil, and wet and dry snow, as examples, produce considerably different results. For purposes of this discussion, these effects are not represented as measurement errors but rather as effects of environmental variations. Chapter 12 discusses the measurement of the environment (ground truth) and its effects in more detail.

## REFERENCES

1. Jasik, Henry, *Antenna Engineering Handbook,* First Edition, p. 13-10, McGraw-Hill Book Co., New York, 1961.
2. Jasik, Henry, *Antenna Engineering Handbook,* First Edition, p. 13-14, McGraw-Hill Book Co., New York, 1961.
3. "Optical Theroy of the Corner Reflector," M.I.T. Radiation Lab Report 433, March 1944.
4. Trebits, R. N., *et al.,* "Millimeter Radar Sea Return Study," Technical Report on Contract N60921-77-C-A168, Georgia Institute of Technology, Atlanta, Georgia, July 1978.

# Chapter 5

# DATA ACQUISITION AND RECORDING SYSTEMS

William J. Steinway
Gulf Applied Research, Inc.

**Authors's Note: The majority of information presented in this chapter has been adapted or extracted from three sources. The material for sections 5.1, 5.3, and 5.4 can be found in more detail in reference [1], and the material in Section 5.2 can be found expounded upon in references [2] and [3]. Please refer to the references at the end of the chapter for exact text and figure credits.**

## 5.1 INTRODUCTION

Today's world functions almost totally on decisions that are based on the gathering of data and its subsequent analysis. There are nearly as many systems for aquiring data as there are types of data. This text, however, confines itself to a particular class of "data aquisition system," defined as:

> an electronic instrument, or group of interconnected electronic hardware items, dedicated to the measurement and recording of analog signals for analysis purposes by either analog or digital processing.

In this context, the Data Acquisition System (DAS) is the interface to the analog/digital electronic world. A graphic representation of where the data acquisition system "fits in" is given in Figure 5-1. Once the parameter to be measured is translated into the analog-electrical domain, the DAS performs the translation to the analog/digital processing/recording domain. In some cases the DAS simply records, or stores, the data, while more sophisticated systems may be capable of analysis or further processing.

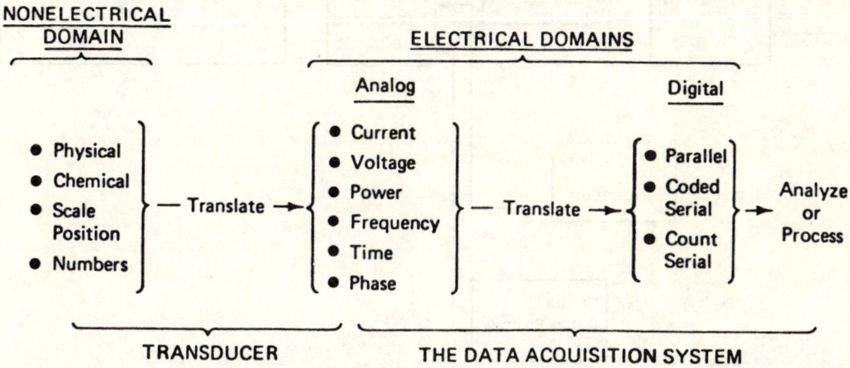


**Figure 5-1.** Where the data acquisition system fits (reprinted from Van Doren, ref. [1]; ©1982 Reston Publishing Company).

Now that we understand the function of this type of system, let us examine the components of a basic data acquisition system. Shown in Figure 5-2 and 5-3 are the basic analog and digital DAS block diagrams. Each of the blocks in the diagrams represents a particular data acquisition function, and each is defined in general terms as follows:

1. Analog Input Multiplexer — The analog input multiplexer permits a number of signal sources to be automatically measured by the same data acquisition hardware.
2. Analog Output Multiplexer — The analog output interface permits a number of signal sources to be expanded to multiple outputs and permits recording on multiple channels.
3. Signal Conditioning — Very often the signal presented to the input of the data acquisition system is not in a form appropriate for recording, and so they must be preconditioned. The required signal conditioning could consist of linear amplifica-

tion, logarithmic amplification, filtering, peak detection, or sample-and-hold.

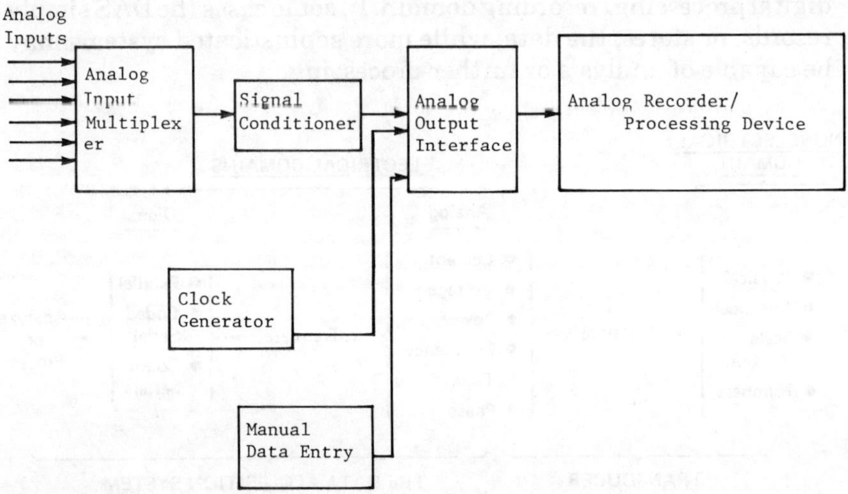


**Figure 5-2.** Analog data acquisition system block diagram.

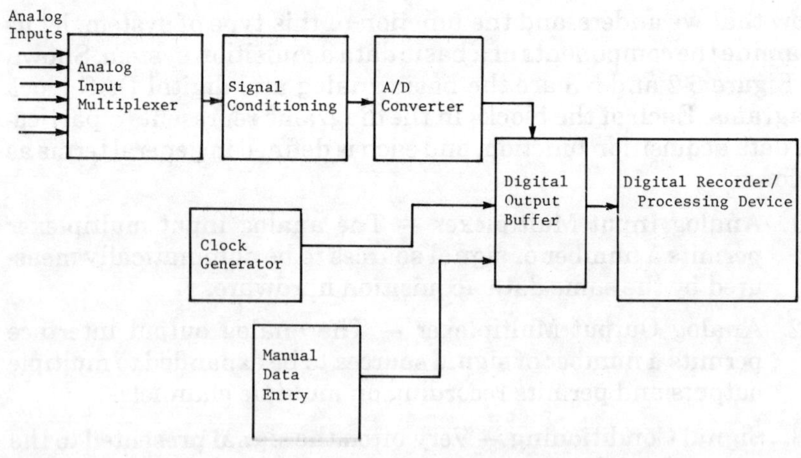


**Figure 5-3.** Digital data acquisition system block dıagram.

4. Analog/Digital Recorder — The analog recorder is a magnetic tape system which records continuous signals on either single or multiple channel tape. Two formats are available; Direct or FM. The digital recorder is either tape or disk.
5. Analog/Digital Converter (A/D) — The analog-to-digital converter actually translates the analog signal into an encoded digital format. The most notable methods are the dual-slope integrating and the successive approximation converters. A/D converters are often refered to by the number of output digits they produce.
6. Clock Generator — The clock provides the master timing for the DAS. It may be a simple multiphased crystal-controlled oscillator, or it may provide the user with a wide selection of multiplexer rates and modes of operation.
7. Manual Data Entry — Many data aquisition systems provide users with a way to tag the data they are accumulating through some sort of manual data entry. They may note such things as the type of data, special conditions surrounding the measurements, and the run number. Such information is generally provided by a microphone/voice input in analog systems, or it may be put onto the "header" that preceeds the measurement data in a digital system.
8. Output Buffer — The output buffer acts as the data collector for the DAS. In an ordered sequence, it gathers up such data as the multiplexer channel number, the signal conditioner gain, A/D converter data, manual data, clock information, and discrete events. The buffer combines the data with with the proper format for entry into the recording or processing system. It also provides the proper buffering and control to interface with the recording or processing device. If the processing device were a microcontroller, the output buffer might be called a "peripheral controller."
9. Processing Device — A number of different equipment types can fill the role of the processing device. These devices can be controlled via a direct interface or through a computer.

## 5.2 DAS CONSIDERATIONS

In the process of determining what type of DAS to use, consideration must be paid to the nature of the data, the required processing in real-time, the storage medium, and cost. Although these are certainly implied by the block diagrams of Figure 5-2 and 5-3, they are not specifically described. Before the actual building of a DAS, these must be explicitly specified as to limits and tolerances. In the following paragraphs, some of the parameters relating to the specification of these items are discussed.

### 5.2.1 Nature of Data

Analog signals are generally described by their bandwidth and maximum amplitude. Clearly, the DAS components must match or exceed the specification of those quantities. In addition, the minimum amplitude meaningful signal must be specified so that the dynamic range of the signal is known, and the number of analog signals must be given. The DAS should come as close as possible to satisfying the dynamic range requirement, as well as recording and processing all signals.

If pre-processing is done on the signals before recording, then some of the signal specifications may change. The components following the processing need only satisfy the changed parameters. Also, the level of error as well as sources of error that can be tolerated must be identified. Calibration signals should be made available to test the actual DAS capability.

### 5.2.2 Real-Time Processing

Real-time processing, or signal conditioning, is done before recording, and can either be before or after A/D conversion for digital systems. Specification of the input/output functions in terms of the bandwidth, dynamic range, and functional manipulation for the processing is important. Ill-defined processing can lead to a system which is too complex in terms of options, or an under-performing DAS.

### 5.2.3 Storage Medium

Tape, either analog or digital, is less expensive and sometimes easier to transport than magnetic disk media. In addition, the data density is higher on tape and highest on analog tape. In the future, disk storage of data will become more advantageous because of advances in data density, cost, and access speed.

In the recording/reproducing business, strip-chart and hot-stylus recorders, recording oscillographs, phonograph records, photographic films, galvanometers, and many other types of recording devices have been used to preserve the output from a transducer. No single one of these devices has all the advantages of the magnetic tape or disk recorder as listed below:[3]

1. Wide frequency range — Magnetic tape or disks permit the recording of information from direct current up into the multimegahertz frequency range.
2. Very wide dynamic range — Magnetic tape or disks permit a range of recording in excess of 40 dB, giving accurate and linear recording from full scale down to 0.3 percent of full scale
3. Low inherent distortion — When an overload occurs with magnetic tape, it occurs gracefully, as contrasted with many electronic devices.
4. Immediate playback — No time is lost in processing magnetic tape or disk. Recordings are available for immediate playback.
5. Multichannel recording — Thousands of bits of information may be recorded simultaneously on a single width of tape using various multiplexing techniques. Additionally, accurate time and phase relationships can be maintained between these simultaneous signal channels.
6. Very high density storage — Several million bits of data can be contained on a single reel of tape or a single magnetic disk.
7. Signal is in its electrical form — the data are preserved in electrical form, so that the event can be re-created at any time. Such a format lends itself to automatic reduction of data.
8. Economic advantage — Magnetic tape or disks can be erased and re-used many times.
9. Repeated playback — Magnetic tape or disks can be played back thousands of times to provide complete recovery of data.

### 5.2.4 Cost

Usually, there is a trade-off between what the data user is willing to pay and the DAS performance. The DAS cannot be the most costly item in the entire experiment/measurement system. The amount of post-processing to be done can be hindered by improper data recording by the DAS. Thus, the DAS should be designed not only to minimize DAS cost, but to minimize total system processing cost. Analog/digital and real-time or post-processing are trade-offs which should be used to minimize total data processing cost.

### 5.2.5 DAS System Parameters

In the design and specification of the DAS, certain parameters are of key importance. Table 5-1 lists and describes a number of the most commonly used parameter terms. An understanding of these is very important since over-specification can increase cost and under-specification leads to poor performance of the DAS.

**TABLE 5-1**
**DAS System Parameters**

| PARAMETERS | DESCRIPTIONS |
|---|---|
| Accuracy | Absolute accuracy is an indication of a system's capability to measure a value referenced to a standard, i.e., National Bureau of Standards values. Relative Accuracy is simply the capability to measure a value relative to full scale. |
| Dynamic Range | The allowable range of inputs over which the system has an accurate output is the dynamic range. |
| Resolution | The minimum discernible level change on the input that causes an output level change is the resolution. |
| Linearity | The capability of a system output to follow the theoretically computed output for a given input. Typically, a constant slope increasing value input yields a constant slope out, or a straight line. |
| Repeatability | If the same input values are repeated many times, the same output values are observed. |
| Drift | Changes in the observed output value for a given input value over a long period of time. These changes are typically due to the component electrical design problems brought on by temperature excursions. Drift affects the linearity, repeatability, and absolute accuracy. |
| Frequency Response | The ability of a system to dynamically respond to changes of the input values. In analog systems, when the output fails to respond to a particular sine wave input frequency the upper limit response has been reached. In digital systems frequency response is specified by the sampling rate. |
| Noise Immunity | The capability of a system to ignore or not respond to high frequency value changes on the input that are noise related. Noise within the frequency response of the system is passed through. |

## 5.3 MAGNETIC TAPE RECORDING

All magnetic recorder/reproducer systems consist of five essential elements:

1. Magnetic heads
2. Magnetic tape
3. Tape transport
4. Record amplifier
5. Reproduce amplifier

Sometimes only a portion of the total system is used, i.e., a record-only or reproduce-only system. But often a magnetic recorder system is made more complex by the addition of accessory equipment such as voice monitors, display cathode ray or meter monitors, and remote controls. Figure 5-4 shows the typical magnetic recorder system and the relationship of one basic element to the other.

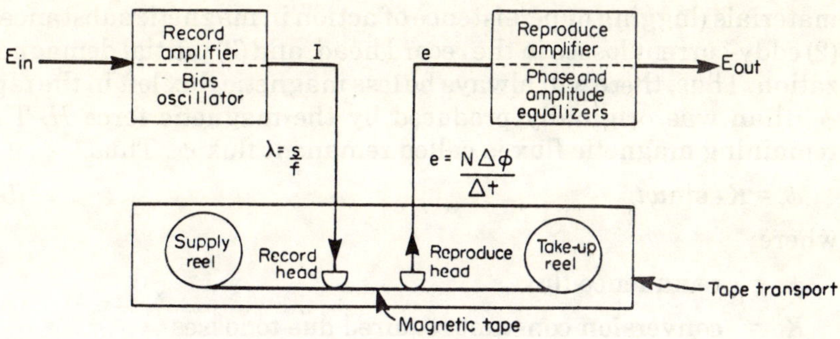


**Figure 5-4.** The basic elements of a magnetic tape recorder/reproducer (reprinted from Bycer, ref. [2]; ©1965 Hayden Book Company).

### 5.3.1 The Functions of the Basic Elements

The magnetic heads may be divided into three categories:

1. Record head
2. Reproduce head
3. Erase head

*The Record Head*

Basically, the function of the record head is to change the current $I$ produced by the record amplifier into magnetic flux $\phi$. As record current varies, so also will the magnetic flux. The change of $\phi$ will be affected by both the amplitude and the direction changes of the record current. The changing current through the head produces a proportional changing magnetic force $H$ in the head.

$$H - I \sin wt \tag{5-1}$$

where

$H$ = magnetic force, oersteds

$I$ = maximum current

$w$ = $2\pi f$

$t$ = time, sec

The magnetic force $H$ in turn produces magnetic flux $\phi$ in the tape. Unfortunately, there are some losses to be considered in the conversion process. These are (1) the hysteresis effect of the head and tape materials (lagging or persistence of action in magnetic substances), (2) eddy-current losses in the record head, and (3) partial demagnetization. Thus, there will always be less magnetic flux left in the tape, $\phi_r$, than was originally produced by the magnetic force $H$. This remaining magnetic flux is called remanent flux $\phi_r$. Thus

$$\phi_r = KI \sin wt \tag{5-2}$$

where

$\phi_r$ = remanence flux

$K$ = conversion constant required due to losses

$w$ = $2\pi f$

$t$ = time, sec

The size (wavelength) of the signal recorded on magnetic tape will be a function of tape speed and frequency. Thus

$$\lambda = \frac{s}{f} \tag{5-3}$$

where

$\lambda$ = wavelength, in

$s$ = tape speed, ips

$f$ = frequency, Hz

*The Reproduce Head*

The basic function of the reproduce head is to change the magnetic field pattern found in the magnetic tape into a voltage *e*. The reproduce head will act as a miniature generator following Faraday's law,i.e.,

$$e = N \frac{\Delta \phi}{\Delta t} \tag{5-4}$$

where

$e$ = instantaneous voltage

$N$ = number of windings around reproduce head core

$\Delta$ = symbol for change

$\phi$ = magnetic flux

$t$ = time, sec

The voltage created in the reproduce head follows a 6 dB/octave curve (within certain limits). By this it is meant that as the frequency doubles, so also will the value of voltage output from the head. Thus, if a constant level of input were fed to the record amplifier at increasing frequencies, the output of the reproduce head would not be constant, but rather would increase at a 6 dB/octave rate (Figure 5-5).

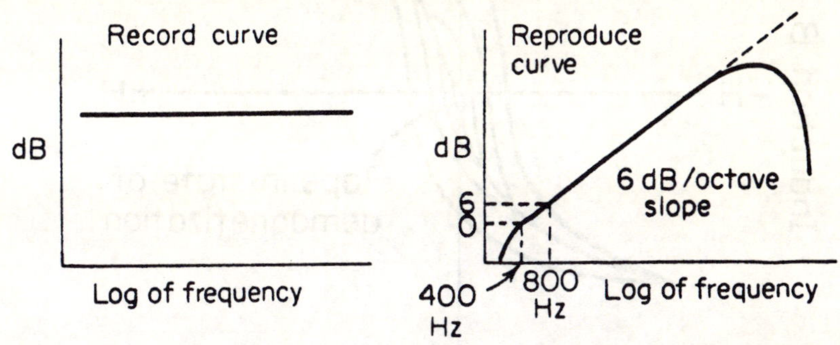


**Figure 5-5.** Record and reproduce curves showing resultant output of a reproduce head to a constant level record signal over a wide frequency range (reprinted from Lowman, ref. [2]; ©1972 McGraw-Hill Book Company).

The deviation from the straight line of the dB/octave curve is caused by losses associated with head construction, magnetic tape characteristics, the speed of tape movement, and many other factors. It is sufficient to say at this time that most of the correction needed to straighten the reproduce curve to one comparable with the record curve will be done by circuits called equalizers, which are mounted in the reproduce amplifiers.

### *The Erase Head*

The basic function of an erase head is to demagnetize the magnetic signal on tape. A very heavy high frequency alternating current (100 mA or more) is fed through the head. This current drives the tape into saturation first in one direction and then in the other. As the tape is pulled away from the head, a slowly diminishing cyclic field is presented to the tape. The diminishing cyclic field leaves the magnetic vectors in the tape in a completely random state. In such a random condition, no single magnetic vector, or group of magnetic vectors, has more strength than the others and the tape is demagnetized, or degaussed. (*See* Figure 5-6 for the demagnetization curve.)

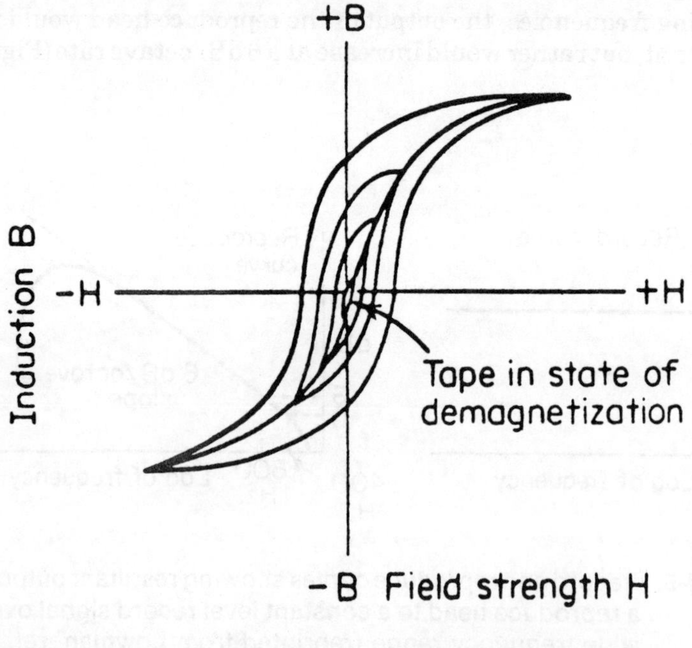


**Figure 5-6.** The demagnetization curve (reprinted from Lowman, ref.[3]; ©1972 McGraw-Hill Book Company).

### 5.3.2 Magnetic Tape

The accepted medium for magnetic recording is a tape consisting of a backing coated with particles of the gamma form of ferric oxide ($Fe_2O_3\gamma$). The backing material, either acetate or polyester film, provides the base to which the magnetic particles are fixed with a synthetic resin binder. Each particle (approximately 25 $\mu in.$ in length) forms an elemental magnet. The north and south poles of these magnets lie along some axis of the iron oxide crystal lattice, the particular axis depending upon the crystalline structure of the oxide used. There will be specific types of oxides used for specific bandwidths of data signals and types of recording techniques, i.e., low, medium, and high frequency response; and digital, audio, video, and longitudinal instrumentation recording techniques.

When the magnetic flux $\phi$ from the record head penetrates into the magnetic tape coating, it reorients the vectors of the magnetic domains (an elemental magnetic volume). With a low value of magnetic flux, a small number of magnetic vectors are aligned with the flux field. When a large magnetizing flux is used, a large number of vectors are aligned. If the tape is moved linearly across the record head gap, the vector pattern will be altered to produce, at any point, a net amount of magnetic vector alignment. This alignment has a magnitude and direction that is a function of the magnetizing field intensity (flux at record gap) that existed at the instant the tape left the gap.

### 5.3.3 The Tape Transport

The tape transport must move tape across the record and reproduce heads at a constant linear velocity, with the least amount of disturbance to tape motion. It must also provide some means of tape storage, either in a form of loop, bin, reel, or magazine. If speed or tension of the tape varies, dynamic time-base errors will be introduced into the system. These will adversely affect the amplitude, phase, and frequency response of the magnetic recorder.

The effort in designing a tape-transport mechanism must be directed toward instantaneous and long term speed errors (drift) that are introduced by inaccuracies of the drive system. They are usually expressed as percentage variation from absolute selected speed. Logical development of each major assembly and careful selection of components, combined with numerous tests and a highly efficient quality control system, will result in good transports with reliable performance. Although tape transports may vary from each other in appearance, size, and operation, the design principles are basically the same. Even the best tape transports fall short of ideal specifications. Further improvements are, however, becoming progressively more difficult and costly to achieve.

### 5.3.4 Important Design Criteria

Generally speaking, the five following points are the most important criteria when designing or purchasing a tape-transport mechanism.

1. It must have reliability and dependability, achieved for the most part by keeping the configuration as simple as possible.
2. It must have high performance; that is to say, the ability to transport tape at the desired speed with no introduction of variables in the signal to be recorded or reproduced and no disturbance of tape motion.
3. It must be standardized and compatible with other equipment. It must have standard speeds, tape widths, reel sizes, track spacing, *et cetera.*
4. It must have low down-time and maintenance cost. It must be easily maintained, and parts must be easily accessible. It must have a long mean-time-between-failure rate.
5. It must have flexibility for modification and freedom from obsolescence. This is generally achieved with modular construction and completely independent major subassemblies, so that modification and updating can be easily accomplished.

### 5.3.5 The Record Amplifier

The record amplifier must change the incoming data signal $E_{in}$ into a form that is suitable for recording on tape. That is, the amplifier must change the $E_{in}$ into current $I$ which, when fed to the magnetic heads, will be converted into magnetic flux $\phi$. The magnetic flux will be used to magnetize the tape so that a given amplitude, a given polarity, and a given point on the tape represent the data $E_{in}$ at a given instant of time.

### 5.3.6 Direct Recording

The direct recording process in Figure 5-7 is, by far, the most familiar, since it is the method used in audio recording. The signal to be recorded is conditioned to activate a recording head through its coil-wound core (Figure 5-8), recording a signal on tape that, essentially, is a direct, unmodified reproduction of the original signal.

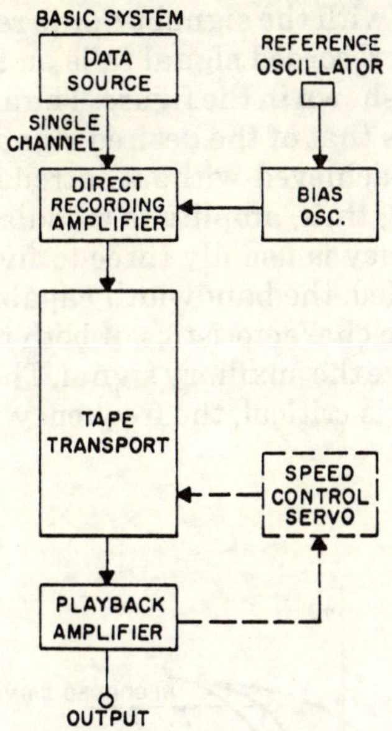


**Figure 5-7.** The direct recording process (reprinted from Bycer, ref. [2]; ©1965 Hayden Book Company).

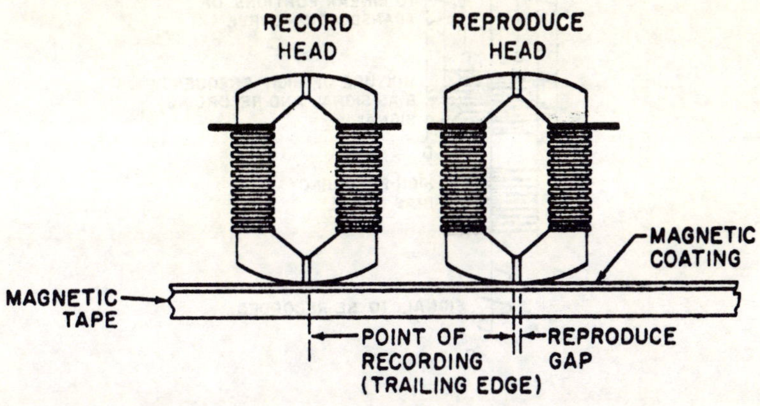


**Figure 5-8.** Magnetic recording and reproducing (reprinted from Bycer, ref. [2]; ©1965 Hayden Book Company).

When a high degree of signal fidelity is required, high frequency bias is linearly mixed with the signal prior to recording, Figure 5-9. With this scheme, the recorded signal falls on the linear portion of the transfer curve, as shown in the figure. The amplitude of the bias signal is several times that of the desired signal. The linear signal mixing of the two is achieved without introducing new sum and difference frequencies; thus, amplitude modulation is avoided. Because the bias frequency is usually three to five times the highest frequency to be recorded, the bandwidth capability of the circuitry and the head response characteristics of both the record and playback operations remove the auxiliary signal. The actual value of the bias signal amplitude is critical, the frequency is not.

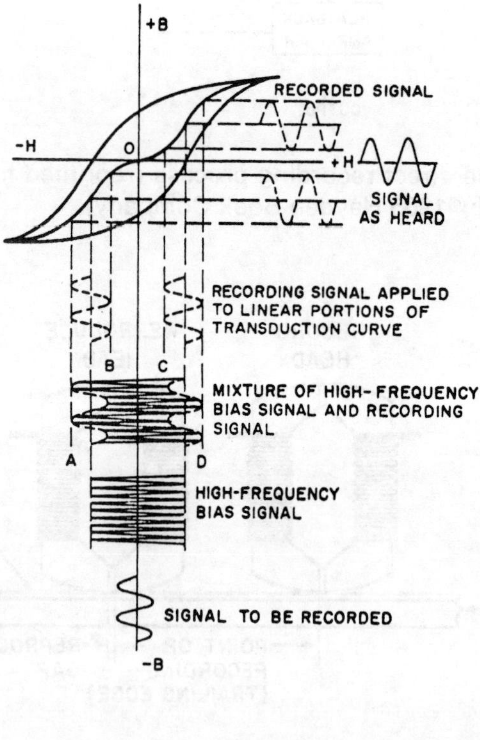


**Figure 5-9.** Linear mixing for direct recording (reprinted from Bycer, ref. [2]; ©1965 Hayden Book Company).

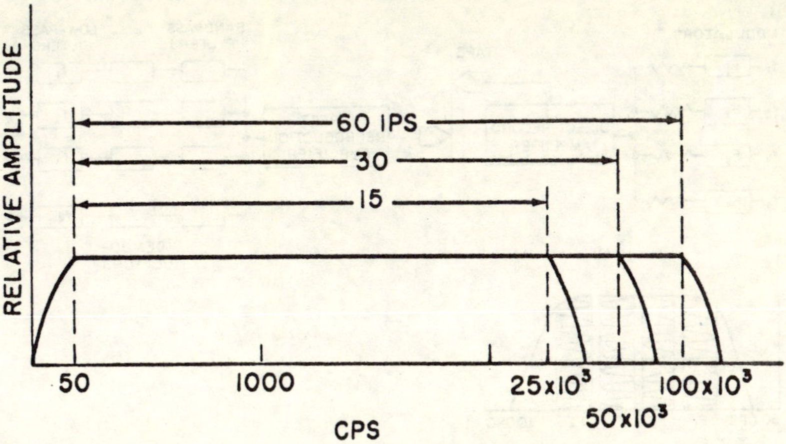


**Figure 5-10.** Direct recording frequency responses at various tape speeds (reprinted from Bycer, ref. [2]; ©1965 Hayden Book Company).

The first requirement of a recording system is that it be capable of recording and reproducing the original signal with a high degree of accuracy in frequency, phase, and amplitude. Good high frequency response is an outstanding advantage of direct recording. Figure 5-10 shows frequency response of the direct recording process for various tape speeds. At 60 inches per second (ips), the effective frequency range within the 3 dB points is 50 cps to 100 kc. The frequency range can be extended proportionally by increasing tape speed or by other means such as rotating heads. The major disadvantage of the direct recording process is its poor low frequency response, especially below 30 cps. The low frequency range could be extended, but only by making the system much more complex. A simpler and more efficient solution is to use frequency modulation techniques when good low frequency response is required.

### 5.3.7 Frequency Modulation (FM) Recording

Direct recording is generally associated with the recording of one signal per recording track. In many cases, however, a wide frequency range may be more efficiently utilized by multiplexing several signals on a number of individual carrier frequencies, each modulated by a separate signal (Figure 5-11). Using a separate subcarrier oscillator, low frequency signals down to direct current modulate

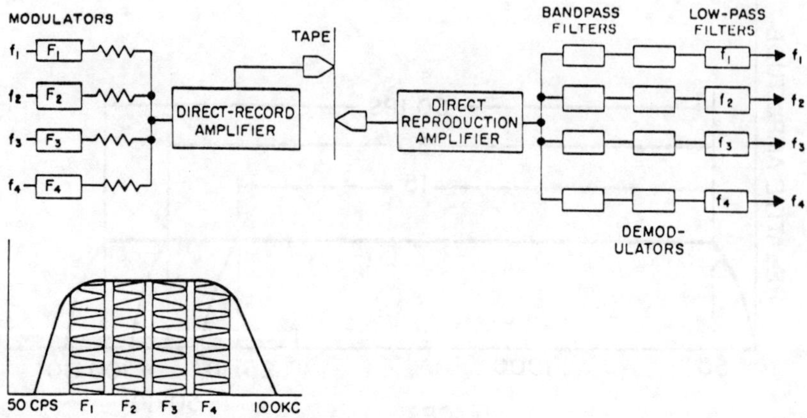


**Figure 5-11.** Basic frequency-division multiplexing of a narrow deviation FM system (reprinted from Bycer, ref. [2]; ©1965 Hayden Book Company).

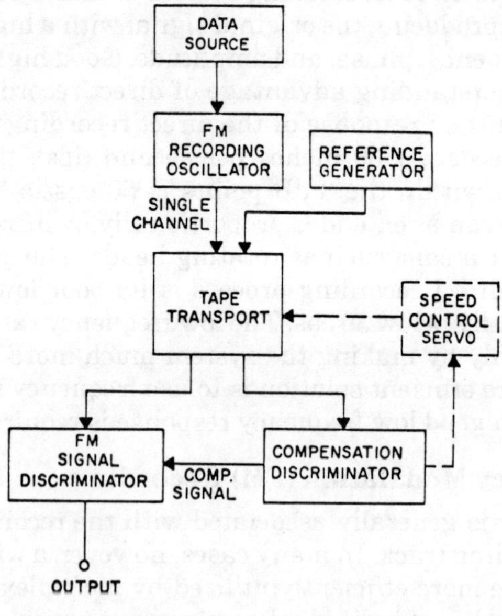


**Figure 5-12.** The FM carrier system (reprinted from Bycer, ref. [2]; ©1965 Hayden Book Company).

each carrier and are then mixed linearly together. The output of the mixer contains the composite signal that is fed into the direct recording system. Here, the full bandwidth and linearity of the direct recording method is utilized to permit a simultaneous recording of a number of signals on a single tape track. The reverse process (demultiplexing) is performed during playback and the original signal is recovered. This modified form of FM recording is known as frequency-division multiplex. In frequency modulation (FM) recording (Figure 5-12), a set of carrier frequencies is chosen with specific allocated sidebands. Each carrier and the amount it may deviate are chosen so that no overlapping of sidebands occurs. The output of the modulator is FM. This composite signal can be used to frequency modulate a center frequency at the midpoint of the direct recording process response. This method, whereby each subcarrier is frequency modulated and the master carrier is frequency modulated, is called FM/FM.

Consider the final FM operation. One signal (composite) is used for wide deviation of a center carrier. For actual values at 60 ips, a center frequency of 54 kc width ±40% deviation has an information capacity of 20 kc (Figure 5-13). Since bandwidth is proportional to tape speed, instead of 20 kc capacity, a 10 kc bandwidth intelligence for 60 cps can be produced with reduced tape speeds, as shown in Figure 5-14, and slower tape speeds will alter bandwidths.

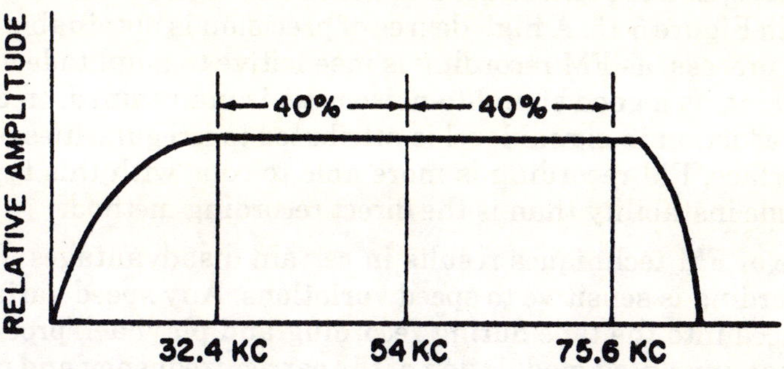


**Figure 5-13.** A wide deviation FM carrier bandwidth at 60 ips (reprinted from Bycer, ref. [2]; ©1965 Hayden Book Company).

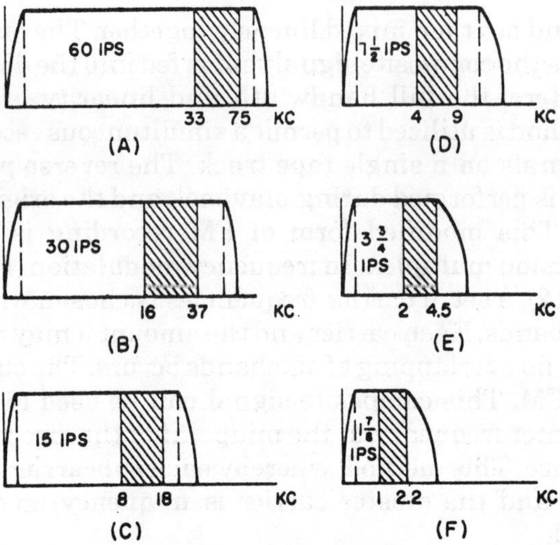


**Figure 5-14** FM carrier and modulation bandwidth at various tape speeds: (a) dc to 100 kHz; (b) dc to 50 kHz; (c) dc to 25 kHz; (d) to 12 kHz; (e) dc to 600 Hz; (f) dc to 300 Hz (reprinted from Bycer, ref. [2];© 1965 Hayden Book Company).

FM recording overcomes two basic limitations associated with the direct recording process: (1) inability to record low frequencies; and (2) amplitude compression. The amplitude characteristics of magnetic tape are only fairly linear. Generally, analog precision is one part in a thousand, or 0.1%. The dynamic range of the direct recording process, and the amount of amplitude and signal distortion, are shown in Figure 5-15. A high degree of precision is obtainable with the FM process, as FM recording is insensitive to amplitude variation as long as a good signal-to-noise ratio is maintained. In cases where reduction in signal level is attributed to irregularities of the tape surface, FM recording is more able to cope with this type of amplitude instability than is the direct recording method.

The use of FM techniques results in certain disadvantages. First, FM recording is sensitive to speed variations. Any speed variation introduced into the tape during recording and playback processes will cause unwanted modulation of the carrier frequency and result in distortion (noise). This is the limiting factor of the FM method. Secondly, FM places stringent requirements upon the tape transport and requires more complex electronic circuitry than the direct recording process. Finally, using FM recording techniques results in reduced information bandwidth relative to the direct recording method.

### 5.3.8 Pulse-Duration Modulation Recording

The third analog recording technique is pulse-duration modulation (PDM), or pulsewidth modulation (PWM), as illustrated in Figure 5-16. The analog information amplitude content exists in the pulse-width and the rate of change of the leading edges of the pulses contains the frequency content. Where FM is a frequency division multiplexing technique for compiling a number of signals onto a single tape channel, PDM accomplishes the same thing by assigning fixed time slots for the same purpose. This later technique is called time division multiplexing and requires a fixed sampling rate of a number of signal channels on a sequential basis.

PDM recording is based on the fact that the frequency and amplitude of any sine wave can be accurately expressed from information obtained by sampling or measuring the amplitude at fixed time intervals. If the information obtained is to be meaningful, a sufficient number of samples must be taken from one cycle so that

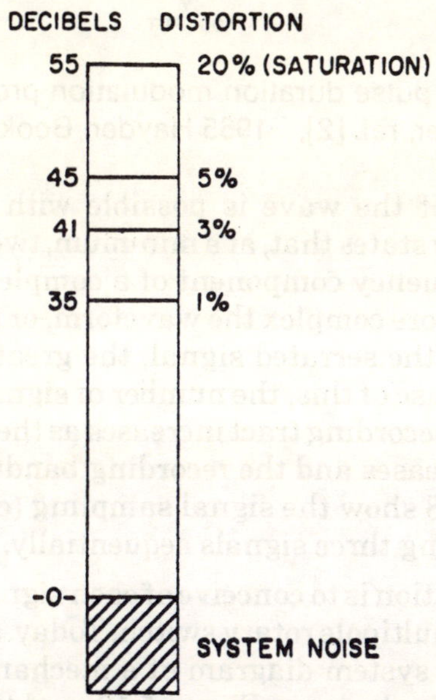


**Figure 5-15.** Direct recording process amplitude tape range (reprinted from Bycer, ref. [2]; ©1965 Hayden Book Company).

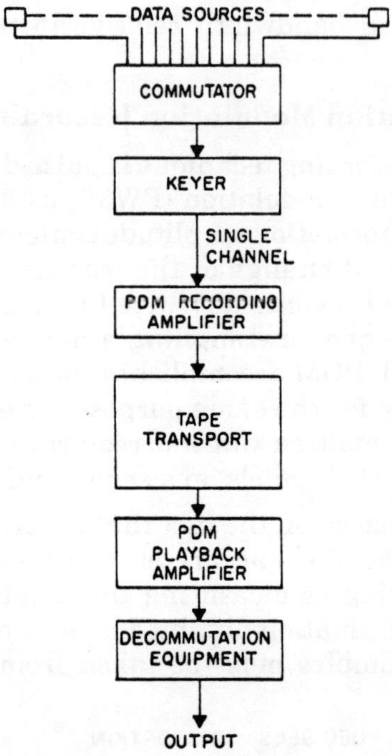


**Figure 5-16.** The pulse duration modulation process (reprinted from Bycer, ref. [2]; ©1965 Hayden Book Company).

reconstruction of the wave is possible with minimal distortion. Sampling theory states that, at a minimum, two samples per cycle of the highest frequency component of a complex wave are required. Therefore, the more complex the waveform, or the better the desired reproduction of the serrated signal, the greater must be the sampling rate. Because of this, the number of signals that can be multiplexed onto one recording tract increases as the frequency content of the signals decreases and the recording bandwidth increases. Figure 5-17 and 5-18 show the signal sampling (chopping) and a time scale of combining three signals sequentially.

A simple illustration is to conceive of each signal source as being fed to one pole of a multipole rotary switch. Today, systems use electronic switching. A system diagram of a mechanical commutator for PDM recording is shown in Figure 5-19, and the signals at various points of the system are shown in Figure 5-20. As the switch rotates, each input contact is sampled and fed to the recording system. Since

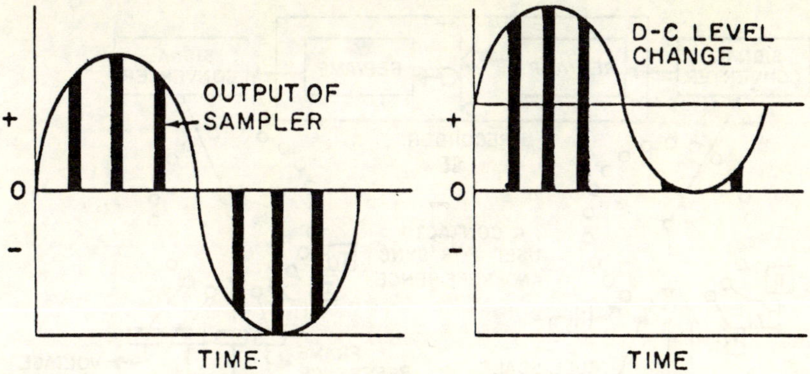


**Figure 5-17.** PDM signal sampling (reprinted from Bycer, ref. [2]; ©1965 Hayden Book Company).

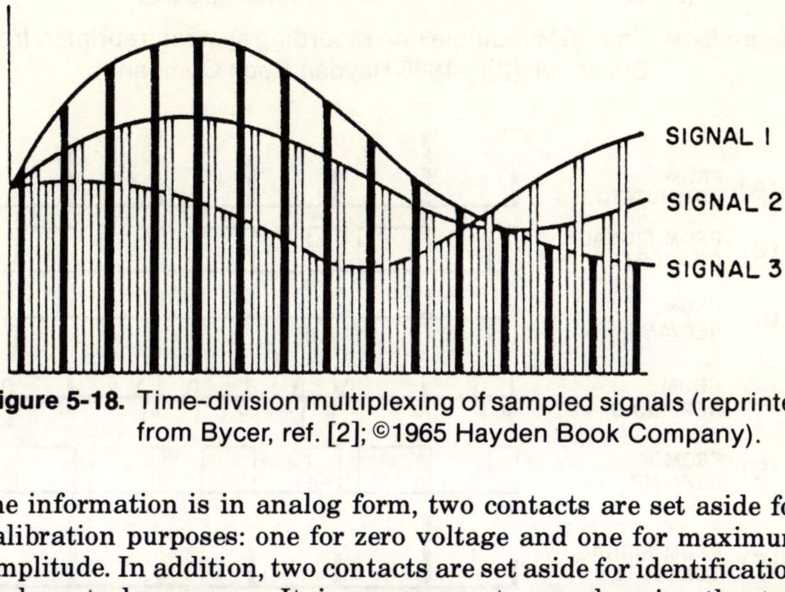


**Figure 5-18.** Time-division multiplexing of sampled signals (reprinted from Bycer, ref. [2]; ©1965 Hayden Book Company).

the information is in analog form, two contacts are set aside for calibration purposes: one for zero voltage and one for maximum amplitude. In addition, two contacts are set aside for identification and control purposes. It is necessary to synchronize the two switches to identify the various signals on playback. For signal recovery, the output of the tape is applied to a similar switching system and synchronization is achieved by means of the timing signals already recorded. The output from the various contacts is fed through suitable low-pass filters that reconstruct, or sine-wave restore, the playback signals. Any final corrections are accomplished by decommutating calibration signals of known voltages and applying the necessary corrections.

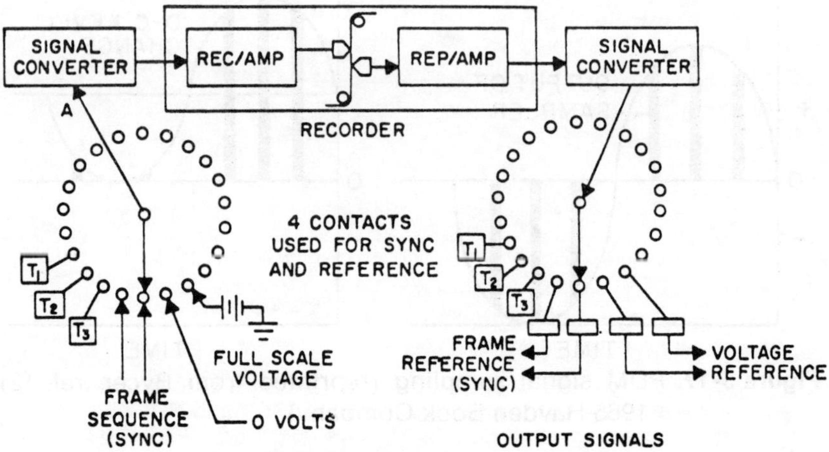


**Figure 5-19.** The PDM multiplexing recording system (reprinted from Bycer, ref. [2]; ©1965 Hayden Book Company).

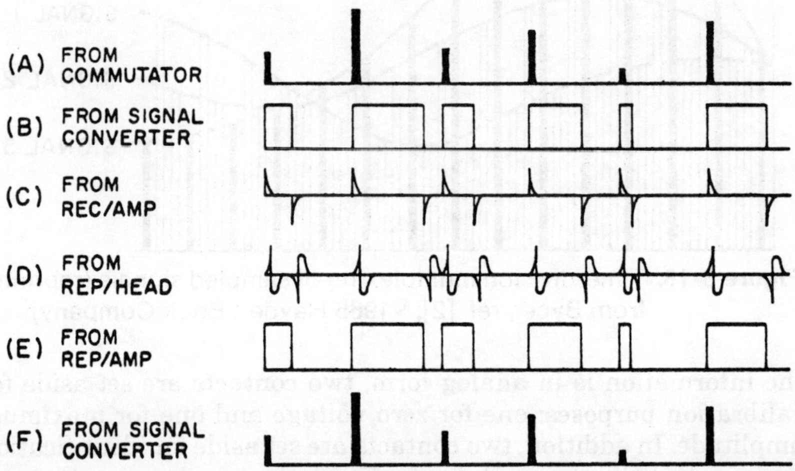


**Figure 5-20.** Signal shapes at various points in the PDM recording system of Figure 5-19 (reprinted from Bycer, ref. [2]; ©1965 Hayden Book Company).

The PDM method requires the conversion of amplitude to pulse-width. For zero voltage, a minimum pulse duration is assigned and, correspondingly, for full signal input, a maximum pulse duration is determined. Any intermediate signal voltage between zero and full scale would be represented by a pulse duration somewhere between these two boundary values. Since the record signal amplitude is constant, PDM is superior to the direct recording process. On the other hand, PDM is sensitive to tape speed variations, but not as much as is FM. Any instantaneous speed variations are integrated over the duration of the pulse.

A variation of PDM is pulse code modulation (PCM). The recorded signal of a PDM process has an infinite number of pulse durations between zero signal input and full scale input. If, during the commutation process, the serrated signal is fed to analog-to-digital converter equipment, a number can be determined and assigned for each signal amplitude. Using a binary code decimal (BCD) numbering system, each decimal column has its own binary representation. Each four-bit weighed column has four discrete pulse durations. Three sets of BCD characters are necessary to read three decimal digits. Here, only four discrete pulsewidths have to be encoded and decoded. With PCM tape, speed variations are a minor problem.

The chief advantage of the PDM recording process is its ability to multiplex a large number of signals onto one information line. Other advantages are its high accuracy, using a self-calibrating method, and good signal-to-noise ratio, resulting from the narrow bandwidth requirements. The PDM process has some serious limitations. It is the most inefficient of the three method processes presented so far, and requires very complex electronic equipment.

### *The Reproduce Amplifier*

The reproduce amplifier should provide an output that is the same as or similar to that fed to the record amplifier, that is $E_{out} = E_{in}$.

### *Direct Reproduce Amplifier*

Since the output of the reproduce head follows Faraday's law, $e = N \Delta\phi/\Delta t$, within specified limits and is nonlinear in other regions, the reproduce amplifier must compensate for this type of irregular output. In direct systems where the frequency response is relatively narrow (up to 100 kHz), only an amplitude equalizer is included as part of the circuitry. As can be seen in Figure 5-21, the equalizer characteristic curve is such that amplitude variations from the reproduce head are compensated for, resulting in a flat reproduce-amplifier output over a specified range of frequencies. In audio reproducing, the curve is not maintained flat, but rather is designed to vary from a straight line in order to be pleasing to the response curve of the human ear.

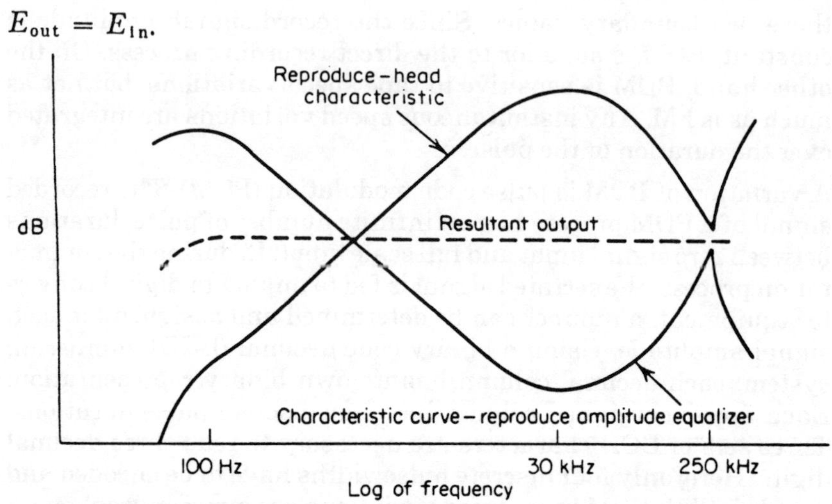


**Figure 5-21.** Resultant of reproduce-head output plus amplitude equalization (reprinted from Lowman, ref. [3]; ©1972 McGraw-Hill Book Company).

When the direct system is one where the frequency response is extended beyond 100 kHz, it is normal that a phase equalizer be included in the reproduce system, in addition to the amplitude equalizer. The need for this arises primarily because phase shifting of the data signal is introduced by the components of the broadband amplitude equalizer. Additionally, if the data are nonsinusoidal in nature (pulse or squarewave), the reactance of the reproduce head, which follows the formula $X_L = 2\pi fL$, tends to distort the wave shape, lengthening rise and fall times, and creating peaks and dips in tops and bottoms. The phase equalizer essentially compensates for the induced delay (phase shift) and reconstructs the reproduced signal into its original form.

### 5.3.9 Limitations and Advantages of Direct Recording

Many of the advantages and disadvantages of the direct record–reproduce system have already been mentioned in earlier sections, but it might be well to reiterate some of the major points:

1. As the frequency decreases, the output voltage from the reproduce head reduces, until it approaches the inherent noise level of the system. Thus, with near-zero frequencies (near dc), there would be no usable output. This is one of the principal limitations of the direct recording process.

2. There is a specific dynamic range for each recording system. The dynamic range is the ratio of the maximum signal which can be recorded with a given level of distortion to the minimum signal that can be recorded (determined by the inherent noise level of the system). As the recording level is increased, the distortion increases correspondingly.
3. Another limitation of the direct record process is amplitude instability. This is normally caused by head-to-tape separation. These losses of signal (dropouts) are relatively unimportant in audio recording, because the human ear tends to integrate the various signal levels and is relatively insensitive to short periods of signal loss. This is not true with instrumentation, since the dropout might occur precisely at the instant that some important transient phenomenon was taking place.

In spite of these limitations, there are many advantages to using the direct recording process, such as:

1. It has the widest frequency bandwidth *versus* tape speed.
2. It has a wide dynamic range and can handle moderate overloads without sudden or drastic increases in distortion.
3. It may be used for frequency-multiplexing a number of signals simultaneously on one track.

### 5.3.10 Frequency Modulation Reproduce

The signal that is recorded on tape in an FM system is a frequency representation of a dc voltage input, or a number of frequencies (carrier and sidebands) representing an ac voltage input. The job of the FM reproduce amplifier is to change these frequencies back into their respective dc or ac values (Figure 5-22). FM parameters for several frequency ranges are listed in Table 5-2.

### 5.3.11 Pulse-Type Reproduce Amplifiers

Where pulse-type recording is used, the reproduce amplifier will be required to reconstitute the time-duration pulses from a series of pulses that are the differentiated forms of the original recorded spike. A triggered multivibrator is most often used for this type of conversion. In the case of frequency shift modulation (FSM), the reproduce amplifier must sense which of the two representative frequencies was recorded on tape (one frequency used to represent a one and another frequency used to present a zero). High-quality filters followed by shaping amplifiers are used for this purpose. With

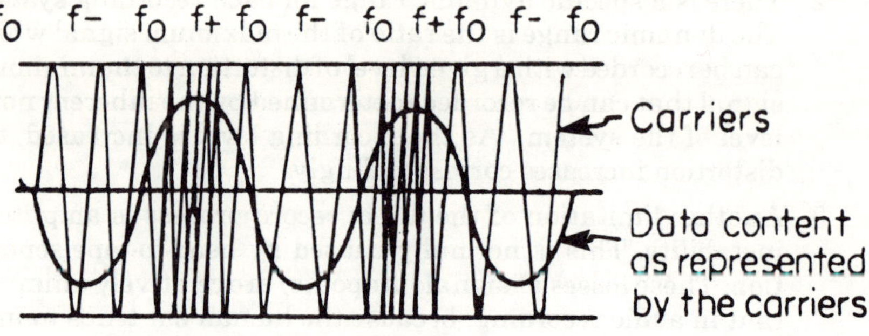


$f_0$ = center carrier frequency = 0 V

f− = CCF − 30% = −1.414 V

f+ = CCF + 30% = +1.414 V

**Figure 5-22.** FM carrier frequencies and the resultant data signal on demodulation (reprinted from Lowman, ref. [3]; ©1972 McGraw-Hill Book Company).

PCM, since, during the recording process, samples of the data are quantized into discrete steps and fed as a code group of ones and zeros to the record amplifier, the reproduce amplifier need only reproduce the recorded digital bits and feed them to the output stages. There the code groups will be used to re-create the original sampled data (Figure 5-23).

### 5.3.12 Advantages and Limitations of the FM Recording Process

The advantages of the frequency modulation recording process are:

1. The ability to record low frequencies down to direct current
2. Freedom from the effects of tape dropouts
3. Excellent phase shift *versus* frequency characteristics and the attendant ability to preserve accurately the waveform of the recorded signal

The disadvantages of the FM process are:

1. Less efficient utilization of the tape, in that at least four times the tape speed is required to give the same upper frequency limit as the direct recording method

**TABLE 5-2**
**Single Carrier and Wideband FM Record Parameters**

| BAND | Tape speed, ips | Center-carrier frequency, kHz | Carrier minus deviation, kHz | Carrier minus deviation, kHz | Modulation frequency |
|---|---|---|---|---|---|
| Low | 60 | 54 | 75.6 | 32.4 | dc to 10 kHz |
| Intermediate | 120 | 216 | 302.4 | 129.6 | dc to 40 kHz |
| | 60 | 108 | 151.2 | 64.8 | dc to 20 kHz |
| Wideband group I | 120 | 432 | 604.8 | 259.2 | dc to 80 kHz |
| Wideband group II | 120 | 900 | 1170.0 | 630.0 | dc to 400 kHz |

*Note*: For details at other speeds the reader is referred to IRIG Telemetry Standards, Document 106-71.

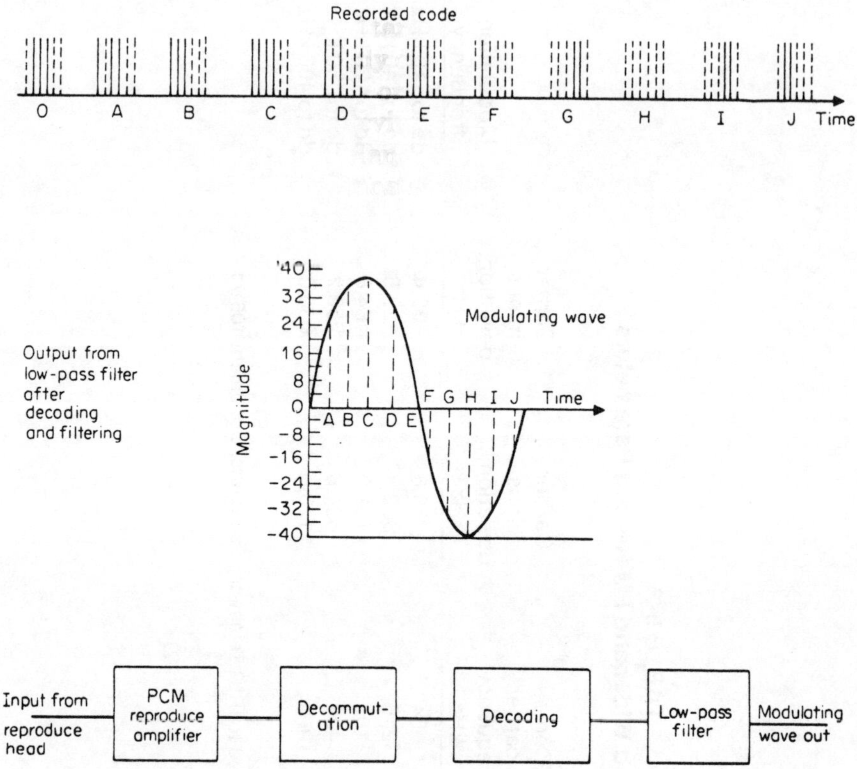


**Figure 5-23.** Transforming a sequency of PCM code into sine-wave output (reprinted from Lowman, ref. [3]; ©1972 McGraw-Hill Book Company).

2. Additional complexity of the electronic circuity required for the modulator, demodulator, and low-pass filters

### 5.3.13 Digital Recording

Digital recording (Figure 5-24) may be considered an extension of, or a variation of, analog recording. A magnetic tape transport which handles digital information on an intermittent basis (stop and go) is constructed in a different fashion than a tape transport handling analog information. The digital tape transport requires some type of tape reservoir to absorb speed differences of the tape reel drive and the tape drive when rapid operational changes occur. Although both recording processes utilize similar signal waveforms and several recording channels, the arrangement of information on tape in each

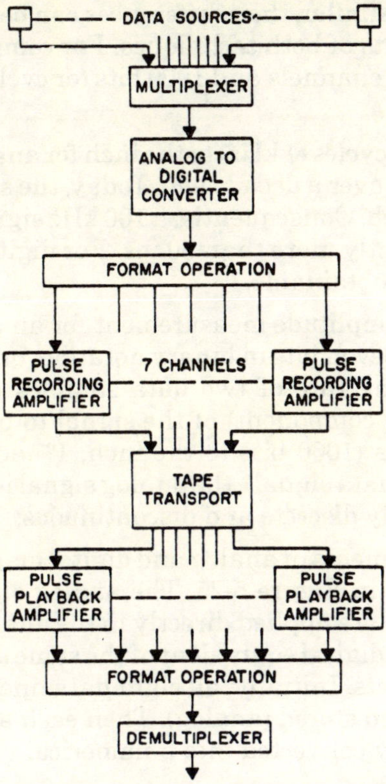


**Figure 5-24.** The digital recording system (reprinted from Bycer, ref. [2]; © Hayden Book Company).

system is not the same. In analog recording, the information to be recorded is generally restricted to a single channel and is referred to as single track or serial recording. In digital recording, the equivalent information is recorded on multiple tape tracks and is referred to as parallel recording.

The field of information storage and retrieval that is concerned with data gathering and preparation, classification, documentation, filing, and retrieval of information, offers extensive opportunities for digital magnetic recording.

### 5.3.14 Comparison of Analog and Digital Recording

A better understanding of both analog and digital magnetic tape recording methods is obtained when a comparison is made. While both methods are similar, their differences can be used effectively to illustrate the merits of both techniques. For comparison, a parallel recording of eight channels and 1000 bits (or cycles) to the inch will be used.

The figure of 1000 cycles (1 kHz) to the inch for analog recording was standard practice over a decade ago. Today, the standard is close to 2000 cycles per inch. Consequently, a 100 kHz signal is recorded at a tape speed of slightly more than 50 ips. For digital recording, 1000 bits to the inch is obtainable.

The precision of amplitude measurement for an analog signal will be limited here to seven bits in binary notation, or 1% accuracy. Also assume a sampling rate of two data samples per sine wave (or highest frequency component) of the signal to be recorded. Using these ground rules (1000 bits to the inch, 1% accuracy, 8 parallel tracks, and 1 kHz data signal), the analog signal is continuous while the data is digitally discrete and discontinuous.

The basic arrangements of analog and digital magnetic tape recording are compared in Figure 5-25. The analog signals (1 kHz sine wave phenomena) are applied directly to the analog tape transport for recording. The digital equivalent of the same data is quite different in many respects. Initially, all eight data lines are interrogated simultaneously and stored (analog). Then each analog-stored data line is sequentially converted into a numerical value and digitally

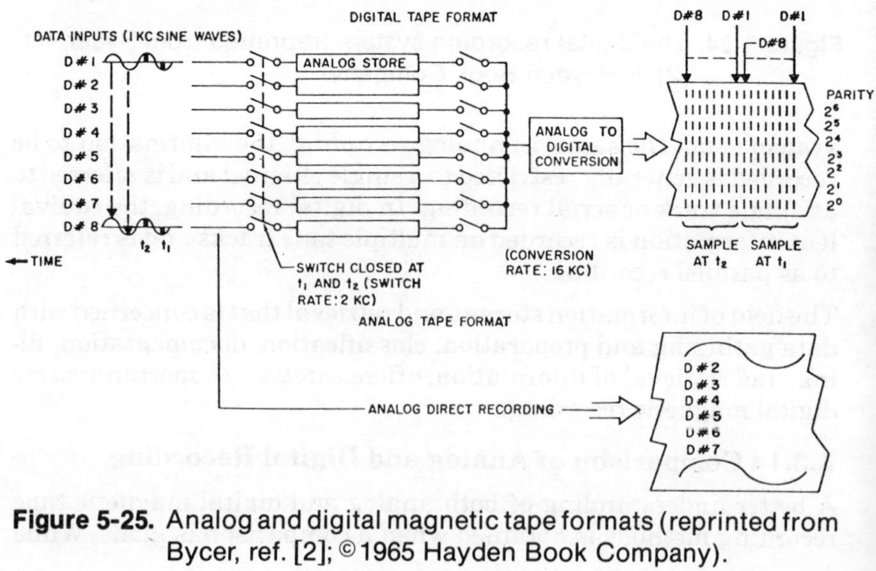


**Figure 5-25.** Analog and digital magnetic tape formats (reprinted from Bycer, ref. [2]; © 1965 Hayden Book Company).

stored. The final process takes the digits, and places each numerical data value in sequential order on tape. The data sample and analog-to-digital conversion processes for each recorded track are performed at twice the highest frequency component (2000 samples per second). Obviously, the analog signal of one complete sine wave occupies 1/100 of an inch of tape. The digital equivalent occupies 16 times (16/1000) as much tape as its analog counterpart, and requires additional conversion hardware and switches.

This simplified illustration of analog and digital magnetic tape recording would imply that analog techniques are far more efficient and, possibly, preferable to the digital method. Actually this is only partially true. In essence, many analog restrictions are not self-evident and need clarification. The parallel recording of analog data requires stringent tape transport speed control and stringent playback amplifier design. For high-quality instrumentation analog recording, each data channel has a carrier, or reference, signal superimposed on the information signal. During playback, this carrier signal provides the amplitude and frequency correction of each recorded track signal for the overall response (record and playback). Also, one of the carrier signals (or a separate reference track) is used in the speed servo-control loop to maintain a constant tape velocity.

Information content is contained in the signal amplitude. Analog recording is sensitive to tape dynamic range, speed variation, spacing between head and tape, and tape surface smoothness. The reference signal amplitude provides compensation, within limits, for all sources of amplitude degradation. Also, any phase and frequency distortion is minimized with good tape speed control.

A few less obvious analog limitations are worth noting in this magnetic tape comparison. In general, the data are limited to one reel of tape because of the continuous signal characteristics. Also, the frequency response is quite wide for the analog method while the digital is fixed frequency. The low-frequency limit is in the vicinity of 50 cps (FM and PDM have a low-frequency limit of 0). Any additional data input requires additional hardware on a per-track basis. High-quality linear magnetic tape characteristics with controlled amplitude compression for equalization are required. Table 5-3 summarizes the advantages and disadvantages of analog and digital recording methods.

### 5.3.15 Digital Magnetic Tape Transport

The function of the tape transport is to move a strip of magnetic tape

**TABLE 5-3**
**Comparison of Analog and Digital Magnetic Tape Recording**

*(Reprinted from Bycer, ref. [2]; © 1965 Hayden Book Company.)*

| TYPE | ADVANTAGES | DISADVANTAGES |
|---|---|---|
| Analog (Direct) | High frequency limit, 100 kc | Low frequency limit, 50 cps |
| | Usually one signal per track | 1% tape speed variation results in 1% error |
| 60 ips | Simplest method | Drop-out effect is serious |
| | | Stringent tape speed control required |
| Analog (FM) | Low frequency limit, 0 cps | High frequency limit, 10 kc |
| | Relatively insensitive to tape dropout | 1% tape speed variation results in 2.5% error |
| 60 ips | One signal for wide deviation or up to 18 inputs (narrow deviation) for one track | Complex circuitry |
| | | High tape transport quality required |
| Analog (PDM) | Low frequency, 0 cps | Limited high-frequency response |
| | Multiplex up to 86 signals on one track | Complex equipment required |
| | Less sensitive to speed variation than FM (1% error for 1% speed change) | Low tape utilization |
| Digital | Low frequency, 0 cps | Limited high frequency (150kc) |
| | High data accuracy | Data must be digitized |
| | Insensitive to amplitude and speed variations | Sensitive to drop-ins and drop-outs |
| | Compatible with computer operations | Requires computer quality tape |
| | Start/stop capability | Low tape utilization |
| | Extremely versatile | |

at constant speed past a magnetic head. Typical elements of a tape transport are listed in Table 5-4. The type of element used depends upon the tape transport application; e.g., the tape reservoir in one application might be a tension-arm type while in another application it might be a vacuum-loop type. No single element is far superior or unquestionably more efficient than another. All of the different types of elements listed in Table 5-4 can be found in equipment currently on the market.

**TABLE 5-4**
**Elements of a Tape Transport**
*(Reprinted from Bycer, ref. [2]; © 1965 Hayden Book Company.)*

| Element | Tape Reservoir | Tape Sensor | Tape Drive |
|---|---|---|---|
| Types of Element | Tension Arm<br>Vacuum Loop | Mechanical<br>Photoelectric<br>Electronic<br>Pressure | Mechanical<br>Pressure Vacuum |

| Tape Brake | Reel Drive | Reel Brake |
|---|---|---|
| Mechanical Pneumatic | Servo Motor; Squirrel Cage; Induction Motor; Synchronous and Induction Motor | Mechanical Magnetic |

The operational elements of a basic tape transport are shown in Figure 5-26. Although the magnetic tape has very little inertia, obtaining controlled motion in conjunction with delicate mechanical handling presents a difficult problem in fast start and fast stop operations. To move the tape past the magnetic head, the tape is trapped against a rotating capstan by mechanical or vacuum pressure. Because of the initial high acceleration of the tape, a period of time must elapse before the tape speed settles down within permissible speed variations. Various methods, such as pressure pads or vacuum hold-down, are used to ensure intimate contact between the tape and the magnetic head during both the transient starting period and steady running.

Since there is a reaction time delay between the motion of tape and its feed from the reel, a tape reservoir is used to supply the tape initially. The reservoir must hold a large enough supply to prevent tape breakage. Before the tape supply is exhausted in the reservoir, the feed reel must be up to speed to supply the tape for continuous tape motion. A complementary operation is performed by a second tape reservoir in conjunction with the take-up reel. This reservoir must be able to absorb the tape spent from the first reservoir, without

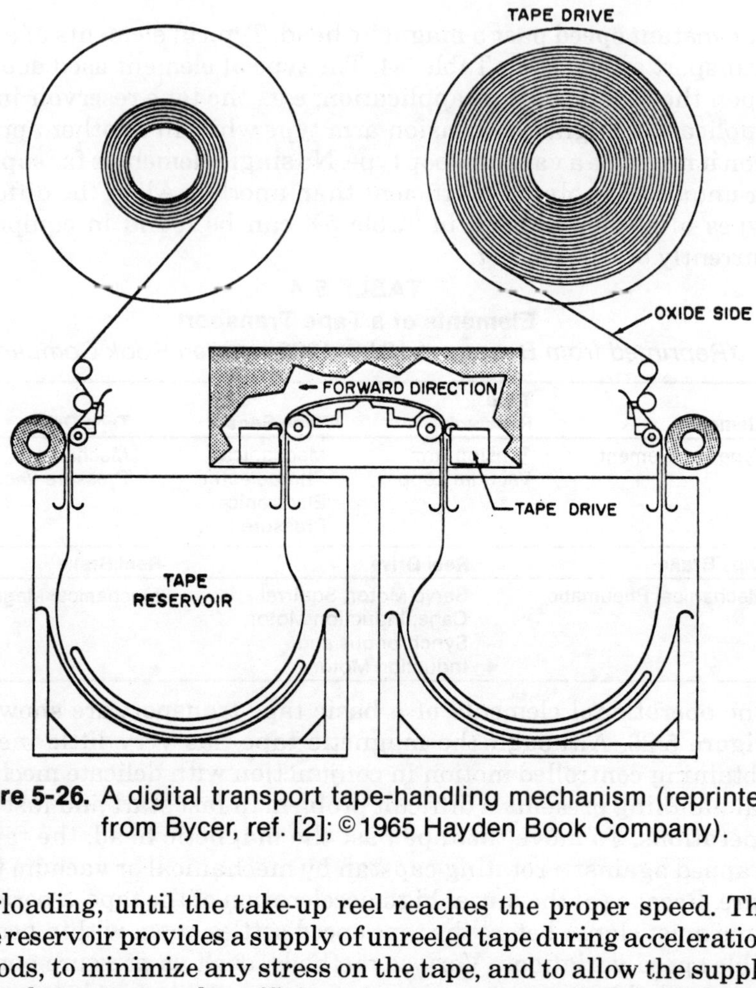


**Figure 5-26.** A digital transport tape-handling mechanism (reprinted from Bycer, ref. [2]; © 1965 Hayden Book Company).

overloading, until the take-up reel reaches the proper speed. The tape reservoir provides a supply of unreeled tape during acceleration periods, to minimize any stress on the tape, and to allow the supply and take-up tape reels sufficient time to overcome their initial inertia. The tape reservoir functions in the same fashion during the braking operation.

Three separate motor drives are needed for these three areas: tape drive, feed drive, and take-up drive. All three motor drives must be controlled by a system of clutches and brakes that facilitate rapid changes in tape speed and direction, and provide sufficient tension to avoid tape slack, slippage, and breakage.

Other transducers are required to ensure proper tape handling and mechanical operation. The tape reservoir supplies signals to the servo control to regulate the speed of the supply and take-up reels.

Tape tension must be monitored and transmitted as input signals to the supply and take-up reel servo system for accelerating the tape from a stationary position, and for the detection of tape breakage. The tape supply is also monitored to supply stop signals to the servo system at each extremity of the tape. In addition, for digital application, the remaining recording time is displayed locally at the tape station control panel or at a remote control console.

In digital recording, the need for fast start-stop operations is paramount, while all other mechanical requirements are secondary. The analog tape recorder is generally used only on a continuous basis while a digital tape transport is operated either on an intermittent or a continuous basis. The analog tape recorder preserves the electrical signals in a linear fashion on magnetic tape and is concerned with the accurate representation and reproduction of the original electrical signal. Tape speed and amplification associated with each process (record and playback) must be identical or compensatory. This is achieved by uniform motion between the heads and the medium. The degree of speed uniformity influences the cost of the equipment and errors due to timing. Flutter or wow must be reduced or eliminated if possible.

For digital applications, problems associated with analog tape recording are present, but are minimized for several reasons. If the digital tape recorder is operated on an intermittent basis, tape speed variations of ±5% from nominal are generally acceptable. Hence, tape speed need not be exactly the same for record and playback operations. The problems associated with flutter or wow are not too important in digital recording unless the playback signal is below the threshold of the playback amplifiers. Amplitude variations and synchronization are treated differently in digital applications and these factors do not have a counterpart in analog recording. The construction and operation of a digital tape transport are very complex.

## 5.4 DIGITAL DATA ACQUISITION SYSTEM

Fundamental to the operation of a digital data acquisition system is the concept of sampling, which may be defined as *the act of measuring a continuous function at discrete time intervals.*

Signals that represent some physical or analog parameter are converted to a series of discrete values. As it turns out, since the data acquisition system is a sampled data system, the overall system transfer function must be examined to achieve a proper solution.

A typical data acquisition/sampled data system is shown in Figure 5-27. As the diagram indicates, three major processes are involved:

1. Sampling
2. Quantization
3. Digital processing

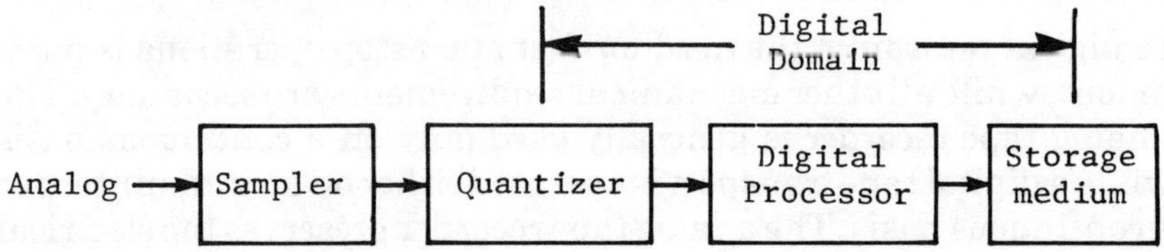


**Figure 5-27.** Typical data acquisition/sampled data system.

A properly designed system generally requires balancing the implementation of one of these processes with the characteristics of the others. For instance, if the required accuracy or system dynamic range is high (a large number of quantizer levels), the sample rate may be affected. The required number of quantization levels may also be dictated by the digital processing algorithm.

### 5.4.1 The Sampling Function

We define sampling as the act of measuring a continuous signal at discrete time intervals. The sampler could be considered as a simple switch. When the switch is closed, it is equivalent to mutiplying the input signal by unity for the duration of the closure period. The results of sampling with a unit-amplitude impulse train is shown in Figure 5-28, where the signal, $f(t)$, is sampled by $h(t)$. The output is the inpulse train, $f^*(t)$, whose amplitudes trace out the envelope of the original signal, $f(t)$.

In actual practice, hardware constraints place severe limitations on the ability to achieve a true impulse sample function. For ease of understanding, however, the impulse function will be used to explain the basic sampling concepts.

The sampling function consists of multiplication in the time domain, as defined by the expression:

$$F^*(t) = f(t)h(t) \tag{5-5}$$

Multiplication in the time domain is convolution in the frequency domain. The effect of sampling in both domains is shown in Figure 5-29. The random time function shown in 5-29a translates to a closed

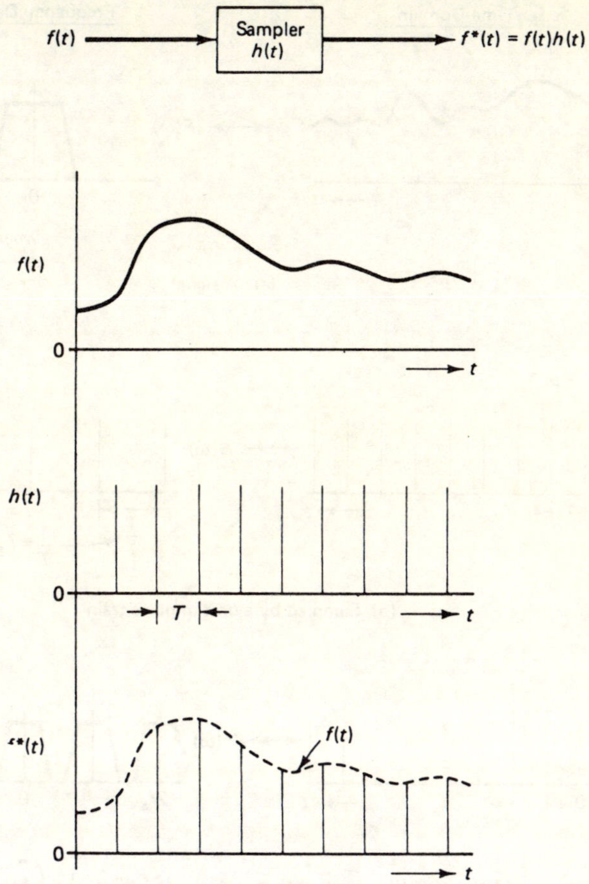


**Figure 5-28.** The impulse samples (reprinted from Van Doren, ref. [1]; © 1982 Reston Publishing Company).

spectrum in the frequency domain. Figure 5-29b shows the sampling inpulse train which is an impulse train in both domains. Finally, in 5-29c, we find that the amplitude-modulated sampled time function, *f*(t)*, appears as a series of spectra as the frequency spectrum, *F(w)*, is convolved with the impulse function, *H(w)*. Each of the spectra of *F*(w)* is centered about multiples of the same frequency, *fs* = *1/T*. Note that these multiple spectra are present in the sampler output and can become a problem later.

Periodic functions in the time domain become sampled spectra in the frequency domain, and sampled functions in the time domain translate into periodic spectra in the frequency domain.

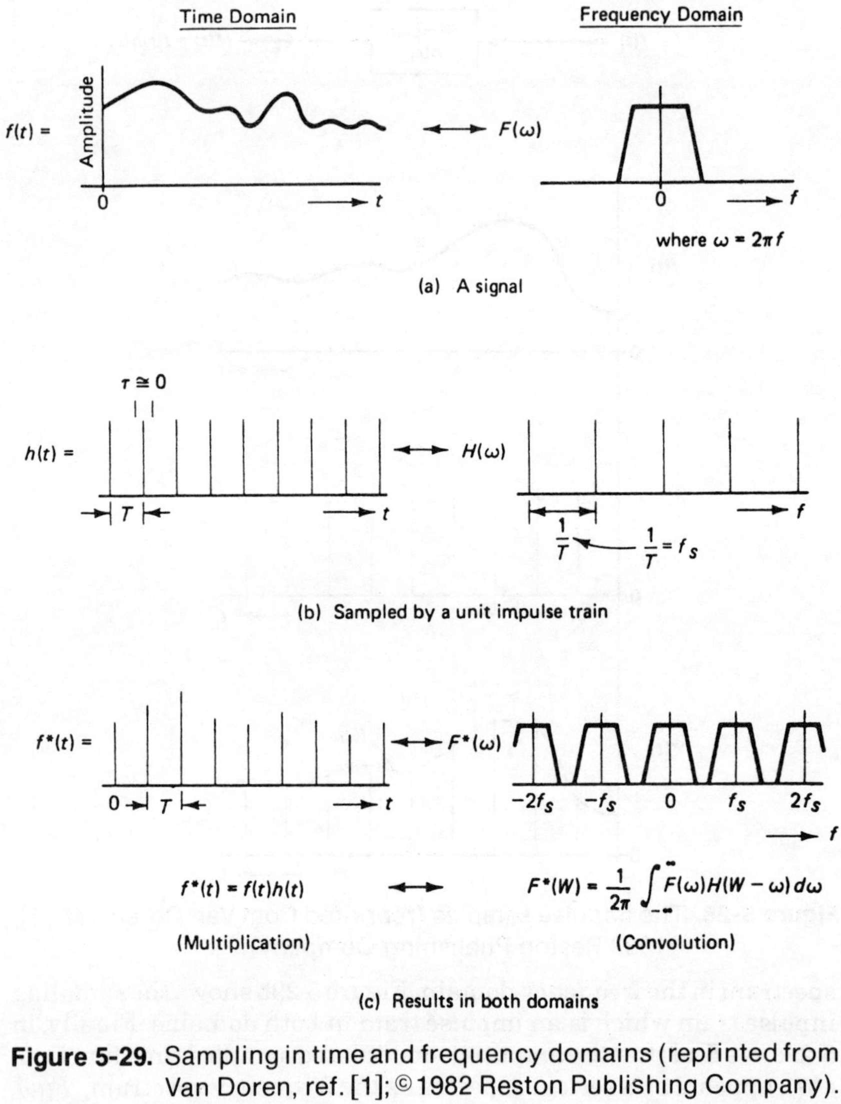


**Figure 5-29.** Sampling in time and frequency domains (reprinted from Van Doren, ref. [1]; © 1982 Reston Publishing Company).

### 5.4.2 The Sampling Theorem

A discussion of sampling would not be complete without reference to the sampling theorem, which states that *signals with a finite bandwidth of f Hertz can be completely described by sampling the time signals at instants separated by $T = 1/2f$ seconds.* In simple terms, the signal must be sampled at a rate at least twice as high as the frequency in the spectrum. This concept is shown in Figure 5-30, where the sinusoid is sampled at its positive and negative peaks. The theory then assumes that the resultant impulse sample train could be passed through an ideal "box" filter and the original sinusoid frequency reproduced. The amplitude of the filter output depends on where the original sinusoid is sampled.

Since the sampling theorem defines the limit for an absolutely perfect situation, it has a number of serious problems. First, suppose we were so unlucky as to choose sample points that fall precisely on the zero crossings. This case, shown in Figure 5-31a, would, of course, produce zero output from the sampler. While the case is trivial, the problem is fundamental.

The second difficulty arises in trying to build a filter that cuts off perfectly above the highest frequency, *f*. In contrast, consider what happens if the reconstruction filter is a simple digital-to-analog converter (zero-order hold). The result, shown in Figure 5-31b, is that

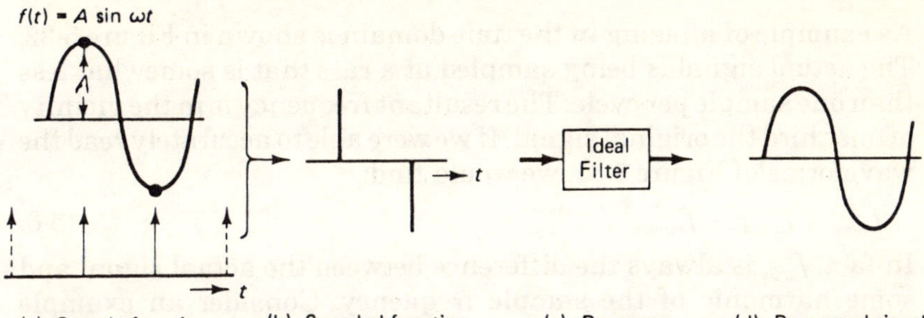


**Figure 5-30.** Basics of the sampling theorem (reprinted from Van Doren, ref. [1]; © 1982 Reston Publishing Company).

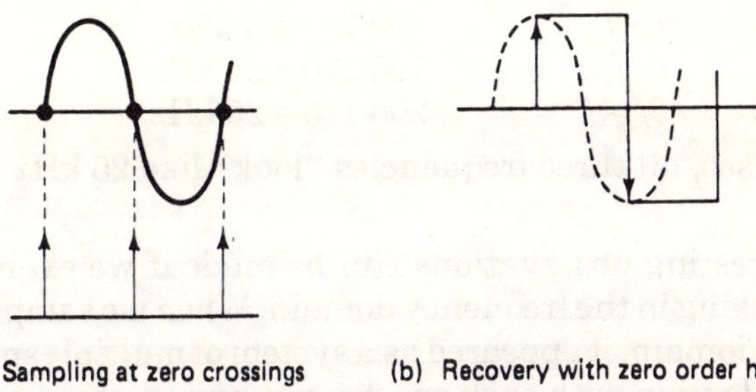


**Figure 5-31.** Obvious difficulties with the sampling theroem (reprinted from Van Doren, ref. [1]; © 1982 Reston Publishing Company).

the square wave contains many more frequencies than the original single sine wave.

The last and perhaps most serious problem is that we tend to become attached to the "highest frequency of interest," and often also consider it to be $f$, the "highest frequency contained." In actuality, no spectrum exists that cuts off perfectly at $f$, and substantial energy exists in the frequencies beyond it. This brings us to the concept of aliasing.

### 5.4.3 Aliasing

Webster's dictionary defines the word alias as "an assumed name" or "to be called by another name." The concept of aliasing, in a sampled data system, means the same thing with reference to frequencies. When a given frequency is sampled at too low a rate, it appears as a totally different lower frequency at the output of the sampler. So we say that the resultant frequency is the "alias" of the original.

As example of aliasing in the time domain is shown in Figure 5-32. The actual signal is being sampled at a rate that is somewhat less than one sample per cycle. The resultant frequency is in the vicinity of one-third the original signal. If we were able to accurately read the waveforms of Figure 5-31, we would find:

$$f_{alias} = f_{actual} - f_{sample} \tag{5-6}$$

In fact, $f_{alias}$ is always the difference between the actual signal and some harmonic of the sample frequency. Consider an example where $f_{sample}$ $(f_s)$ = 100 kHz and the signal contains the frequencies 75 kHz, 125 kHz, and 175 kHz. The sample results are:

| $f$ | Formula | Result |
|---|---|---|
| 75 kHz | $f_s - f$ | 100-75 = 25 kHz |
| 125 kHz | $f - f_s$ | 125-100 = 25 kHz |
| 175 KHz | $2f_s - f$ | 200-175 = 25 kHz |

As we can see, all three frequencies "look" like 25 kHz as they are aliased.

Other interesting observations can be made if we examine the effects of aliasing in the frequency domain. When we sampled a signal in the time domain, it appeared as a system of multiple spectra in the frequency domain with each spectrum centered about a multiple of the sample frequency, $f_s$.

Consider what happens if we reduce the sample rate. In the time

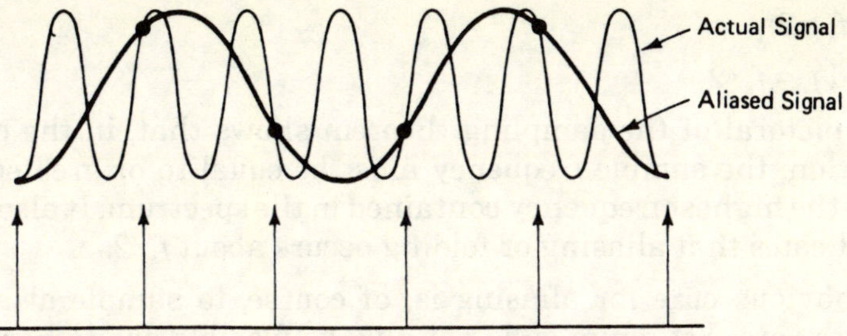


**Figure 5-32.** Aliasing in the time domain (reprinted from Van Doren, ref. [1]; ©1982 Reston Publishing Company).

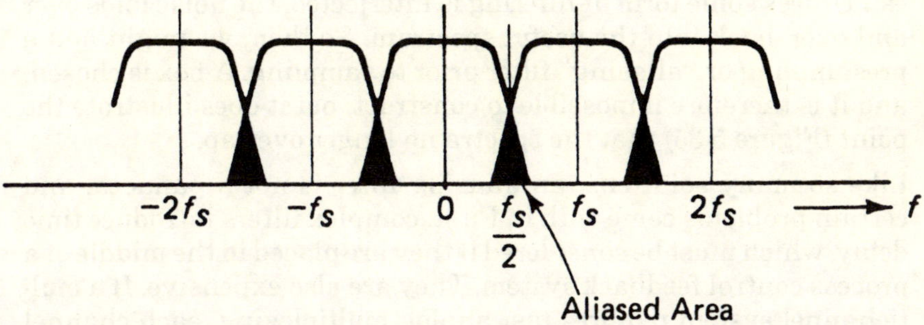


**Figure 5-33.** Aliasing in the frequency domain (reprinted from Van Doren, ref. [1]; ©1982 Reston Publishing Company).

domain, the sample pulses spread apart. In the frequency domain, however, the spectra move closer together. Eventually, if we continue to reduce the sample rate, the spectra touch and finally overlap. Once the spectra have overlapped, aliasing has occurred. Such a situation is found in Figure 5-33. An important item to note is that the aliasing occurs about a point that is equal to one-half the sample frequency, or $f_s/2$. This point is often referred to as the folding frequency because all frequencies above $f_s/2$ are "folded" back into the spectrum below it. As an interesting side note, consider the rectangular spectra of Figure 5-34. The original spectrum and the first harmonic spectrum can be butted tightly together at the very point of folding, $f_s/2$.

The following expression may be written:

$$f_s - f_c = f_c \tag{5-7}$$

where

$f_c$ = the cut-off frequency

or $f_s = 2f_c$

and $f_c = f_s/2$

This pictoral of the sampling theorem shows that, in the perfect situation, the sample frequency must be equal to or greater than twice the highest frequency contained in the spectrum. It also clearly indicates that aliasing or folding occurs about $f_s/2$.

The obvious cure for aliasing is, of course, to sample at a high enough rate. Yet doing so is not always feasible due to hardware constraints. Also recognize that signals usually contain noise, which is much wider in its bandwidth than the frequencies of interest. Unless some form of filtering is interjected, the noise folds over and over, back into the useful spectrum. So then, we might add a presampling or "aliasing" filter prior to sampling. A box is chosen, and it is therefore impossible to construct, but it does illustrate the point (Figure 5-35) that the spectra no longer overlap.

Like so many solutions, an aliasing filter is not a panacea, and certain problems come with it. First, complex filters introduce time delay, which must be considered if they are placed in the middle of a process control feedback system. They are also expensive. If a multichannel system requires fast analog multiplexing, each channel must have a separate filter prior to the multiplexer. The cost per filter must then be multiplied by the number of channels. The solution to a given problem is usually a combination of sample rate selection and presample filtering.

### 5.4.4 Frequency Resolution

Often the data acquisition system is required to determine which frequencies are present in a spectrum and how much energy each frequency contains, relative to the others. Such an application is called spectral, or spectrum, analysis. Since this application means the system must be able to separate one frequency from another, we

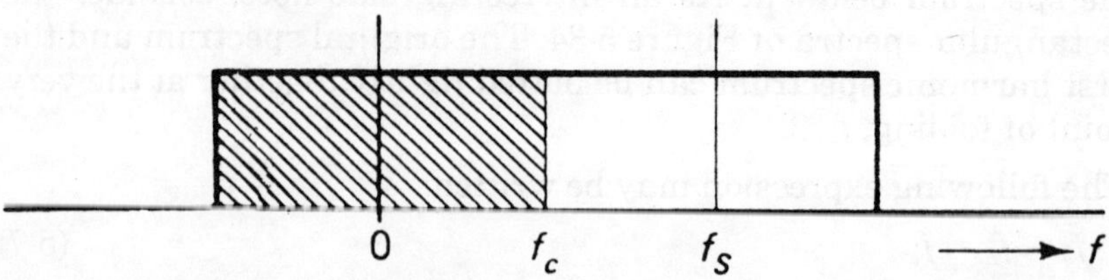


**Figure 5-34.** Sample theorem with ideal spectrum (reprinted from Van Doren, ref. [1]; © 1982 Reston Publishing Company).

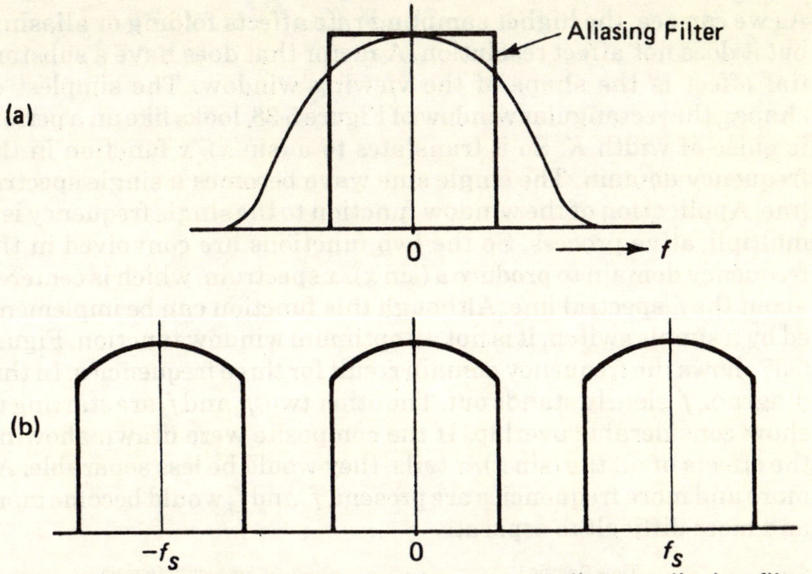


**Figure 5-35.** Removal of aliasing with a presampling or aliasing filter (a) and resultant (b) (reprinted from Van Doren, ref. [1]; © 1982 Reston Publishing Company).

are interested in frequency resolution. In a general sense, the frequency resolution postulate states that if a function is available for $K$ seconds, $1/K$ cycles per second (cps) may be resolved. At least one complete cycle of the frequency must be seen, in other words. Although spectrum analysis requires adequate sample rate selection, like any other sampled data process, the resolution is a function of the viewing period and not of the sampling frequency. Two examples:

*Example 1*

Observe for: 100sec

Sample rate: 1,000 samples/sec

Foldover frequency: 500 cps

Resolution: 1/100 cps

*Example 2*

Observe for: 100 sec

Sample rate: 5,000 samples/sec

Foldover frequency: 2,500 cps

Resolution: 1/100 cps

As we can see, the higher sampling rate affects folding or aliasing, but it does not affect resolution. A factor that does have a substantial effect is the shape of the viewing window. The simplest of shapes, the rectangular window of Figure 5-36, looks like an a periodic pulse of width $K$. So it translates to a $(\sin x)/x$ function in the frequency domain. The single sine wave becomes a single spectral line. Application of the window function to the single frequency is a multiplicative process. So the two functions are convolved in the frequency domain to produce a $(\sin x)/x$ spectrum, which is centered about the $f_o$ spectral line. Although this function can be implemented by a simple switch, it is not an optimum window function. Figure 5-37 shows the frequency domain result for three frequencies. In this diagram, $f_1$ clearly stands out. The other two, $f_2$ and $f_3$ are starting to show considerable overlap. If the composite were drawn showing the effects of all the $(\sin x)/x$ tails, they would be less separable. As more and more frequencies are present, $f_2$ and $f_3$ would become more and more difficult to separate.

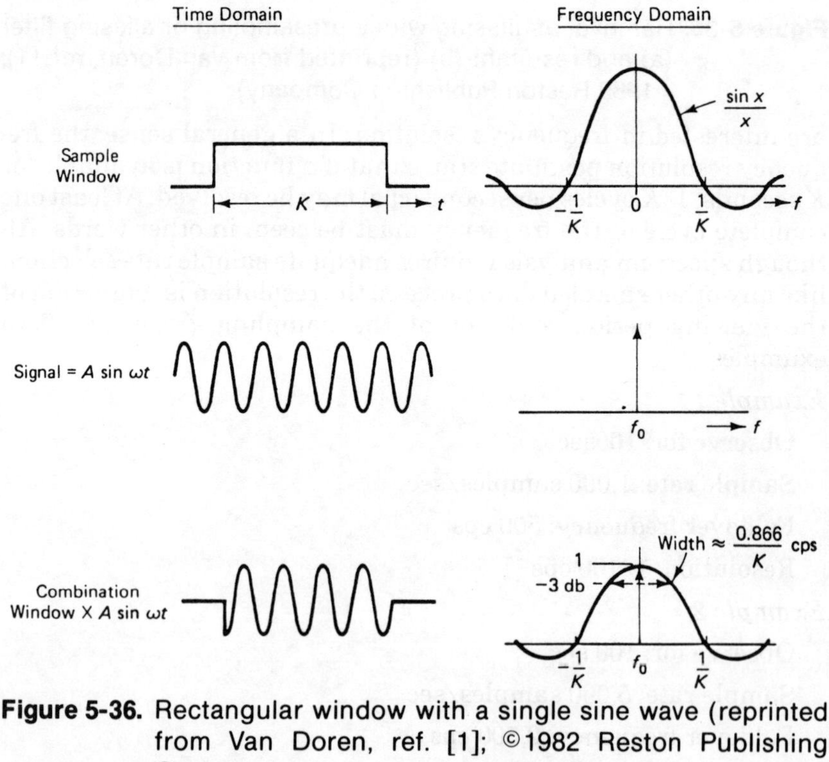


**Figure 5-36.** Rectangular window with a single sine wave (reprinted from Van Doren, ref. [1]; ©1982 Reston Publishing Company).

Fortunately, several other very tailored windows have been developed to solve this problem. The most popular of these are found in Figure 5-38. Each has less ringing or residue in its frequency plot tails than the rectangular (sin $x$)/$x$ combination. While the separation is cleaner for these window functions, several penalties are associated with them. Besides being difficult to implement in hardware, the resolution is no longer $1/K$, but ranges from $1.5K$ to $2.0K$.

### 5.4.5 Quantization

The major premise of the digital data acquisition system is that signals are converted from the analog domain to the digital domain, so that a digital computer or some other digital process may operate on them. This process, called quantization, is defined as *the conversion of an input function that has values in a continuous range to an output that has only discrete values.*

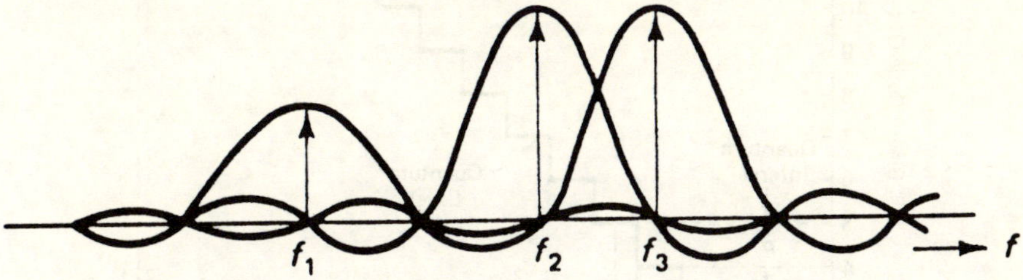


**Figure 5-37.** Rectangular window with three frequencies (reprinted from Van Doren, ref. [1]; ©1982 Reston Publishing Company).

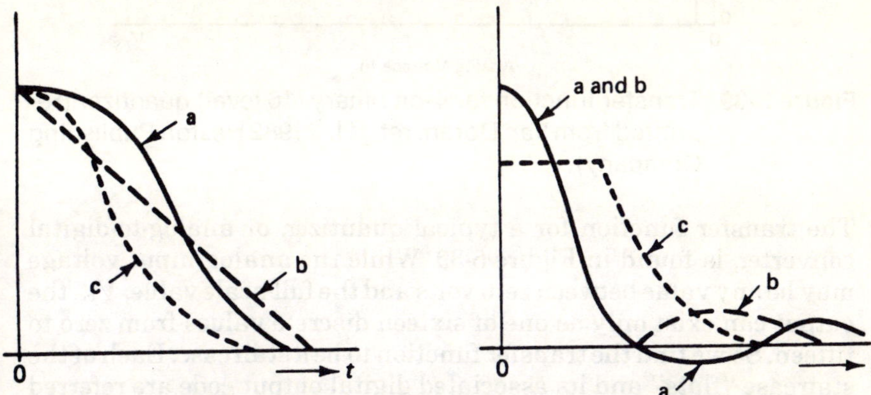


**Figure 5-38.** Other windows: (a) Hanning or raised cosine, (b) Barlett, and (c) Parzen (reprinted from Van Doren, ref. [1]; ©1982 Reston Publishing Company).

Sometimes, quantization is combined with sampling and the two operations occur simultaneously. In other implementations, such as when a sample-and-hold is used, sampling and quantization are two very distinct and separate operations. Here, the sample-and-hold makes the basic measurement and holds the value while the quantizer converts it into a digital format.

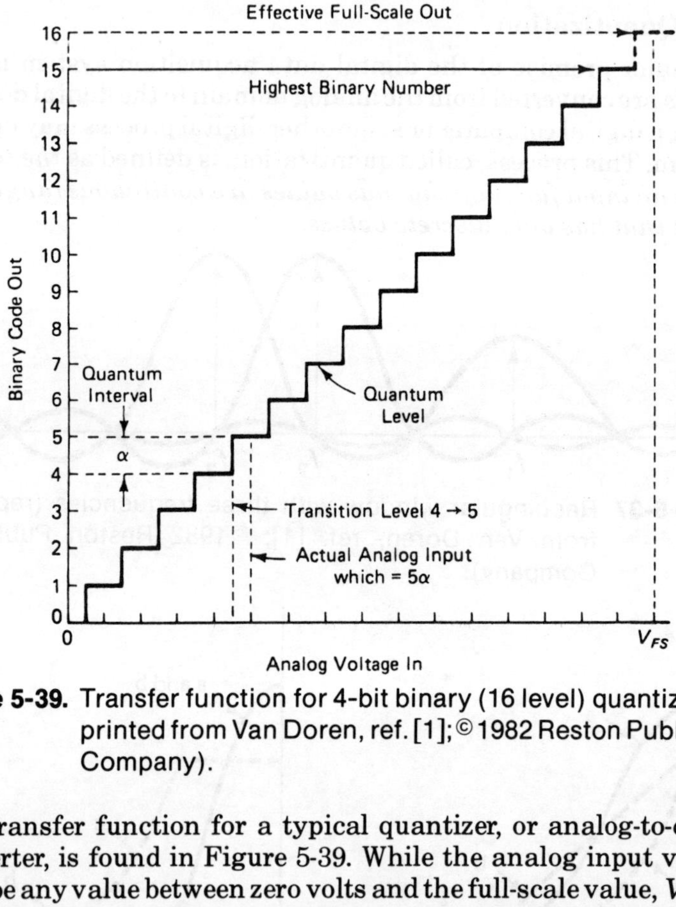


**Figure 5-39.** Transfer function for 4-bit binary (16 level) quantizer (reprinted from Van Doren, ref. [1]; © 1982 Reston Publishing Company).

The transfer function for a typical quantizer, or analog-to-digital converter, is found in Figure 5-39. While the analog input voltage may be any value between zero volts and the full-scale value, $V_{FS}$, the output can exist only as one of sixteen discrete values from zero to fifteen. So we find the transfer function to be a staircase. Each of the staircase "flats" and its associated digital output code are referred to as quantum levels. The distance between any two quantum levels is called a quantum interval. For a binary (or power of 2) based quantizer, the quantum interval, $\alpha$, is defined as:

$$\alpha = \frac{V_{FS}}{2^n} \tag{5-8}$$

where

$V_{FS}$ = full scale voltage input

2 = number base for binary

$n$ = number of bits (binary digits)

$2^n$ = number of quantizing intervals

For the quantizer of Figure 5-39, we find that:

$$\alpha = \frac{V_{FS}}{2^n} = \frac{V_{FS}}{2^4} = \frac{V_{FS}}{16} \tag{5-9}$$

Should five bits have been required, $\alpha$ would be:

$$\frac{V_{FS}}{2^5} \text{ or } \frac{V_{FS}}{32} \tag{5-10}$$

and the transfer function staircase would contain twice as many steps as Figure 5-39, and they would be half-sized. The steps continue to double in number and halve in size as the number of quantizer bits increase, as follows:

| $n$ | Steps |
|---|---|
| 1 | 2 |
| 2 | 4 |
| 3 | 8 |
| 4 | 16 |
| 5 | 32 |
| 6 | 64 |
| 7 | 128 |
| 8 | 256 |
| 9 | 512 |
| 10 | 1,024 |
| 11 | 2,048 |
| 12 | 4,096 |
| 13 | 8,192 |
| 14 | 16,384 |
| 15 | 32,768 |
| 16 | 65,536 |

Referring to Figure 5-39, note that level 16 does not actually exist and that the highest binary number output available is level 15. It should be mentioned that the voltage range of 0 volts to $V_{FS}$ was selected for simplicity. In actuality the quantizer input could range from any arbitrarily low value to any arbitrarily high value.

Several binary codes are commonly used. The most common are the straight binary and offset binary, which are found in Table 5-5. Straight binary is a unipolar code where 0000 is equivalent to level 0 (also analog zero), while 1111 is equivalent to level 15. On the other hand, offset binary is a bipolar code with 0000 equal to -7, and 1111 equal to +7. For this situation, the 23 bit (most significant) is the sign bit, with 0 = minus and 1 = plus. There are two "zero" levels in the offset binary code, each half a quantum interval from true analog zero.

Two other codes are most compatible with standard digital computers (Table 5-6). They are the one's and two's complement binary codes. Both bipolar, they differ primarily in the positioning of the true analog zeros. The one's complement code is identical with the offset binary code, with the exception that the sign bit is inverted. In this case, 1 is negative and 0 is positive. Like offset binary, this

**TABLE 5-5**
**Straight and Offset Binary Codes**
*(Reprinted from Van Doren, ref. [1]; © 1982 Reston Publishing Company.)*

| | Bit Position | | | | Level | |
|---|---|---|---|---|---|---|
| | $2^3$ | $2^2$ | $2^1$ | $2^0$ | Straight Binary | Offset Binary |
| Straight Binary Zero → | 0 | 0 | 0 | 0 | 0 | -7 |
| | 0 | 0 | 0 | 1 | 1 | -6 |
| | 0 | 0 | 1 | 0 | 2 | -5 |
| | 0 | 0 | 1 | 1 | 3 | -4 |
| | 0 | 1 | 0 | 0 | 4 | -3 |
| | 0 | 1 | 0 | 1 | 5 | -2 |
| | 0 | 1 | 1 | 0 | 6 | -1 |
| | 0 | 1 | 1 | 1 | 7 | -0 |
| Offset Binary Zero → | | | | | | |
| | 1 | 0 | 0 | 0 | 8 | +0 |
| | 1 | 0 | 0 | 1 | 9 | +1 |
| | 1 | 0 | 1 | 0 | 10 | +2 |
| | 1 | 0 | 1 | 1 | 11 | +3 |
| | 1 | 1 | 0 | 0 | 12 | +4 |
| | 1 | 1 | 0 | 1 | 13 | +5 |
| | 1 | 1 | 1 | 0 | 14 | +6 |
| | 1 | 1 | 1 | 1 | 15 | +7 |

system has two zero levels. The two's complement code, however, has only one zero level, and it is coincident with the 0000 code. There is also one more negative level (-8) than positive (+7). The sign is also the complement of that for offset binary. Careful examination shows that all three bipolar codes discussed thus far are simple extrapolations from straight binary requiring either offsetting of analog zero with respect to the code levels or simple inversion of the sign bit.

An expansion of a segment of Figure 5-39 is found in Figure 5-40. Here we see the transitions between the given levels $K - 1$ and $K$ and $K + 1$. Of interest here is the fact that, once the output code has made the transition from level $K - 1$ to level $K$, it does not change again until the analog input has progressed from $V_K - \alpha/2$ to $V_K + \alpha/2$, or a value equivalent to a full quantum interval. So obviously, when the quantizer output level is $K$, all we can say about the input is that it is somewhere between $V_K - \alpha/2$ and $V_K + \alpha/2$. This element of uncertainty produces an irreducible error situation. The only time the output code is accurate is when the input is exactly $V_K$. Having the output code remain constant, while the input ranges over all possible

**TABLE 5-6**
**Ones and Twos Complement Binary Codes**
*(Reprinted from Van Doren, ref. [1]; © 1982 Reston Publishing Company.)*

| | Bit Position | | | | Level | |
|---|---|---|---|---|---|---|
| | $2^3$ | $2^2$ | $2^1$ | $2^0$ | Two's | One's |
| | 1 | 0 | 0 | 0 | -8 | -7 |
| | 1 | 0 | 0 | 1 | -7 | -6 |
| | 1 | 0 | 1 | 0 | -6 | -5 |
| | 1 | 0 | 1 | 1 | -5 | -4 |
| | 1 | 1 | 0 | 0 | -4 | -3 |
| | 1 | 1 | 0 | 1 | -3 | -2 |
| | 1 | 1 | 1 | 0 | -2 | -1 |
| | 1 | 1 | 1 | 1 | -1 | -0 |
| Ones Zero → | | | | | | |
| Twos Zero → | | | | | | |
| | 0 | 0 | 0 | 0 | 0 | +0 |
| | 0 | 0 | 0 | 1 | +1 | +1 |
| | 0 | 0 | 1 | 0 | +2 | +2 |
| | 0 | 0 | 1 | 1 | +3 | +3 |
| | 0 | 1 | 0 | 0 | +4 | +4 |
| | 0 | 1 | 0 | 1 | +5 | +5 |
| | 0 | 1 | 1 | 0 | +6 | +6 |
| | 0 | 1 | 1 | 1 | +7 | +7 |

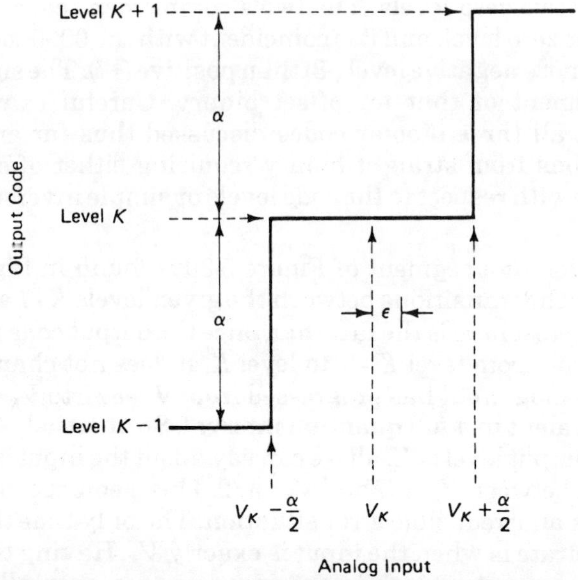


**Figure 5-40.** Uncertainity at quantizer output (reprinted from Van Doren, ref. [1]; © 1982 Reston Publishing Company).

values $V_K \pm \alpha/2$, is the same as adding noise to the quantizer input. This effect is called quantization error or quantization noise. If we consider any given analog input, *f(t)* to occur over a long period, it is probable that all values of $V_K \pm \alpha/2$ will be present and will cause the *K*th level to occur in a random fashion. The amplitude of the effective noise is established by the number of quantizer bits or quantum intervals. If the input deviation from the voltage value equivalent to the *K*th level, $V_K$, is designated (Figure 5-41), the RMS quantization error is calculated as follows:

$$\sqrt{\frac{1}{\alpha}\int_{-\alpha/2}^{\alpha/2} \epsilon^2 \, d\epsilon} = \sqrt{\frac{\alpha^2}{12}} = \frac{\alpha}{2\sqrt{3}} \tag{5-11}$$

The character of the quantization error function is found in Figure 5-41. The error is triangular in nature, and it cycles back and forth between $+\alpha/2$ and $-\alpha/2$ as the analog input proceeds up or down the scale. Quantization error has a probability function that is rectangular in nature since, over a large period, the input is as likely to occur at any one value of $\epsilon$ as any other, while the output remains fixed. This situation is true for the perfect quantizer. Yet quantizers themselves add noise from a number of internal sources. This noise, generally Gaussian in character, has a definite impact on the effective quantization error rate.

Figure 5-42 depicts this situation when the analog input is centered in the quantum interval. This figure shows that the noise peaks above $\pm 2.5\sigma$ extending into the adjacent quantum intervals and that 1.25 percent of the time the output code is in error. Unfortunately, as discussed, the input does not always occur at the center of the interval, and it is as likely to be at any other value.

Figure 5-43 gives an indication of what happens when the input is offset from the center. Although the error output favors the $K + 1$ side, it has also substantially increased in value. The quantizer output will be in error 6.7 percent of the time under these conditions. This type of result occurs when the RMS quantizer noise is only 20 percent of a quantum interval. Therefore, the noise figure of the analog-to-digital converter (quantizer) should be considered when making a selection.

## 5.5 SYSTEM ERRORS

Since most systems are digital today, the error analyses briefly presented here are oriented toward the Digital DAS. Figures 5-44 and 5-45 illustrate the common ac and dc error sources in the front-end elements of the DAS.

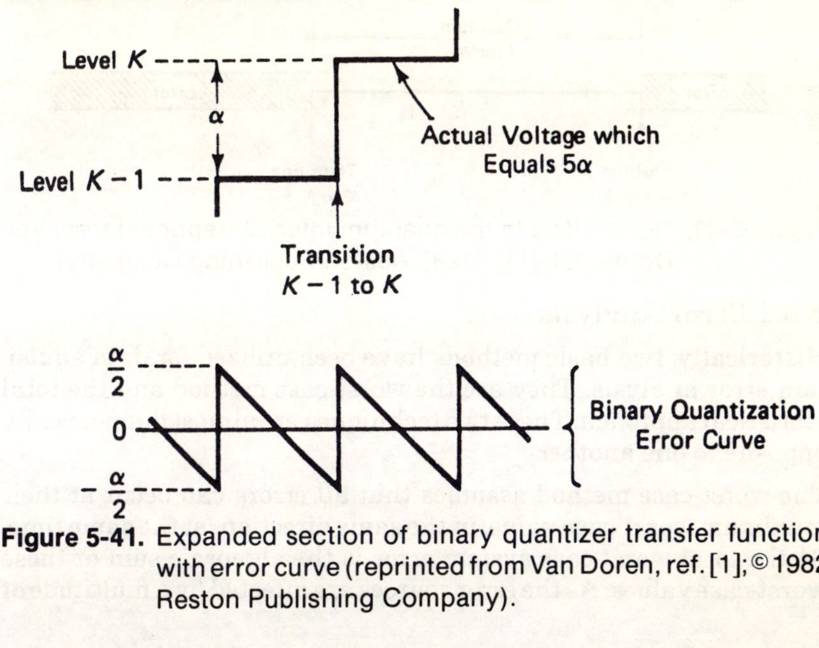


**Figure 5-41.** Expanded section of binary quantizer transfer function with error curve (reprinted from Van Doren, ref. [1]; © 1982 Reston Publishing Company).

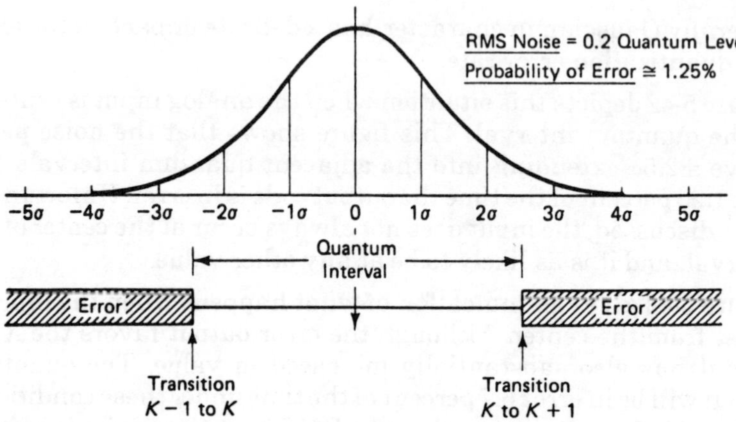


**Figure 5-42.** Noise centered in the quatum interval (reprinted from Van Doren, ref. [1]; © 1982 Reston Publishing Company).

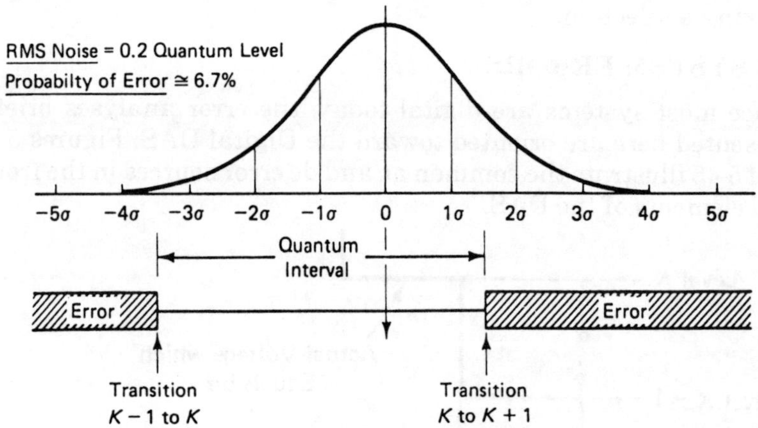


**Figure 5-43.** Noise offset in the quantum interval (reprinted from Van Doren, ref. [1]; © 1982 Reston Publishing Company).

### 5.5.1 Error Analysis

Historically, two basic methods have been utilized for data aquisition error analysis. They are the worst-case method and the total statistical approach. These two techniques are almost diametrically opposite to one another.

The worst-case method assumes that all errors can occur at their maximum worst-case value, in the same direction, at the same time. So the total worst-case system error is the algebraic sum of these worst-case values. As the error sources are affected by a multitude of

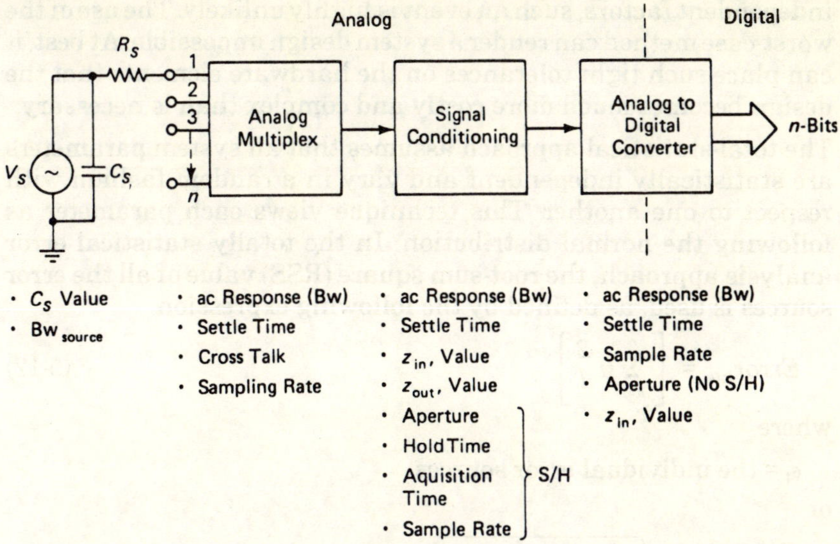


**Figure 5-44.** Data Acqisition system block diagram with ac error sources (reprinted from Van Doren, ref. [1]; © 1982 Reston Publishing Company).

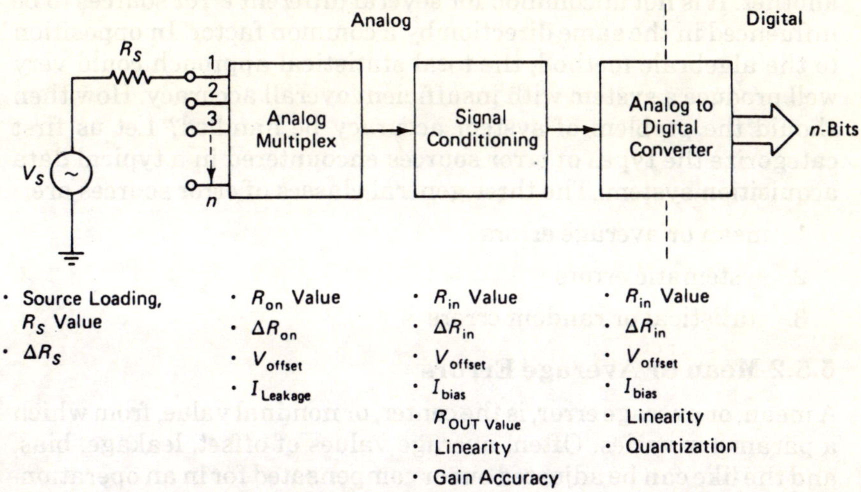


**Figure 5-45.** Data acquisition system block diagram showing dc error sources (reprinted from Van Doren, ref. [1]; © Reston Publishing Company).

independent factors, such an event is highly unlikely. The use of the worst-case method can render a system design impossible. At best, it can place such tight tolerances on the hardware elements that the design becomes much more costly and complex than is necessary.

The total-statistical approach assumes that all system parameters are statistically independent and vary in a random fashion with respect to one another. This technique views each parameter as following the normal distribution. In the totally statistical error analysis approach, the root-sum square (RSS) value of all the error sources is used, as defined by the following expression:

$$Error_{total} = \left[ \sum_{n=1}^{n} \epsilon_{I\,1}^{\,2} \right]^{1/2} \qquad (5\text{-}12)$$

where

$\epsilon_I$ = the individual error sources

or

$$Error_{total} = \sqrt{\epsilon_1^2 + \epsilon_2^2 + \epsilon_3^3 + \ldots \epsilon_n^2}$$

The values used are usually either the $3\sigma$ values or the maximum value from the manufacturer's data sheet.

Unfortunately, not all error sources are independent from one another. It is not uncommon for several different error sources to be influenced in the same direction by a common factor. In opposition to the algebraic method, the total statistical approach could very well produce a system with insufficient overall accuracy. How then should the problem of system accuracy be handled? Let us first categorize the types of error sources encountered in a typical data acquisition system. The three general classes of error sources are:

1. mean or average errors
2. systematic errors
3. statistical or random errors

### 5.5.2 Mean or Average Errors

A mean, or average error, is the center, or nominal value, from which a parameter varies. Often, average values of offset, leakage, bias, and the like can be adjusted out or compensated for in an operational system.

### 5.5.3 Systematic Errors (Parameters)

Systematic errors are those which always vary as some definite function of the system operating conditions. Some of the causes for these variations are temperature, power supply voltage change, input voltage change, input voltage levels, and age. Some systematic errors may appear random with respect to one another because their causes are independent.

### 5.5.4 Statistical or Random Errors

Statistical or random errors are usually a product of an arbitrary selection of a manufacturer's components and parts. For example, out of a thousand operational amplifiers of a given type, some will have positive offset voltages while some will have negative offset voltages. If not tampered with, the distribution of parameters will follow the normal curve. Like systematic errors, random error effects can be caused by changes in operating conditions. The difference is that the result is postive for some components and negative for others, rather than always in one direction. The practical way to combine all three types of errors is as follows:

1. Add all average value or mean value errors algebraically.
2. Add all systematic error values due to specific causes algebraically to form parametric error functions.
3. RSS all random errors with the parametric values calculated in entry 2 using the maximum values of entry 2 as $3\sigma$ values. Utilize the $3\sigma$ values of the random errors or the maximum deviations where the $3\sigma$ values are not available.
4. Finally, combine the error sources in the following manner:

$$Error_{total} = \sum_{n=1}^{n} \epsilon_{an} + \left[ \sum_{m=1}^{m} \epsilon_{rm}^{2} + \sum_{s=1}^{s} \epsilon_{ps}^{2} \right]^{1/2}$$

where

$\epsilon_{an}$ = average error due to cause $n$

$\epsilon_{rm}$ = random error due to cause $m$

$\epsilon_{ps}$ = parametric error due to cause $s$

The one error source we have not dealt with as yet is quantization error. Although random in nature, when not coupled with noise, its distribution is rectangular rather than normal. If the quantization error is significant with respect to the other system errors, the distributions would have to be added point by point. Since doing so is difficult, the recommended approach is to combine all the analog errors, as shown, and to carry the quantization error as a separate

**TABLE 5-7**
**Potential Error Sources**
*(Reprinted from Van Doren, ref. [1]; © 1982 Reston Publishing Company.)*

| Error Term | Average Errors | Random Errors | Systematic Errors: Temperature | Systematic Errors: PS Volts | Systematic Errors: Input Volts [Differential] | Systematic Errors: Frequency |
|---|---|---|---|---|---|---|
| 1. *Source of Transducer:* | | | | | | |
| a. Source loading | √ | √ | | | | |
| b. Deviation of $R_S$ | √ | √ | √ | | | |
| c. Source cpacitance | | √ | | | | |
| d. ac responce | | √ | | | | √ |
| e. Input volts — cumulative | | | | | | |
| 2. *Mux or Scanner:* | | | | | | |
| a. $R_{on}$ | √ | √ | √ | √ | √ | |
| b. $V_{offset}$ | √ | √ | √ | √ | | |
| c. $I_{leakage}$ | √ | √ | √ | | | |
| d. ac response | | √ | | | √ | √ |
| e. Settle time | | √ | | | √ | |
| f. Crosstalk | | √ | | | √ | √ |
| g. Sampling accuracy | √ | | | | | √ |

| | | | | | | |
|---|---|---|---|---|---|---|
| 3. *Signal Conditioning:* | | | | | | |
| a. $R_{in}$ | √ | √ | | | | |
| b. $V_{offset}$ | √ | √ | | | | |
| c. $I_{bias}$ | √ | √ | | | | |
| d. $R_{out}$ | √ | √ | | | | |
| e. Linearity | √ | | √ | √ | √ | |
| f. Gain | √ | | √ | √ | √ | |
| g. Settle time | | √ | | | | |
| h. ac response | | √ | | | | √ |
| i. $Z_{in}$ | | √ | | | | |
| j. $Z_{out}$ | | √ | | | | |
| k. Aperture (S/H) | √ | | | | | |
| 4. *A/D Converter:* | | | | | | |
| a. $Z_{in}$ | | √ | | | | |
| b. $I_{bias}$ | √ | | √ | √ | | |
| c. $V_{offset}$ | √ | | √ | √ | | |
| d. Linearity | | | √ | √ | √ | |
| e. Quantization | | √ | | | | |
| f. ac response | | √ | | | | √ |
| g. Settle time | | √ | | | | |
| h. Sample accuracy | √ | | | | | √ |
| i. Aperture | √ | | | | | √ |

factor. The final result is of the form:

| $y\% \pm$ | $x\% \pm$ | $1/2$ LSB |
|---|---|---|
| Average Error | Random and Systematic Error | Quantization Error |

The potential error sources are summarized in Table 5-7.

## REFERENCES

1. Van Doren,. H., *Data Acquisition Systems*, Reston Publishing Company, Inc., Reston, Virginia, 1982. Figures 5-1 and 5-27-45; Tables 5-5, 5-6, and 5-7; Text for Sections 5.1, 5.3 and 5.4.

2. Bycer, B. B., *Digital Magnetic Tape Recording: Principles and Computer Applications*, Hayden Book Company, Inc., New York, 1965. Figures 5-7-20, 5-24, 5-25, 5-26; Tables 5-3, 5-4; Text for Section 5.2.

3. Lowman, C. E., *Magnetic Recording*, McGraw-Hill Book Company, New York, 1972. Figures 5-4, 5-5, 5-6, 5-21, 5-22, 5-23; Table 5-2; Text for Section 5.2.

# CHAPTER 6
# DATA ANALYSIS PROCEDURES

Margaret M. Horst
Engineering Experiment Station
Georgia Institute of Technology



## 6.1 OVERVIEW

The analysis of radar reflectivity data is as important as the actual measurement of the data if results are to be presented in a useful, concise form. Sound analysis techniques usually can salvage some worthwhile information from the most disastrous measurement program, even if only an indication of trends or relative values is obtained. On the other hand, poor data analysis techniques can butcher the results from the most careful and well-executed measurement program. The conclusion is, obviously, that data reduction and analysis must be as scrupulously planned and carried out as the actual data measurement process.

For the successful acquisition of meaningful radar reflectivity information, the data analysis procedure to be used and the specific quantities sought should be firmly established before the measurement program begins. This helps to ensure that parameters needed for later data reduction will be recorded during the measurements. Preliminary data analysis and spot checks of results in the field are desirable, although the bulk of data reduction and analysis is usually accomplished off-line after the measurement program has been completed. At a minimum, calibration closure should be verified in the field before data are recorded and data should be monitored and kept within the calibrated range of the receiver and the recording device during the entire course of the measurements. (*See* Chapters 3 and 4 for detailed information on calibration closure procedures and recording techniques.)

The first task faced by the analyst is the retrieval and calibration of stored data, usually on a pulse-by-pulse basis. Reducing these data to amplitude distributions is widely employed as straightforward presentation of the statistical description of the fluctuating radar return. (*See* Chapter 13 for a more detailed discussion of the statistics of the return.) Other useful temporal quantities of the radar return are obtained from the autocorrelation and autocovariance of the return while power spectral densities and Doppler spectra provide important information in the frequency domain.

Relatively new measurement techniques offer greatly increased resolution for radar reflectivity data, enabling researchers to locate individual scattering centers on targets, for example. The analysis of radar imagery obtained with beam-sharpened (inverse synthetic aperture radar) or range-sharpened (frequency code) systems can provide detailed signature information.

Techniques for extracting information in these forms from raw video output data are presented in the remainder of this chapter, along with an indication of some of the most common data analysis problems and suggestions for their solutions.

## 6.2 DATA RETRIEVAL AND CALIBRATION

### 6.2.1 Digital vs. Analog Data

Since most data analyses sooner or later require the use of a digital computer, one of the first problems faced by the analyst deals with the transformation necessary to input the data to the computer. Different sets of problems arise depending on whether the radar data are recorded in analog or digital form, and these considerations should be addressed early in the design of the experiment. Direct digital recording usually simplifies the problem of later input to a computer, although sometimes analog-to-digital convertor (ADC) resolution must be sacrificed for speed to record pulse-by-pulse radar data on several channels. Care must also be taken in the design of the header block format for each record to ensure that all necessary descriptive information about the data run is recorded. Some provision should also be made for operator comments to cover unexpected situations. This could be included on the digital tape or in an ancillary written log, depending on the circumstances.

If the data are recorded in analog form, slightly different considerations drive the choice of the ADC. Sample rate is still important, and the rate achievable in practice may be much slower than the ADC maximum sample rate, since the software to control the ADC for the number of channels being sampled may be slower than the sample time of the ADC. At least eight bit resolution is desirable for most radar reflectivity data with nine or ten bits preferable. More resolution is needed if logarithmic data are digitized with the goal of linearizing the data at a later time, due to the nonlinear relationship between voltage and the ADC bits. The signal input to the ADC should cover the voltage range of the ADC as fully as possible. For example, if the ADC covers a ±5 volt input range, the signal on a typical FM analog tape should be amplified from its ± 1.5 volt nominal range using variable gain and offset for each channel to cover the allowed input range to the ADC. This minimizes any errors due to ADC inaccuracy, and provides the full available resolution.

Three channels on an analog recorded data tape are somewhat unique; they are time code, voice log, and synchronization pulse (sync). Using the sync pulse to trigger the ADC samplers ensures that all samples are taken at the same place on the radar pulse, but investigation has shown no practical difference in the statistics of the return when a simple sine or square wave generator is used, set for any frequency between the *prf* and twice the *prf*. A frequency slightly higher (say, 10%) than the *prf* is suggested when a pulse generator is used instead of a sync pulse, to avoid the mischance of always sampling between pulses.

The time code and voice log provide some redundancy in information, but both should always be recorded. Without at least one of these channels, a tape of otherwise valuable radar data becomes useless. With time code, voice log, and a written log, data can accurately and efficiently be identified on a tape. Experience has also shown that often a valuable piece of information mentioned on the voice log will be entirely omitted from the written log.

### 6.2.2 Data Calibration

Table 6-1 presents a list of the quantities needed for accurate radar system calibration. They are not presented in order of importance. Most are not expected to vary over the course of the measurement program, but several (transmitter power, waveguide losses) should be measured daily to ensure that they have not changed.

**TABLE 6-1**
**Quantities Needed for Accurate Calibration**

- Peak transmitted power
- Power level out of signal generator
- Radar frequency
- Pulse repetition frequency (*prf*)
- Pulsewidth and shape
- Antenna gain and antenna pattern
- Miminum detectable signal (MDS) (receiver noise)
- Waveguide losses from antenna to injected signal port
- Range reference for zero range
- Antenna height

As discussed in Chapter 4, two distinct types of calibrations should be recorded. The first is designed to describe the receiver and data acquisition system transfer function. It consists of an injected signal, first at a level high enough to induce receiver saturation, followed by succeeding levels attenuated by some constant amount (usually 5 or 10 dB), down to receiver noise. For a perfect logarithmic receiver, equally spaced attenuation levels should yield equally spaced voltage levels on tape, except for the effects of receiver saturation and noise. A voltage distribution analysis (VDA) performed on the step calibration would yield results similar to those shown in Figure 6-1. For the data reduction system used to produce Figure 6-1, the voltage range from saturation (0 dB) to receiver noise (−60 dB) is divided into 128 bins. The tape is sampled at the radar *prf*, the voltage level of each sample is determined, and the count in the bin

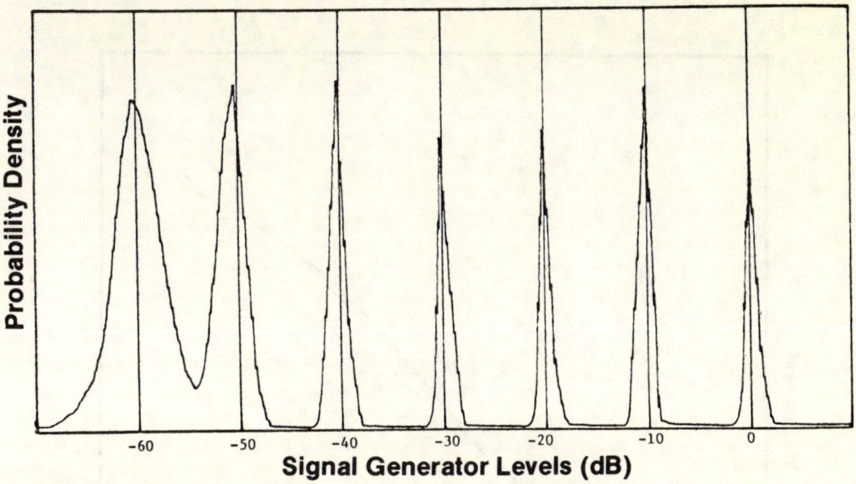


**Figure 6-1.** Typical step calibration density plot.

corresponding to that voltage level is incremented by one. The mode, or most often occurring value, of each level is chosen as the bin corresponding to that level, and bin number is plotted *versus* attenuation level as in Figure 6-2. In this case, for the region between -10 dB and -50 dB, the calibration can adequately be described by a linear fit to the points. If radar data of interest occur above -10 dB or below -50 dB, a cubic fit better describes the effects of noise and saturation. An *rms* error of less than 0.5 dB over the range of signal generator levels where data occur should be the goal. If this accuracy cannot be obtained with a linear fit to the transfer function, it can usually be achieved with a cubic or piece-wise linear fit. Even the cubic fit will not allow accurate calibration of signals above receiver saturation or below receiver noise, so care must be taken to keep the radar data in the appropriate region.

Using the measurement of the absolute power level injected by the signal generator, the measurement of waveguide losses between the antenna and the coupler used for injecting the calibration signal, and the calibration described above, injected power may be related to the received power at the antenna. As an example, suppose that the power out of the signal generator set at 0 dB is measured to be -3.2 dBm; the 20 dB coupler used to inject the signal is measured as -20.2 dB; and losses from the antenna to the coupler are measured as 2.7 dB. Then, if a received signal corresponds to the same output value as a signal generator setting of -36.4 dB, the actual received power at the antenna is - 36.4 - 3.2 - 20.2 + 2.7 = - 57.1 dBm.

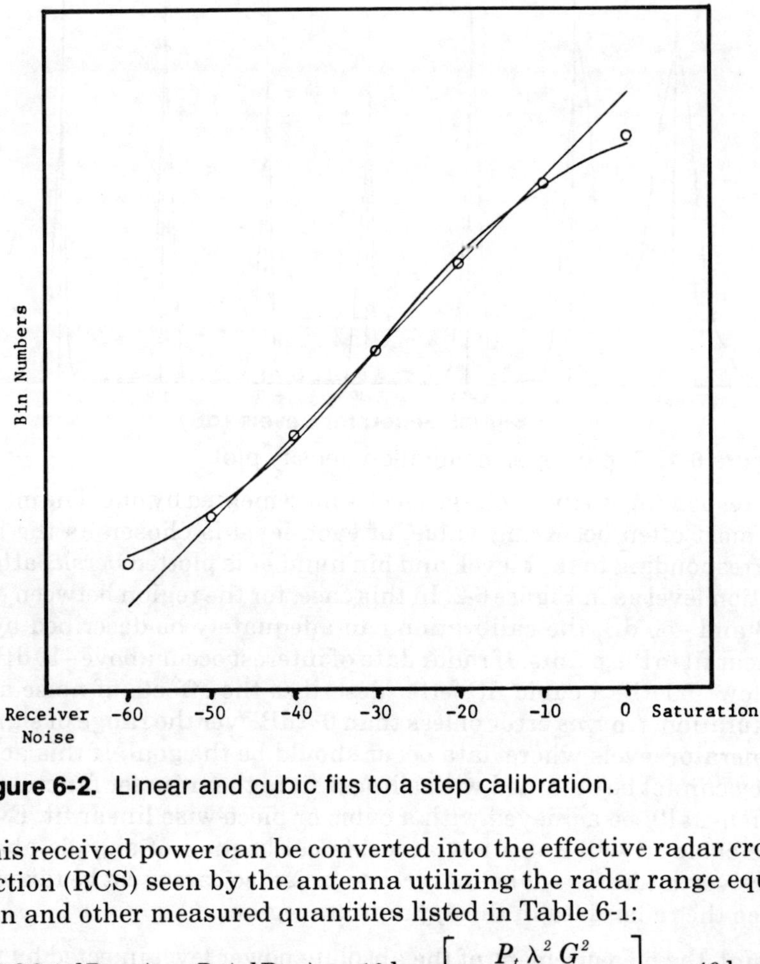


**Figure 6-2.** Linear and cubic fits to a step calibration.

This received power can be converted into the effective radar cross section (RCS) seen by the antenna utilizing the radar range equation and other measured quantities listed in Table 6-1:

$$RCS\,(\,dBsm\,) \;=\; P_r\,(\,dBm\,) \;-\; 10\log\left[\frac{P_t\,\lambda^2\,G^2}{(\,4\,\pi\,)^3}\right] \;+\; 40\log R \tag{6-1}$$

where

$P_r$ = received power in dBm

$P_t$ = peak transmitted power in milliwatts

$\lambda$ = radar wavelength in meters

$G$ = antenna gain

$R$ = range in meters

With this information, the analyst is ready to verify that the system is accurately calibrated by computing the measured RCS of some standard reference target, as discussed in Chapter 4. The actual RCS of the reference target must be known, preferably from measurements (e.g., on a compact range) as well as from theory. A "free space" measurement of the reference target is desirable; if not possible, then the multipath field of the target must be as accurately known as possible. An amplitude distribution of the reference target return is calculated (see next section), and the peak of this distribution, usually the 90% or 99% point, is compared with the known RCS of the target. An absolute calibration run should be made at least once a day, or more often if possible. Figure 6-3 is an example of how even an unfortunate choice of a reference target in a multipath field can still serve the purposes of absolute calibration. The reference target was a corner reflector atop a piling in the Gulf of Mexico.[1] The geometry of the scenario resulted in the top of the piling being

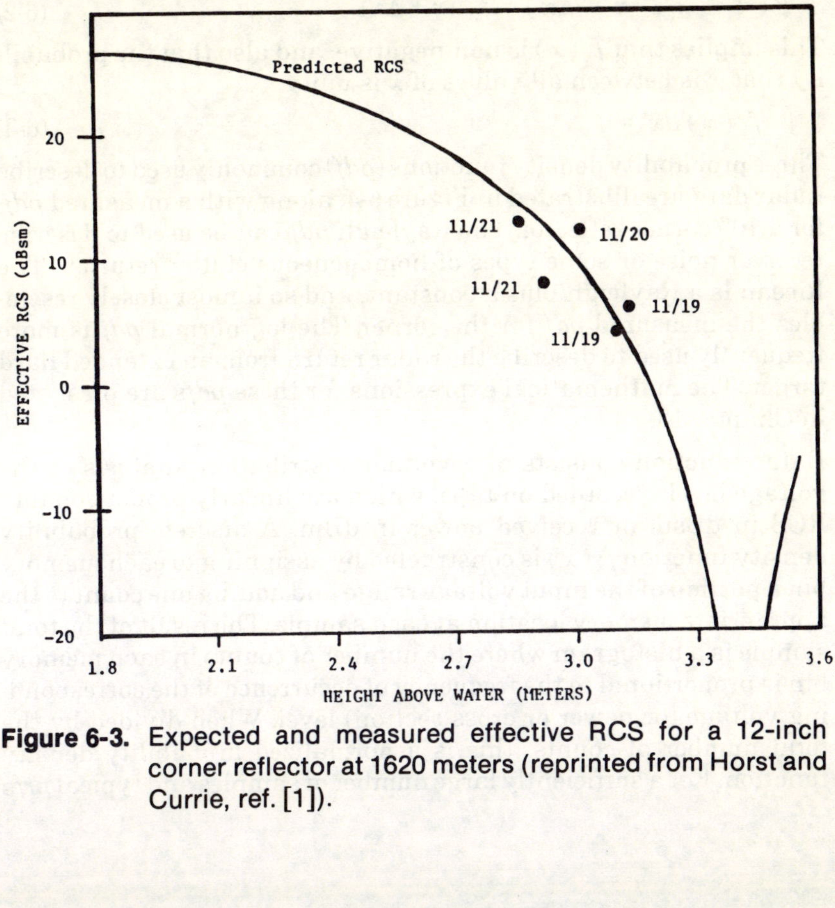


**Figure 6-3.** Expected and measured effective RCS for a 12-inch corner reflector at 1620 meters (reprinted from Horst and Currie, ref. [1]).

very nearly in a null of the multipath field. Even so, with tidal variations taken into account, measured values for the corner showed the calibration to be closed and reliable data were obtained on the targets of interest.

## 6.3 AMPLITUDE ANALYSIS

### 6.3.1 Amplitude Distributions

As is discussed in Chapter 13, radar echoes from targets typically fluctuate in amplitude with time, leading easily and naturally to the treatment of the RCS as a random variable.[2] Since received power and RCS are linearly related through the radar range equation, either could be represented by the random variable $\underline{x}$. The probability that the value of the random variable $\underline{x}$ lies on the interval between $x$ and $x + \Delta x$ is given by the probability density function, $f_x(x)$, times the width of the interval, $\Delta x$:

$$P(x \leq \underline{x} \leq x \times \Delta x) = f_X(x + \Delta x) \tag{6-2}$$

This implies that $f_x(x)$ is non-negative, and also that the probability that $\underline{x}$ is between all values of $x$ is unity:

$$\int_{-\infty}^{\infty} f_x(x)\, dx = 1 \tag{6-3}$$

Three probability density functions (*pdf*) commonly used to describe radar data are illustrated in Figure 6-4, along with a measured *pdf* for a 10″ corner reflector. The Rayleigh *pdf* can be used to describe receiver noise or some types of homogeneous clutter returns. The Ricean is a Rayleigh plus a constant, and so it most closely resembles the measured *pdf* for the corner. The log normal *pdf* is more frequently used to describe the radar return from an extended hard target. The mathematical expressions for these *pdf*s are presented in Chapter 13.

Data reduction consists of a voltage distribution analysis of the voltage levels recorded on tape, which are linearly proportional to RCS in dBsm or received power in dBm. A discrete probability density function $f_x(x)$ is constructed by assigning to each memory bin a portion of the input voltage range and adding one count to the appropriate memory location at each sample. The result of the total sample is a histogram where the number of counts in each memory bin is proportional to the frequency of occurrence of the corresponding voltage (or power or cross section) level. When divided by the total number of counts, this is a normalized probability density function. For a sufficiently large number of samples and typical *prf*s

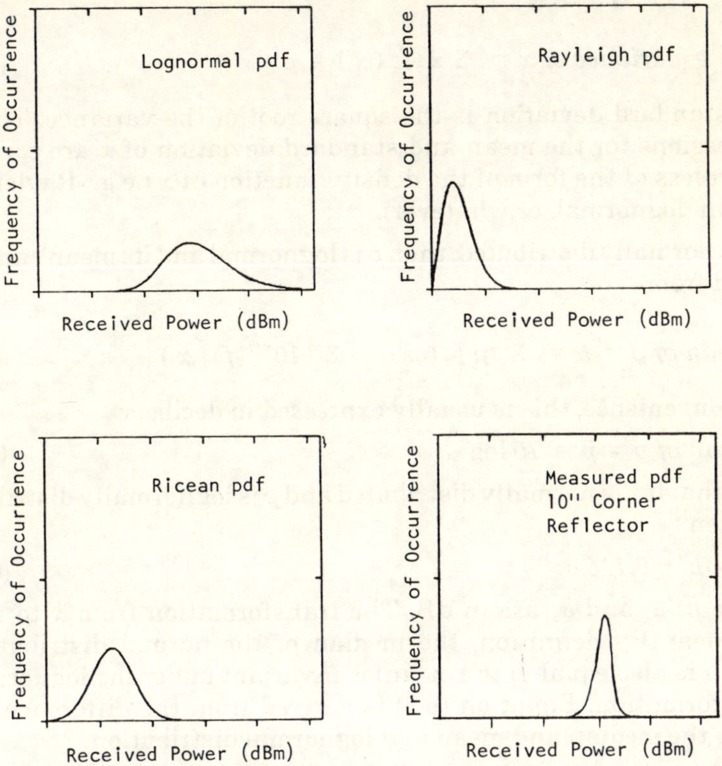


**Figure 6-4.** Probability density functions (*pdf*) commonly used to describe radar return.

in the 1 to 2 kHz range, this implies total sample times of 15 seconds to 2 minutes for most applications.

### 6.3.2 Logarithmic vs. Linear Data

Other quantities which may be useful in describing the radar return can be calculated from the density function. Note that $x$ is either received power in dBm or RCS in dBsm; that is

$$x = 10 \log p \tag{6-4}$$

where $p$ is either received power in milliwatts or RCS in square meters. The mean of $x$ is given by

$$\mu = mean\ of\ x = \sum_{-\infty}^{\infty} x f_x(x) \tag{6-5}$$

and the variance is

$$\sigma_x^{\,2} = variance\ of\ x = \sum_{-\infty}^{\infty} x^2 f_x(x) - \mu^2 \tag{6-6}$$

The standard deviation is the square root of the variance. These expressions for the mean and standard deviation of $x$ are correct regardless of the form of the density function of $x$ ( e.g., Rayleigh, Ricean, lognormal, or whatever).

If $x$ is normally distributed, then $p$ is lognormal and its mean can be found from

$$mean\ of\ p = \alpha = \sum_{-\infty}^{\infty} p f_p(p) = \sum_{-\infty}^{\infty} 10^{x/10} f_x(x) \tag{6-7}$$

For convenience, this is usually expressed in decibels:

$$mean\ of\ p = \overline{p} = 10 \log \alpha \tag{6-8}$$

Note that if $x$ is normally distributed and $p$ is lognormally distributed, then[3]

$$\overline{p} = \mu + 0.115\, \sigma_x^{\,2} \tag{6-9}$$

where $\overline{p}$, $\mu$, and $\sigma_x$ are in dB. The transformation from $x$ to $p$ is nonlinear. By definition, the median of the normal distribution (which is also equal to the mean) is invariant under the lognormal transformation. Equation (6-9) is derived from the difference between the median and mean of a lognormal distribution.

The quantity characterizing the spread of the lognormal distribution, similar to the way that $\sigma_x$ characterizes the spread of the normal distribution, is not the standard deviation of $p$, but rather the standard deviation of the ratio of $p$ to its mean.[4] Let

$$r = \frac{p\ (\text{in } m^2)}{\alpha\ (\text{in } m^2)} \tag{6-10}$$

Then

$$variance\ of\ r = \sigma_r^{\,2} = \sum_{-\infty}^{\infty} (10^{r/10})^2 f_x(x) - \alpha^2 \tag{6-11}$$

$$standard\ deviation\ of\ p = \sigma_p = 10 \log \sigma_r \tag{6-12}$$

A summary of the variables used to denote the different statistical quantities discussed in this section is given in Table 6-2.

The probability distribution function $F(x)$ is defined

$$F(x) = \int_{-\infty}^{x} f(\xi)\, d\xi \tag{6-13}$$

**TABLE 6-2**
**Summary of Statistical Quantities**

| VARIABLE NAME | UNITS | |
|---|---|---|
| $x$ | dBsm | A value in the distribution of RCS in dBsm |
| $\mu$ | dBsm | Mean of RCS distribution in dBsm |
| $\sigma_x$ | dB | Standard deviation of the RCS distribution in dBsm |
| $p$ | $m^2$ | A value in the distribution of RCS in $m^2$ |
| $\alpha$ | $m^2$ | Mean of the RCS distribution in $m^2$ |
| $\overline{p}$ | dBsm | 10 log $\alpha$ |
| $\sigma_r$ | -- | Standard deviation of $r = p/\alpha$ |
| $\sigma_p$ | dB | 10 log $\sigma_r$ |

Obviously, $F(-\infty) = 0$ and $F(+\infty) = 1$. The distribution function is a non-decreasing function of $x$, and is obtained by numerical integration of the density function. In practice, the scale on the amplitude distribution plot is usually chosen as exponential (so that a Rayleigh distribution will appear as a straight line) or normal probability. Figure 6-5 presents a typical amplitude distribution plot on normal probability paper. The curve is nearly linear between 0.001 and 0.9999, indicating that the distribution is normal in dBsm, or lognormal in square meters. Figure 6-6 illustrates an amplitude distribution of receiver noise, which is Rayleigh distributed, plotted on a normal probability scale.

Often, target RCS amplitude data are recorded as a function of target orientation. For example, the target may be placed on a turntable and rotated at a constant speed while the radar measurements are made as a function of target aspect angle. There are several good illustrative ways to present such data. One way is to divide the pulse-to-pulse RCS data into small regions (say, 2° aspect coverage), and plot the mean (in linear or logarithmic space) or median of each 2° region in polar format. Another possibility is to employ a sliding window grouping specified by a window length and slide interval. For instance, in the case of a 100-sample window length and a slide increment of 50 samples, the first group would consist of samples 1-100, the second of samples 51-150, the third of samples 101—200, *et cetera.* These data groups could be analyzed statistically and polar plot displays constructed for the mean (in linear or logarithmic space), median, mode, variance, or standard deviation. Amplitude distributions could be generated and specific

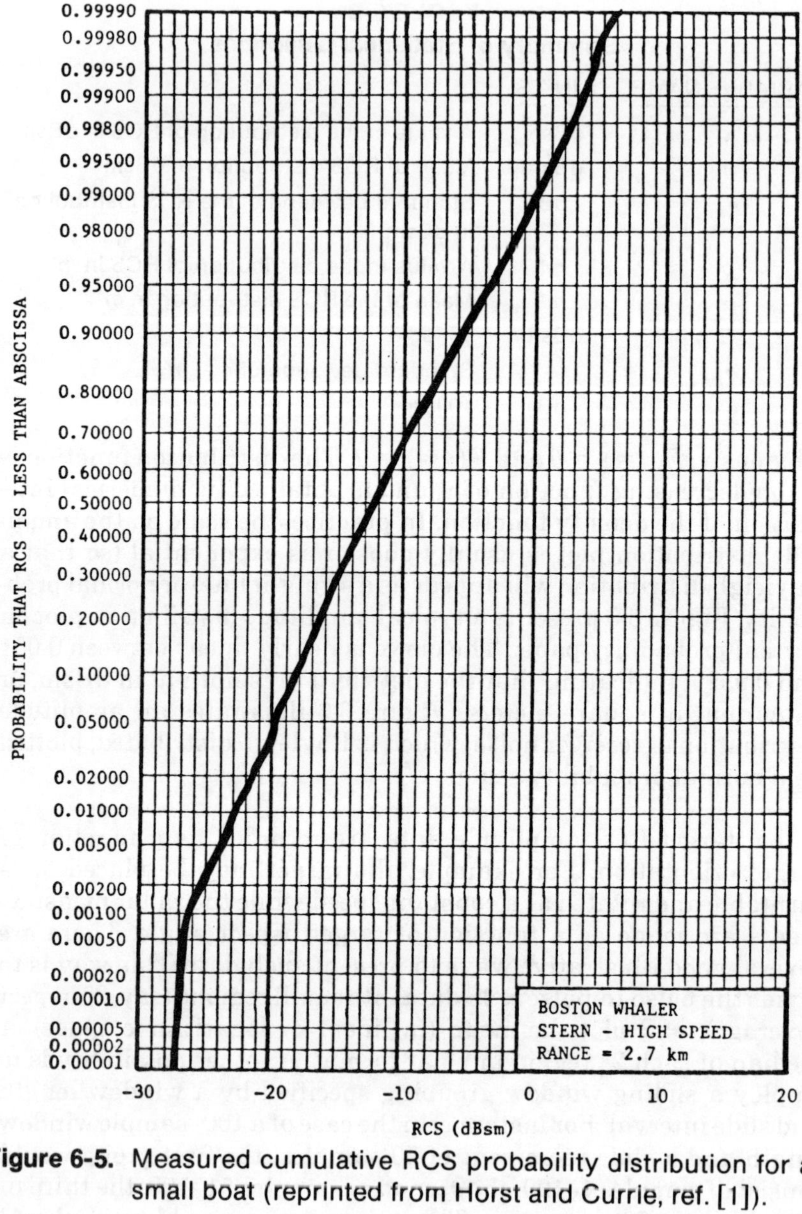


**Figure 6-5.** Measured cumulative RCS probability distribution for a small boat (reprinted from Horst and Currie, ref. [1]).

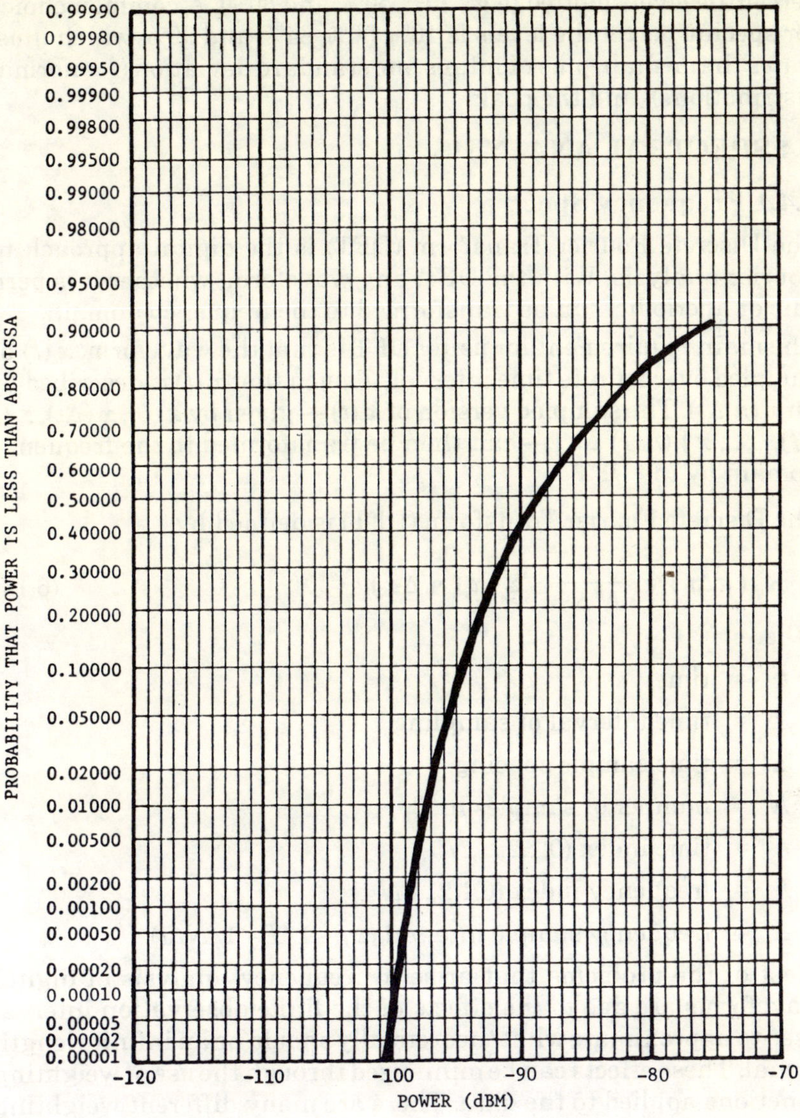


**Figure 6-6.** Cumulative distribution plot for receiver noise power.

percentile levels plotted (e.g., 20%, 50%, and 80%). A common choice for specific percentile levels is 16%, 50%, 84%, and 99%, which illustrates the median, plus or minus one standard deviation (for normal distributions), and the peak.

## 6.4 SPECTRAL ANALYSIS

### 6.4.1 Frequency Spectra

The Discrete Fourier Transform (DFT) is the digital approach to Fourier analysis. With the DFT, any waveform, whether it is periodic or aperiodic, can be transformed to the frequency domain.

The main requirement for using DFT is that the waveform, $x\,(t)$ be sampled over a finite time interval, $T$, such that $x\,(t)$ is described by an $x\,(n\,\Delta t)$. The sampled version of $x\,(t)$ — given by $x\,(0)$, $x\,(\Delta t)$, $x\,(2\,\Delta t)$, . . . , $x\,[\,(n\text{-}1)\,\Delta t\,]$ — can then be transformed to the frequency domain by the DFT.

The Discrete Fourier Transform (DFT) is defined by:

$$X_d\,(k\Delta f) = \frac{1}{N\,\Delta\,t} \sum_{n=0}^{N-1} x\,(n\,\Delta\,t)\,e^{-j2\,kn/N} \tag{6-14}$$

Where:

$t$ = time

$T$ = time interval of sampling

$\Delta t$ = time sample spacing

$N$ = number of samples in $T$

$n$ = time index (0,1,2,...,$N$-1)

$k$ = frequency index (0,1,2,...,$N$-1)

$\Delta f$ = frequency sample interval ($\Delta f = 1/T = 1/N\Delta t$)

Most of the problems that arise in frequency analysis of digital waveforms are due to the effects of the finite observation interval used to approximate what theoretically should be an infinite length signal. These effects can be minimized through the use of weighting functions applied to the data. There are many different weighting functions, called windows, and each has its own special advantages and disadvantages. Table 6-3 provides a summary of commonly used windows and some of their performance characteristics. For radar data in general, the choice has not been found to be critical. A simple window, such as the cosine-squared or Hanning window, has usually been found to provide adequate results. An excellent summary of data windows and their pertinent performance characteristics is given in reference[5].

### 6.4.2 Correlation Properties

The autocorrelation $R(t_1, t_2)$ of a process $\underline{x}(t)$ is the joint moment of the random variables $\underline{x}(t_1)$ and $\underline{x}(t_2)$:

$$R(t_1, t_2) = E[\underline{x}(t_1)\underline{x}(t_2)] \tag{6-15}$$

If the process is stationary, at least weakly, the autocorrelation depends only on the difference $\tau = t_1 - t_2$:

$$R(\tau) = E[\underline{x}(t+\tau)\underline{x}(t)] \tag{6-16}$$

If the process is ergodic, the autocorrelation may be expressed as

$$R(\tau) = \lim_{T\to\infty} \frac{1}{2T}\int_{-T}^{T} \underline{x}(t+\tau)\underline{x}(t)\,dt \tag{6-17}$$

The autocovariance is defined as

$$C(\tau) = E\{[\underline{x}(t+\tau) - \eta][\underline{x}^*(t) - \eta^*]\} = R(\tau) - |\eta|^2 \tag{6-18}$$

where $\eta$ = mean of $x = E[\underline{x}(t)]$. The Fourier transform of the autocorrelation is the power spectrum:

$$S(\omega) = \int_{-\infty}^{\infty} e^{-j\omega\tau} R(\tau)\,d\tau \tag{6-19}$$

In one method of calculating measured autocorrelation or autocovariance, the radar data are sampled at the *prf* rate or higher. The sampled values are calibrated to form a time history $x(t)$ of received power in milliwatts. A Fast Fourier Transform (FFT)[6] is used to transform the time history to the frequency domain. The size of the FFT is usually a power of 2; for $2^{10}$, or 1024 points, sampled at 1 kHz, only one second of data would be considered. To increase the data base for the autocorrelation, time histories of several consecutive data intervals can be constructed, transformed to frequency domain, and their frequency spectra averaged. If the zero frequency (dc) component of the averaged frequency spectrum is set to zero, the means are removed and autocovariance rather than autocorrelation is computed. Note that the dynamic range of the ADC used may require dropping the calibration for the means in favor of increased resolution for the smaller fluctuations around the mean.

The averaged frequency spectra for radar data often indicate that the majority of the energy is concentrated in the low frequencies, usually below 20 or 30 Hz. Sometimes averaged spectra will show

**TABLE 6-3**
**Parameters of Data Windows**

| WINDOW | EQUATION | GAIN | HIGHEST SIDELOBE LEVEL (dB) | SIDELOBE FALL-OFF (dB/OCT) |
|---|---|---|---|---|
| Rectangle | $A = 1$ for $t = 0$ to $T$ | 1.00 | -13 | - 6 |
| Triangle | $A = 2t/T$ for $t = 0$ to $T/2$<br>$A = -2t/T$ for $t = T/2$ to $T$ | 0.50 | -27 | -12 |
| Hanning | $A = 0.5\,[1 - \cos(2\pi t/T)]$ for $t = 0$ to $T$ | 0.50 | -32 | -18 |
| Hamming | $A = 0.08 + 0.46[1 - \cos(2\pi t/T)]$ for $t = 0$ to $T$ | 0.54 | -43 | - 6 |
| Blackman | $A = 0.42 + 0.50 \cos\left(\frac{2\pi t}{T}\right) + 0.08 \cos(4\pi t/T)$ for $t = 0$ to $T$ | 0.42 | -58 | -18 |

| | | | | |
|---|---|---|---|---|
| Minimum 3-Sample Blackman-Harris | $A = a_o - a_1 \cos\left(\frac{2\pi t}{T}\right) + a_2 \cos\left(\frac{4\pi t}{T}\right) - a_3 \cos\left(\frac{6\pi t}{T}\right)$<br>for $t = 0$ to $T$<br>$a_o = 0.42323$<br>$a_1 = 0.49755$<br>$a_2 = 0.07922$<br>$a_3 = 0$ | 0.42 | -67 | - 6 |
| Minimum 4-Sample Blackman-Harris | as above with<br>$a_o = 0.35875$<br>$a_1 = 0.48829$<br>$a_2 = 0.14128$<br>$a_3 = 0.01168$ | 0.36 | -92 | - 6 |

contributions at multiples of the 60 Hz line frequency. Since the source of these extraneous frequencies cannot always be identified and the unwanted frequencies removed on input, it is as effective and more efficient to add a software option to low-pass or selectively bandpass filter the frequency spectrum by zeroing out those components above a specified cutoff frequency or around line frequency multiples. Any energy legitimately belonging to the spectrum at these frequencies is assumed to be an extremely small percentage of the total energy, so that its absence does not appreciably affect the autocorrelation or autocovariance.

Conjugate multiplication is then carried out on the averaged and edited frequency spectrum, and an inverse FFT is performed to carry the autocorrelation back to the time domain. Usually the autocorrelation or autocovariance is then normalized by dividing all terms by the magnitude of the first ($\tau = 0$) term. Figure 6-7 presents a typical normalized autocovariance plot showing differences in the decorrelation of a target and clutter.

Decorrelation time is a useful concept employed to relate the decorrelation statistics of radar return under varying conditions. There is no "standard" definition of decorrelation time; commonly used definitions include the time for the normalized auto covariance to drop to zero, or the time for the normalized auto covariance to drop to $1/e = 0.368$.

Since the autocorrelation and the frequency spectrum are Fourier transform pairs, they contain identical information about the radar signal. However, visualizing the Fourier transform of an arbitrary function is not always easy, so it is sometimes useful to present power spectral densities (PSD) instead of (or as well as) autocorrelations or autocovariances. For example, it is much easier to extract dominant frequencies from a PSD than from an autocorrelation.

## 6.5 SPECIAL WAVEFORMS

### 6.5.1 Coherent Data Analysis

When coherent (amplitude and phase) radar measurements are made, Doppler spectra can be computed for the complex radar return. The quantities needed are the amplitude ($A$) of the return and the in-phase ($I$) and quadrature ($Q$) components of the phase angle or the in-phase and quadrature linear components. These can be recorded as the sine and cosine of the phase angle. The calibrated logarithmic amplitude is antilogged and multiplied by the complex

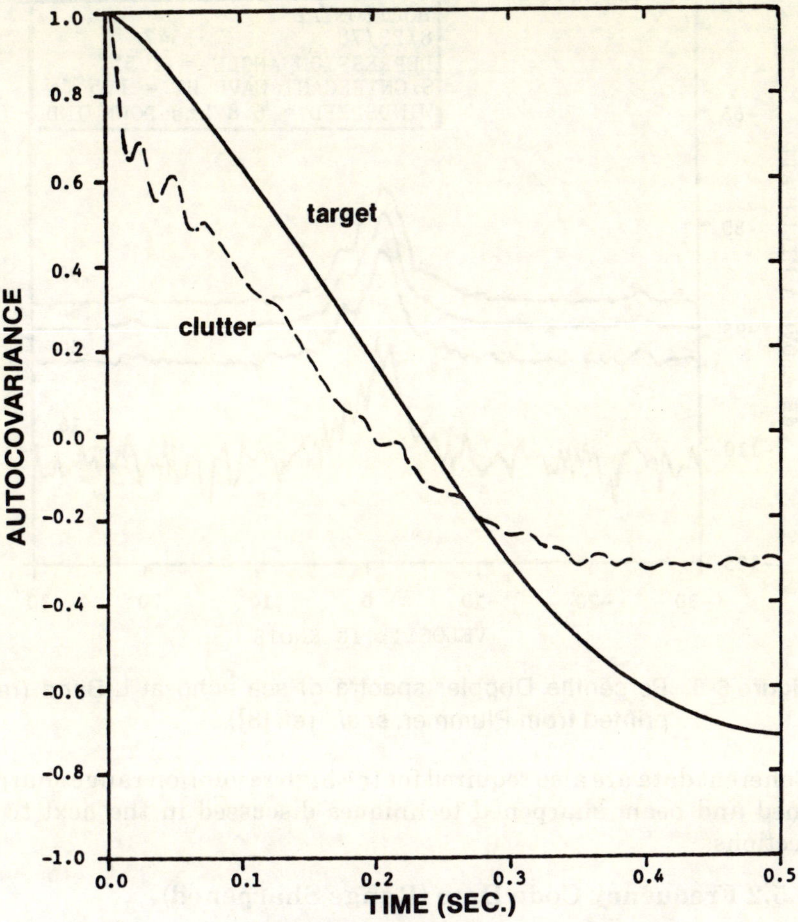


**Figure 6-7.** $K_u$ band autocovariance of a target in sea clutter; 1° depression angle and horizontal polarization.

combination of $I$ and $Q$ channels to yield the complex signal $10^{A/20}$ $(I + jQ)$, which is the output of an ideal linear receiver. Conversely the complex signal can be computed from the linear components to yield $\sqrt{I^2 + Q^2}\,[\tan^{-1}(Q/I)]$. A DFT is performed on the complex points, using an algorithm such as the modified Cooley-Tukey,[7] and the power spectrum is computed. Figure 6-8 illustrates percentile Doppler spectra for sea echo at L-band, where the four curves represent the 16th, 50th, 84th, and 99th percentile levels of the amplitude probability distribution.[8]

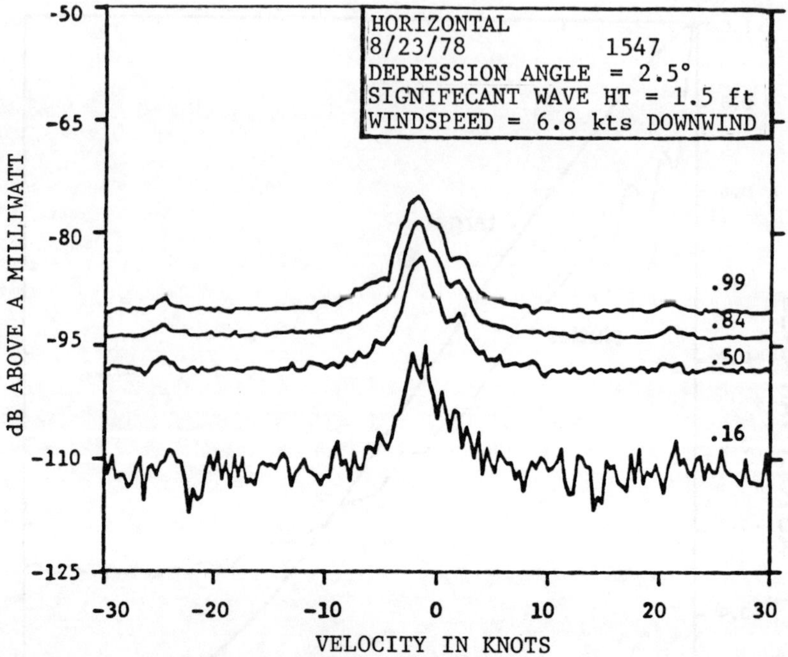


**Figure 6-8.** Percentile Doppler spectra of sea echo at L-Band (reprinted from Plummer, *et al.*, ref. [8]).

Coherent data are also required for the high resolution range sharpened and beam sharpened techniques discussed in the next two sections.

### 6.5.2 Frequency Code Data (Range Sharpened)

High range resolution can be obtained by using very short pulsewidths and correspondingly large bandwidths, but is more easily implemented using a linear frequency modulation in a CW or long pulse system. The coherent return signal is sampled at equal frequency step intervals, windowed as appropriate, and then an inverse Fourier transform is applied to accomplish range-sorting. A good development of the theory is found in reference [9].

Figure 6-9 illustrates an example of the range resolution obtainable with a frequency stepped, range sharpened system. The figure shows a range profile of the return from two 100 $m^2$ corner reflectors separated by a distance of 2 feet. The range profile was generated by illuminating the two reflectors with a 100 *ns*, 35 GHz pulse and stepping the transmitted frequency coherently for 128, 5 MHz steps

(640 MHz total frequency excursion). The complex Fourier transform was then computed from the resulting in-phase and quadrature signals and the amplitude of the Fourier transform plotted. The signal-to-noise ratio of the recorded signal was 8 dB, so approximately 20 dB of compression gain was achieved. (The noise level in Figure 6-9 is approximately 28 dB below the peak signal from the first reflector.)

Since range is directly related to frequency, no relative motion between the radar and target is required for range sorting. To achieve high resolution in the cross range direction (normal to the radar line-of-sight), this relative motion will become necessary.

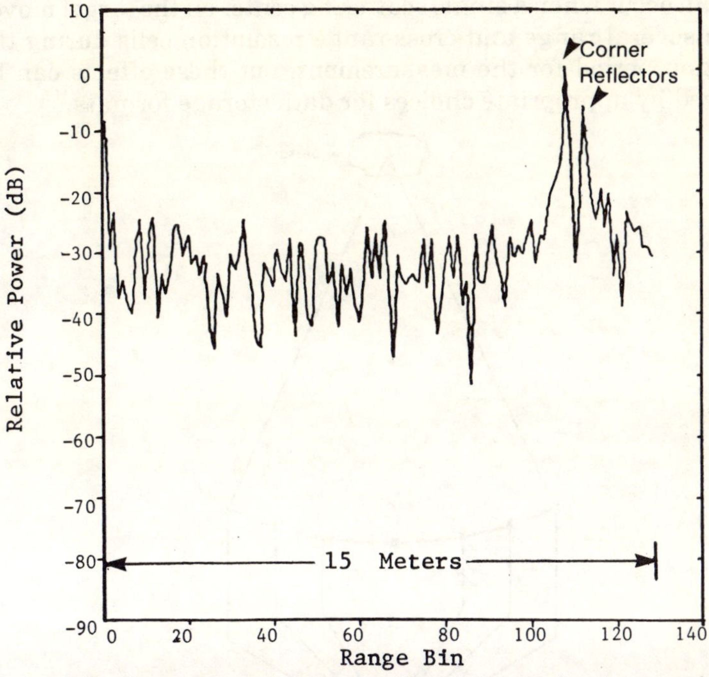


**Figure 6-9.** Compressed range profile of the received power from two 100 $m^2$ corner reflectors (courtesy of the Georgia Institute of Technology).

### 6.5.3 Inverse Synthetic Aperture Radar (ISAR) Data (Beam Sharpened)

High resolution radar reflectivity image data can be obtained by using synthetic aperture ("beam sharpening") techniques. In these techniques, increased resolution is obtained in the cross range dimension by the relative motion between target and radar. This relative motion can arise from the motion of the radar system, as in traditional SAR applications, or it can be due to the motion of the target, as in inverse SAR or ISAR applications. The latter is most commonly used to characterize the reflectivity of targets, such as missiles, small aircraft, and tanks that can be rotated on a turntable to produce the relative motion.

At a given range from the radar, consecutive azimuthal samples of the return are coherently summed, as illustrated in Figure 6-10. Most of the data processing later employed deals with the geometrical transformations to construct undistorted linear images from data acquired along circular arcs of constant range. Smearing of the image can occur when a single scattering center on the target moves through several range and cross-range resolution cells during the integration period for the measurement, but these effects can be minimized by appropriate choices for data storage formats.[10]

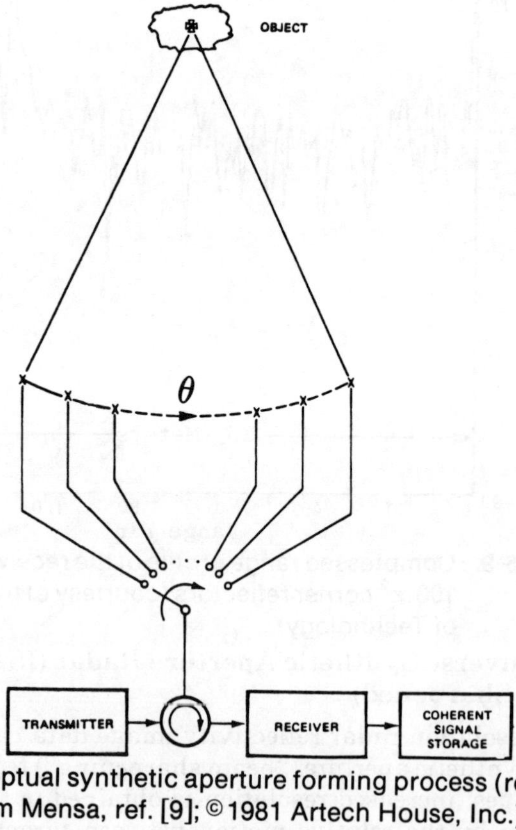


**Figure 6-10.** Conceptual synthetic aperture forming process (reprinted from Mensa, ref. [9]; © 1981 Artech House, Inc.).

The ISAR image is constructed by applying an appropriate window to the data (*see* reference [5]) and performing a DFT. The process is repeated for each range resolution cell and the results stored in a two-dimensional (range *versus* cross range) array. The results can be displayed as waterfall-type plots, or as picture-type images, where the brightness of each picture element (pixel) corresponds to the RCS level of that range-cross range position.

## 6.6 TYPICAL PROBLEMS

The single biggest problem faced by the analyst is usually insoluble after the measurements are over. It is, of course, the absence of a vital piece of information necessary for calibration. The best way to avoid such a disaster is to close the calibration manually in the field, at least roughly. An error of a dB or two can be straightened out at home later, or even accepted gracefully, but a corner that measures 20 dB above its true value can invalidate a complete data set. If the calibration does not close in the field, the first step is to verify all the measured quantities needed. Measure the peak transmitter power, waveguide losses, power out of the signal generator, radar operating frequency, and antenna gain if possible. If the target is in a multipath field, are the field structure and the target's position accurately known? Is the range to the target accurately pointed at the radar? If calibration is difficult in one part of the multipath field, try repositioning the target. Is the target accurately known? Note that most of the remedies for a calibration that does not close are impossible once the field tests are over, again underscoring the importance of closing the calibration in the field.

Most errors in measurements will tend to lower the measured value for the standard target. For example, if peak transmitted power should change, it is more likely to decrease than to increase. Similarly, waveguide losses are more likely to increase, antenna gain to decrease, *et cetera.* Therefore, a measured value much larger than the theoretical is potentially more serious than an error on the low side.

If worst comes to worst and the absolute calibration cannot be reconciled, all may not be lost. At least relative measurements can be reported, e.g., upwind/downwind ratios of sea clutter cross section. Some statistics, such as decorrelation time or Doppler velocity, do not depend on absolute amplitude calibration and do provide valuable data on the processes contributing to radar return.

One of the principal uses of published reflectivity data is to provide inputs to radar system models for performance evaluation. Usually, if a number of similar data sets are available, then these sets are compared to provide assurance that the data used in a model are realistic and correct. Unfortunately, all too often, data sets taken under supposedly similar conditions appear to be quite different. This does not necessarily mean that the data are invalid since there can be a number of reasons why the data may appear to be different. These include the following:

1. Two data sets may be expressed in terms of similar but different parameters. For example, the reflectivity of land may be expressed in terms of $\sigma^\circ$ (radar cross section per unit area of the radar beam projection on the ground) or in terms of $\Gamma$ (radar cross section per unit area of the perpendicular beam area) where $\Gamma = \sigma^\circ \sin E$ and $E$ is the depression angle. A large body of early data is expressed in terms of $\Gamma$, while recently, $\sigma^\circ$ has been more popular.
2. Two sets may use the same parameter, such as $\sigma^\circ$, but may define it differently. For instance, some experimenters use the median value of radar cross section of clutter to calculate $\sigma^\circ$ rather than the mean value. This can result in values of $\sigma^\circ$ being several dB lower for some types of clutter than when the mean values are used. Also, often one way or two way beamwidths are used indiscriminately to calculate the radar beam area (*see* Chapter 2).
3. If two data sets are measured with radar systems that are significantly different in characteristics, then the results can be quite different. For example, the effective radar spot size (beamwidth times pulse length) has a strong influence on both the average values and statistics which will be measured for clutter. Large spot sizes tend to yield Rayleigh statistics while small spot sizes tend to result in lognormal statistics.
4. Differences in geometry between two measured data sets can result in reflectivity differences. For example, for sea clutter the alignment of the radar look direction with the wind/wave direction strongly affects the backscatter.[11] This effect is seen also in alignment of radar look direction with crop row directions.
5. Environmental differences can affect measured data. This tends to be more important at millimeter wave frequencies.

For example, the reflectivity of snow-covered ground at 35 GHz can vary 15 dB over a few hours of time when the snow surface freezes or melts.[12] Also, surface moisture on foliage and on the ground has been shown to affect the reflectivity.

6. Finally, there is always the possibility that a data set was not measured carefully. This can be due to a number of reasons including: problems with the measurement system, improper calibration, or incorrect data reduction.

The following criteria should be helpful in assessing data set validity.

1. The measurement radar system parameters should be described accurately.
2. The data parameters should be accurately defined so that they can be related to other parameters.
3. The method of calibration should be described and any multipath effects should be discussed.
4. The experimental geometry should be described, and the experiment should last over a long enough period so that temporal effects are averaged out. (Beware of data which are all taken within a few minutes on one day.)
5. Whenever it is likely to be a factor (almost always, above 35 GHz), adequate ground truth measurements should be performed and correlated with the reflectivity data.
6. Data more than 10 years old should be used with care as they may not be valid. This is not because experimenters were less careful in the past, but because measurement and analysis tools have taken a giant leap in performance in the last few years. In some of the early experiments, experimenters were measuring the characteristics of their measurement systems rather than the environment, through no fault of their own.
7. Finally, it is only common sense to compare several similar data sets when they are available, to generate confidence in their validity. If the preceding criteria have been met then it should be possible to relate and compare different data sets successfully.

## REFERENCES

1. Horst, M. M., and N. C. Currie, "RCS Measurements of WIDS Targets," Memorandum Report No. 1 on Contract N00612-79-

C-8004, HR-13, Georgia Institute of Technology, Atlanta, Georgia, January 1980.

2. Papoulis, A., *Probability, Random Variables, and Stochastic Processes,* McGraw-Hill Book Company, New York, 1965.
3. Long, M. W., *Radar Reflectivity of Land and Sea,* Artech House, Dedham, Massachusetts, 1983.
4. Atchinson, J., and J. A. C. Brown, *The Log-Normal Distribution,* Cambridge University Press, London, 1969.
5. Harris, F. J., "On the Use of Windows for Harmonic Analysis with the Discrete Fourier Transform," *Proc. IEEE*, Vol. 66, No. 1, January 1978, pp. 51-83.
6. Bringham, E. O., *The FFT*, John Wiley and Sons, New York, 1975.
7. Cooley, J. W., and J. W. Tukey, "An Algorithm for the Machine Calculation of Complex Fourier Series," *Mathematics of Computation,* April 1965.
8. Plummer, D. K., *et al.*, "Some Measured Statistics of Coherent Radar Sea Echo and Doppler," Final Report on Contract N62269-78-C-0328, Georgia Institute of Technology, Atlanta, Georgia, December 1979.
9. Mensa, D. L., *High Resolution Radar Imaging*, Artech House, Dedham, Massachusetts, 1981.
10. Walker, J. L., "Range-Doppler Imaging of Rotating Objects," *IEEE Transactions on Aerospace and Electronic Systems*, Vol. AES-16, No. 1, January 1980, pp. 23-52.
11. Currie, N. C., F. B. Dyer, and R. D. Hayes, "Analysis of Radar Rain Return at Frequencies of 9.375, 35, 70, and 95 GHz," Technical Report No. 2 on Contract DAAA25-73-C-0256, Georgia Institute of Technology, Atlanta, Georgia, February 1975.
12. Currie, N. C., F. B. Dyer, and G. W. Ewell, " Characteristics of Snow at Millimeter Wavelengths," *IEEE AP-S International Symposium Digest,* October 1976.

# Chapter 7

# BISTATIC RADAR CROSS SECTION MEASUREMENTS

G. W. Ewell
Engineering Experiment Station
Georgia Institute of Technology

## 7.1 OVERVIEW

While many radar systems utilize the same antenna for transmission and for reception, this is not the only configuration which can be utilized. Separate antennas may be used for transmission and reception, and such a radar system is termed a "bistatic radar." Bistatic radar may be grouped into two general types: the narrow-angle bistatic systems where the transmit and receive antennas are immediately adjacent to each other, and the wide-angle bistatic systems where the transmitting and receiving antennas are separated by a substantial distance. A simplified sketch of the bistatic geometry is given in Figure 7-1, which shows the bistatic angle and separation of the transmitting and receiving antenna.

As in the case for monostatic radar systems, when discussing bistatic radar system performance, some measure of the scattering strength of targets and clutter is a necessary item of information. The measure used for the bistatic radar case is the bistatic radar cross section defined as[1]

$$\sigma = 4\pi R_o^2 \lim_{R_o \to \infty} \left[ \frac{\overline{H}_R^2}{\overline{H}_o^2} \right] \qquad (7\text{-}1)$$

where

$\sigma$ = radar cross section ($m^2$)

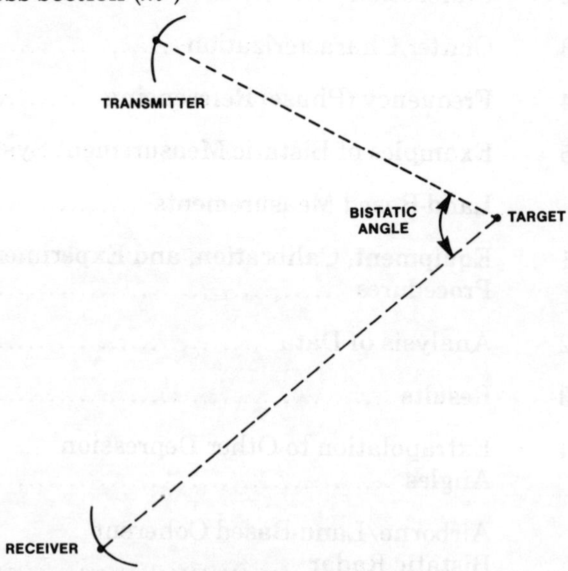


**Figure 7-1.** Simplified bistatic radar geometry.

$\overline{H}_o$ = magnetic vector of incident plane wave

$\overline{H}_R$ = magnetic vector of scattered wave at observer

$R_o$ = range from target to observer

In the bistatic case, the radiation source and the receiver are at different locations, and the cross section is not only a function of the target orientation and frequency, but of the bistatic angle between the transmitter and receiver as well.

In general, monostatic and bistatic radar cross sections are not equal. It can be shown[1] that for small bistatic angles, the bistatic RCS is closely approximated by the monostatic RCS measured on the bisector of the bistatic angle, $\beta$, and measured at a frequency lower than the frequency of interest by a factor

$$\frac{F_m}{F_i} = \cos \beta/2 \tag{7-2}$$

where

$F_i$ = frequency of interest

$F_m$ = measurement frequency

$\beta$ = bisector of bistatic angle

Unfortunately, for larger bistatic angles such a simple equivalence is no longer valid. The bistatic radar cross section may be calculated for simple geometric shapes, and symmetry properties may be invoked[2] for symmetrical bodies. For more complex bodies and for clutter a measurement program is often required if meaningful bistatic cross section data are to be obtained.

## 7.2 THE BISTATIC RADAR EQUATION

A bistatic radar equation can be developed using procedures which are analogous to those used to develop the conventional monostatic radar equation. One form of this bistatic radar equation is given by[3]

$$P_r = \frac{P_t G_t G_r \lambda^2 \sigma}{(4\pi)^3 R_1^2 R_2^2 L} \tag{7-3}$$

where

$P_r$ = received power

$P_t$ = transmitted power

$G_t$ = transmitting antenna gain

$G_r$ = receiving antenna gain

$\lambda$ = wavelength

$\sigma$ = bistatic radar cross section

$R_1$ = range from transmitter to target

$R_2$ = range from target to receiver

$L$ = system losses

Alternate forms of the radar equation, which separately account for propagation effects, are sometimes employed. They are discussed later in this chapter.

## 7.3 BISTATIC CLUTTER GEOMETRY

Bistatic measurements may be carried out for either CW or pulsed cases. For the CW case where one of the terminals is in motion, range resolution may be obtained by Doppler filtering which extracts data from a range element given by the intersection of the hyperbolas of constant Doppler frequency return and the antenna pattern on the earth's surface; one such experiment is described by Pidgeon.[4] Unfortunately, for a pulsed bistatic radar system the identification of scattering elements, which contribute to the return from distributed targets and clutter, is not as straightforward as is the case for a monostatic system. Usually, a means of synchronization and extraction of information which arrives with a constant delay between the transmitter and receiving site is used. In a bistatic system, however, lines of constant delay become ellipses as shown in Figure 7-2. The power from clutter arriving at the receiver becomes the sum of all elements which contribute to the received power at the radar at a given time. The power contributed by each element is a function of the position, and the illumination of that element by the transmitting antenna, the transmitted pulse, and the receiving antenna gain as shown in Figure 7-3. The concept of $\sigma°$ can be utilized in such geometry, but needs to be applied with care. In the event that the transmitted pulse is much longer than the area bounded by the intersection of the transmitting and receiving beams, one can associate an effective area with the geometrical intersection of the transmit and receive antenna beams (Figure 7-3). However, if the portion of the surface effectively illuminated by the transmitted pulse is smaller than the intersection of the beams, then more complex integration will be required to define the effectively illuminated clutter area, and to permit calculation of a meaningful value of $\sigma°$.

In general, for the case of distributed clutter characterized by its radar cross section per unit area, one can express the received power as

$$P_r = \frac{P_t \lambda^2 \sigma^\circ}{(4\pi)^3 L} \, A \int \frac{G_t G_r \, dA}{R_1^2 R_2^2} \tag{7-4}$$

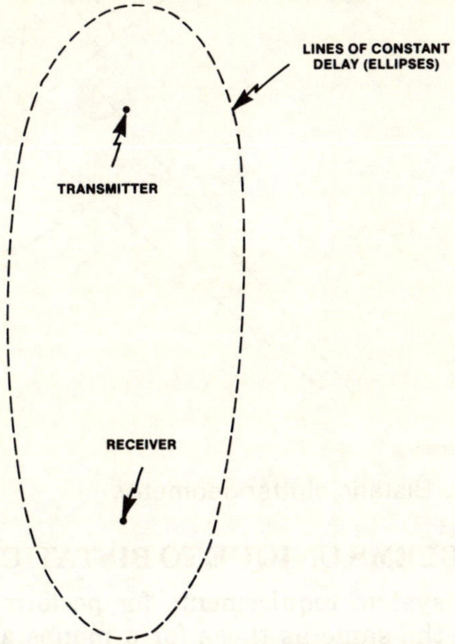


**Figure 7-2.** Lines of constant delay.

where the integration area $A$ is defined by the region over which both antenna patterns have appreciable response, and by the ellipses of constant range delay which bound the transmitted pulse. If desired, similar expressions explicitly calling out the propagation factor may be developed. In general, an analytic evaluation of the integral is required, but numerical methods are also often required. It should be emphasized that for relatively narrow beams, clutter is present at the receiver for only a small fraction of the time between transmitted pulses, corresponding to the simultaneous illumination of the intersecting antenna beams by the transmitted pulse (Figure 7-4). If additional receiving antennas and receivers are employed, or if the receiving antenna is scanned rapidly, returns from a number of different clutter or target areas may be monitored simultaneously.

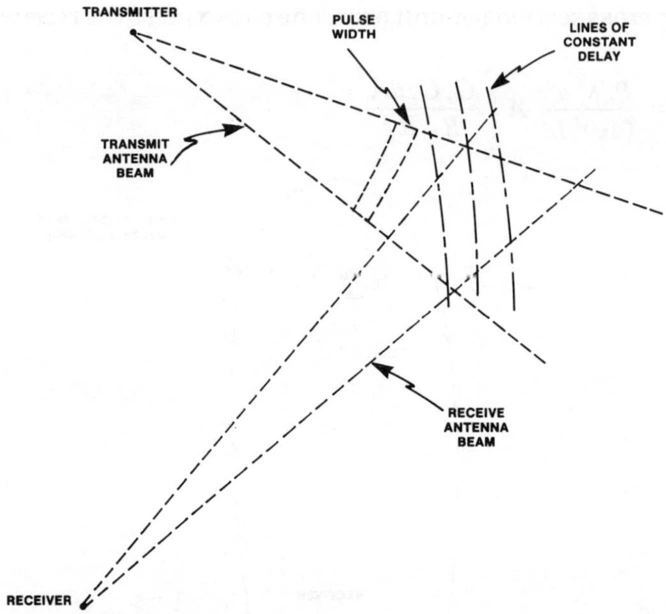


**Figure 7-3.** Bistatic clutter geometry.

## 7.4 PROBLEMS UNIQUE TO BISTATIC MEASUREMENTS

The basic system requirements for performing bistatic measurements are the same as those for a monostatic measurement program. The target returns must be extracted, the desired information must be removed from these signals, and the data must be adequately calibrated. Unfortunately, the separation of the transmitting and receiving sites in a bistatic system introduces additional complexities not present in a conventional monostatic radar cross section measurement program. The more complex geometries involved make a careful description of the target doubly important.

### 7.4.1 Synchronization

To sample or identify returns from the areas of interest, a time synchronization must be established between the transmitting and receiving site. Such synchronization may be established in a number of ways, perhaps the simplest being the transmission of a portion of the radiated pulse directly toward the receiving site, using a high gain and high directivity antenna. For widely separated sites, and for appropriate bistatic angles, such a signal may provide an excellent method for synchronizing a data acquisition system at

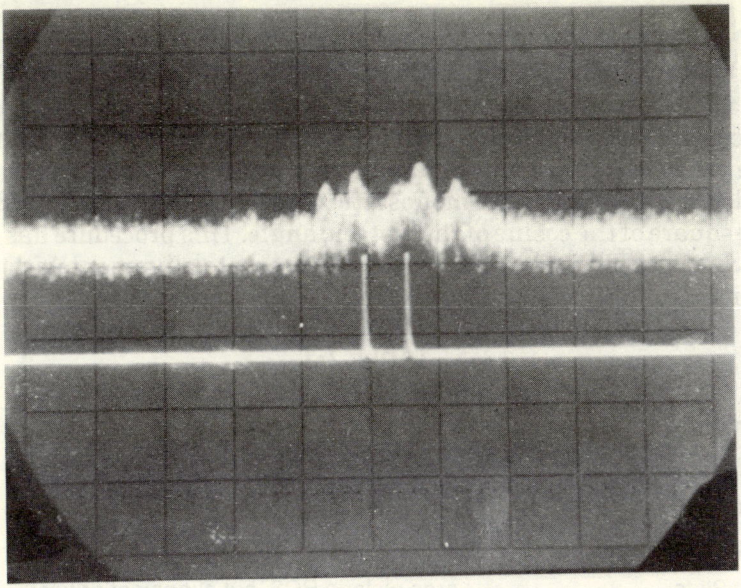

**Figure 7-4.** A-scope representation of the sea clutter return in a pulsed bistatic radar showing the relatively small period of time the two narrow antenna beams and the transmitted pulse coincide. Horizontal scale in 0.4$\mu s$/cm and the start of the sweep is delayed 22$\mu s$ from the synchronization pulse. The two markers indicate the location of sampling gates.

the receiving site. However, there are other cases for which this procedure may not be suitable, and in such instances the use of pretriggers for the transmission of synchronization information by RF link, land lines, or transmission in a different frequency band may be necessary.

### 7.4.2 Calibration

Relative calibration of a bistatic system may be readily accomplished by the injection of standard test signals in the conventional manner. Indirect calibration may be carried out using measured radar parameters as in the monostatic case. Absolute calibration, on the other hand, presents a different problem since a number of standard targets have either a bistatic cross section which may be difficult to calculate, or extremely small bistatic RCS, making measurements at extended ranges difficult.

Table 7-1 lists a few standard targets whose bistatic cross section is both relatively large and easily calculable. A dihedral corner reflector with its seam horizontally oriented is particularly useful. It can be continuously rotated to alleviate alignment problems; the peak of the received signal as the diplane is rotated provides a signal whose radar cross section is the monostatic radar cross section reduced by the square of the cosine of the bistatic angle. This procedure has the advantage that it does not require an accurate alignment with the transmitter or receiver.

**TABLE 7-1**
**Some Possible Calibration Targets For Bistatic Radars**

| STANDARD TARGET | COMMENTS |
|---|---|
| Sphere | Already small RCS |
| Cylinder | Gets smaller (see above) |
| Trihedral Corner | Highly retro-reflective |
| Flat Plate | Large bistatic RCS, but alignment critical |
| Dihedral | Alignment critical; may be rotated with seam horizontal to ensure alignment |

### 7.4.3 Clutter Characterization

As mentioned earlier, identification of a resolution cell in the bistatic case is more complex than the monostatic case, requiring knowledge of the time delay and the two antenna positions. The illuminated clutter area is a function of the specific geometry, and numerical integration must often be used to determine the illuminated area. Since the illuminated area changes, and may be difficult to physically visualize, care must be taken when defining the observation area to ensure an accurate description of its physical composition. Once such precautions have been taken, clutter characterization may proceed as with a monostatic system.

### 7.4.4 Frequency (Phase) Referencing

If coherent measurements of the bistatic cross section are required, a phase reference must be provided at the receiver. As was the case for a noncoherent system, a reference signal in pulsed form may be transmitted from the transmitter to the receiver, and the received pulse can be used to phase lock a coherent oscillator (as in a coherent-on-receive MTI system). If available at the transmitter, a CW coherent local oscillator signal may be transmitted to the receiving site. If such transmissions are undesirable, a CW phase reference at the transmitter can be heterodyned in frequency, with the two frequencies transmitted to the receiving site and mixed to form the original frequency. While this approach will provide the desired phase reference to phase detect the received signals, such an approach is awkward, to say the least. Using suitable atomic frequency standards instead of coherent reference signals has been suggested, but such an approach has never been tried to the knowledge of the author.

### 7.4.5 Examples of Bistatic Measurement Systems

Data describing the bistatic backscatter from ground targets and clutter are not commonly encountered in the available radar literature, and descriptions of measurement systems are rarer still. One bistatic radar measurement system assembled to measure simultaneous monostatic radar cross section of ship targets and sea clutter will illustrate a number of features any such bistatic radar cross section measurement system must possess.[6,7] Portions of the following section are an adaptation of a paper given at the RADAR 82 Conference,[7] and are reproduced with the permission of the Institution of Electrical Engineers.

## 7.5 LAND-BASED MEASUREMENTS

The recent interest in bistatic radar, multi-static radar, and electronic support measure (ESM) systems has led to a need for information concerning the bistatic reflectivity of clutter viewed by such systems. Unfortunately, little information is available describing land or sea clutter for a bistatic radar system,[4] particularly when operating near grazing incidence. Hence, the development of realistic theoretical models describing bistatic reflectivity has been severely handicapped by the lack of experimental data. This section discusses a set of experimental data describing simultaneous bistatic radar cross section (BRCS) and monostatic radar cross section (MRCS) measurements of sea clutter taken with a moderately high resolution X-band radar system operating near grazing incidence for horizontal and vertical polarizations over a range of bistatic angles from 23 to 85 degrees.

### 7.5.1 Equipment, Calibration, and Experimental Procedures

The transmitter and monostatic receiver for this experiment was the X-band radar located at the Georgia Tech Field Site, Boca Raton, Florida. The bistatic receiver was located 3.5 km (1.9 mi) north of the Boca Raton inlet. Major system parameters are summarized in Table 7-2. Figure 7-5 is a photograph of the receiving site, and Figure

**TABLE 7-2**
**Characteristics of Land-Based Monostatic and Bistatic Radar Systems**

| PARAMETER | MONOSTATIC SYSTEM | BISTATIC RECEIVER |
|---|---|---|
| Frequency | 9.38 GHz | 9.38 GHz |
| Peak Power | 250 kW | — |
| Pulsewidth | 0.2*μs* | — |
| *prf* | 1000 *pps* | — |
| Antenna | Cut Paraboloid (1° Az. x 3° El.) | Paraboloid (1.5° x 1.5°) |
| Polarizations Used | *HH* or *VV* | *HH* or *VV* |
| Antenna Gain | 38 dB | 37 dB |
| IF Center Frequency | 60 MHz | 60 MHz |
| IF Bandwidth | 5 MHz | 20 MHz |
| IF Response | Logarithmic | Logarithmic |
| Dynamic Range | 80 dB | 75 dB |
| Antenna Height | 75 ft (MSL) | 27 ft (MSL) |
| Site Separation | — | 1.9 nmi |

7-6 is a simplified block diagram of the equipment. A transmitted pulse sample was radiated through a separate antenna toward the receiving site. This signal, received through a dedicated antenna (the small dish visible in Figure 7-5) and receiver, was used to establish a timing reference between the sites. Range-gated sample-and-hold circuits were used to sample the monostatic (at the transmitter site) and the bistatic (at the receiver site) logarithmic video return signals. The transmitting antenna was normally fixed at a constant bearing of 90 or 150 degrees, while the directions of the receiving antenna and range delay were varied to change the bistatic angle. Bistatic data were taken with the range gate sampling the peak of the signal formed by the intersection of the transmitting and receiving antenna beams. Monostatic data were acquired using a sampler positioned to sample the monostatic return corresponding to this intersection point. The resulting stretched signals, along with voice annotation and time code, were recorded on magnetic tape, using instrumentation FM tape recorders.

Calibration of recorded data was accomplished using several techniques. Known RF test signals were injected into the receiver front ends and varied in known increments over the dynamic ranges of the receivers while the resultant video signal was sampled and

**Figure 7-5.** View of the bistatic receiving site located approximately 3.5 km south of the transmitting site. The large dish is the target receiving dish while the smaller dish visible below was used to receive the synchronization pulse.

recorded on tape. This information was used to establish the transfer function of the receiver, data acquisition, and recording system at each site. Next, the received power from standard targets was used to establish levels corresponding to signals received from targets of known radar cross section. A rotating diplane (dihedral corner reflector) was used to calibrate both the monostatic and the bistatic systems, and a trihedral corner reflector was used as an additional aid in calibrating the monostatic system. The rotating diplane was oriented such that the seam was horizontal and the axis of rotation was vertical. This permitted observation of the specular cross section, which is approximately the monostatic cross section reduced by the square of the cosine of half the bistatic angle. The reduction from monostatic cross section was small, since the bistatic angle was kept small for the calibration data runs.

The next step in calibration involved using the radar equation to determine expected values of received power from these standard targets. A useful form of the bistatic radar equation is

$$P_r = \frac{P_t G_t G_r \sigma \lambda^2 F_t^2 F_r^2}{(4\pi)^3 R_1^2 R_2^2} \tag{7-5}$$

where

$P_r$ = received power

$P_t$ = transmitted power

$G_t$ = transmitting antenna gain

$G_r$ = receiving antenna gain

$\sigma$ = target radar cross section

$\lambda$ = wavelength

$F_t$ = propagation factor for transmitting antenna-to-target path

$F_r$ = propagation factor for receiving antenna-to-target path

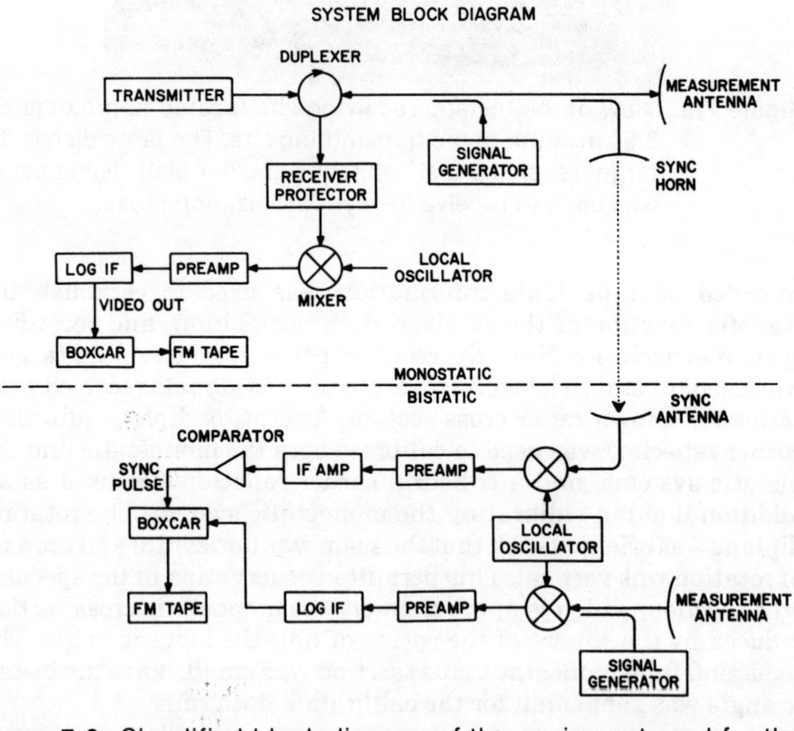


**Figure 7-6.** Simplified block diagram of the equipment used for the measurement of bistatic radar cross sections of ship targets and clutter.

$R_1$ = range from transmitter to target

$R_2$ = range from target to receiver

The propagation factors $F$ are included to account for the effects of multipath returns. When the angle between the direct and reflected rays is small, the function $F$ may be written as

$$F = 1 + \rho D \exp(-j\alpha) \tag{7-6}$$

where

$\rho$ = reflection coefficient of the ocean surface

$D$ = divergence factor to account for the curved reflecting earth

$\alpha$ = phase difference between the direct and reflected waves at the target

In the monostatic case $R_1 = R_2 = R$ and $F_t = F_r = F$, and Equation (7-5) reduces to the familiar monostatic radar equation:

$$P_r = \frac{P_t G_t G_r \sigma \lambda^2 F^4}{(4\pi)^3 R^4} \tag{7-7}$$

Comparisons of measured and calculated received powers from the standard targets were used as measures of the quality of the radar cross section data. The expected value of received power was calculated using Equations (7-5) and (7-7), including the propagation factor given in Equation (7-6). Radar cross sections of the standard reflectors were calculated using measured physical dimensions; these radar cross sections were used with measured values of gain, transmitted power, and range to calculate expected values of received power.

The corner reflectors were elevated at a height which would place them near the peak of the interference lobe. Since accurate alignment of a corner reflector may be difficult from a small boat, values of received power equal to the largest value received (the value corresponding to the 0.99 point on the cumulative probability distribution) were associated with the received power from the reflectors. This procedure was justified due to the expectation that the +18 dBsm corner reflector would dominate the return from the small boat supporting the corner reflector (this was verified experimentally); then the maximum return would occur when the reflector was aligned with the receiver and when the reflector was also at the peak of the interference lobe.

Comparisons of the calculated and measured returns from numerous calibration runs conducted throughout the experiment showed excellent agreement between measured and calculated values; worst-case errors were less than ±2 dB, with ±1 dB differences being more typical. This excellent agreement between measured and calculated received powers from standard targets gives a high degree of confidence in data bracketed by such calibrations. (*See* Chapters 3 and 4 for a more detailed discussion of calibration techniques.)

### 7.5.2 Analysis of Data

Data describing monostatic and bistatic clutter returns were returned to Georgia Tech and used, along with calibration information, to generate distributions of received power at each of the receivers.[5] Since the multipath propagation factor is difficult to interpret in the case of distributed clutter returns, and since multipath effects are normally included in values of normalized clutter return, the free space radar equations

$$P_r = \frac{P_t G^2 \lambda^2 \sigma}{(4\pi)^3 R_1^4} \ [monostatic] \tag{7-8}$$

and

$$P_r = \frac{P_t G_t G_r \lambda^2 \sigma}{(4\pi)^3 R_1^2 R_2^2} \ [bistatic] \tag{7-9}$$

were used to convert received power to radar cross section. The impact of separately accounting for the propagation factor is treated in section 6.4.

Since illuminated cell sizes varied widely, the use of the normalized radar cross section per unit area, $\sigma^\circ$, was selected as the measure of clutter reflectivity. It was calculated by dividing the clutter cross section by the illuminated area. For the monostatic case, the calculation of effective illuminated area is rather straightforward and is given by

$$kR_1 \left(\frac{c\tau}{2}\right) [BW] \tag{7-10}$$

where

$k$ = a constant close to unity, dependent upon beamshape[8]

$BW$ = 3 dB antenna beamwidth, expressed in radians

$c$ = velocity of light

$\tau$ = width of the transmitted pulse

$R_1$ = range to the illuminated area from the radar

for incidence angles near grazing.

Calculation of the effective illuminated area for the bistatic case is more complex since the illuminated area is determined by the intersection of the two antenna beams and the portions of the constant delay ellipses illuminated by the transmitted pulse. Calculation of the illuminated area was accomplished using a digital computer; the assumed geometry is shown in Figure 7-7. In Figure 7-7, the nominal intersection of the beam is indicated by the "aimpoint." A clutter cell

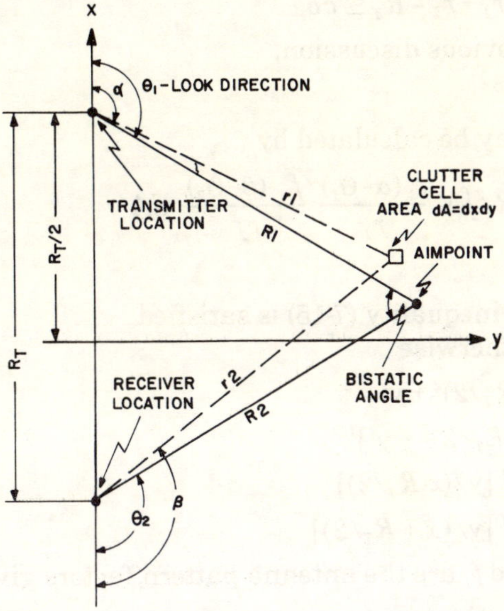


**Figure 7-7.** Geometry and nomenclature for the experiment and for calculation of bistatic illuminated area (reprinted from Ewell and Zehner, ref. [7]; © 1982 IEE).

under consideration is indicated by the small square. The clutter return is obtained by integrating all incremental clutter areas which contribute to the return. $BW_1$ and $BW_2$ are the assumed beamwidths of the antennas. The delay at the bistatic receiver relative to the timing pulse received from the transmitter, $\delta$, is

$$\delta = \frac{1}{c}\,(R_1 + R_2 - R_T) \tag{7-11}$$

and the coordinates of the aimpoint in rectangular coordinates are

$$(R_1 \cos \Theta_1 + \frac{R_T}{2}, R_1 \sin \Theta_1) \tag{7-12}$$

Then one may write

$$\Theta_2 = \tan^{-1} (R_1 \sin \Theta_1 / R_1 \cos \Theta_1 + R_1) \tag{7-13}$$

and

$$R_2 = (R_1^2 + R_T^2 + 2R_1 R_T \cos \Theta_1)^{1/2} \tag{7-14}$$

If a rectangular pulse of width $\tau$ is transmitted, clutter elements will make a contribution if

$$c(\delta - \tau) \leq r_1 + r_2 - R_T \leq c\delta \tag{7-15}$$

From the previous discussion,

$$area = \sigma / \sigma^\circ$$

This area may be calculated by

$$area = R_1^2 R_2^2 \int \Delta \frac{f_t(\alpha - \Theta_1)}{r_1^2} \frac{f_r(\beta - \Theta_2)}{r_2^2} dA \tag{7-16}$$

where

$\Delta$ = 1, if inequality (7-15) is satisfied
0, otherwise

$r_1 = [(x - R_T/2)^2 + y^2]^{1/2}$

$r_2 = [(x + R_T/2)^2 + y^2]^{1/2}$

$\alpha = \tan^{-1} [y/(x - R_T/2)]$

$\beta = \tan^{-1} [y/(x + R_T/2)]$

and the $f_t$ and $f_r$ are the antenna pattern factors given by

$$f(\alpha) = [\sin(u)]/u \tag{7-17}$$

where

$$u = 2.783\ \alpha / BW$$

The integration was performed numerically starting at the aimpoint and increasing until either $\Delta$ went to zero or until $u$ for both antennas was greater than $\pi$. The resulting values of illuminated area were used to convert values from bistatic cross section to bistatic radar cross section per unit area.

### 7.5.3 Results

Data are presented as the ratio of median values, $\sigma^\circ_{\text{monostatic}} / \sigma^\circ_{\text{bistatic}}$, as a function of bistatic angle. Since the results are a function of look direction, environment, and polarization, results are separated by look direction and by day. Figures 7-8 to 7-10 show the ratio of

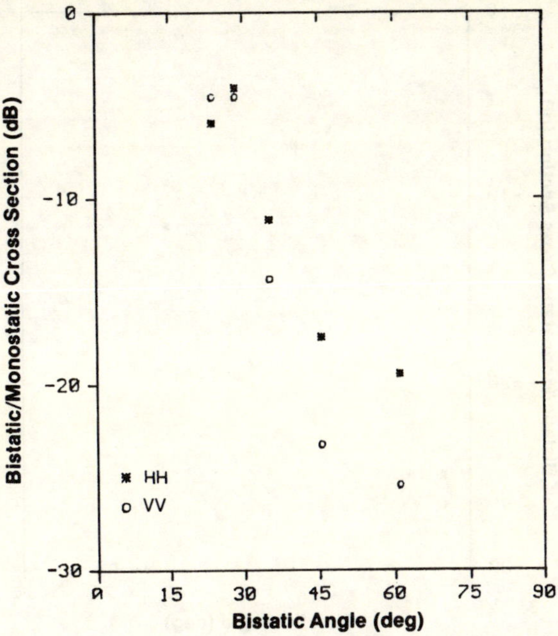


**Figure 7-8.** Ratio of bistatic to monostatic radar cross section per unit area as a function of bistatic angle for horizontal (HH) and vertical (VV) polarizations; data measured August 3 with transmitter look direction of 90 degrees (reprinted from Ewell and Zehner, ref. [7]; © 1982 IEE).

median bistatic to monostatic $\sigma^\circ$ expressed in dB as a function of bistatic angle, indicating a rather abrupt decrease in BRCS from MRCS, which generally continues to decrease with increasing bistatic angle. The weather conditions during these measurements were typical of southern Florida during the summer months; i.e., a fairly steady sea from the east or southeast with occasional local frontal activity and associated showers. Visual estimates of sea conditions were 1.2 - 1.8 m (4-6 ft) significant waveheight on August 3 and 4, and 0.9 m (3 ft) on August 5. The seas were somewhat disorganized on August 4, which may contribute to the relatively broad scatter of the data in Figure 7-10.

Most amplitude distributions were nearly lognormal in shape. The lognormal standard deviation (LSD) was calculated as the difference between the cumulative distribution values in dB at the 50 and 84 percent points. Figure 7-11 is a scatter diagram of monostatic LSD *versus* bistatic LSD. There appears to be a tendency for the

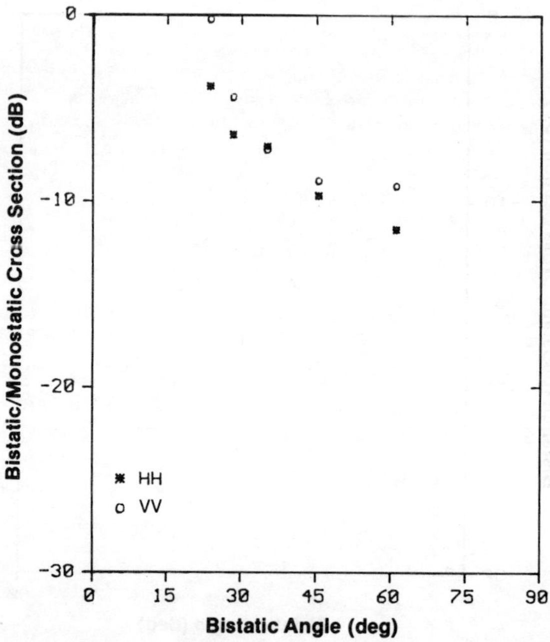


**Figure 7-9.** Ratio of bistatic to monostatic radar cross section per unit area as a function of bistatic angle for horizontal (HH) and vertical (VV) polarizations; data measured August 5 with transmitter look direction of 90 degrees (reprinted from Ewell and Zehner, ref. [7]; © 1982 IEE).

monostatic return to have a larger LSD than the bistatic, and the horizontal polarization to have a larger LSD than the vertical. A number of days and bistatic angles are lumped together in Figure 7-11, since significant daily differences, or a dependence upon bistatic angle, could not be identified.

### 7.5.4 Extrapolation to Other Depression Angles

The cross section values reported in Figures 7-8 to 7-10 are "effective" cross section values, in which the effect of the propagation factor are implicity included in the calculated cross section values. These data were taken at relatively low angles, and the effects of multipath may be significant when attempting to apply these results to other geometries. Since propagation effects may be significant, a brief review of the propagation factor behavior for sea clutter return at low angles is appropriate.

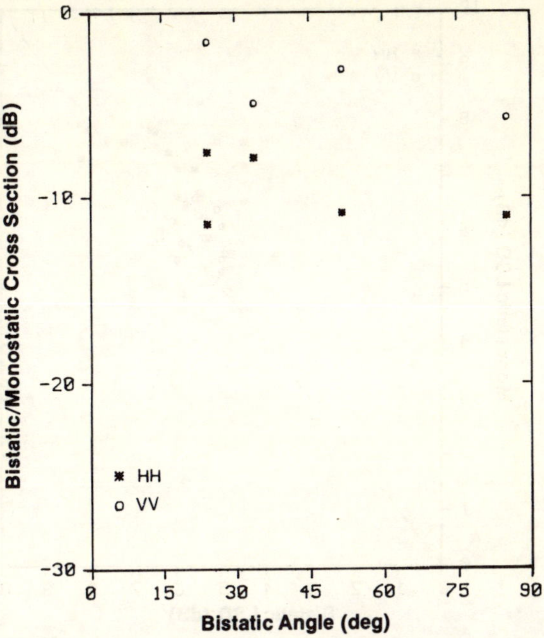


**Figure 7-10.** Ratio of bistatic to monostatic radar cross section per unit area as a function of bistatic angle for horizontal (HH) and vertical (VV) polarizations; data measured August 4 with transmitter look direction of 150 degrees (reprinted from Ewell and Zehner, ref. [7];© 1982 IEE).

The behavior of the propagation factor $F$, as a function of depression angle $\psi$, may be drastically different for different values of $\psi$. For relatively large $\psi$ (a few degrees), interference effects are not dominant, and $F \sim 1$; this would lead to a dependence of monostatic received power with range as $R^{-3}$. At relatively small values of $\psi$ (typically small fractions of a degree), interference effects are important and $F$ varies as $\psi^{-1}$, giving rise to a range dependence of monostatic received power as $R^{-7}$. The value of $\psi$ which separates these two regions is called the "critical angle," $\psi_c$. Values of $\psi \geq \psi_c$ are considered to the "plateau region," while values of $\psi \leq \psi_c$ are considered to be in the "interference region." The value of $\psi_c$ is given by

$$\psi_c = \frac{\lambda}{K\,\sigma_h} \tag{7-18}$$

where

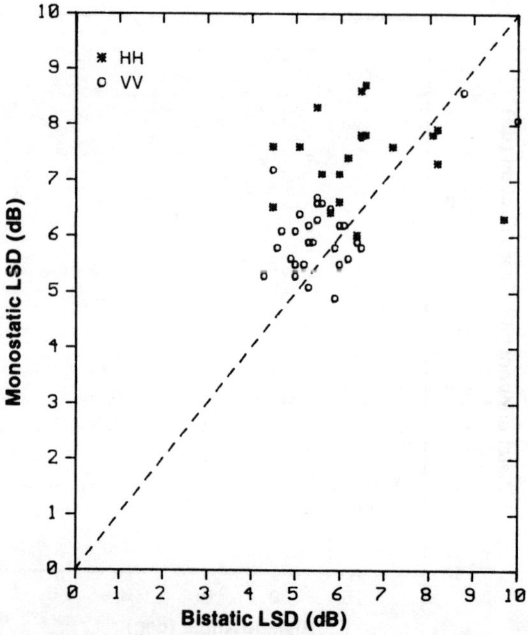


**Figure 7-11.** Scatter diagram for monostatic versus bistatic log normal standard deviations (LSD) in dB. Data for horizontal (HH) and vertical (VV) polarizations are included, and the dotted line is the loci of equal LSD (reprinted from Ewell and Zehner, ref. [7]; © 1982 IEE).

$\lambda$ = operating wavelength

$\sigma_h$ = *rms* surface roughness, typically about one-fourth the significant wave height

$K$ = a constant, close to $4\pi$

These relationships may be used to correct data, for example, to the plateau region as follows. If one denotes $\sigma_m{}^\circ$ as the monostatic cross section per unit area and $\sigma_b{}^\circ$ as the bistatic cross section per unit area, one may write

$$F_1^4\, \sigma_m^{\circ\prime} = \sigma_m{}^\circ$$

and

$$F_1{}^2 F_2{}^2\, \sigma_b^{\circ\prime} = \sigma_b^\circ$$

where

$\sigma_m^{\circ\prime}$ = monostatic cross section per unit area calculated, with propagation effects separately accounted for

$\sigma_b^{m\prime}$ = bistatic cross section per unit area calculated, with propagation effects separately accounted for

Then one may write

$$\sigma_b^{\circ\prime} / \sigma_m^{\circ\prime} = (F_1/F_2)^2 \, \sigma_b^{\circ}/\sigma_m^{\circ} \qquad (7\text{-}19)$$

Unfortunately, it is difficult to employ the relationships set forth in

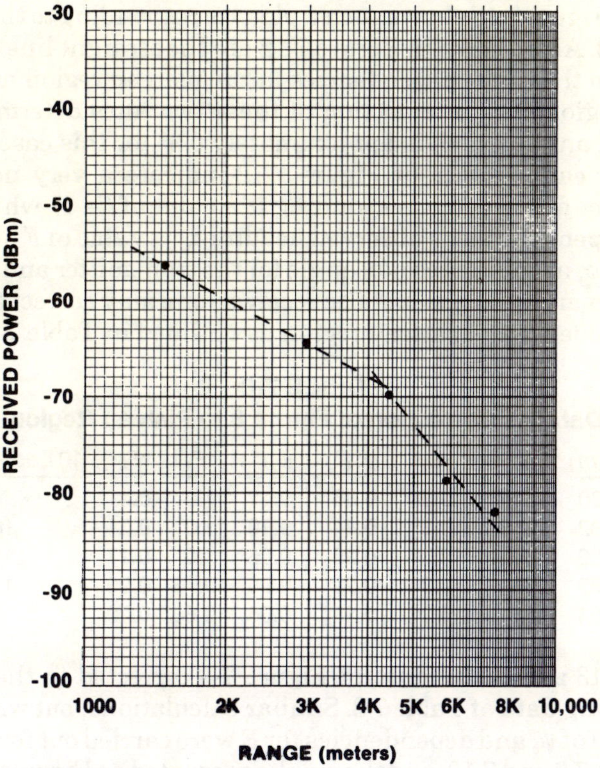


**Figure 7-12.** Received power as a function of range for the monostatic data from August 3, bearing 090, and horizontal (HH) polarization. The dotted lines are estimates of the true behavior and their intersection corresponds to the critical angle.

the preceeding paragraph to correct sea clutter data to other depression angles for two principal reasons. The first is the measured range laws often deviate from the ideal because $\sigma^\circ$ may be depression angle dependent, or the presence of ducting or anomalous propagation may affect the dependence of $F$. The second problem is that the value of $\psi_c$ calculated by Equation (7-18) may differ substantially from the value observed experimentally, particularly when visual estimates of sea state are employed to determine $\sigma_h$.

One approach to determine $\psi_c$ and the functional dependence of $F$ is to examine experimental values of monostatic received power as a function of range, assuming $\sigma_m^{\circ\prime}$ is a constant, and that $F$ is a function of $\psi$. Figure 7-12 shows such a plot, corresponding to the data of Figure 7-8. Also shown in Figure 7-12 are the straight line approximations to the range dependencies in the plateau region and interference region. The intersection of these two lines determines the transition angle, which is approximately 0.3° for this case (assuming a flat earth). Data of Figure 7-12 indicate a very nearly $R^{-3}$ dependence above $\psi_c$ (corresponding to a value of $F = 1$), while below $\psi_c$ a $R^{-6}$ dependence was observed, leading to a value of $F = (\psi/\psi_c)^{.75}$. Denoting $\psi_1$ as the depression angle for the transmitter and $\psi_2$ as the depression angle for the bistatic receiver, one may correct the cross section values to the plateau region as outlined in Table 7-3 below.

**TABLE 7-3**
**Data for Correction of Figure 8 to Plateau Region**

| $\beta$ (°) | $R_1$ (m) | $R_2$ (m) | $\psi_c$ | $\psi_1$(°) | $\psi_2$(°) | $F_1$ | $F_2$ | 10 $\text{Log}_{10}$ $(F_1/F_2)^2$ |
|---|---|---|---|---|---|---|---|---|
| 61 | 1500 | 3982 | 0.3° | 0.86 | 0.12 | 1.0 | 0.4 | 8.0 |
| 46 | 2982 | 4889 | 0.3° | 0.44 | 0.10 | 1.0 | 0.3 | 10.5 |
| 35 | 4482 | 6019 | 0.3° | 0.29 | 0.08 | 1.0 | 0.27 | 11.4 |
| 29 | 5982 | 7297 | 0.3° | 0.22 | 0.06 | 0.73 | 0.20 | 11.2 |
| 24 | 7464 | 8514 | 0.3° | 0.18 | 0.06 | 0.60 | 0.20 | 9.5 |

Figure 7-13 shows the data of Figure 7-8 corrected to the plateau region using data of Table 7-3. Similar calculations, but with different values of $\psi_c$ and dependencies for $F$ were carried out for the data of Figures 7-9 and 7-10, and the results presented as Figures 7-14 and 7-15. While the dependence on bistatic angle is reduced and essentially eliminated for the data of Figure 7-15, a substantial amount of data still evidence a significant dependence on bistatic angle when extrapolated to the plateau region.

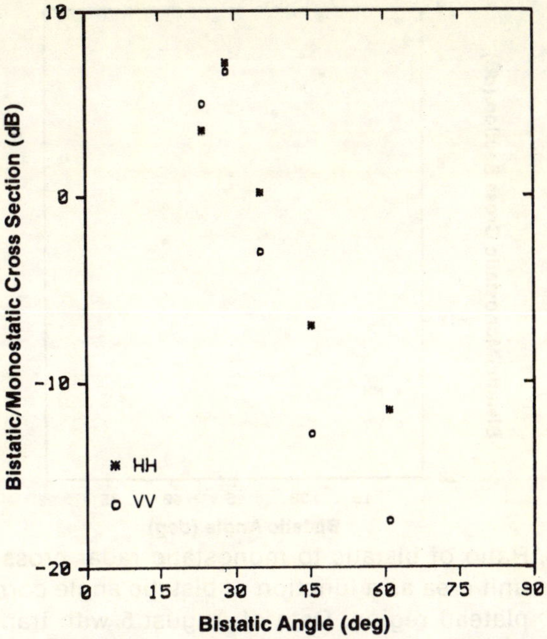


**Figure 7-13.** Ratio of bistatic to monostatic radar cross section per unit area as a function of bistatic angle corrected to the plateau region. Data of August 3 with transmitter look direction of 90 degrees.

## 7.6 AIRBORNE/LAND-BASED COHERENT BISTATIC RADAR

Bistatic radar cross section measurements are not limited to ground-based systems; in addition to the CW system mentioned earlier,[4] pulsed airborne experiments have permitted bistatic measurements to be performed for a variety of clutter types.[9,10] The Environmental Research Institute of Michigan conducted a bistatic measurement program which was interesting not only because one terminal was airborne and the other on the ground, but also because coherent data were acquired. The transmitter was mounted in an aircraft and not only illuminated the ground area being measured, but also sent a local oscillator signal directly to the receiving station on the ground. A simplified system block diagram is given as Figure 7-16, and a summary of the basic system parameters is given as Table 7-4.

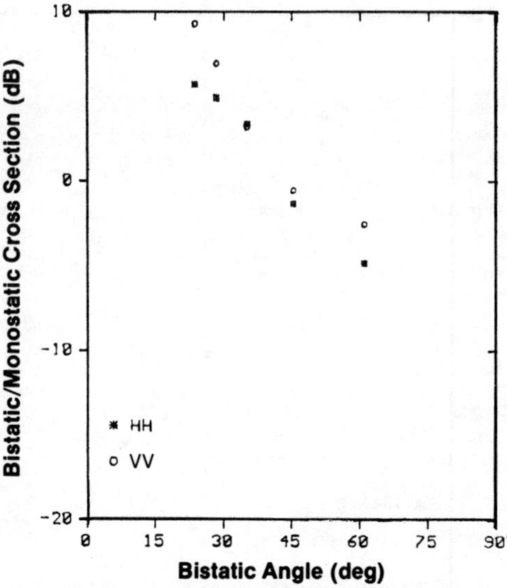


**Figure 7-14.** Ratio of bistatic to monostatic radar cross section per unit area as a function of bistatic angle corrected to the plateau region. Data of August 5 with transmitter look direction of 90 degrees.

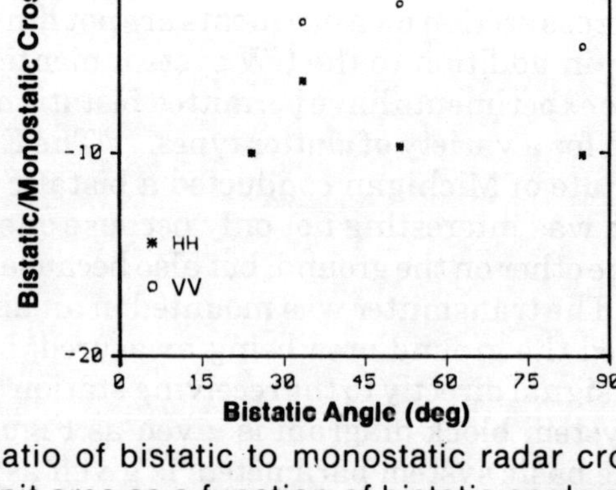


**Figure 7-15.** Ratio of bistatic to monostatic radar cross section per unit area as a function of bistatic angle corrected to the plateau region. Data of August 4 with transmitter look direction of 150 degrees.

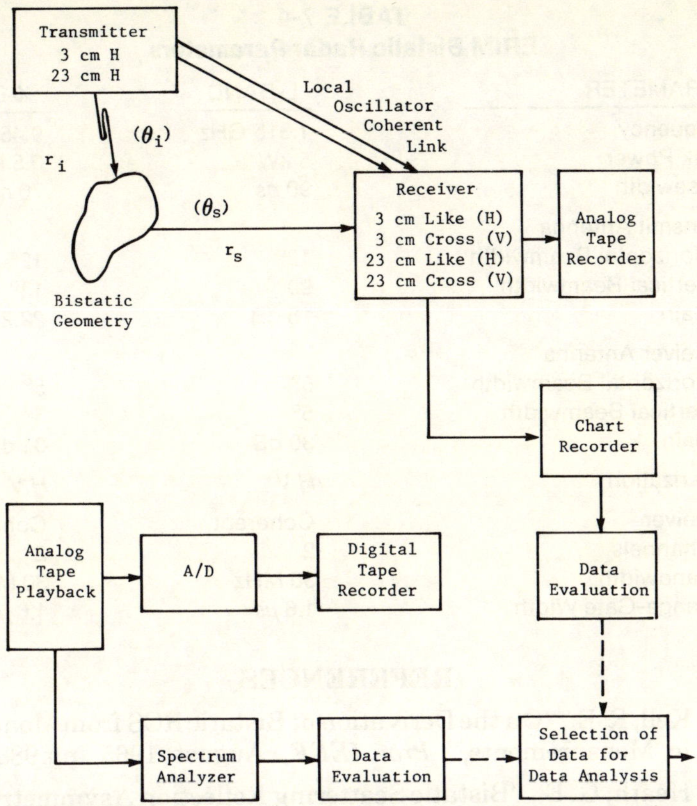


**Figure 7-16.** Functional block diagram of ERIM coherent bistatic measurement system (adapted from Larson, ref. [9]).

In operation, the area near the receiver was illuminated constantly by the airborne transmitting antenna, and the receiving antenna scanned the observation area several times during an aircraft flight. This approach provides a number of samples from each clutter cell, along with samples from a number of different clutter areas. These series of airborne tests illustrate clearly the requirements for use of standard targets of known radar cross section as a means of validating proper system performance. Lacking such calibration targets, and observing a substantial discrepancy between results obtained during this and earlier measurement programs, the hesitancy which one might have in using results of these measurements would be substantially alleviated if more comprehensive calibration procedures had been employed.[11]

**TABLE 7-4**
**ERIM Bistatic Radar Parameters**

| PARAMETER | L-BAND | X-BAND |
|---|---|---|
| Frequency | 1.315 GHz | 9.450 GHz |
| Peak Power | 5 kW | 0.5 kW |
| Pulsewidth | 90 *ns* | 70 *ns* |
| Transmit Antenna | | |
| Horizontal Beamwidth | 12° | 12° |
| Vertical Beamwidth | 90° | 13° |
| Gain | 16 dB | 22.2 dB |
| Receiver Antenna | | |
| Horizontal Beamwidth | 5° | 5° |
| Vertical Beamwidth | 5° | 5° |
| Gain | 30 dB | 31 dB |
| Polarization | *H,V* | *H,V* |
| Receiver | Coherent | Coherent |
| Channels | 2 | 2 |
| Bandwidth | 30 MHz | 30 MHz |
| Range-Gate Width | 1.6 $\mu s$ | 1.6 $\mu s$ |

## REFERENCES


1. Kell, R. E., "On the Derivation of Bistatic RCS from Monostatic Measurements," *Proc. IEEE*, August 1965, pp.983-988.
2. Heath, G. E., "Bistatic Scattering Reflection Asymmetry, Polarization Reversal Asymmetry, and Polarization Reversal Reflection Asymmetry," *IEEE Transactions on Antennas and Propagation*, Volume AP-29, Number 3, May 1981, pp. 429-434.
3. Skolnik, M. I., *Introduction to Radar Systems*, McGraw-Hill Book Company, New York, 1962.
4. Pidgeon, V. W., "Bistatic Cross Section of the Sea," *IEEE Transactions on Antennas and Propagation*, Volume AP-14, Number 3, May 1966, pp. 405-406.
5. Ewell, G. W., and S. P. Zehner, "Bistatic Radar Cross Section of Ship Targets," *IEEE AP-S International Symposium Digest*, June 1978.
6. Ewell, G. W., and S. P. Zehner, "Bistatic Radar Cross Section

of Ship Targets," *IEEE Journal of Oceanic Engineering*, Volume OE-5, No. 4, October 1980, pp. 211-215.

7. Ewell, G. W., and S. P. Zehner, "Bistatic Sea Clutter Return Near Grazing Incidence," *RADAR-82*, IEE Conference Publication No. 216, October 1982, pp. 188-192.
8. Barton, D. K., *Radar System Analysis*, p. 96, Artech House, Dedham, Massachusetts, 1976.
9. Larson, R. W., *et al.*, "Measurements of Bistatic Clutter Cross Section," RADC Report RADC-TR-79-15, May 1979.
10. Larson, R. W., *et al.*, "Bistatic Clutter Measurements," *IEEE Transactions on Antennas and Propagation*, Vol. AP-26, No. 6, November 1978, pp. 801-804.
11. Vander Schur, R. E., and P. G. Tomlinson, "Bistatic Clutter Analysis," RADC Report RADC-TR-79-70, April 1979.

# CHAPTER 8

# FAR FIELD RCS MEASUREMENT RANGES

**Eugene F. Knott**

**Engineering Experiment Station**
**Georgia Institute of Technology**

## 8.1 GENERAL CONSIDERATIONS

### 8.1.1 The Far Field Requirement

As was discussed in Capter 1, the radar cross section of a target can be defined as

$$\sigma = \lim_{r \to \infty} 4\pi R^2 \left| \frac{E_s}{E_o} \right|^2 \tag{8-1}$$

The limiting process essentially requires that the target be illuminated by a plane wave, yet the wave incident on a target in practical measurement situations is nearly always a spherical wave, due to the finite separation between the target and the instrumentation system. As a matter of practical interest, one wonders how "spherical" the incident wavefronts are, while being reasonably good approximations of plane wave incidence.

One way to resolve the question is to assume the source radar to be a point source, and to examine the deviation of the incident phase fronts from perfect uniformity over an aperture having the same width as the target. From the geometry of Figure 8-1, we can find the distance *h* in terms of the range *r* (or *r*-*h*) and the transverse width.

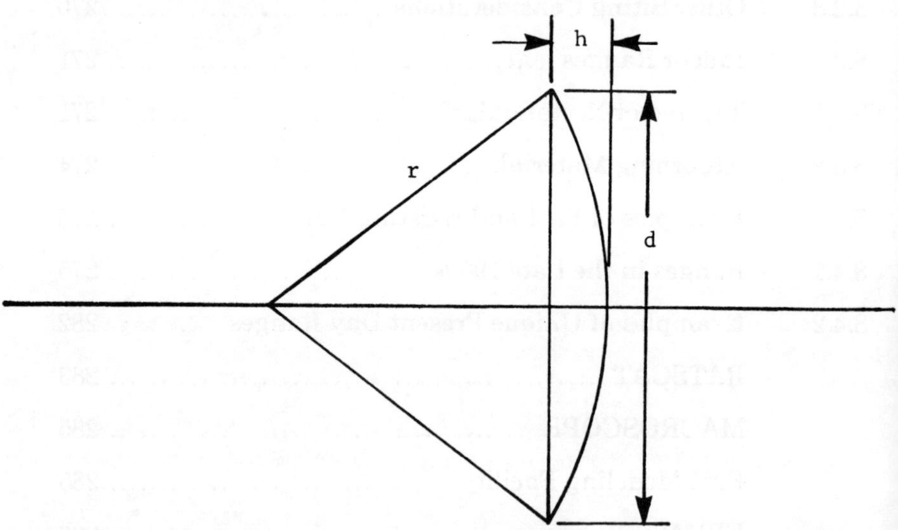


**Figure 8-1.** Phase deviation over a transverse aperture due to spherical wavefronts.

$$h = R \left\{ 1 - \left[ 1 - \left( \frac{d}{2R} \right) \right]^2 \right\}^{1/2} \tag{8-2}$$

which, assuming $d << 2R$,

$$h \cong \frac{d^2}{8R} \tag{8-3}$$

Thus, the phase of the incident wave at the center of the aperture is different from the phase at the extremes of the aperture by the amount $kh$, where $k = 2\pi/\lambda$ is the free space wave number. If we arbitrarily require that this phase deviation be less than $\pi/8$ radians (22.5 degrees), then the familiar far field condition is obtained,

$$R > 2 \frac{d^2}{\lambda} \tag{8-4}$$

Note that the requirement $kh \leq \lambda/8$ is quite arbitrary, and does not consider any accuracy statements about the measurement of the far scattered field. Despite its arbitrary nature, (8-2) has become a widely used yardstick for "ensuring" the accuracy of the far field RCS measurements.

Kouyoumjian and Peters considered the range requirement in far greater detail,[1] including the axial decay in incident field strength, as well as the transverse phase deviation. They concluded that a measurement can "be optimized with respect to minimum range and maximum sensitivity by a proper choice of antenna size," although no general statements could be made on the accuracy of the measurements. Knott and Senior later conducted a numerical study of errors on very simple targets,[2] and did assign values to the errors. They concluded that even the standard far field criterion [Equation (8-4)] might not be sufficient for certain classes of targets. Some of the large errors noted were due to near field cancellation of the returns from various scattering centers on the targets that would not, in fact, occur for true plane wave incidence. Nevertheless, Equation (8-4) remains a good rule of thumb for establishing the minimum range at which targets should be measured.

A plot of the far field distance prescribed by Equation (8-4) is given in Figure 8-2 for a variety of frequencies and target sizes. The far field requirement is not too hard to meet at low frequencies, but typical radar sensitivities make it difficult to meet at high frequencies. For example, a 10-foot target should be measured at a range of not less than 200 feet at 1 GHz, but not less than 2000 feet at 10 GHz.

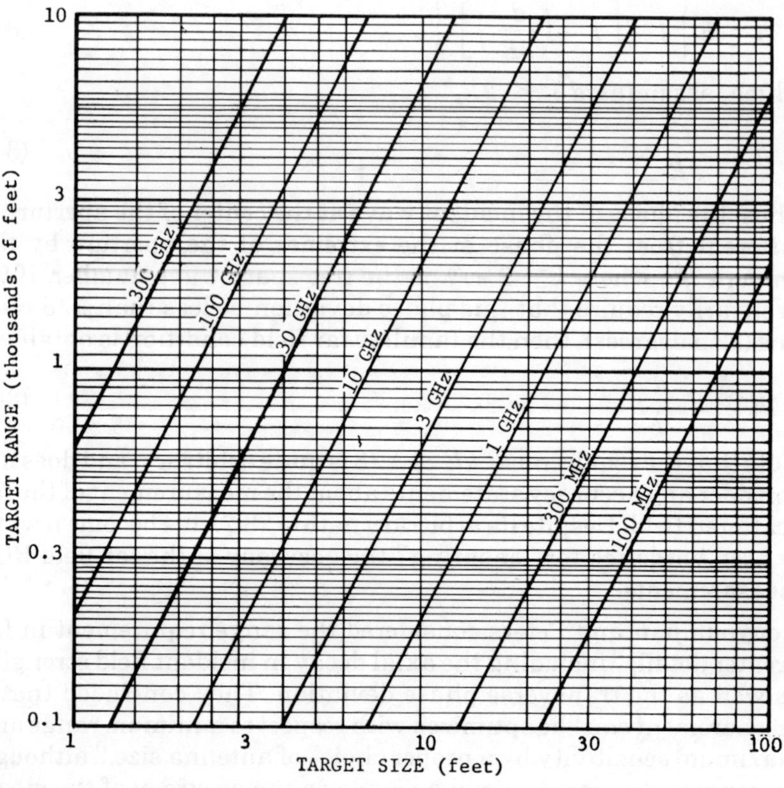


**Figure 8-2.** The far field distance.

In some cases the radar sensitivity is not good enough for the target to be measured at the far field distance, and a range considerably less than that may have to be selected to ensure adequate received signal strengths. For very simple targets, such as cylinders and flat plates, the first symptom of near field effects on the RCS patterns is the filling in of nulls and the reduction of the amplitudes of the sidelobes. As the range is further shortened, the nulls become entirely filled in, the sidelobes become shoulders in the pattern, and the main lobe amplitude decreases significantly.

The effects of measuring more complex targets at less than the standard far field distance are less obvious. At high frequencies each feature of the target scatters energy more or less independently of other target features, and each feature is usually significantly smaller than the overall target. Such features, which could be tail fins, engine inlets, nose tips, or other target surfaces, could each be in the far field when its own characteristic size is considered, while the composite target may not be. Thus the amplitude of the scattering from each feature, as well as the locations of peaks and nulls in its own pattern, are less sensitive to the measurement range. The primary effect of a near field measurement in this case is the slight shifting of the lobes and nulls of the composite pattern as compared with the true far field pattern. Therefore, measurements performed at less that the standard far field distance can often be defended.

Moreover, high accuracy in RCS measurements is often unnecessary unless the data are to be used as a basis for detailed analytical work or the development of scattering predictions. More often than not, users of test data need only mean or median values which are averages of the return over an aspect angle window that is moved across the RCS pattern. The mean or median is often used in assessments of weapons systems performance, in which many other approximations of physical processes are made that could be worse than the errors in the data. Thus, the end use of the data should be considered in determining how important the near field effects may be.

Finally, it should be pointed out that it is the transverse target dimension which figures predominately in calculating the far field range. In the nose-on or end-on aspect angle region, the transverse target dimension of missiles and slender bodies allows the target to be brought much closer to the radar. Since the end-on returns are usually much smaller than the broadside return, this helps improve system RCS sensitivity. RCS patterns in the broadside aspect angle region can usually be measured comfortably at a greater range. Although it requires two separate measurements and the "splicing together" of separate RCS patterns valid over separate aspect angle regions, this is often the price that must be paid for high quality data.

### 8.1.2 Target Support Structures

Several target support techniques are available. These include:

low density plastic foam columns
high strength non-metallic support lines
absorber-covered metal pylons
hollow solid plastic support columns

Of the four, the first has been the most widely and successfully used, while the last has been used the least. Each has its advantages and limitations; plastic foam columns offer relatively low background return when properly designed and used, but have limited weight handling qualities. Suspending targets from filamentary lines is attractive because support lines can have smaller returns than foam plastic columns, but target attitude control is hard to maintain. Metal pylons have excellent load bearing properties, but require special attachments to mate with the target, and the target may have to be disfigured at the mounting point. Moreover, an absorber-covered metal pylon increases the target/support coupling and is *very* expensive. Hollow plastic columns are probably useful only over a narrow frequency range where the front and rear column returns can be "tuned" to cancel each other, but outside these narrow limits the returns can be nearly as high as that of a bare metal column.

Freny[3] discusses the four methods, but offers no data by which to assess metal or solid plastic support columns. Data are presented on dielectric support lines, and some data are given on foamed plastic columns. However, more detailed information on foamed plastic supports is available in other references.

As pointed out by Plonus[5] and Senior, *et al.*,[6] the radar echo of a plastic foam column can be ascribed to two mechanisms. One is a surface reflection due to the abrupt change in the medium the incident wave finds itself propagating in, and the other is an "incoherent" volume return due to millions of tiny internal scatterers (i.e., the cell structure). The volume contribution is conceptually independent of the shape of the column and therefore irreducible; that is, column tuning can have no effect. Plonus' result for the incoherent return is

$$\sigma = \frac{\pi}{2}\, t^2 k^4 a\, |\epsilon_r - 1|^2\, V \tag{8-5}$$

where

| | |
|---|---|
| $t$ | = the mean cell wall thickness |
| $a$ | = the mean cell radius |
| $V$ | = the volume of the column |
| $\epsilon_r$ | = the dielectric constant of the base polymer |
| $k$ | = $2\pi/\lambda$, the free space wave number |

Extensive experiments conducted at the University of Michigan did not confirm this prediction. Because of the very small returns from test blocks and ogives, the measured data show large deviations.

**TABLE 8-1**
**Comparison of Cylindrical Pedestals at X-Band**
(*Adapted from Senior, et al., ref. [6]; © 1964 Microwaves*)

| MATERIAL | DIAMETER (INCHES) | SPECULAR CROSS-SECTION, $s$ (dBsm) | THEORETICAL INCOHERENT CROSS SECTION, $s_1$ (dBsm) | "MEASURED" INCOHERENT CROSS-SECTION (dBsm) |
|---|---|---|---|---|
| Tyrilfoam | 26.04 | -15.4 | -20.2 | -33.3 |
| Pelaspan Escofoam | 18.98 | -12.5 | -35.7 | -48.0 |
| Styrofoam FB | 14.66 | - 9.9 | -55.2 | -19.0 |
| Styrofoam DB | 15.40 | - 7.5 | -43.0 | -49.3 |
| Styrofoam FR | 13.34 | - 9.3 | -61.5 | -53.2 |
| Thursue | 19.13 | - 9.7 | -60.4 | -50.2 |

**TABLE 8-2**
**Properties of Foamed Plastics**
(*Adapted from Senior, et al., ref.* [6]; © *1964 Microwaves*)

| MATERIAL | MANUFACTURER | TYPE | BASE PLASTICS Density, $\bar{p}$ (lbs/ft$^3$) | DIELECTRIC CONSTANT, $\bar{r}$ | DENSITY $P$ | FOAMED PLASTICS ELAS. MED., $E$ (lbs/In$^2$) | CELL DIAMETER $Za$ (In) |
|---|---|---|---|---|---|---|---|
| Tyrilfoam | Dow Chemical Co. Midland, MI | Styrene-acrylonitrile | 66.8 | 2.57 at 1 G | 0.70 | 207 | 0.461 |
| Pelaspan | Dow Chemical Co. | Polystyrene | 66.5 | 2.55 | 1.15 | 733 | 0.125 |
| Escofoam | Emerson & Cuming Canton, MA | | | | | | |
| Styrofoam FB | Dow Chemical Co. | Polystyrene | 66.5 | 2.55 | 1.76 | 2061 | 0.025 |
| Styrofoam DB | Dow Chemical Co. | Polystyrene | 66.5 | 2.55 | 1.80 | 1692 | 0.057 |
| Styrofoam FR | Dow Chemical Co. | Polystyrene | 66.5 | 2.55 | 1.97 | 3000 | 0.011 |
| Thursue | Wyandotte Chem. Co. Wyandotte, MI | Polyurethane | 70 | 2.06 | 2.04 | 710 | 0.019 |

Nevertheless, a value of about -58 dBsm per cubic foot seems typical of the averaged data for several classes of plastic foams. Table 8-1 is a comparsion of the electrical performance of several materials abstracted from reference [6], and Table 8-2 is a list of the physical properties of the materials. Expandable polystyrene bead foam is currently in wide use because of its superior all-around performance. It is available in large billets (17 x 48 x 192 inches) and is therefore amenable to column fabrication with minimum effort.

Supporting a target with dielectric lines requires the erection or availability of a pair of towers straddling the target region. The towers should be placed outside the range gate and well outside the mainlobe of the radar antennas; *see* Figure 8-3. A sling or harness must be made to accommodate the target, and it must be guyed to a

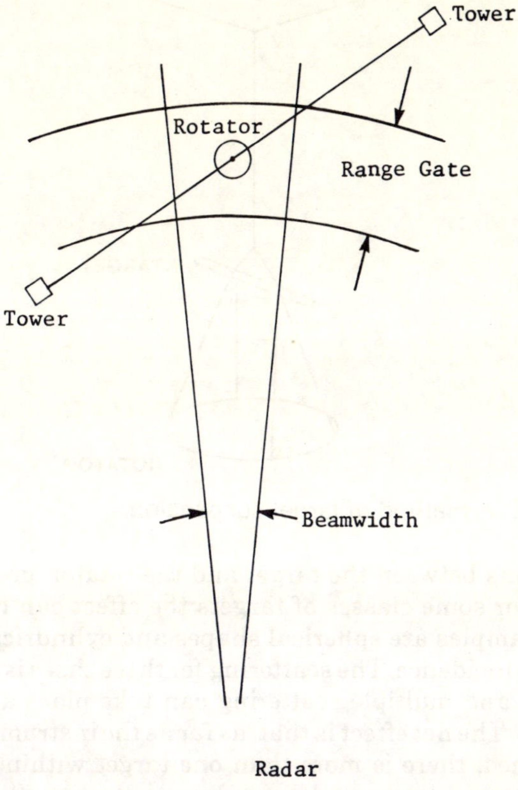


**Figure 8-3.** Tower placement for target suspension of the target rotator.

rotator in order to control the aspect angle of the target; *see* Figure 8-4. The support lines must be strong enough to bear the tension of these guy lines in addition to the weight of the target. The line tension is usually such that the target will oscillate in aspect angle even with the rotator moving at constant speed. It requires calm conditions, very careful attention to target alignment, and a great deal of patience to obtain RCS patterns using this suspension technique, yet the effort may often be well worth it. The experimenter should be aware that all the guy lines to the rotator will be normal to the incident wave at two points in a 360 degree azimuth cut, and their returns, although small, will likely be visible in the recorded data.

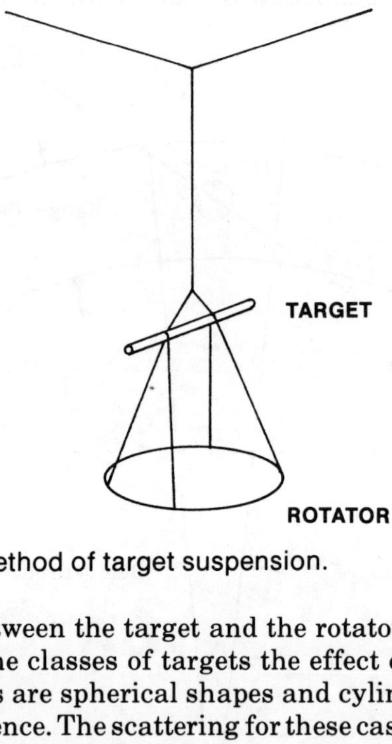


**Figure 8-4.** A method of target suspension.

Interactions between the target and the rotator/ground are possible, and for some classes of targets the effect can be surprisingly large. Examples are spherical shapes and cylindrical shapes near broadside incidence. The scattering for these cases is nearly omnidirectional, and multiple scattering can take place as suggested in Figure 8-5. The net effect is that, as far as the instrumentation radar is concerned, there is more than one target within the beam and within the range gate, and depending on the actual geometry, they seem to be aligned one behind the other. The intensities of the reflections from these "ghost" targets is not as high as that of the

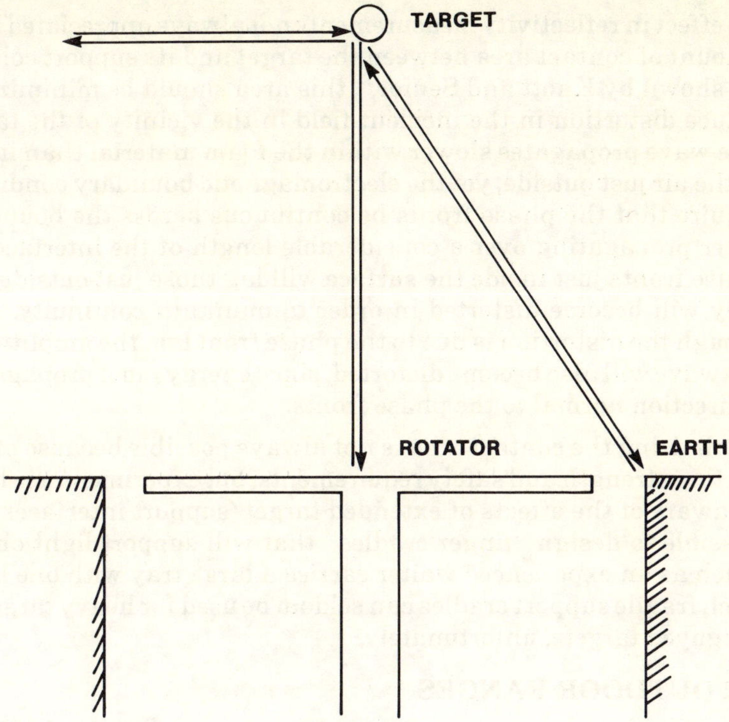


**Figure 8-5.** Target/rotator interactions.

real target, yet they can be large enough to add constructively or destructively (depending on the target height), introducing measurement errors. Since metallic spheres are in very common use as calibration standards, the potential for error is more prevalent than might be supposed.

Obviously the most expedient remedy is to cover the rotator and the ground in the vicinity of the target with radar absorbing material (RAM). This is not as easily accomplished as it seems, since the target support column (if one is used) must somehow be mounted on the RAM. Moreover, especially on outdoor ranges, the rotator, and therefore, the RAM coating, is exposed to the weather. Consequently the material must be rigid, must be able to withstand the environment, and must be rugged enough to support the target as well as personnel who may have to stand on the rotator during target mounting procedures. Despite these harsh requirements, radar absorbing coatings can be designed and manufactured having the necessary qualities.

An effect in reflectivity measurements not always appreciated is the amount of contact area between the target and its support column. As shown by Knott and Senior,[7] this area should be minimized to reduce distortion in the incident field in the vicinity of the target. The wave propagates slower within the foam material than it does in the air just outside, yet the electromagnetic boundary conditions require that the phase fronts be continuous across the boundary. After propagating over a considerable length of the interface, the phase fronts just inside the surface will lag those just outside, and they will become distorted in order to maintain continuity. Even though the distortion is due to the phase front lag, the amplitude of the wave will also become distorted, since energy must propagate in a direction normal to the phase fronts.

Minimizing the contact area is not always possible because of conflicting strength and safety requirements, but experimenters should be aware of the effects of extended target/support interfaces. It is possible to design "finger cradles" that will support light objects much as an experienced waiter carries a large tray with one hand. Such fragile support cradles can seldom be used for heavy targets or for guyed targets, unfortunately.

## 8.2 OUTDOOR RANGES

### 8.2.1 Ground Plane or Not?

The proximity of the ground to the antenna and the target is sometimes hard to avoid, and one solution is to exploit the reflection from the ground. This is done at the RATSCAT site at Holloman Air Force Base, NM, and at the McDonnell Douglas Corporation Facility near Palmdale, CA, otherwise known as the Grey Butte site. The exploitation of the ground plane effect requires maintenance in the form of grading and leveling of the soil unless an asphalt or concrete ground plane has been constructed.

The ground plane effect is illustrated in Figure 8-6. Its performance depends on the reflection of an indirect wave at the specular point *P*. For a flat, smooth earth the location of this point can be found by placing an image antenna directly below the actual antenna just as far below the ground surface as the real antenna is above it. Transmitted energy and received target returns propagate along four distinct paths (*see* Figure 8-6):

| | |
|---|---|
| Direct path | ATA |
| "Image" path | APTPA |

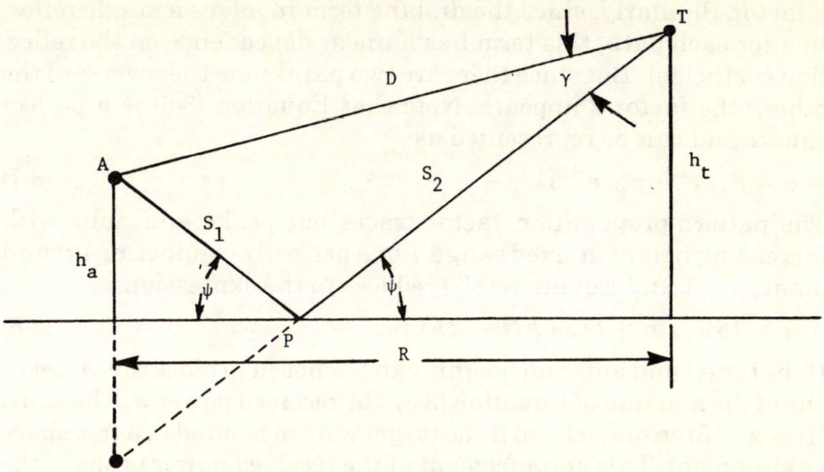


**Figure 8-6.** Ground plane geometry. Radar antenna is at point A, radar target at point T, and the specular point is P.

Diplane path #1 ATPA

Diplane path #2 APTA

Of the four, the last two are actually bistatic returns, since the angles of incidence and scattering at the target are different. For most measurement situations, the bistatic angle is insignificant, however.

If the bistatic angle is small and the target return is insensitive to small differences in the arrival angle of the direct and indirect paths, the effective radar cross section as seen from the radar is

$$\sigma = \sigma_o [ e^{i2kD} + \rho^2 e^{i2kI} + 2\rho e^{ik(D+I)} ]^2 \tag{8-6}$$

where

$\sigma_o$ = free space radar cross section of the target

$\rho$ = effective reflection coefficient of the ground plane

$D$ = length of the direct path

$I$ = $S_1 + S_2$ = length of the indirect path

$k$ = $2\pi/\lambda$ = free space wave number

The first term in Equation (8-4) is the free space return, the second is the image return and the third is the diplane return. Since there are two ground plane reflections involved in the total propagation path for the image term, the square of the reflection coefficient appears as a factor. Similarly, since the diplane term involves a single reflection for each path, this term has a linear dependence on the reflection coefficient. But since there are two paths (one the reverse of the other), the factor 2 appears. Note that Equation (8-6) is a perfect square and can be represented as

$$\sigma = \sigma_o' ( e^{ikD} + \rho e^{ikI} )^4 \tag{8-7}$$

The pattern-propagation factor traces out peaks and nulls with increasing height at fixed range. For a perfectly conducting ground plane, $\rho = -1$ and Equation (8-7) reduces to the expression

$$\sigma = 16 \sigma_o \sin^4 [ 1/2 ( kD - kI ) ] \tag{8-8}$$

If the target and antenna heights can be chosen to place the target at one of the maxima of Equation (8-8), the received power will be 12 dB ($10 \log_{10} 16$) stronger than if the target were measured in a free space environment. This enhancement of the received power is one of the ground plane's most attractive advantages. However, it does require that antenna and target heights be controllable, since the target must be placed at a maximum in the vertical interference pattern created by the ground plane. Almost invariably, the first maximum above the ground plane is the one used.

An illustrative example will show the application of Equation (8-8) to a measurement situation. The maxima of Equation (8-8) occur when the electrical angle is an odd multiple of $\pi/2$ radians, or when

$$1/2 \, k ( D - I ) = m \frac{\pi}{2} \tag{8-9}$$

where $m$ is an odd integer. Therefore, the path length difference must be

$$\delta = D - I = m \frac{\lambda}{2} \tag{8-10}$$

From the geometry of Figure 8-6,

$$D = [ ( h_t - h_a )^2 + R^2 ]^{1/2} \cong R + 1/2 \frac{(h_t - h_a)^2}{R} \tag{8-11}$$

$$I = [ ( h_t - h_a )^2 + R^2 ]^{1/2} \cong R + 1/2 \frac{(h_t + h_a)^2}{R} \tag{8-12}$$

and the path length difference is

$$\delta = D - I \cong 2 \frac{(h_a h_t)}{R} \tag{8-13}$$

The requirement [ Equation (8-9) ] is therefore approximately satisfied by

$$h_a h_t = m \frac{R\lambda}{4} \tag{8-14}$$

This relationship is illustrated in Figure 8-7 for a radar frequency of 3 GHz, a target range of 1400 feet, and $m = 1$ (the first lobe).

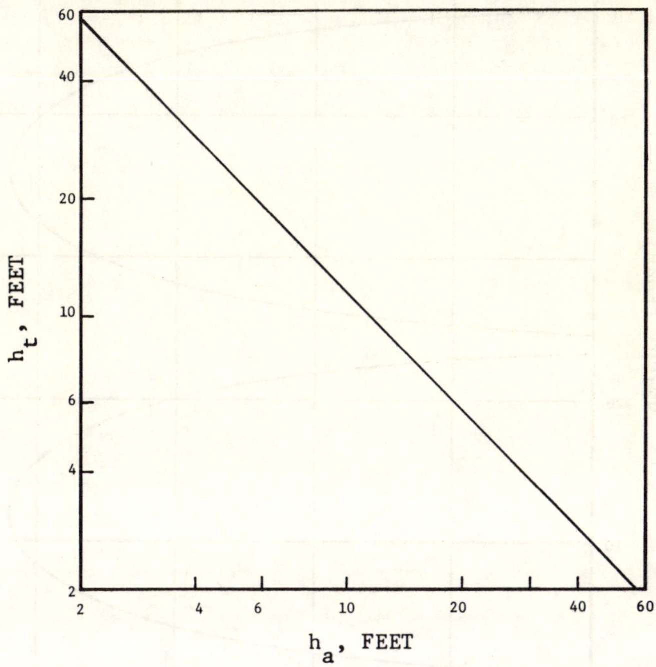


**Figure 8-7.** Relation between $h_a$ and $h_t$ for an ideal ground plane at 3 GHz for $R$ = 1400 feet.

Selection of appropriate antenna and target heights depends on several factors, such as the maximimum antenna height available, the maximum safe target height, whether an existing support column can be used, and the capabilities of target handling equipment. Suppose, for example, that a target height of 12 feet is a compromise among all these factors. Then Equation (8-14) or Figure 8-7 would show that an antenna height of 9.5 feet is required. This can be verified by raising or lowering a reference target at the desired range. A plot of the relative received power for an ideal ground plane is shown in Figure 8-8. Note that the first lobe does indeed attain its maximum value 12 feet above the ground plane. If the vertical profile were to be measured using a small pickup antenna, the same kind of pattern would be measured, except that the abscissa would be halved because it would be the one-way (transmitted) power being measured instead of the two-way (scattered) power.

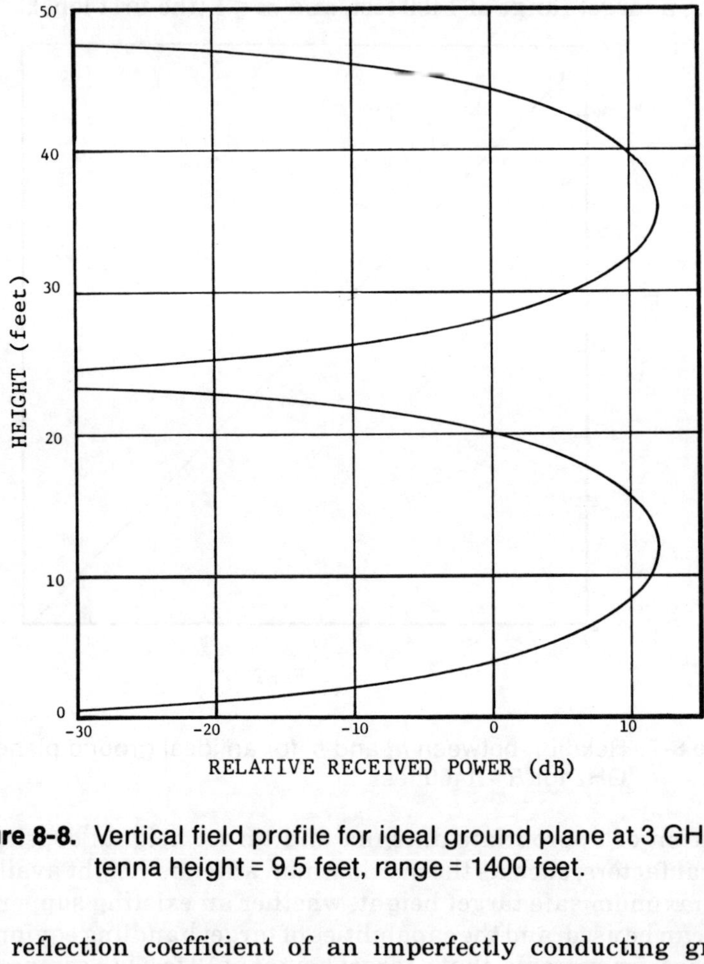


**Figure 8-8.** Vertical field profile for ideal ground plane at 3 GHz. Antenna height = 9.5 feet, range = 1400 feet.

The reflection coefficient of an imperfectly conducting ground plane, such as formed by actual soil, is not −1. At higher microwave frequencies, for example, 1.5 to 15 GHz, the amplitude may be 0.9 and the phase angle may be 170°. Nevertheless, the vertical location of the first lobe of the profile in Figure 8-8 is not substantially different and its amplitude drops only by 1 dB. If the ground can be characterized by a complex dielectric constant

$$\bar{\epsilon}_r = \epsilon_r + i\,\epsilon_r'' \tag{8-15}$$

Then the classical Fresnel reflection coefficients are

$$\rho_h = \frac{\sin\psi - (\bar{\epsilon}_r - \cos^2\psi)^{1/2}}{\sin\psi + (\bar{\epsilon}_r - \cos^2\psi)^{1/2}} \tag{8-16}$$

$$\rho_h = \frac{\bar{\epsilon}_r \sin\psi - (\bar{\epsilon}_r - \cos^2\psi)^{1/2}}{\bar{\epsilon}_r \sin\psi + (\bar{\epsilon}_r - \cos^2\psi)^{1/2}} \tag{8-17}$$

where $\psi$ is the grazing angle (*see* Figure 8-6). Unfortunately, the dielectric properties of the ground depend strongly on the moisture content and can change from day to day, or even hour to hour. Moreover, the moisture content is not uniform with depth and this is not accounted for in Equation (8-16) or Equation (8-17). The Physical Science Laboratory (PSL) of New Mexico State University measured the soil properties as a function of moisture content [9] at RATSCAT at 1.5 GHz, for example, and fitted a fourth order polynomial to the data. (The empirical relationship is shown in Figure 8-9.) Based on these data and a theoretical description for the frequency and temperature dependence, estimates of $\epsilon_r$ can be made at other frequencies, and the Fresnel reflection coefficients can be estimated. Since these data are for the specific soil found at the RATSCAT site, the estimates do not necessarily hold for other sites with different kinds of soil.

Since the reflection coefficients differ for horizontal and vertical polarization, ground plane measurements made for circular polarization are subject to error. The net incident wave at the target is, in general, elliptically polarized, and the ellipticity can be accentuated by the second ground reflection when the target echo reaches the instrumentation radar. There is no easy way to avoid this. The effect becomes more pronounced as the frequency decreases and can be particularly troublesome for frequencies in the UHF band.

It is possible for the reflection coefficient to be precisely zero for certain combinations of grazing angle and dielectric constant. This occurs for vertical polarization and all of the energy is propagated into the ground. It corresponds to the case when the numerator of (8-10) vanishes; the angle at which this occurs is called the Brewster angle. The geometry and dielectric constants of ground plane ranges are such that this rarely occurs, however.

The surface roughness of the ground plays an important role because it reduces the effective reflection coefficient. The effect of surface roughness can be estimated by factor $\rho_s$,[21, 22]

$$\rho_s = \exp[-2(k\,\sigma_h \sin\psi)^2] \tag{8-18}$$

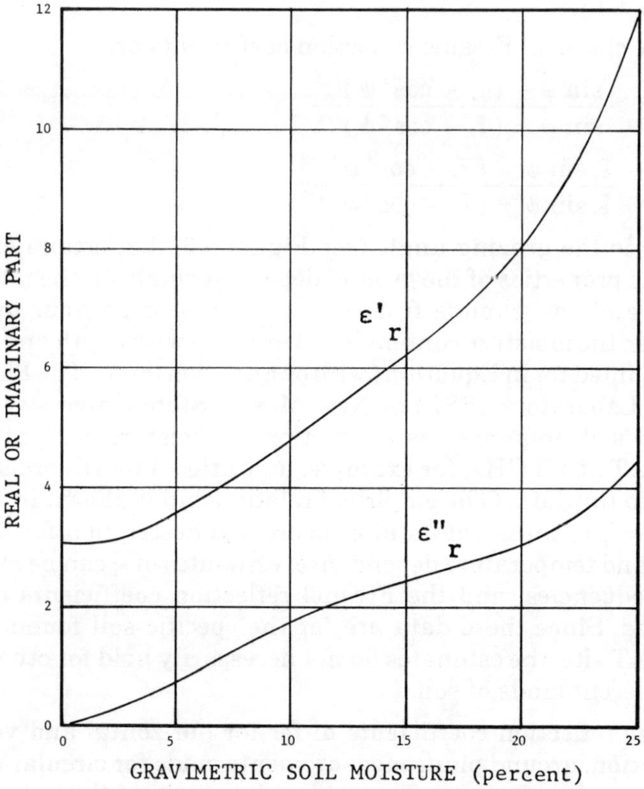


**Figure 8-9.** PLs fitted polynominal to measurements of the complex dielectric constant of RATSCAT soil at 1.5 GHz (adapted from Bush, ref. [9]).

where $\sigma_h$ is the RMS surface height variation. The roughness factor accounts for the dispersion of energy in directions other than the specular direction, thereby decreasing the energy along the specular. The surface roughness also creates an incoherent forward scattered field whose relative power is proportional to $(1-\rho_s^2)$.

This effect can be estimated, but before measurements of targets are made, the net incident field structure in the target zone should be probed to verify that the incident field is acceptably uniform. Thus, an effective reflection coefficient, which is the product of the classical Fresnel reflection coefficient and the roughness factor, should be used

$$\rho_e = \rho_s \tag{8-19}$$

where $\rho$ is either of Equation (8-16) or Equation (8-17), depending on the transmitted polarization. Ground vegetation will modulate the ground plane reflection because brush and grass blades are moved by the wind, and there is hardly ever a perfectly calm day. Vegetation will also reduce the reflected signal, hence a good ground plane should be bare.

The properties of $\rho_s$ are illustrated in Figure 8-10 for small grazing angles. The grazing angle for typical outdoor operation seldom exceeds 5 degrees, and 1 degree or less is quite common. If $\sigma_h = 2\lambda$ is established as an upper limit on the tolerable RMS surface roughness (which corresponds to $\rho_s = .91$ at $\psi = 1$ deg.), the ground plane can be quite rough and still give good performance at low frequencies. For example, this translates to nearly 16 inches at 1.5 GHz, which is quite rough. On the other hand, the requirement becomes less than an inch at millimeter wavelengths, and unless the ground

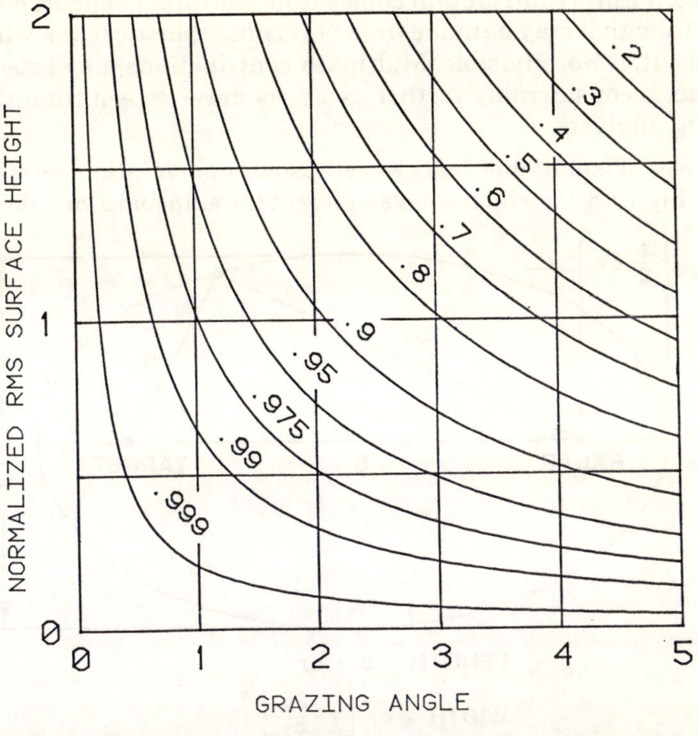


**Figure 8-10.** The surface roughness factor $\rho_s$ for small grazing angles.

plane is a concrete or asphalt runway, maintaining such smoothness would be difficult. Thus, for millimeter wavelengths it may be necessary to *suppress ground reflections* using such techniques as high gain antennas (to minimize ground illumination), large antenna and target heights, and perhaps radar fences erected between the radar and the target.

### 8.2.2 Passive Clutter and Multipath Reduction

Clutter and multipath can introduce measurement errors unless they can be suppressed or exploited. As mentioned above, a ground plane range actually exploits the multipath effect by providing a smooth surface between the radar and target. Other multipath effects can occur in addition to the ground plane reflection, and examples are nearby buildings, trees, and shrubbery. Multipath returns differ from clutter returns in that clutter or background signals arise from scatterers located at or near the target distance. The target positioner is a clutter source unless adequately shielded and, of course, a finite contribution comes from the target support column, even though it may be made from "invisible" plastic foam, which, it turns out, is *not* invisible. Multipath contributions may arise from obstacles considerably farther away, as may be seen from the following analysis.

Surfaces of equal time delay are ellipsoids of revolution, as suggested by Figure 8-11, which shows a slice of the ellipsoid along its major

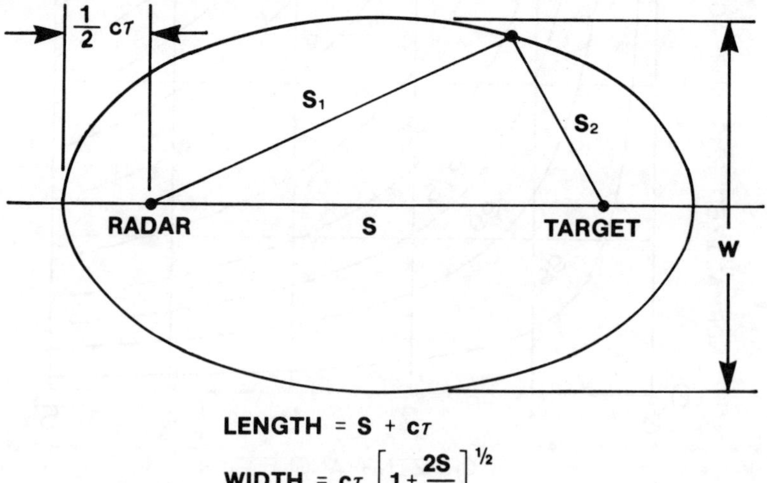


$$\text{LENGTH} = S + c\tau$$

$$\text{WIDTH} = c\tau\left[1 + \frac{2S}{c\tau}\right]^{1/2}$$

**Figure 8-11.** Ellipse of constant delay.

axis. The radar and test target lie at the foci, and the axis of revolution coincides with the line of sight. The indirect path $S_1 + S_2$ is constant for any point on this surface, hence the path length difference $\delta$ is constant. The maximum path length difference cannot exceed $c\tau$, where $c$ is the speed of light and $\tau$ is the shorter of either the transmitted pulse width or the width of the range gate. If the path length difference exceeds this value, the multipath return will not be detected.

If $S$ is the distance between the radar and target, the total length of the ellipsoid is

$$\ell = S + c\,\tau \tag{8-20}$$

and its width is

$$w = c\,\tau\left(1 + \frac{2S}{x\,\tau}\right)^{1/2} \tag{8-21}$$

By way of example, if $\tau = 0.2\ \mu s$ then $c\,\tau = 197$ feet; if the target range is $S$ = 1400 feet, then

$$\ell = 1400 + 197 = 1577\ feet \tag{8-22}$$

$$w = 197\left(1 + \frac{2800}{197}\right)^{1/2} = 768\ feet \tag{8-23}$$

Hence the clear (obstacle-free) width of this test range should be at least half its length. Multipath returns can be generated by an obstacle or scatterer which shares the range gate with the target.

Clutter can be reduced by illuminating as little ground area as possible in the vicinity of the target. The ground itself should be smooth and bare. Target supports should be carefully designed, and target rotators should be well shielded. If necessary, the rotator and the surrounding ground should be covered with radar absorbing material to minimize interactions between the target and the ground.

The use of radar clutter fences can eliminate the specular return from the ground in the event the ground plane effect is undesired, as might be the case at millimeter wavelengths. However, PSL concluded in its study of the RATSCAT ground plane range[9] that the clutter fences offer no distinct advantages. This is because the top edge of the fence diffracts incident energy toward the target as well as back toward the radar, and edge diffraction is very hard to suppress. One way to reduce the diffraction is to tighten up the transmitted beam and to increase the target height so as to minimize the illumination of the edge — but this is precisely what would be done to eliminate the specular reflection from the ground. Consequently the cost and effort of deploying radar clutter fences seems hard to justify. However, if the ground has vegetation which cannot be eliminated or controlled, the use of fences should be necessarily be ruled out.

### 8.2.3 Other Siting Considerations

Most radar reflectivity measurements are performed in connection with Department of Defense programs and many of the targets, as well as the data collected, constitute classified material or information. Moreover, test targets and data may be regarded as proprietary even if not classified, hence a reflectivity range must often satisfy security requirements. These requirements are not always specific, but in general, direct access to the site must be subject to rigid control. This could require security fences, guards, and patrols.

Physical access is not the only aspect of control; remote observation of measurement activities is a possibility that must be considered, and this often requires that the site be well removed from urban areas. In addition, measurement activity is subject to observation from satellites, and operations may have to be suspended during transits of certain space objects. Government agencies provide information on these satellite passes to authorized personnel.

The government also has restrictions on the power radiated by radio frequency systems. FCC regulations are complicated and change from year to year. Range operators should become acquainted with frequency and power limitations; the cognizant local FCC office should be identified and consulted whenever there is a question on operating limits.

Favorable environmental conditions are important, but not always within management's means to acquire. Local corporate real estate may have to be used for a reflectivity range even if local weather is poor for outdoor work. Management must also reconcile the cost of maintaining a distant range in a more favorable climate, as well as travel costs of range users back and forth to the home office. The Southwestern United States is a good place for an outdoor range because of the possibility of year-round operation. General weather conditions are good there because of the scant rainfall. However, winds, which tend to be strong in the flat lands, are the most common cause for suspending measurement operations. Vegetation tends to be sparse or easily cleared. On the other hand, the summer heat tends to be hard on personnel.

Southeastern sites also tend to have favorable year-round weather, with calmer winds but more frequent rain. The average annual precipitation for the state of Georgia, for example, is 55 inches, which is almost exclusively rain (as opposed to snow) in the southern two-thirds of the state. The Northeast and Midwest have more moderate rainfall but much harsher winters. The Pacific Northwest is calm but wet and moderate. Further inland, the Northwest climate is drier but winters are very cold. Thus, site selection involves many factors, not the least of which are the climate and cost of operation.

## 8.3 INDOOR RANGES

### 8.3.1 Tapered or Rectangular?

Emerson's survey of radar absorbents and anechoic chambers[10] is recommended as further reading for anyone with more than a passing interest in radar reflectivity or antenna measurements. Indoor measurement ranges offer immunity from weather, and therefore higher productivity due to greatly reduced weather outages. In addition, security is much easier to implement, and there is less likelihood of violating any FCC regulations on radio emissions. On the other hand, indoor ranges require a considerable capital investment for the building or enclosure the range is housed in, as well as the shielding and radar absorbing materials that need to be mounted on the inner walls.

Early indoor ranges were simply rooms partitioned off from the rest of a laboratory and lined with absorbing material. Early uses appear to have been for antenna pattern measurements, and later uses were for radar cross section measurements. The anechoic chamber range requirements are more difficult to satisfy for reflectivity measurements than for antenna measurements, and the users of indoor ranges found themselves trying to reduce interior reflections to lower levels by the use of better and more costly absorbents. Most of the modern indoor chambers use high performance pyramidal materials with reflectivities of -50 dB or better; these ranges have a useful bandwidth that often exceeds 100 GHz.

A great deal of attention was paid to the rear wall of these indoor ranges because the rear wall intercepts the main beam of energy from the transmitting antenna and residual reflections from the rear wall can have a large effect. Efforts to reduce the reflection included the use of high quality absorbents and a novel tilt-wall concept, which was supposed to allow the wall to be angled so as to place a null in its scattering pattern at the target or at the instrumentation radar. It is doubtful, however, if the tiltable wall ever worked as well as its developers thought it would.

A less obvious source of measurement error was the reflections from other interior walls, and two ideas were presented to reduce the reflections. Assuming the reflections emanate from the specular points of a rectangular chamber, as shown in Figure 8-12 (a), one can

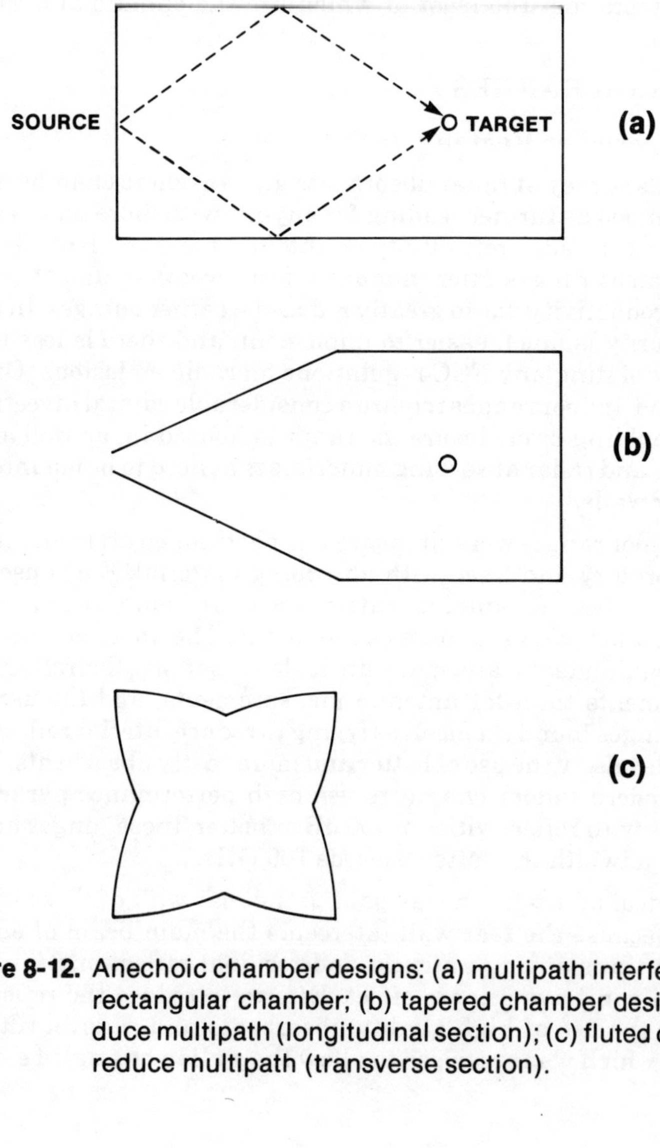


**Figure 8-12.** Anechoic chamber designs: (a) multipath interference in rectangular chamber; (b) tapered chamber design to reduce multipath (longitudinal section); (c) fluted design to reduce multipath (transverse section).

use a tapered design along the longitudinal chamber axis or a fluted design in the transverse dimension. The fluted concept was used in many of the anechoic chambers built by Emerson and Cuming, Inc., of Canton, Massachusetts, and the tapered design was evolved at the B. F. Goodrich Sponge Products Division in Shelton, Connecticut.[13]

The tapered design in Figure 8-12(b) is fairly effective in reducing the multipath effects shown in Figure 8-12 (a) and has been well documented. King, *et al.*,[11] built small models of rectangular and tapered rooms and probed the internal fields using a small horn as a transmitter. They found that the multipath fields were essentially eliminated, although they also discovered that the decay in field strength with propagation distance deviated from the normal free space value. Thus, absolute gain measurements made in a tapered anechoic chamber might be in error unless a known gain standard antenna can be used for calibration. Nevertheless, tapered chambers are superior for radar cross section measurements, especially if the measurements must be made at low frequencies for which high gain antennas (to reduce sidewall illumination) cannot be used.

In investigating the capabilities of tapered chambers, Dybdal and Yowell[12] reported that, with careful attention given to the absorbent material, reflectivity measurements can be made at frequencies as low as 120 MHz and as high as 93 GHz in a tapered anechoic chamber. At the millimeter wavelengths (which are about ⅛ inch at 93 GHz), the sharp tips on the pyramidal absorbing material in the chamber must be maintained, or else the effectiveness of the material is degraded. Apparently the paint ordinarily applied to absorbent materials cannot be used at these very short radar wavelengths, and the rounding or breaking of the tips due to handling, or use during installation, must be avoided to preserve the high performance.

The largest anechoic chamber believed to have been built was at the NASA Manned Spacecraft Center in Houston, Texas. It is 175 feet long and has a transverse cross section of 50 x 50 feet. The rear wall of this huge chamber is hung on rails and can be slid out of the way so that large test objects can be brought in to the chamber, or so that tests can be made on objects up to a quarter mile a way. Such large chambers require tall support columns (in this case, up to 25 feet long), and special attention must be given to target handling and target mounting problems.

### 8.3.2 Absorbing Materials

Early absorbent materials are discussed by Emerson.[10] Among them were absorbers made from animal hair (pig bristles), which is still marketed commercially today. The loss mechanism in most modern radar absorbents is carbon and/or some form of iron (carbonyl iron or ferrites) dispersed in a suitable matrix. Many forms of radar absorbers are resonant, but those used in anechoic chambers are broad-band materials, for obvious reasons.

Broadband performance is achieved by matching the impedance at the outer surface of the material to the impedance of free space and by gradually tapering the impedance to zero at the rear of the absorber layer where it is mounted to the chamber walls. This tapering can be achieved in one of two ways: a geometrical taper allows constant bulk properties to be used, while a "taper" in the characteristics from one layer to the next of a layered material allows flat sheets to be used. Geometrical tapering produces the most effective boardband material and the familiar pyramidal absorbers are popular for anechoic changers because of their high performance. The graded (layered) absorbers do not perform as well and find their most useful applications in diagnostic studies or experimental testing rather than in lining anechoic chamber walls.

The graded dielectric and tapered absorbers are usually made of carbon-loaded urethane foams. The basic method of manufacture is to compress a block or layer of foam between a pair of plates, immerse the compressed piece in a bath containing a concentration of carbon particles in suspension, and then to release the compression and allow the piece to expand like a sponge, drawing the solution and carbon particles into the open-cell structure. This may be done more than once. The piece is squeezed a final time to force out excess liquid and is allowed to dry or cure. The carbon density within the foam matrix is controlled by the bath concentration and other steps in the manufacturing process. The cured absorber block is typically sprayed with a low reflectivity paint.

Table 8-3 lists the properties of Emerson and Cuming's commercial line of radar absorbing materials.[23] The presentation of this information should not be construed to represent endorsement of these products, however. The lists are included because of the succinct form given in Emerson and Cuming's brochure and because of the wide variety of materials and applications encompassed in the list.

Except for Emerson and Cuming, absorber manufacturers have not been very successful. The B. F. Goodrich Sponge Products Division in Shelton, Connecticut, suffered a disastrous fire in the 1970s, and the decision was made not to resume production of high quality radar absorbents. The Plessy Microwave Materials Company, whose headquarters are in the United Kingdom, shut down its North American production facilities early in 1980. The McMillan Company, one of the pioneers in the production of resonant absorbers, has been defunct for several years.

## 8.4 EXAMPLES OF PAST AND EXISTING RANGES

### 8.4.1 Ranges in the Late 1960s

Many large corporations operate their own reflectivity ranges in support of internal development projects or government contracts for weapons systems research and development. Examples are the facilities of large aerospace companies such as McDonnell Douglas, Boeing, and General Dynamics. A large number of indoor ranges are in operation, some of them fairly small and many of them opeated by academic institutions such as Georgia Tech, the University of Michigan, and Ohio State University. A few are government owned and operated, such as the MACROSCOPE range at Ft. Belvoir, Maryland, and the RATSCAT range at Holloman Air Force Base, New Mexico. A few modern ranges are special-purpose facilities not necessarily used to measure the free space radar cross section of test objects.

It is interesting to note the similarities and differences among large outdoor ranges of 10 to 15 years ago. In a 33-month study of ranges conducted by the University of Michigan [14, 15] in the late 1960s five ranges were selected in an evaluation of large target capabilities. Of the five, two have gone out of business, and like the manufacturers of radar absorbent materials, one might surmize that the commercial market for radar reflectivity measurements is very competitive. The five ranges participating in the evaluation are listed in Table 8-4.

Of the five ranges listed, conductron and Radiation Service ceased operations in the 1970s. All except Micronetic (now Teledyne Micronetics) use the ground plane in their measurements; the Micronetics range is unusual in that an asphalt-coated berm in the shape of an inverted vee runs from the transmitter to the target turntable. The purpose of this berm is to deflect ground reflected waves away from the target and away from the receiver, and it is therefore intended to

**TABLE 8-3**
**Characteristics of Commercial Radar Absorbents**
(*Reprinted from ref.* [23])

| ECCOSORB ABSORBER | VHP | SPY | WG |
|---|---|---|---|
| Claim to Fame | Best Performance for Anechoic Chambers | Best Weatherproof Use | Low Cost for Non-critical Use |
| Frequency Range | 50 MHz & Up | 500 MHz & Up | 300 MHz & Up |
| Basic Composition | Loaded Urethane Foam | Loaded Open Cell Urethane Foam | Loaded Urethane Foam |
| Basic Geometry | Solid Pyramids on Flat Base | Solid Pyramids on Flat Base | Solid Wedges on Flat Base |
| Cost/Sq. Ft. | $4.50-$50.00 | $12.00-$17.50 | $25 |
| Service Temp. | To 250°F | To 250°F | To 250°F |
| Color of Surface | Light Blue | Light Blue | Light Blue |
| Power Handling | 2.0 watts/sq. in. | 10 watts/sq. in. | 2 watts/sq. in. |
| Fire Retardancy | Self-Extinguishing | Self-Extinguishing | Self-Extinguishing |
| Outdoor Usage | Use SPY | Excellent | Must be treated |
| Fuel Oil Resistance | Must be specially treated | Poor | Must be protected |
| Installation | Mech. Mount Large Pcs ECCOBOND 87H(NF) Small Pcs | Mech. Mount or ECCOBOND 87H(NF) | ECCOBOND 87H(NF) |
| Base Dimension | 24" x 24" | 24" x 24" | 24" x 24" |
| Type | 2 4 8 12 18 26 45 70 110 150 | 9 12 18 24 | 2 4 8 12 18 26 |
| Height or Thickness, In. | 2 4 8 12 18 26 45 70 110 150 | 9 12 18 24 | 2 4 8 12 18 26 |
| Weight, lb/sq ft | 0.5 0.5 1.0 1.2 1.5<br>2.0 3.5 6.0 8.0 11.0 | 5.5 4.8 5.8 6.2 | 1.5 2.6 4.5<br>6.0 8.5 10 |
| Lowest Frequency in GHz for Maximum Reflectivity of -17dB | - - - - - - - - - - | - - - - | Consult Emerson & Cuming, Inc. Canton, MA for Recommendations |
| -20dB | - - - - - - - - - - | 1 0.5 0.4 0.3 | |
| -30dB | 5.0 3.0 1.0 0.8 0.5 0.3<br>0.2 0.1 0.1 0.06 | 1.5 1 0.5 0.5 | |
| -40dB | 10.0 5.0 3.0 3.0 1.0 1.0<br>0.5 0.3 0.2 0.12 | 3 3 1 1 | |
| -50dB | 20.0 15.0 10.0 5.0 5.0<br>3.0 3.0 1.0 0.5 0.30 | 9 9 9 9 | |
| Technical Bulletin | 8-2-3 | 8-2-15A | 8-2-3B |

| HPY & HPY-W | CV & CV-W | OCF | SF | MX410 |
|---|---|---|---|---|
| Lightweight<br>High Performance<br>for Chambers | Electrically<br>Tapered, Best for<br>20 GHz & Up | Weatherproof<br>High<br>Frequency | Resonant<br>Very Thin<br>Flexible | Resonant<br>Thinnest<br>Flexible |
| 50 MHz & Up | 1 GHz & Up | 1 GHz & Up | 1-16 GHz | X-Band |
| Loaded<br>Urethane Foam | Loaded<br>Urethane Foam | Loaded Open-Call<br>Urethane Foam | Loaded Silicone<br>or Urethane | Resistive Front<br>Metal Foil Back |
| Hollow Pyramids<br>on Flat Foam Base | Foam Blocks<br>Convoluted Front | Foam Blocks<br>Convoluted Front | Flat Sheets | Flat Sheets |
| $4.50-$20.00 | $4.50-$11.25 | $7.50-$12.50 | $24.00-$49.00 | $48.00 |
| To 250°F | To 250°F | To 250°F | To 350°F SF<br>To 250°F SF-U | |
| White, Black<br>or Colors | White, Black<br>or Colors | White, Black<br>or Colors | Gray | Black |
| 2 watts/sq. in. | 1.5 watts/sq. in. | 10 watts/sq. in. | 10 watts/sq. in. | 3 watts/sq. in. |
| Self-Extinguishing | Self-Extinguishing | Self-Extinguishing | Non-Burning | |
| Use HPY-W | Use CV-W | Excellent | OK | OK |
| Use HPY-W | Use CV-W | Poor | OK | OK |
| Mech. Mount<br>Large Pcs<br>ECCOBOND<br>87H(NF)<br>Small Pcs | Stack or Adhere<br>ECCOBOND<br>87H(NF) | Stack or Adhere<br>ECCOBOND<br>87H(NF) | ECCOSIL TP50 | ECCOBOND<br>87H(NF) |
| 24" x 24" | 24" x 24" | 24" x 24" | 12" x 12" | 22" x 22" |
| 12 18 24 30 54 72 | 4 6 9 | 4 6 9 | 24 types | 410 |
| 12 18 24 30 54 72 | 4 6 9 | 4 6 9 | 0.035 to 0.22 | 0.025 |
| 1.0 1.2 1.3<br>2.5 3.2 3.7 | 0.8 0.9 1.8 | 1.1 1.2 2.5 | 0.6 to 6 | 0.156 |
| 0.5 0.4 0.3 0.2<br>0.12 0.08<br>1 0.8 0.6 0.4 0.3 0.2 | 2.5 1.5 1<br><br>5 3 1.5 | 2.5 1.5 1<br><br>5 3 1.5 | -20dB or better<br>at resonant freq. | -20dB or better<br>9.375 GHz |
| 2 1.5 1.2 0.8<br>0.6 0.5 | 8 6 3 | 8 6 3 | | |
| 10 7 6 5 4 3 | 25 25 25 | 25 25 25 | | |
| 8-2-2 | 8-2-14 | 8-2-15 | 8-2-18 | 8-2-20 |

| ECCOSORB ABSORBER | RF | NZ | FGM |
|---|---|---|---|
| Claim to Fame | Resonant<br>Thin<br>Flexible | Broadband<br>Low Frequency<br>+ High Temperature | Thin,<br>Very Broadband<br>2-12 GHz<br>Flexible Too |
| Frequency Range | 3-16 GHz | 50 MHz & Up | 2-12 GHz |
| Basic Composition | Reinforced<br>Urethane<br>Plastic & Foam | Fired Ferrite | Ferrites, etc.<br>in Silicon Rubber |
| Basic Geometry | Flat Sheets | NZ-1-Pyramid<br>NZ-2, NZ-31, NZ-41 and<br>NZ-51 = Flat Tile | Thin<br>Flat Sheets |
| Cost/Sq. Ft. | $10.00 | $60.00-$100.00 | $24.00-$49.00 |
| Service Temp. | To 300° F | To 500° F | To 500° F |
| Color of Surface | Green | Black | Brown |
| Power Handling | 1 watt/sq. in. | 20 watts/sq. in. | 10 watts/sq. in. |
| Fire Retardancy | | Non-Burning | Non-Burning |
| Outdoor Usage | Seal Edges | OK | OK |
| Fuel Oil Resistance | Seal Edges | OK | OK |
| Installation | ECCOBOND<br>87H(NF) | Mechanically mount,<br>or adhere with<br>ECCOSIL 1776 | ECCOSIL TP50 |

| Base Dimension | 24" x 24" | | | | 1" x 1" | 2.25" x 2.25" | | | | 12" x 36" | |
|---|---|---|---|---|---|---|---|---|---|---|---|
| Type | S | C | X | Ku | 1 | 2 | 31 | 41 | 51 | 40 | 125 |
| Height or Thickness, in. | 0.75 | 0.32 | 0.17 | 0.11 | 1 | 0.25 | 0.29 | 0.275 | 0.20 | 0.040 | 0.125 |
| Weight, lb./sq ft | 1.5 | 0.6 | 0.3 | 0.2 | 8.5 | 4.5 | 7.7 | 6.4 | 5.1 | 1.0 | 2.5 |
| Lowest Frequency in GHz for Maximum Reflectivity of -17dB -20dB -30dB -40dB -50dB | -25dB or better at 3 GHz | 5.5 GHz | 9.375 GHz | 16 GHz | -15dB Ave from 50MHz to 12GHz | -15dB at 0.2 GHz<br>-30dB<br>-10dB at 1GHz<br>-5dB at 5GHz | -35dB Resonance at 0.25 | 0.4 GHz | -35dB 0.8 | -10dB Ave from 4GHz to 10GHz | -12dB Ave from 2GHz to 12GHz |
| Technical Bulletin | 8-2-19 | | | | 8-2-17 | | | | | 8-2-23 | |

| UPF | H | FR | CV-CB | AN & AN-W | AN-P |
|---|---|---|---|---|---|
| Best Broadband<br>Weatherproof | Original Free<br>Space<br>Absorber<br>Low Cost | Old Standby<br>Where Flat,<br>White Surface<br>is Needed | Rigid Walkway<br>and Flooring<br>Absorber | Thin, Flexible<br>Easily Cut<br>Wrap Around | Flexible<br>Weatherproof<br>Fuelproof |
| 100 MHz-12 GHz | 300 MHz<br>& Up | 455 MHz<br>& Up | 300 MHz<br>& Up | 455 MHz<br>& Up | 2.4 MHz<br>& Up |
| Sintered<br>Ferrite &<br>Urethane | Curled<br>Rubbermaid<br>Fibers | Rigid Poly-<br>Styrene<br>Foam | Rigid Poly-<br>Styrene<br>Foam | Loaded<br>Urethane<br>Foam | Closed Cell<br>Poly-Ethylene<br>Foam |
| Hollow<br>Pyramids<br>on Flat Base | Flat<br>Sheets | Smooth Faced<br>Blocks | Smooth Faced<br>Blocks | Flat Sheets | Flat Sheets |
| $125.00 | $1.50-$7.50 | $2.50-$4.25 | $11.25-$21.25 | $3.50-$16.00 | $5.00-$6.75 |
| To 250°F | To 200°F | To 170°F | 170°F | To 300°F | To 200°F |
| Gray-Black | Light | White | White | White | White, Black<br>or Colors |
| 3 watts/sq. in. | 1.5 watts/sq in | 1 watt/sq in | 1 watt/sq in | 1.5 watts/sq in | 1 watt/sq in |
| Non-Burning | Self<br>Extinguishing | Self<br>Extinguishing | Self<br>Extinguishing | Self<br>Extinguishing | |
| Excellent | Must be<br>covered | OK | Must be<br>treated | Use AN-W | Excellent |
| Good | OK | NO | Poor | Use AN-W | Excellent |
| Mech. Mount<br>or ECCOSIL 1776 | Hooks or<br>ECCOBOND<br>45 | Stacking,<br>Anchors or<br>ECCOBAND<br>45 | Lay Loose | Drape, Tie or<br>Adhere with<br>ECCOBOND<br>87H(NF)<br>or large<br>blankets | Mech. Mount<br>or<br>ECCOBOND<br>90H<br>or large<br>blankets |
| 11.25" x 11.25" | 24" x 24" | 12" x 48" | 24" x 24" | 24" x 24" | 24" x 24" |
| UPF | 1 2 4 8 12 | 330 340 | 9 12 18 24 | 72 73 74 75<br>77 79 | 74 75 |
| 1.25 | 1 2 4 8 12 | 2 4 | 9 12 18 24 | 1/4 3/8 3/4<br>1 1/8 2 1/4 4 1/2 | 3/8 3/4 1 1/8 |
| 10 | 0.33 0.45 1.1<br>1.7 2.3 | 0.7 1.25 | 1.4 2 3 3.9 | 0.1 0.2 0.3 0.5<br>0.9 0.3 | 0.15 0.3 0.5 |
| -15dB<br>Ave<br>from<br>100MHz<br>to<br>12GHz | - - - - - | - - | - - - - | - - - - - | 7.5 3.5 2.4 |
| | 5 2.5 1 0.5 0.3 | 2.3 0.94 | - 0.5 0.4 0.3 | 20 7.5 3.5 2.4<br>1.2 0.6 | - - - |
| | - - - - - | - - | 3 2.5 1.5 1 | - - - - - | - - - |
| | - - - - - | - - | 10 10 7 4 | - - - - - | - - - |
| | - - - - - | - - | - - - - | - - - - - | - - - |
| 8-2-17A | 8-2-1 | 8-2-5 | 8-2-14B | 8-2-4&8-3-4A | 8-2-4B |

| ECCOSORB ABSORBER | RM | RMP-S & RMP-V | FBC | HT | | PHD |
|---|---|---|---|---|---|---|
| Claim to Fame | Flexible High Temp | Comformable for Outdoors & High Temp | Lightweight Structural Core Material | Lightweight Use to 650°F High Power | | Premium Reflectivity Highest Power |
| Frequency Range | 2.4 GHz & Up | 2.4 GHz & Up | 3 GHz & Up | 940 MHz & Up | | 3 GHz & Up |
| Basic Composition | Open-Cell Sil Rubber Foam | Sill Rubber (-S) Vinyl (-V) | Coated Fabric | Closed-Cell Ceramic Foam | | Fiberglass Honeycomb |
| Basic Geometry | Flat Sheet | Array of Pyramids | Array of Pyramids | Flat-Faced Block | | Array of Pyramids |
| Cost/SQ FT | \$60.00 | \$29.00-\$39.00 | \$14.00 | \$20.80-\$21.70 | | \$50-\$100 |
| Service Temp. | To 400°F | To 400°F (-S) To 300°F (-V) | To 200°F | To 650°F | | To 500°F |
| Color of Surface | White | Black | Yellow | Silver | | Black |
| Power Handling | 3 watts/sq in | 5 watts/sq in | 2 watts/sq in | 25 watts/sq in | | 200 watts/sq in |
| Fire Retardancy | Self Extinguishing | Self Extinguishing | Self Extinguishing | •Non-Burning | | |
| Outdoor Usage | OK | Outstanding | OK | OK | | Excellent |
| Fuel Oil Resistance | OK | Outstanding | OK | OK | | Excellent |
| Installation | ECCOSIL 1776 | ECCOSIL 1776 | Drape or ECCOSIL 1776 | Stack, Mortar or ECCOSIL 1776 | | Mech. Mount or ECCOSIL 1776 |
| Base Dimension | 24" x 24" | 12" x 12" | 12" wide rolls | 12" x 18" | | 24" x 24" |
| Type | 75 | 75 | 75 | 98 | 99 | 3 |
| Height or Thickness, In. | 1 | 1 | 1 | 2 | 3 | 3½ |
| Weight, lb/sq ft | 1 | 1 | 1 | <0.15 | 0.7 | 2 |
| Lowest Frequency in GHz for Maximum Reflectivity of 17dB | - | 3 | 3.0 | 2.7 | 0.94 | 0.4 |
| 20dB | 2.4 | - | - | - | - | - |
| 30dB | - | - | - | - | - | - |
| 40dB | - | - | - | - | - | - |
| 50dB | - | - | - | - | - | - |
| Technical Bulletin | 8-2-9 | 8-2-9A | 8-2-24 | 8-2-7 | | 8-2-7A |

**TABLE 8-4**
**Reflectivity Ranges Evaluated in the Late 1960s**

| SITE | TRANSMITTER | MINIMUM RANGE USED | GEOMETRY | DATA RECORDED Type | DATA RECORDED Dynamic Range | DIGITAL EQUIPMENT |
|---|---|---|---|---|---|---|
| Conductron | CW | 200 | Ground Plane | Amplitude | 40 dB | Paper Tape |
| Radiation Service | Pulse | 1000 | Ground Plane | Amplitude | 40 dB | Punch Cards |
| General Dynamics | Pulse | 1800 | Ground Plane | Amplitude<br>Phase | 50 dB<br>360° | Paper Tape |
| RATSCAT | Pulse | 1200 | Ground Plane | Amplitude<br>Phase (L-Band) | 50 dB<br>360° | Paper Tape |
| Micronetics | Pulse | 600 | Direct "Λ" | Amplitude | 40 dB | Magnetic Tape |

provide freespace target illumination conditions. Micronetics was in the forefront of data recording in those days, being the only range to take advantage of the high speed features of magnetic tape as a recording medium. The other four recorded data on paper tape and punched cards, which was a very slow process. Very few modern ranges, if any, record data on these media now.

Conductron and Micronetics both made measurements at much less than the far field range for the largest target and the highest frequencies in the test program. The remaining three ranges also used less than the far field range, although not as foreshortened as the ranges used by Conductron and Micronetics. The data collected by Conductron suffered severe near field distortion, with the broadside lobe of a metal cylinder being split into a pair of lobes with a shallow null between them, and with the amplitude several dB below the true far field value. This was a consequence of using a CW instrumentation system that radiated very low power, probably much less than a watt. The other ranges used more conventional pulsed systems, with pulsewidths ranging from 0.1 to 1 microsecond.

Only two ranges (RATSCAT and General Dynamics') were able to measure the phase of the backscattered signal. General Dynamics, it might be noted, had the prime contract to build and instrument the RATSCAT range; hence, it is not unexpected that both ranges had the same capability of coherence. General Dynamics used the greatest range in the measurement of the test objects, which were a collection of precision-made cylinders, the largest being 32 feet long and 5 feet in diameter. Of particular interest is the fact that the now-defunct Radiation Services range turned in the most accurate results for the lowest cost in the least time, a tribute to the company and its range crew. Not all of these virtues are apparent in Table 8-5, which is a summary of how the five ranges were rated on their measurements of the precision cylinders.[15]

### 8.4.2 Examples of Unique Present-Day Ranges

The world, and primarily North America, is dotted by several dozen radar reflectivity ranges of both the indoor and outdoor kinds. Many of them share common methods of instrumentation and calibration, and many have similar sensitivities and equipment. But rather than catalog the capabilities and features of these ranges, which do not differ in large measure from one facility to the next within a given class, it is of interest to examine the features of some of the more unusual ranges.

**TABLE 8-5**
**Range Ratings**
*(Adapted from Hiatt, et al., ref. [15]; © 1965 IEEE)*

| Range | Facilities, Techniques, and Procedures | CYLINDER DATA ACCURACY Direct Polarization | Cross Polarization |
|---|---|---|---|
| Conductron Corporation | B– | D | C |
| Radiation Services | B | B | C |
| General Dynamics, Ft. Worth | B+ | C | D |
| RATSCAT | B | C | C |
| Micronetics | B– | D | D |

A selection of such ranges is presented below. The list is not intended to be exhaustive, nor to endorse the particular virtues of the facilities discussed. It was assembled only with the idea of demonstrating the variety of approaches to determining radar reflectivity characteristics.

### *RATSCAT*

RATSCAT is probably the most impressive outdoor reflectivity range in current operation. Located on several square miles of alkali flats on the White Sands Missile Range in south central New Mexico, the site is remote and secure. Full scale aircraft and missiles weighing up to 50,000 pounds can be mounted on a rotator 40 feet in diameter at a range of 5600 feet. Several other small rotators (17 feet in diameter) dot the site at fixed ranges from 458 to 2458 feet. Large mobile target shelters can be towed over the target and rotator for protection against weather and satellite observation. Large mobile cranes are used to mount heavy targets on the support columns, while smaller targets are handled with hoists built into the mobile target shelters.

Figure 8-13 shows a relatively modest-size target (a TALOS missile) in place for measurement at Pit #2. The rotating turntable supporting the plastic foam columns is 17 feet in diameter and can be raised or lowered by a hydraulic ram. The rotator and inner surface of the pit are covered with radar absorbing material to suppress interactions between the target and its environment. The Operations Building can be seen in the background, flanked by three antenna hoist-

**Figure 8-13.** TALOS missile mounted on support columns at Pit 2. The RATSCAT operations building visible in the background is 1158 feet away (courtesy of Ratscat).

ing and positioning frames. Such massive support columns as shown in the photograph are necessary for heavy targets.

Frequency coverage is continuous from 30 MHz to about 24 GHz, and four millimeter wave bands through 95 GHz are covered by special millimeter wave instrumentation radars. Typical transmitter outputs are 1 kW for the continuous frequencies, although high power transmitters of 10 and 100 kW are available for certain frequency bands. The phase of the target echo can be measured as well as the amplitude, and digital data collection is being provided more frequently than in the past because of instrumentation upgrading projects at the site. Data are processed in a variety of ways to suit the range users' needs.

The RATSCAT reflectivity range is a division of the U.S. Air Force 6585th Test Group, with headquarters at Holloman Air Force Base, New Mexico. A 48-page brochure describing the facilities and capabilities of RATSCAT is published annually. The range is operated as a service for government agencies only. A survey of the RATSCAT concept is given by Marlow, *et al.*[16]

### MACROSCOPE

MACROSCOPE is an unusual facility operated by the U.S. Army Mobility Equipment Research and Development Command at Ft. Belvoir, Virginia. This CW range uses a homodyne detection scheme in which a sample of the transmitter signal is used as the local oscillator signal. The frequency of operation is about 98.6 GHz and the facility is used primarily in diagnostic studies of the returns from scale models of ships and military vehicles. The transmitted signal is amplitude modulated at a rate of 100 kHz.

The range can be operated in one of two modes. In one mode, the target is illuminated by a broad beam and is oscillated back and forth at a rate of a few Hertz, thereby shifting the frequency of scattered waves. This allows for discrimination against background return. A second detector extracts the amplitude of the shifted signal from the 100 kHz modulation. In the other mode of operation, a microwave lens is installed and the target (on its ground plane) is scanned in range and cross range through the focus of the lens. The resolution of the system is purported to be about 0.5 inch in range and cross range. A signal processing system displays the imagery on a television screen and the image reveals the strong scattering centers on the target as bright spots. The computer used with the facility and the software developed for it allows close studies to be made of small areas on the target. Although the amplitudes of the various contributions to the radar echo can be obtained in a coarse sense, the facility is most useful in the location and identification of dominant target scatterers.

In its early configuration, the system oscillated the target and ground plane together, but currently only the target is moved. The data are calibrated in amplitude by replacing the target with a metallic calibration sphere. Although an imperfectly conducting ground plane could be used, or a rough one to simulate the reflection from the sea or terrain, this has not been done. The operators of the range have not had the requirement to perform this simulation, and they contend that the diagnostic purposes of the facility are served just as well with a "shiny" ground plane, since the target images in the ground plane are brighter. A brief description of the facility is given by Paddison, *et al.*[17] in their paper on ship cross sections.

### *EMI Modeling Facility*

EMI Electronics Limited has developed a radar modeling capability at the United Kingdom National Radio Modeling Facility. Emphasis at this facility is on the development of instrumentation systems and the collection and interpretation of radar scattering data at frequencies up to 2000 GHz (2 THz). Since virtually all of the measurements and testing are performed on scale models, EMI has fabricated many models over the years, ranging from missiles and shells to ships and aircraft. EMI has also developed state of the art components, such as RF sources and detection systems.

All the measurements are conducted indoors. As of 1978, nine different radar systems were operable in conjunction with seven different model support systems. Unlike most indoor facilities, this one makes limited use of radar absorbing material and relies instead on range gating to eliminate background reflections. As many as seven different systems can be operated simultaneously, although it is doubtful this ever occurs.

Targets are usually suspended from overhead carriages by means of thin electric lines, although tapered support columns may also be used. The line of sight is inclined such that the support lines are not viewed at normal incidence. For some measurement systems the target is slowly rotated by the support carriage while being viewed by a stationary radar. For others the target is held stationary while the instrumentation is rolled past the target on circular or straight tracks. Obviously the tracks must be very smooth and precisely laid for coherent measurements at these frequencies, which EMI has the capability to perform.

A succinct survey of the modeling facility, including photographs and diagrams, is given by Cram and Woolcock.[18]

*ERIM*

The Environmental Research Institute of Michigan (ERIM) has operated a radar imaging system for several years in which the test target is rotated in azimuth while a coherent radar is trained upon it. The system, known as the Rotating Platform Radar Imaging Facility, allows radar parameters to be varied in a controlled manner, and changes in the imagery can be observed at less expense than with an airborne radar.

The radar antenna illuminates a target on the ground-based platform which rotates about a vertical axis, as shown in Figure 8-14. The rotation of a rigid target introduces a Doppler gradient which can be sensed by a coherent radar and used to obtain fine cross range resolution. In effect, a synthetic aperture is generated in the cross range direction.

**Figure 8-14.** A field piece in position of ERIM's platform. The cans on either side of the targeted are calibration standards which remain on the platform with the target during measurements (courtesy of ERIM).

To obtain fine range resolution, the transmitted signal must have a very wide bandwidth. Although the required bandwidth could be obtained with a narrow pulse, the ERIM system uses an alternate method well suited to two-dimensional optical processing and digital recording. The frequency of the transmitted pulse is varied linearly with time during the pulse, producing a chirped pulse. After the amplitude and phase information in the return signal is recovered by the radar and recorded, a processor produces a two-dimensional Fourier transform (i.e., an image) of the target scene. The image displays the radar cross section in slant range and Doppler coordinates.

Two types of processing are used with the facility. For optical processing, the data are recorded on film which is then developed and placed in a coherent optical processor. An image of the target scene appears at the output plane of the processor where it can be studied or photographed. Optical processing allows data from a full 360° target rotation to be superimposed on a single photograph. For digital processing, the data are recorded on analog tape, which is later digitized and packed on a compatible magnetic tape. The one-dimensional Fourier transform of the data provides radar cross section as a function of slant range. To form a range-Doppler image, the computer arranges the data in a special polar format and forms a two-dimensional Fourier transform. The amplitude of digitally processed images is proportional to the square root of the radar cross section, hence the returns of individual scattering centers can be measured if there are calibrated reflectors in a scene. The cylindrical cans on either side of the field piece in Figure 8-14 are such calibrated targets.

Data currently can be collected at either of two center frequencies (10.0 and 34.1 GHz) and a 94 GHz capability is under development. At 10 GHz any linear polarization can be transmitted and there are two orthogonal, linearly polarized receiving channels. Normally, both parallel and cross-polarized returns are recorded on analog tape for subsequent digitization, and one channel can also be recorded on film for optical processing. Both monostatic and bistatic data can be collected at X-band. The X-band antennas are mounted on an antenna tower 90 feet from the platform center, and the depression angle is continuously variable form 0 to 24.8°. At 34.1 GHz, the available transmitter polarizations are horizontal, vertical, and circular, and both parallel and cross polarized returns are received. Only monostatic data are collected at 34. 1 GHz, and the depression angle is fixed at 5 degrees.

The Air Force Avionics Laboratory has been the primary sponsor for the facility. Because the output imagery is classified, the use of the facility is restricted to the government and government contractors. The facility provides (1) high resolution range-Doppler images from which the cross sections of individual scattering centers can be determined and (2) radar cross section as a function of slant range. It does not provide total radar cross section. The facility is also useful for simulating airborne SAR radars so that the effect of parameter changes can be studied. Data have been collected on targets such as vehicles, artillery pieces, radar camouflage nets, space objects, and scale models of large targets.

*The University of Michigan*

The Radiation Laboratory at the University of Michigan operates a unique surface field measurement range in which the induced fields on a target are probed. The facility cannot be strictly classed as a reflectivity range, yet it is sufficiently unusual to be discussed briefly. Developed in the mid-1960s in support of U.S. Air Force studies of the scattering from re-entry bodies, the facility is currently being used to investigate surface fields as might be induced by the electromagnetic pulses generated by nuclear blasts.

The range is housed in an indoor tapered anechoic chamber of more or less conventional design, except that a hole has been cut in the ceiling of the receiving end of the chamber. The test object is placed under this hole on a foamed plastic column in the desired orientation and remains stationary while the surface fields are probed. A broadband transmitting antenna has been developed and installed at the transmitting end (the apex) of the chamber, and the frequency capability of the system is continuous from 100 MHz to 5 GHz.

Two kinds of surface probing can be performed. A "free space" unbalanced loop probe, typically from 3 to 10 mm in diameter, can be moved over the surface of the target, thereby sampling the tangential magnetic field. The probe loop is formed at the end of a small diameter, semi-rigid coaxial line which projects downward from the hole in the chamber ceiling. The probe and its lead are supported by a low-reflectivity support structure attached to a carriage above the ceiling, and the carriage is driven by small, remotely controlled dc motors. Since the metallic probe lead projects vertically into the chamber, the incident wave is alway chosen to be horizontally polarized to minimize interactions with the probe lead and to reduce stray pick-up.

In the other mode of operation, small loop probes or monopoles are attached to the target surface at selected stations and their leads brought outside the target at some point of geometric symmetry. The leads are run vertically up to the hole in the ceiling, as with the movable probe, to minimize stray induced fields. The fixed probes are oriented to sample the tangential magnetic field or the normal electric field, depending on the desired data to be collected.

The RF sources are a pair of sweep oscillators that are switched into or out of operation by a control computer. These sources can cover the frequency range from 10 MHz to 18 GHz but not the entire range. A set of four power amplifiers, each covering a portion of the entire system, boosts the 10 mW source output to 1 watt of input to the broadband antenna. The amplifiers are switched in and out of operation by the control computer as needed, depending on the particular frequency demanded.

The computer is programmable and the user can specify the frequencies and frequency intervals to be used in the measurements. The data are collected at a sequence of discrete frequencies, rather than continuously over the entire spectrum. The probe output signals are detected and amplified by a system of preamplifiers and a microwave network analyzer extracts the phase and amplitudes of the signals at each frequency. These are then converted to digital form in A/D coverters and stored or written on tape by the control computer for subsequent analysis and processing.

A brief description of the facility in its early configuration is given by Knott, *et al.*, [19] and a discussion of later improvements may be found in Liepa.[20, 21]

*Georgia Tech*

Georgia Tech operates what is called a Compact Range, in which antenna and reflectivity measurements are made in the extreme near field of a large parabolic reflector. The concept originated at Georgia Tech and the facility has been used for nearly a decade in support of research at the Engineering Experiment Station. When fed at its focus by a low gain feed horn, the reflector provides a collimated beam of parallel wavefronts, thereby approximating a quasi-planar wave within a very short distance, hence the name of the facility. This facility is described in greater detail in the next chapter.

## REFERENCES

1. Kouyoumjian, R.G. and L. Peters, Jr., "Range Requirements in Radar Cross Section Measurements," *Proc. IEEE*, Vol. 53, August 1965, pp. 920-928.
2. Knott, E.F. and T. B. A. Senior, "How Far Is Far?" *IEEE Transactions on Antennas and Propagation*, Vol. AP-22, pp. 732-734, September 174.
3. Freeny, C. C., "Target Support Parameters Associated with radar Reflectivity Measurements," *Proc. IEEE*, Vol. 53, August 1965, pp. 922-936.
4. Senior, T. B. A., Knott, E. F., Freeny, C. C. and Smith, C. H., "Radar Cross Section Target Supports — Plastic Materials," Rome Air Development Center, Griffiss AFB, N. Y., Report RADC-TDR-64-381, June 1964.
5. Plonus, M. A., "Thcorctical Investigation of Scattering from Plastic Foams, "*IEEE Transactions on Antennas and Propagation*, Vol. AP-13, January 1965, pp. 88-93.
6. T. B. A, Senior, M. A., Plonus and E. R., Knott, "Designing

Foamed-Plastic Materials," *Microwaves*, December 1964, pp. 38-43.

7. Knott, E. F. and T. B. A. Senior, " Studies of Scattering by Cellular Plastic Materials," University of Michigan Radiation Laboratory Report 5849-1-F, April 1964.
8. Kerr, D. E., ed., *Propagation of Short Radio Waves*, Volume 12 of the MIT Radiation Laboratory Services, Boston Technical Publishers, Inc., Boston 1964, page 116.
9. Bush, T. F., ed., "Evaluation of the RATSCAT Ground Plane Range," Report PE00911, Physical Science Laboratory, New Mexico State University, Las Cruces, New Mexico, August 1978.
10. Emerson, W. H., "Electromagnetic Wave Absorbers and Anechoic Chambers Through the Years," *IEEE Transactions on Antennas and Propagation*, Vol. AP-21, July 1973, pp. 484-490.
11. King, H. E., F. I. Shimabukuro, and J. L. Wong, "Characteristics of a Tapered Anechoic Chamber," *IEEE Transactions on Antennas and Propagation*, Volume AP-15, May 1967, pp. 488-490.
12. Dybdal, R. B., and C. O. Yowell, "VHF to EHF Performance of a 90-foot Quasi-Tlapered Anechoic Chamber," *IEEE Transactions on Antennas and Propagation*, Vol. AP-21, July 1973, pp. 579-581.
13. Emerson, W. H., and H. B. Sefton, Jr., "An Improved Design for Indoor Ranges," *Proc. IEEE*, Vol. 53, August 1965, pp. 1079-1081.
14. Hiatt, R. E., E. F. Knott, and T. M. Smith, "Evaluation of Selected Radar Cross Section Measurement Ranges — I," University of Michigan Radiation Laboratory, Report 7462-1-F(I), March 1969.
15. Hiatt, R. E., E. F. Knott, and T. M. Smith, "Evaluation of Selected Radar Cross Section Measurement Ranges — II: Cylinder Tests and Range Evaluation Procedures," University of Michigan Radiation Laboratory, Report 7462-1-F(IIa), March 1969.
16. Marlow, H. C., D. C. Watson, C. H. Van Hoozer, and C. C. Freeny, "The RATSCAT Cross Section Facility," *Proc. IEEE*, Vol. 53, August 1965, pp. 946-954.

17. Paddison, F. C., C. A. Shipley, A. L. Maffett, and M. H. Dawson, "Radar Cross Section of Ships," *IEEE Transactions on Aerospace and Electronic Systems*, Vol. AES-14, January 1978, pp. 27-34.

18. Cram, L. A., and S. C. Woolcock, "Review of Two Decades of Experience Between 30 GHz and 900 GHz in the Development of Model Radar Systems," *Millimeter and Submillimeter Wave Propagation and Circuits, AGARD (Advisory Group for Aerospace Research and Development) Conference Proceedings* No. 245, September 1978.

19. Knott, E. F., V. V. Liepa, and T. B. A. Senior, " A Surface Field Measurement Facility," *Proc. IEEE*, Vol. 53, August 1965, pp. 1105-1107.

20. Liepa, V. V., "Sweep Frequency Surface Field Measurements," Sensor and Simulation Note 210, Air Force Weapons Laboratory, July 1975.

21. Liepa, V. V., "Measurements of Surface Fields on Scale Model E-4B Aircraft," University of Michigan Radiation Laboratory, Report 01678-1F, October 1979.

22. Beckmann P. and A. Spizzichino, *The Scattering of Electromagnetic Waves from Rough Surfaces*, Pergammon Press, New York, 1963.

23. "Microwave Absorbers," Brochure 313673, Emerson & Cuming, Inc., issue date unknown, but probably published in the late 1970s.

# CHAPTER 9
# COMPACT RANGE MEASUREMENTS

**Michael T. Tuley**
**Engineering Experiment Station**
**Georgia Institute of Technology**

## 9.1 INTRODUCTION

Accurate measurements of radar cross section require that the target be illuminated by a uniform plane wave; that is, a wave having constant amplitude and phase. The most common way to achieve quasi-plane wave conditions is to place the target at a sufficient distance from the source so that the incident spherical wave, over the dimensions of the target, can be considered plane. A common rule of thumb for achieving that condition is to require a $2d^2/\lambda$ separation between the target and the source where $d$ is the largest target dimension and $\lambda$ is the wavelength. Relaxed conditions are often possible and are detailed in Chapter 8.

Satisfying the rule-of-thumb condition for a 5-foot target requires that a 500-foot range be used at X-band, a sometimes impractical separation between source and target. To reduce the separation requirement, several "compact range" techniques have been proposed and implemented. One realization uses a dielectric lens to produce a quasi-plane wave, and that technique will be discussed briefly. A second technique, utilizing a parabolic reflector, has been more widely implemented and will be discussed in detail.

Reference [1] is the basic paper describing the use of a dielectric lens in a compact range configuration, and reference [2] describes the use of parabolic reflectors in a compact range. While this chapter will focus discussion on the reflector compact range, general comments concerning signal generation, target support, feed requirements, and background signal rejection apply equally well to either configuration.

### 9.1.1 Description

The major advantage of a compact range over the conventional reflectivity range lies in its smaller size. For that reason, compact ranges will generally be indoor installations, often constructed in anechoic chambers. All consist of the following set of generic components, although the details may vary.

Plane wave converter (lens or reflector)

Signal generation and feed system

Target support structure

Receiving and recording electronics

Reference targets

The component which defines the compact range is a device to convert a spherical or cylindrical wave from the feed into a quasi-plane wave. This may be done using a lens or a reflector. Only point source feeds, and thus, spherical-to-plane-wave conversion are discussed in this chapter. Reference [2] contains an excellent discussion of the use of a line source feed with a parabolic cylinder reflector to convert cylindrically diverging rays into a plane wave.

The power generation and feed system must produce a stable signal at the desired frequency with enough power to give adequate signal-to-noise ratios and must transmit with a radiation pattern and placement designed to reduce the problems of stray radiation. The target support structure is normally made of a low backscatter material such as styrofoam and is mounted on a positioner allowing target RCS to be measured as a function of aspect. If the compact range is not built in an anechoic chamber, absorber should be placed at strategic locations within the enclosure to reduce stray radiation.

The associated electronics require a calibrated receiver, a recording device, and a method of discriminating against residual stray radiation. Discrimination may be accomplished by nulling the background in a CW system or by temporal discrimination of the return of a pulsed or swept frequency system.

As in other reflectivity ranges the compact range basically measures relative power. Thus, for absolute RCS to be provided, reference targets of known radar cross section are required. Spheres, cylinders, and flat plates, whose RCS may be calculated analytically, are most often used as reference targets.

### 9.1.2 Theory of Operation

*Lens Range*

The lens type compact range typically uses a low dielectric constant, plano-convex lens to adjust the phase of the transmitted wave so that it is effectively plane. While the use of a metal or high dielectric constant lens has been proposed, use of those types for RCS measurements is not usually feasible, since the backscatter from the lens can exceed that from the target.[1] Thus, materials with a dielectric constant below 1.05, such as styrofoam, are normally prescribed. A recent investigation[3] proposed use of a "reversed" lens configuration to reduce lens return and thereby allow use of solid plastics with moderate dielectric constants ($\epsilon_r \approx 2.5$) rather than foams for lens construction.

The lens is used to transform the spherical wavefront radiated by the source to a planar phase front. To do this, path lengths to a plane on the target side of the lens must be matched, which for the conventional lens configuration in Figure 9-1 leads to an equation for the lens surface of:

$$R = \frac{f(n-1)}{n \cos \Theta - 1} \tag{9-1}$$

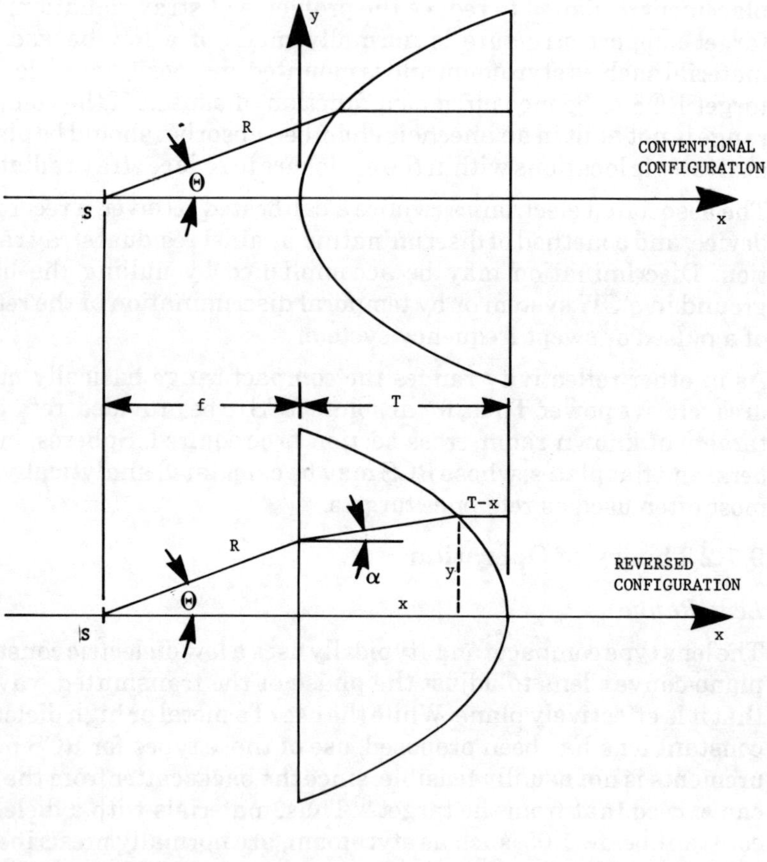


**Figure 9-1.** Two lens configurations. The lens contours are *not* the same (reprinted from Knott, ref. [3]).

where $n$ is the index of refraction of the lens, $f$ is the focal length, and the other variables are labeled in Figure 9-1. This is the equation of a hyperbola with the focus at the feed point.[4] Equation (9-1) was derived from geometric optics considerations and makes use of the fact that the effective path length of a ray in a medium with an index of refraction, $n$, is increased by the factor $n$ over its value in free space. Thus, the thickness of the lens adjusts the effective path length of all the rays leaving the feed to make them equal. The RF source is placed at point $S$ in Figure 9-1 and the phase, but not the amplitude, of the wave is theoretically constant in the y-z plane to the right of the lens. The target is normally placed one to two lens diameters from the lens.

A major problem with lens type compact ranges is the power reflected from the lens. For the conventional configuration all of the rays entering the lens arrive at the flat rear face in phase and normal to the face, providing the maximum coherent power reflection back to the source. The ratio of the reflected to transmitted power in that case is just the square of the voltage reflection coefficient, $\Gamma$, of the lens material and is given by

$$|\Gamma|^2 = \left| \frac{1 - n}{1 + n} \right|^2 \tag{9-2}$$

Reflected power from the lens is minimized by minimizing the refractive index of the material from which the lens is constructed, driving the choice toward low density plastic foams.

The reversed configuration, also shown in Figure 9-1, reduces the reflection problem by avoiding any physical surface which coincides with a constant phase contour. The equation for that surface cannot easily be written in the same form as that for the conventional lens. One convenient form of the equation for the surface is

$$x = \frac{\cos\alpha}{n - \cos\alpha} \left[ T(n - 1) - f\,\frac{1 - \cos\Theta}{\cos\Theta} \right] \tag{9-3}$$

where by Snell's law

$$\sin\Theta = n \sin\alpha \tag{9-4}$$

The maximum thickness, $T$, of a reversed lens is related to the lens diameter, $D$, by

$$T = \frac{D}{n - 1} \left[ \sqrt{(f/D)^2 + 1/4} - f/D \right] \tag{9-5}$$

The major advantage of the reversed lens is the reduced lens reflection level. Table 9-1 provides formulas for the ratio of reflected to transmitted powers for the apex and face of conventional and reversed lenses, where G is the power gain of the transmit/receive horn illuminating the lens. Figure 9-2 plots the Table 9-1 results for the case of a 12 foot diameter, 90 foot focal length lens at 35 GHz using a 35 dB gain feed antenna. For comparison purposes, the return from a 1 $m^2$ target and from a rear wall covered with typical pyramidal absorber are also plotted. Note that even for relatively low indices of refraction the lens return will exceed target return, but that the return from the reversed configuration lens is typically 20 dB or more below that of the conventional lens.

**TABLE 9.1**
**Formulas for $P_r/P_t$**
(*Adapted from Knott, ref.* [3])

| RETURN | FORMULA |
|---|---|
| flat face, conventional lens | $\left[\frac{n-1}{n+1}\right]^2$ |
| apex, conventional lens | $\left[\frac{G\lambda(n-1)}{8\pi n f}\right]^2$ |
| flat face, reversed lens | $\left[\frac{G\lambda(n-1)}{8\pi f(n+1)}\right]^2$ |
| apex, reversed lens | $\left[\frac{G\lambda(n-1)}{8\pi f(nf+T)}\right]^2$ |

Because of problems of dielectric nonuniformities and complications of construction, lenses of foamed plastic have generally been limited to diameters around one meter, thereby severely limiting target size. Solid plastics can easily be manufactured with uniform dielectric constants, but as seen, use of them significantly increases lens reflection. A pulsed system can be employed to gate out the lens return, but the problems of receiver recovery time must be solved. Employment of the reversed lens configuration can significantly reduce lens reflection, making the receiver recovery time problem manageable, even for solid plastic lenses. Thus, if a large lens system is desired, a reversed lens configuration using a solid plastic lens, combined with a short pulse source and fast recovery receiver, might be considered.[3]

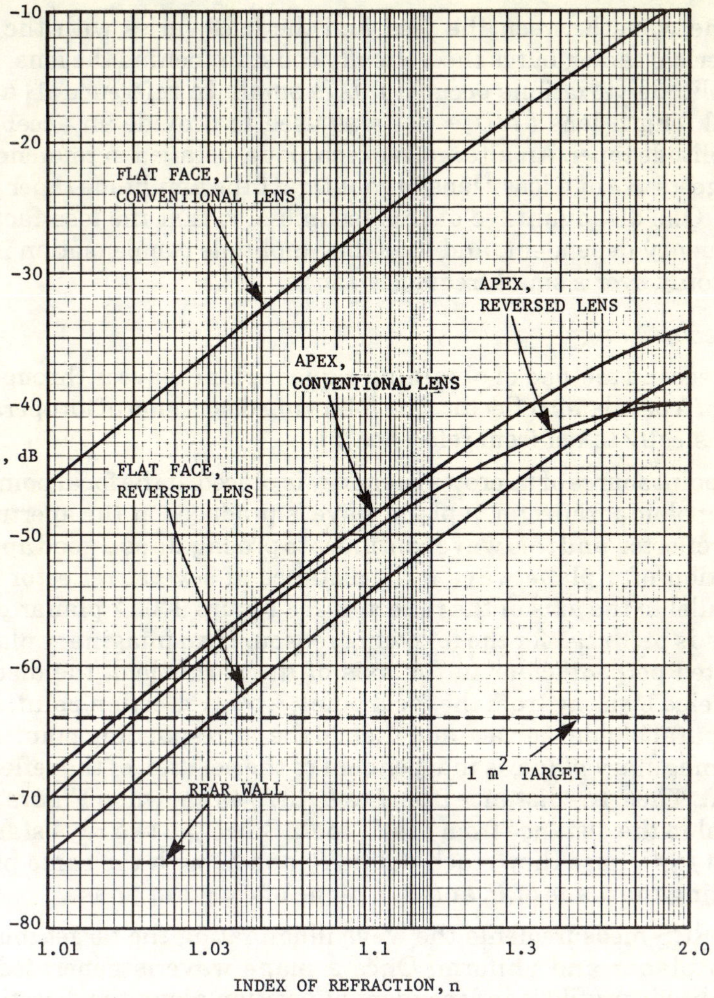


**Figure 9-2.** Comparison of front and rear lens return at 35 GHz for a 12 foot diameter lens with a focal length of 90 feet (reprinted from Knott, ref. [3]).

The maximum allowable target dimensions are clearly a function of lens size but also depend on other design factors. The size of the useful measurement area is often controlled by the amplitude taper, and production of the maximum flat amplitude area relative to the lens diameter can be accomplished in two ways. Since the amplitude of the wave between the feed and the lens varies with the $1/R^2$ spherical factor, but is theoretically constant behind the lens, large amplitude variations occur if $f/D$ is small. In reference [1] a long focal length lens ( $f/D \approx 9$ ) was used to provide an acceptable amplitude taper. An alternative technique, detailed in reference [4], suggests use of a lossy lens to counteract the amplitude taper problem. One disadvantage of this technique is that the loss factor is frequency dependent, and thus the amplitude compensation is correct only over a small frequency range.

*Reflector Range*

The second method of obtaining a quasi-plane wave is through the use of a parabolic reflector. The following discussion of its operation is basically taken from reference [5].

If a point source of microwave energy is placed at the focal point of a paraboloidal reflector, a plane wave is produced at the aperture of the reflector, and this wave continues to propagate as a good approximation of a plane wave for a distance of several reflector radii parallel to the axis of the reflector (*see* Figure 9-3). A planar phase front is produced for the following reason. If an imaginary plane is erected perpendicular to the axis of the paraboloid, the distance traveled by a ray from the focal point, $f$, to a reflection point, $R$, on the paraboloidal surface and then to the corresponding point, $P$, on an imaginary plane, is independent of the position of the reflection point. Thus, the distance from $f$ to $R_1$ to $P_1$ as shown in Figure 9-3 is equal to the distance from $f$ to $R_2$ to $P_2$. Since the two rays start out from $f$ in phase they reach points $P_1$ and $P_2$ with the same phase, making the plane of $P_1$ and $P_2$ a plane of constant phase.

For RCS measurements the wave illuminating the target must be both planar and uniform. Once a plane wave is generated, the amplitude profile is independent of position along the direction of propagation. That is to say, if the measurement plane is moved along the axis of the paraboloid to some other position, the amplitude of each ray is the same as it was in the original position. However, referring again to Figure 9-3, as the wave propagates from the focal point to the reflector surface, it diverges as a spherical wave. Therefore, if rays 1 and 2 are emitted from $f$ with equal amplitudes, the ray at $R_1$ has a smaller amplitude than the ray at $R_2$, simply because it has traveled a greater distance. Also, the ray at $P_1$ has a smaller amplitude than the ray at $P_2$ because the amplitude profile does not vary after the wave becomes a plane wave. For rays 1 and 2 to have equal amplitudes at points $P_1$ and $P_2$, it is necessary that ray 1 be initiated with a larger amplitude.

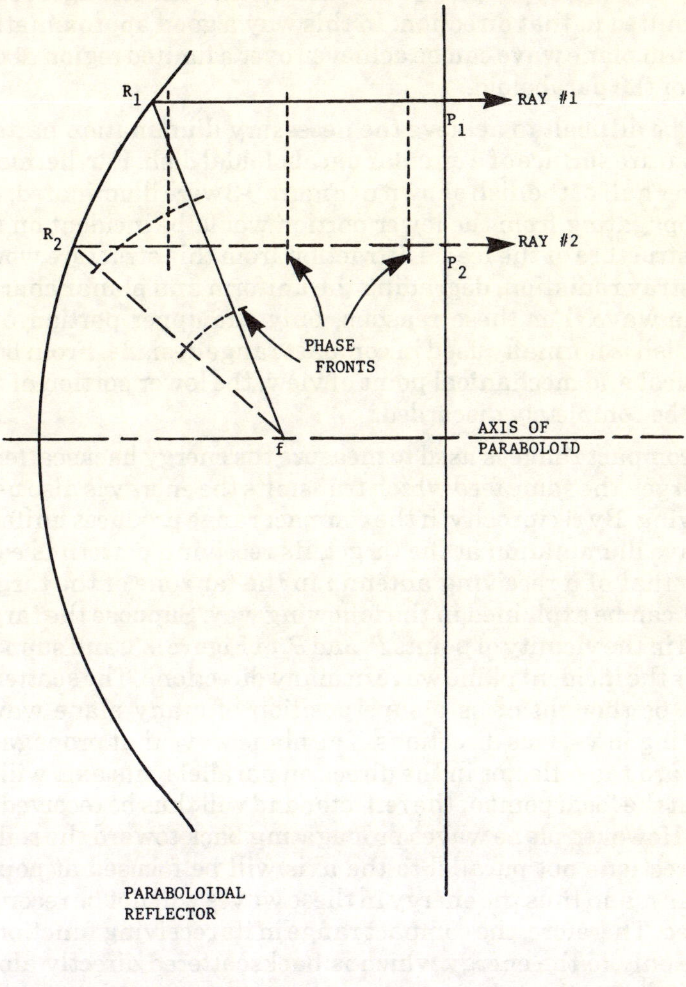


**Figure 9-3.** Principle of plane wave generation (reprinted from Bodnar, *et al*., ref. [5]).

Any microwave feed placed at the focal point, *f*, will have an amplitude pattern with some taper to it. Through proper design, the shape of the pattern can be used to compensate for the fact that the different rays must travel different distances to reach the reflector surface. Practically, it is necessary that the peak of the feed pattern be aimed toward the upper part of the dish so that the strongest rays will be emitted in that direction. In this way a good approximation to a uniform plane wave can be achieved over a limited region above the axis of the paraboloid.

It would be difficult to achieve the necessary illumination pattern over the entire surface of a circular paraboloidal dish. Furthermore, if the lower half of the dish shown in Figure 9-3 were illuminated, the wave propagating from the lower portion would be incident on the support structure of the feed. Diffraction from this structure would provide stray radiation, degrading the uniform and planar character of the wave. For these reasons, only the upper portion of a circular dish is normally used in compact range designs. From both an electrical and mechanical point of view, the lower portion of the dish can be completely discarded.

When a compact range is used to measure the energy backscattered from a target, the same feed which transmits the energy is also used for receiving. By reciprocity, if the compact range produces uniform plane wave illumination at the target, its receiving pattern is equivalent to that of a receiving antenna in the far zone of the target. This fact can be explained in the following way. Suppose the target is located in the vicinity of points $P_1$ and $P_2$ of Figure 9-3, and suppose it scatters the incident plane wave in many directions. The scattered field can be thought of as a superposition of many plane waves propagating in various directions. The plane wave that propagates back toward the reflector in the direction parallel to its axis will be focused at the focal point of the reflector and will thus be received by the feed. However, plane waves propagating back toward the reflector in directions not parallel to the axis will be focused at points other than *f*, and thus the energy in these waves will not be received by the feed. Therefore, the compact range in its receiving function is sensitive only to the energy which is backscattered directly along the original direction of propagation, that is, the direction parallel to the axis of the paraboloid.

### 9.1.3 Advantages and Disadvantages of a Compact Range

The most obvious advantage of a compact range compared to a conventional far field range is its reduced size. This, in turn, leads to several other advantages. However, compact ranges are not without their accompanying disadvantages.

The fact that a compact range is small allows it to be constructed in an enclosed area. Thus, it can be used in all weather and provides more security than can typically be obtained on an outdoor range. If the RF environment in the vicinity of the test facility is cluttered, the compact range enclosure can be shielded to provide a quiet area. Similarly, if it is desired that radiation from the measurement area not be available for outside intercept, the shielding can do double duty. Because of the small distance between the source and target, often a lower power frequency source may be used than on an outdoor range. Fewer problems are encountered in the areas of support deterioration, positioner maintenance, and range maintenance, since the range is indoors.

Because dielectric lens ranges are not well proven for use with medium to large targets, disadvantages of the more common reflector-type compact range are emphasized here. The major disadvantage is reflector cost. The reflector surface must be precise if a plane wave is to be achieved. The size of the reflector for a compact range is primarily determined by the size of the largest target which is to be measured. The cost of a compact range reflector is directly proportional to the area of the reflector, and, therefore, an increase in target size by a factor of two results in an increase of cost by a factor of four. The cost of the reflector also increases directly with the frequency of operation, because of tighter tolerance requirements on the reflector surface. Therefore, a facility capable of measuring 10-foot targets at 100 GHz could be as much as 40 times more expensive than a facility capable of measuring 5-foot targets at 10 GHz (assuming that the total cost is the product of the frequency cost and the size cost).[5] Clearly, a compromise between cost and measurement capability must be made. As a reference point, a commercial 12-foot compact range reflector with feed and and feed positioner, manufactured by Scientific Atlanta and shown in Figure 9-4, currently (1983) costs $208,000. The prototype of that reflector has been used at Georgia Tech to make measurements on targets with maximum dimensions around six feet at frequencies between 2 and 18 GHz.

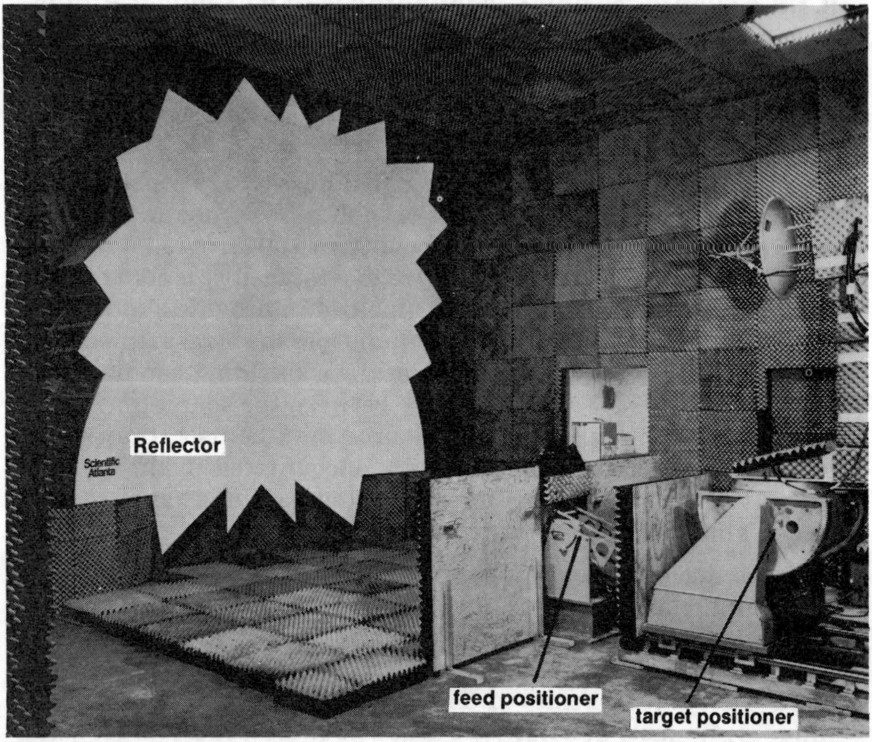


**Figure 9-4.** Compact range shown set up for antenna testing (courtesy of Scientific Atlanta, Inc.).

A second problem, which can also plague conventional ranges, is reduction of stray radiation. Primary sources of stray radiation for a compact range are inadequate absorber in the enclosure, diffraction from the reflector edges, reflection from the feed support and positioner, and radiation from the back lobes of the feed. Note the placement of absorber in Figure 9-4 to reduce stray reflections, and the serration of the reflector edge to reduce the effects of edge diffraction.

A final problem is range sensitivity due to vibration and temperature variations. This limit arises when a CW system with a nulling loop is used, and the problem is quantified in section 9.3.

## 9.2 COMPACT RANGE EQUIPMENT

Figure 9-5 provides a block diagram of the electronic and microwave equipment necessary for implementation of a dual-channel CW compact range. A stable frequency source is used to produce power at the desired frequency. A majority of the power is transmitted, via couplers and a magic tee, through the feed. The couplers provide a reference signal for the receiver, a nulling loop signal, and a source locking signal for the synchronizer. The return signal from the target is routed through the magic tee into the receiver path. A nulling signal, whose amplitude and phase have been adjusted to provide a minimum signal with no target present, is added to the return in a 3 dB coupler. A receiver and chart recorder or, for the case of an automated system, a receiver and computer, are used to provide storage and output of received power as a function of target position. Return from a standard target of known RCS must be used if absolute RCS is to be obtained. This section will discuss the components described, beginning with requirements for the reflector, feed, and source.

### 9.2.1 Reflector Configuration

Perhaps the most important component of a compact range is the collimating reflector, and the most critical factor affecting the performance of the compact range is the surface shape of the reflector. Below, the matter of surface tolerance is considered in more detail.

Suppose, for example, that a range reflector has small ($< \lambda/100$) random deviations from the correct paraboloidal shape, and that the areas of the individual deviations are less than a square wavelength. The amplitude of the field at a point in the test region of the compact range will be the result of integrating contributions from all parts of the range reflector. In this case, the errors due to the deviations will tend to average out, leaving a fairly uniform amplitude distribution across the test region. Suppose, on the other hand, that the area of the individual deviations is comparable to the area of the range reflector. In this case, the effect of the error will be similar to that of a slightly defocused reflector, and the amplitude distribution will still be fairly uniform across the test region.

Deviations with areas of intermediate sizes are more troublesome than either small or large ones. The worst situation will be realized when the sizes of the deviations are comparable to a Fresnel zone. So in the fabrication of the range reflector, one must control not only

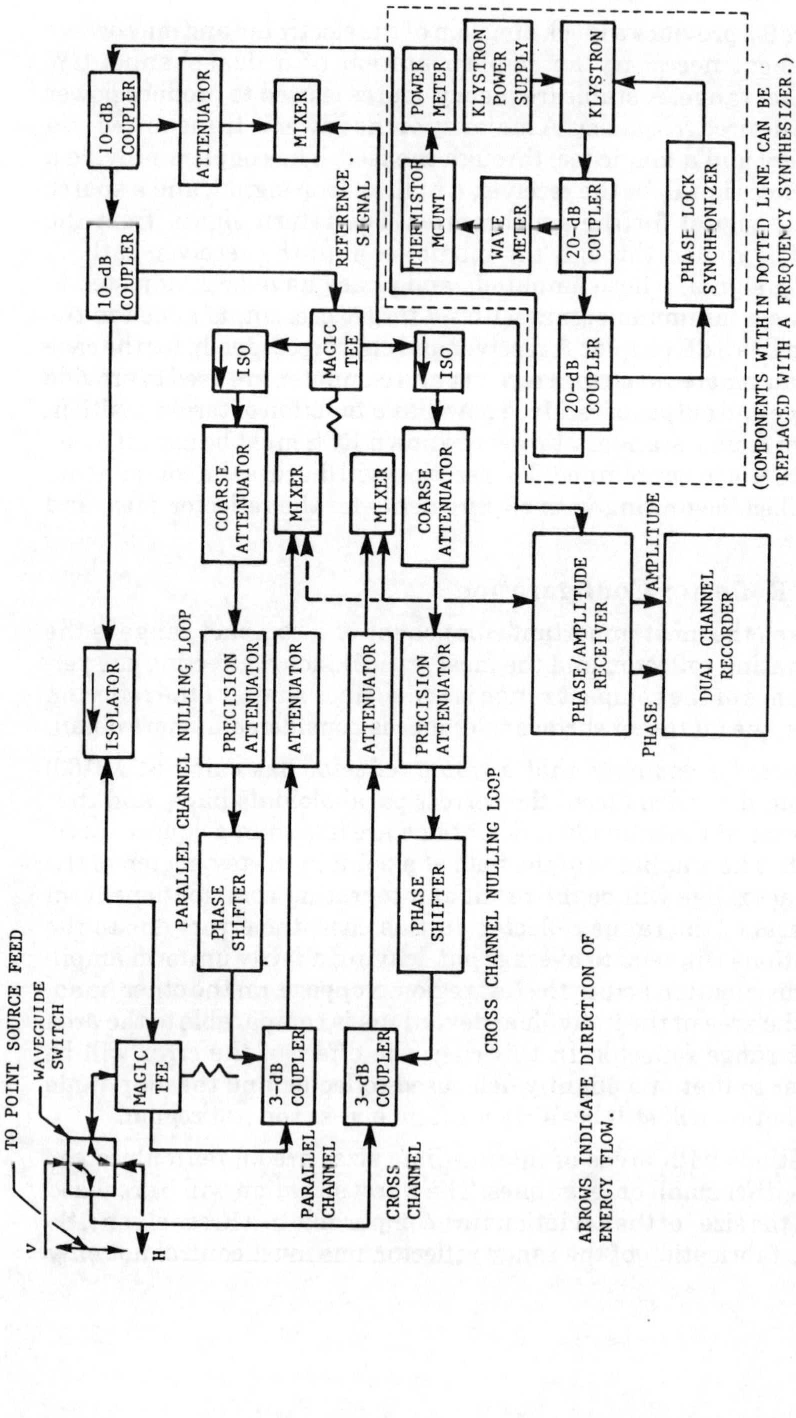


**Figure 9-5.** Transmitter, receiver, and nulling loop for a dual channel CW compact range (reprinted from Bodnar, *et al.*, ref. [5]).

the amplitude of deviations from the desired surface, but also the size of the areas over which the deviations are allowed to extend.

Another factor affecting the performance of a compact range is that of polarization purity. Even for an ideal range, cross polarized components will be generated by the reflection process. These components can tend to zero along the principal planes of the reflector, but in other directions, cross polarized components occur even in theory. By increasing the focal length of the reflector and thus decreasing its curvature, the cross polarized components can be reduced; however, they cannot be completely eliminated throughout the test region.

Diffraction is another problem affecting the operation of a compact range. Aperture diffraction occurs when the principal dimensions of the reflector are only a few wavelengths. Thus, it is necessary for the diameter of the reflector to be at least several tens of wavelengths. Edge diffraction also occurs due to the discontinuity of illumination produced at the edges of the reflector. As mentioned before, this problem can be reduced by serrating the edges of the reflector.

The use of an off-axis cut paraboloid reduces the amount of energy which is incident on the test object after being diffracted or scattered by the feed and its supports, because the main beam of collimated energy passes above the feed. Such diffracted or scattered energy can be reduced further by locating the test object essentially over the feed and by placing absorbing material between them.

### 9.2.2 Feed Design

The feed used at the focus of the reflector strongly affects compact range performance. Direct radiation from the feed to the target can be a source of serious problems; however, it can be reduced to an acceptable level by designing a feed with low radiation in the direction of the test object and by placing high-quality absorbing material in a position to intercept this radiation. Thus, concern should be given to the backlobes of the feed horn, and placement of the feed should allow space for absorber behind it. Furthermore, the feed amplitude taper should give as wide an active area as possible while reducing illumination at the reflector edge to a minimum to reduce diffraction effects.

### 9.2.3 Source Requirements

Selection of a frequency source for a compact range is critical because of stringent requirements for power output and stability over the range of frequency coverage. Ideally, high power, broad frequency range, and excellent stability should be available simultaneously in the selected frequency source. Unfortunately, compromises must be made. Broadband, stable sources for the common microwave frequency bands (2-18 GHz) are available from several manufacturers. Hewlett Packard and Watkins Johnson, among others, manufacture digital frequency synthesizers which cover this frequency range. Power output is currently around 10 dBm maximum, which will limit the sensivity of the compact range at X-band and higher frequency bands. Medium power TWT amplifiers are available from several manufacturers, and in combination with a frequency synthesizer can provide the desired power output, stability, and frequency coverage.

A second option for frequency sources is the use of phase-locked klystron oscillators. Table 9-2 lists klystron types suitable for compact range use. The required power output of either type of source, as a function of frequency and measurement accuracy, is discussed in the next section.

**TABLE 9.2**
**Klystron or BWO Sources**
(*Reprinted from Bodnar, et al., ref.* [5].)

| MODEL | FREQUENCY COVERAGE | POWER OUTPUT (mW) |
|---|---|---|
| Varian VA-183GA | 1-2 GHz | 100 |
| Varian VA-184T | 2-4 GHz | 100 |
| Varian EM-1137L | 3.6-4.3 GHz | 150 |
| Varian V-259 Series | 4.4-8.5 GHz (6 Tubes) | 1000 |
| Varian X-13 | 8.1-12.4 GHz | 150 |
| Varian EM-1149 Series | 12-16 GHz (7 Tubes) | 1000 |
| Varian VA 308 Series | 15.9-18 GHz (3 Tubes) | 800 |
| Varian EM-1188 Series | 18-26.5 GHz (9 Tubes) | 650 |
| Varian VA-322 Series | 26.5-40 GHz (34 Tubes) | 500 |

*Transmitter Power Requirements*

The power required from the transmitter for a compact range depends on a number of factors including receiver sensitivity, minimum cross section to be measured, accuracy to which the measurement is to be made, the compact range geometry, and the frequency of operation. Curves of received power *versus* frequency can be obtained using the radar equation,

$$P_r = \left( \frac{G^2 \lambda^2 \sigma}{(4\pi)^3 R^4} \right) P_t \tag{9-6}$$

where

$G$ = the feed horn gain

$\sigma$ = the target cross section

$\lambda$ = the wavelength

$R$ = the distance from the feed to the reflector surface

$P_t$ = the transmitted power, and

$P_r$ = the received power

Figure 9-6 is a plot obtained from the above equation with $G = 8$, $R = 4$ meters, and $P_t = 10$ mW. Also shown in this figure is a curve of CW sensitivity for a typical superheterodyne instrumentation receiver. For measurement accuracy on the order of 0.2 dB, the received signal should typically be 30 dB above receiver noise, as shown in Figure 9-7. Therefore, from about 1 to 5 GHz, the transmitter power must be 10 mW in order to measure a –40 dBsm target with an accuracy of 0.2 dB. At 10 GHz, a transmitted power of 200 mW is required. At 20 GHz, 1 watt is required, and at 30 GHz, 100 watts are required. Typical frequency synthesizers can produce a power output of +10 dBm. Thus, they can provide adequate power for measuring –40 dBsm targets up to 5 GHz. Above this frequency, some sort of power amplification, such as a TWT, is required. Klystrons, on the other hand, typically produce on the order of hundreds of milliwatts, and so could be used directly to about 10 GHz. Above 10 GHz, power amplification would be required even with klystron sources. Thus, the block diagram shown in Figure 9-5 must also include TWT amplifiers for higher frequencies, if a measurement accuracy of 0.2 dB at –40 dBsm is required.

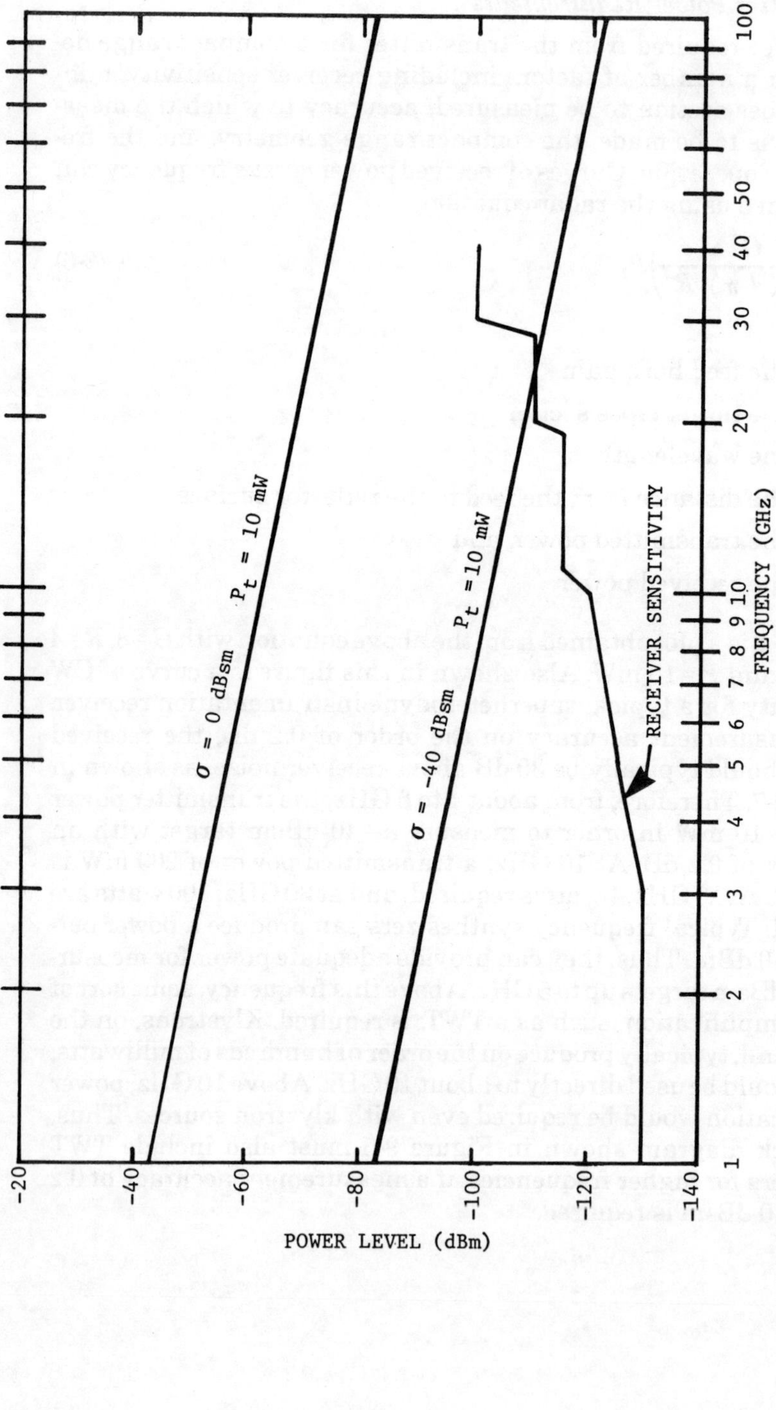


**Figure 9-6.** Power received on a compact range versus frequency for a transmitted power of 10 mW, G = 8, and R = 4 meters (reprinted from Bodnar, *et al.*, ref. [5]).

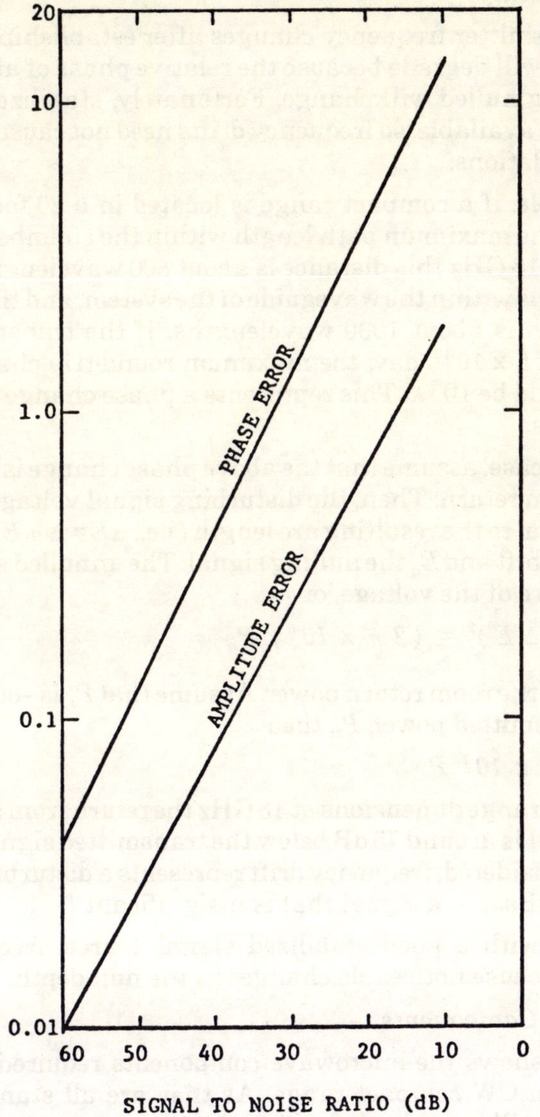


**Figure 9-7.** Phase and amplitude measurement errors versus signal to noise ratio (reprinted from Bodnar, *et al.*, ref. [5]).

*Source Stability*

If the transmitter frequency changes after establishing a null, the null depth will degrade because the relative phase of all the components being nulled will change. Fortunately, stabilized frequency sources are available, so frequency drifts need not cause appreciable null degradations.

For example, if a compact range is located in a 20-foot by 36-foot chamber, the maximum path length within the chamber is about 14 meters. At 18 GHz this distance is about 800 wavelengths. Add 200 wavelengths within the waveguide of the system, and the maximum path length is about 1000 wavelengths. If the transmitter has a stability of $5 \times 10^{-10}$/day, the maximum round-trip change in path length would be $10^{-6}\,\lambda$. This represents a phase change of only $2\pi \times 10^{-6}$ radians.

As a worst case, assume that the above phase change is experienced for the room return. Then, the disturbing signal voltage is approximately equal to the resulting arc length (i.e., $\Delta E = \Delta\phi E_n$) where $\phi$ is the phase shift and $E_n$ the nulling signal. The unnulled signal power is the square of the voltage, or

$$\Delta P = (\Delta E)^2 = (2\pi \times 10^{-6})^2 P_n \tag{9-7}$$

where $P_n$ is the room return power. Assume that $P_n$ is -60 dB relative to the transmitted power, $P_t$, then

$$\Delta P \approx 4 \times 10^{-17} P_t \tag{9-8}$$

For typical range dimensions at 18 GHz the return from a one square meter target is around 75 dB below the transmitted signal. Thus, for the case considered, frequency drift represents a disturbing signal of about -90 dBsm — a signal that is insignificant.

Therefore, with a good stabilized signal source, frequency drift should not cause noticeable changes in the null depth.

*Microwave Components*

Figure 9-5 shows the microwave components required for a dual-polarization CW compact range. As they are all standard items, little space will be spent describing them.

The major choice to be made concerning components is whether a coaxial or waveguide system will be used within a particular band. Coaxial components are generally smaller and cover a wider bandwidth than waveguide components. However, coaxial ranges generally suffer in sensitivity because of problems in background nulling.

That is, a good null is harder to achieve and more difficult to hold over a long period of time with coaxial systems than with waveguide systems. Some of the reasons that are postulated for this are that coaxial systems are less rigid and have a higher thermal coefficient of expansion than waveguide. Another reason for the difficulty in achieving a deep null with coaxial systems are that variable coaxial attenuators are linear (in dB) over their entire attenuation range, while their waveguide counterparts stretch the first dB (i.e., 0 to 1.0) over a large mechanical movement of the control knob. This means that waveguide systems can have one attenuator for coarse settings in series with another that is operated only over the region 0 to 1 dB for precise nulling (*see* Figure 9-5).

Georgia Tech has assembled a coaxial component reflector range which operates over the region 2.4 - 8 GHz. Its utility is satisfactory in the lower portion of the band but decreases at the higher frequencies. Below C-band, a coaxial system is likely to be preferable simply from size and weight considerations. Above C-band, waveguide is certainly preferred.

## 9.3 OPERATION

### 9.3.1 Field Probe Requirements

Prior to using a compact range, or any other range, it is necessary to probe the field in the measurement area to ensure that a uniform plane wave is present. The normal method for doing so is to move a wide beamwidth antenna in transverse horizontal and vertical cuts across the measurement area while recording received power and phase. Figure 9-8 illustrates a field probe pattern made on Scientific Atlanta's compact range.

Several factors must be kept in mind when setting up for field probe measurements. The first is the pattern of the probe antenna. If the probe beamwidth is small, energy scattered in the vicinity of the target position from wide angles will not be received. Thus, the field amplitude across a cut may appear to be more constant than it is. A second problem is ensuring that the probe is moved in a plane parallel to the phase front. The mechanism carrying the probe must be mechanically stable, and it must be rigid. Note that for the patterns in Figure 9-8, a 12° change in relative phase at 15 GHz requires a probe movement toward or away from the reflector of only 0.026 inches. Obviously, allowable mechanical tolerances become greater as the frequency decreases.

The limits which should be set on field probe amplitude and phase variation depend on the same factors discussed in Chapter 8 for conventional ranges. In general, a good rule of thumb is the one presented in reference [6] stating that echo patterns of reasonable accuracy can be obtained with phase variations across the target of less than $\pi/8$ radians and amplitude variations of less than 1 dB. For precise measurements of essentially flat targets, requirements may need to be tightened to $\pi/16$ radians and 0.2 dB. Note that even for the more stringent requirements, a target with dimensions in the 3 to 4 foot range could be measured, based on the field probe of Figure 9-8.

### 9.3.2 Calibration Target Selection

Because the compact range provides only relative amplitude outputs, calibration targets must be used to convert those outputs to absolute RCS. Calibration targets can also be helpful in assessing the linearity of range response and the level of the stray background field.

The usual calibration target for any range is either a metal sphere or plate. Metal spheres are particularly appropriate because analytical formulations exist for their RCS as a function of radius and frequency, and because they are essentially isotropic scatterers. For a sphere whose radius, $a$, is large enough that $2\pi a/\lambda > 10$, the RCS is simply the projected area, or

$$\sigma = \pi a^2 \tag{9-9}$$

Plates larger than several wavelengths on a side have an RCS given by

$$\sigma = 4\pi A^2/\lambda^2 \tag{9-10}$$

where $A$ is the area of the plate and $\lambda$ the wavelength. Reference [7] provides curves which can be used to correct the calculated RCS of very small plates.

Horizontally mounted cylinders also make convenient calibration targets, since it is easy to ensure that the peak of the pattern is observed. Cylinders whose radii, $a$, are such that $2\pi a/\lambda > 10$ have a peak RCS given by

$$\sigma = 2\pi a L^2/\lambda \tag{9-11}$$

where $L$ is the cylinder length.

One useful measurement procedure requiring multiple calibration targets is a linearity check of the compact range. At least two targets

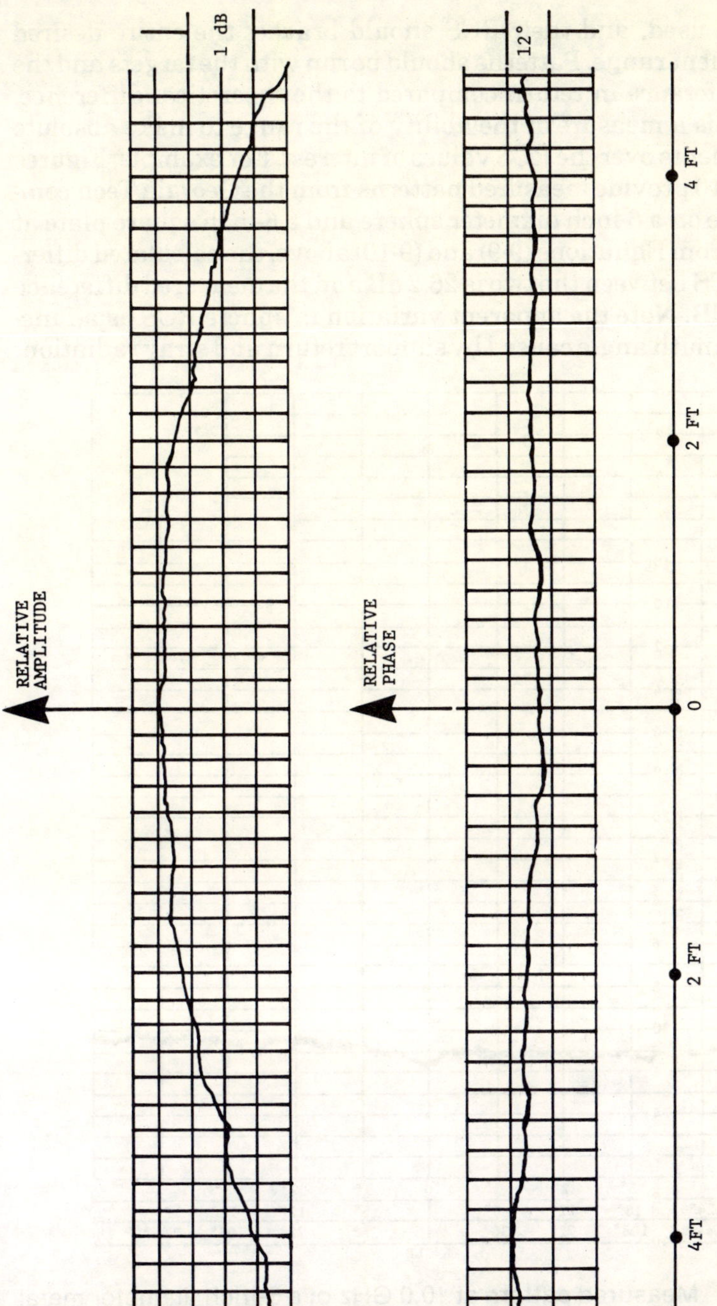


**Figure 9-8.** Horizontal cut through the center of Scientific Atlanta compact range showing the amplitude and phase variations over a 9-foot region at 15 GHz (reprinted from Bodnar, *et al.*, ref. [5]).

should be used, and their RCS should bracket the entire desired measurement range. Patterns should be run with the targets and the actual difference in return compared to the theoretical difference. The error is a measure of the ability of the range to make absolute measurements over the RCS values of interest. For example, Figures 9-9 and 9-10 provide measured patterns from the Georgia Tech compact range on a 6-inch diameter sphere and a 6-inch square plate at 10 GHz. From Equations (9-9) and (9-10) above, the calculated difference in RCS between the two is 26.2 dB and the measured difference about 26 dB. Note the apparent variation in sphere RCS as a function of azimuth angle caused by support return and stray radiation.

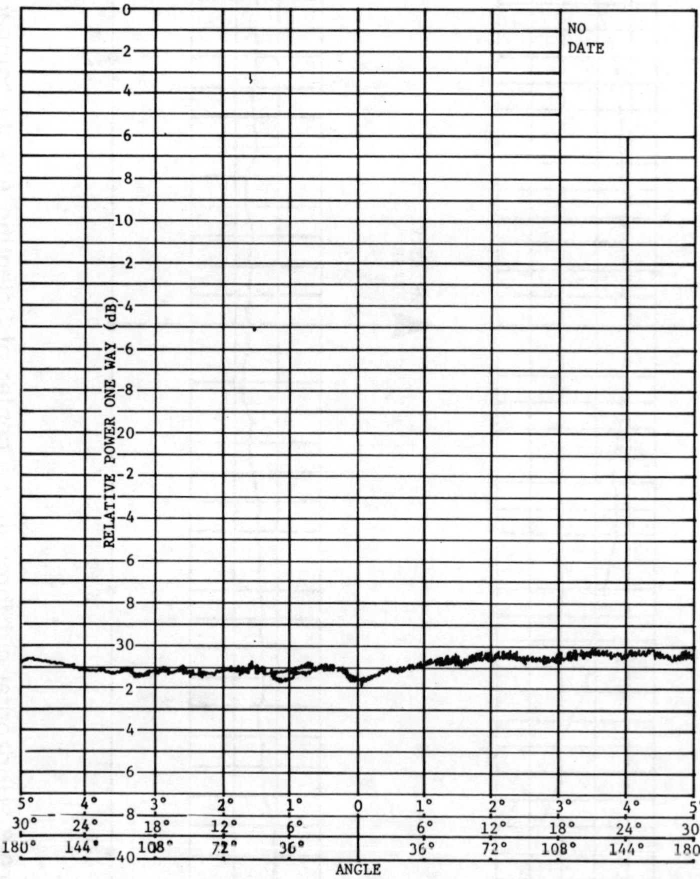


**Figure 9-9.** Measured pattern at 10.0 GHz of a 6-inch diameter metal sphere.

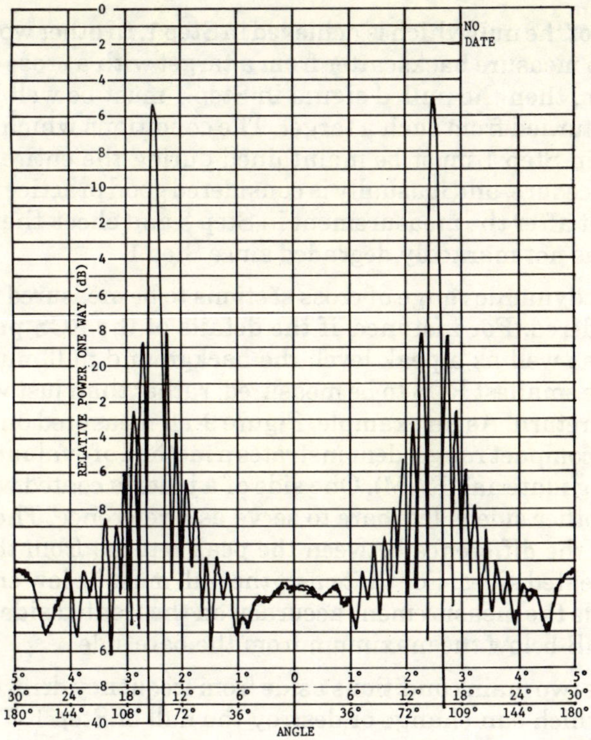


**Figure 9-10.** Measured pattern at 10.0 GHz of a 6-inch square metal plate.

### 9.3.3 Null Level and Range Sensivity

The sensitivity of a compact range is generally controlled by the null level which can be obtained, rather than by receiver sensitivity. The following material, much of it taken from reference [5], provides an analysis of the primary contributions to null degradation.

When making measurements on CW reflectivity ranges, the procedure generally is (1) to null out the returned signal in the absence of a target, (2) to place the target in the incident field, and (3) to measure the return signal with the target in place. The signal measured in Step 3 is assumed to be the target return,* since the return from other objects was presumably nulled out in Step 1. One can easily see that the sensivity of the range depends upon the depth and

*There are some inherent errors due to shadowing of the background by the target and due to multipath return but these errors generally are small and will not be discussed here.

stability of the null which is achieved in Step 1. In other words, if one wishes to measure backscatter from a target with a cross section of -30 dBsm, then the nulled signal in Step 1 must be well below the signal returned from such a target. The conditions which produced the null in Step 1 must be maintained during the entire measurement procedure, and it usually is considered good practice to remove the target after the measurement in Step 3 and check the null to be sure it has not markedly degraded since Step 1.

The total dynamic range of cross sections to be measured must also be considered. For instance, if the details of the RCS pattern are desired, as well as a peak level, the background null must be well below the smallest RCS to be measured, rather than just well below the peak return. As an example, Figure 9-11, measured on the Georgia Tech compact range, demonstrates a method of evaluating radar absorbing material (RAM). One side of a plate is coated with RAM, and the other side is left bare to serve as a reference. The RCSR is given by the difference between the peak returns from the treated and untreated sides. For that case, the null must be low enough not to degrade the measurement accuracy on the treated side peak — a level 20 dB below the maximum from the bare side.

There are two main conditions aside from frequency drift, discussed earlier, which can change or destroy the null of Step 1. The first is physical movement of range components (for example, movement of the transmit/receive antenna relative to its surroundings or vibrations of an attenuator or phase shifter which cause its value to fluctuate); the second is environmental changes (for example, changes in temperature or humidity). Each of these conditions will be discussed briefly in this section.

One should understand that the above conditions are critical on any CW reflectivity range whether it is an outdoor range, a range within an anechoic chamber, or a compact range. The conditions will be discussed as applied to a compact reflectivity range, but similar conclusions would apply to other ranges. Use of a pulsed or FM-CW system removes the necessity for a nulling loop. However, either system carries penalties in increased equipment cost and complexity, plus design problems related to resolution requirements and receiver recovery time.

*Physical Movement of Range Components*

In a CW compact reflectivity range there are three main signals that must be nulled out before placing a target on the support structure.

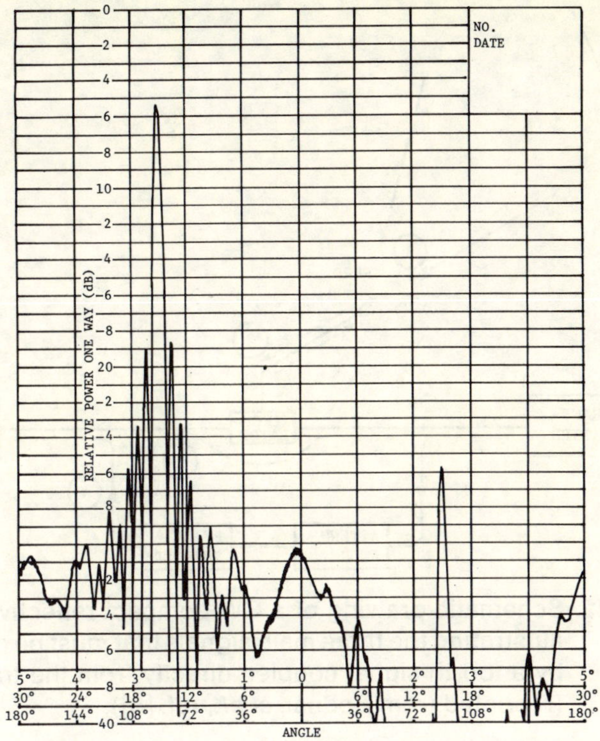


**Figure 9-11.** Measured pattern at 10.0 GHz of a 6-inch square metal plate with Emerson and Cuming AN 75 absorber on one side.

They are (1) leakage from the transmitter to the receiver at the circulator (or magic tee), (2) reflections within the waveguide or coaxial transmission lines, and (3) reflections from the room and surrounding equipment. The resultant of the three signals is nulled out by a fourth signal which can be coupled directly from the transmitter line as indicated in Figure 9-12. It is convenient to reference the depth of the null to the signal that would be returned by a target having a cross section of one square meter. To estimate the signal returned from a target, one needs only to use the radar equation which was given earlier as Equation (9-6).

At first glance, one might assume that the returned signal on a compact range will be much larger than on a far field range because of the small value for $R$. This is not necessarily so because the feed gain is typically much less than the antenna gain on a far field

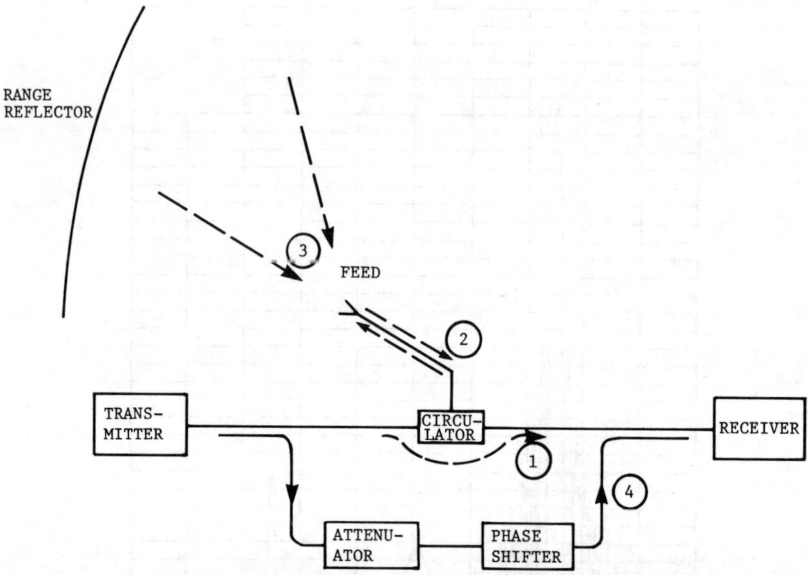


**Figure 9-12.** Schematic drawing of a CW compact reflectivity range illustrating the three main signals that must be nulled out by a fourth signal coupled directly from the transmitter (reprinted from Bodnar, *et al*., ref. [5]).

range. Although in general the return signal strengths on the two types of ranges will be approximately the same, for physically small targets a high gain feed (with an attendant higher amplitude taper) may be used to increase the return signal or decrease transmitted power requirements.

The four signals indicated in Figure 9-12 are illustrated as vectors in Figure 9-13. Signals 1 and 2 are confined to the waveguide (or coaxial) plumbing. Their phases at the receiver can vary with frequency and temperature changes, but they will be insensitive to movements of range components. Signal 3, on the other hand, radiates into the test chamber, is scattered, and then is received again by the feed. The phase of Vector 3 can be changed by small movements of the feed, of the range reflector, or of surfaces of the test chamber.

Figure 9-14 illustrates radiation within a compact range chamber. Typically, about 65 percent of the radiated energy strikes the reflector, is collimated, and then stikes the back wall. About 10 to 12 percent of the energy strikes the ceiling, and the remainder is essen-

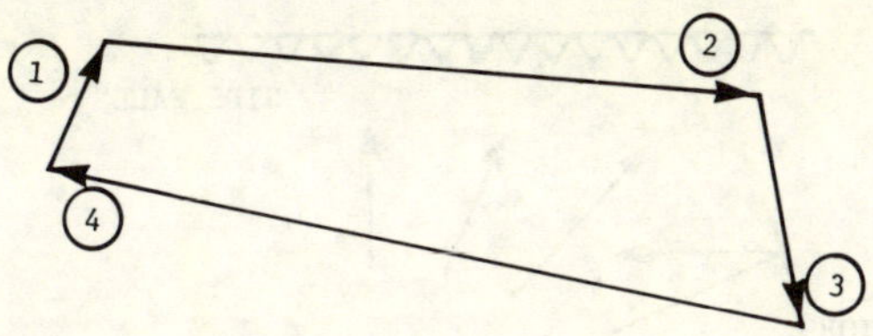


**Figure 9-13.** Vector representation of the four signals indicated in Figure 9-12. The magnitude and phases of the first three vectors have been chosen arbitarily; the magnitude and phase of the fourth vector is adjusted so that it terminates at the origin of vector 1 (thus producing a null signal) (reprinted from Bodner, *et al*., ref. [5]).

tially divided equally among the two side walls and the floor. In an absorber-lined chamber one can consider that all energy leaving the feed eventually strikes absorber, so the magnitude of Vector 3 essentially depends on backscatter reflections from the absorber. In this respect, the signal represented by Vector 3 can be called "room return" in a manner similar to the custom of using "target return" to refer to the signal returned from a target. Thus, in a well-designed chamber, the room-return power probably would be about -50 to -60 dB relative to the transmitted power (assuming a typical reflection coefficient of the absorber of -45 to -55 dB).

Keep in mind that Vector 3 is actually the resultant of several vectors, each of which represents a contribution from a particular part of the chamber. For example, if the back wall moves while the rest of the chamber is stationary, then only the components representing reflections from the back wall will be changing phase.

The various parts of a compact reflectivity range must be examined to estimate the null-depth sensitivity to movements. The most sensitive part is probably the feed, because all energy radiated and received passes through this device. The second most sensitive part is the range reflector, because the largest portion of radiated energy strikes the reflector and because rays are reflected from it twice —once on transmission and once on reception. The third most sensitive part is the back wall, because essentially all of the energy reflected off of the reflector strikes the back wall. Of lesser sensitivities will be the side walls and floor.

Since the feed appears to be the most sensitive component consider it first. The feed actually radiates energy in many directions; so when it moves in a particular direction, all reflected rays are not

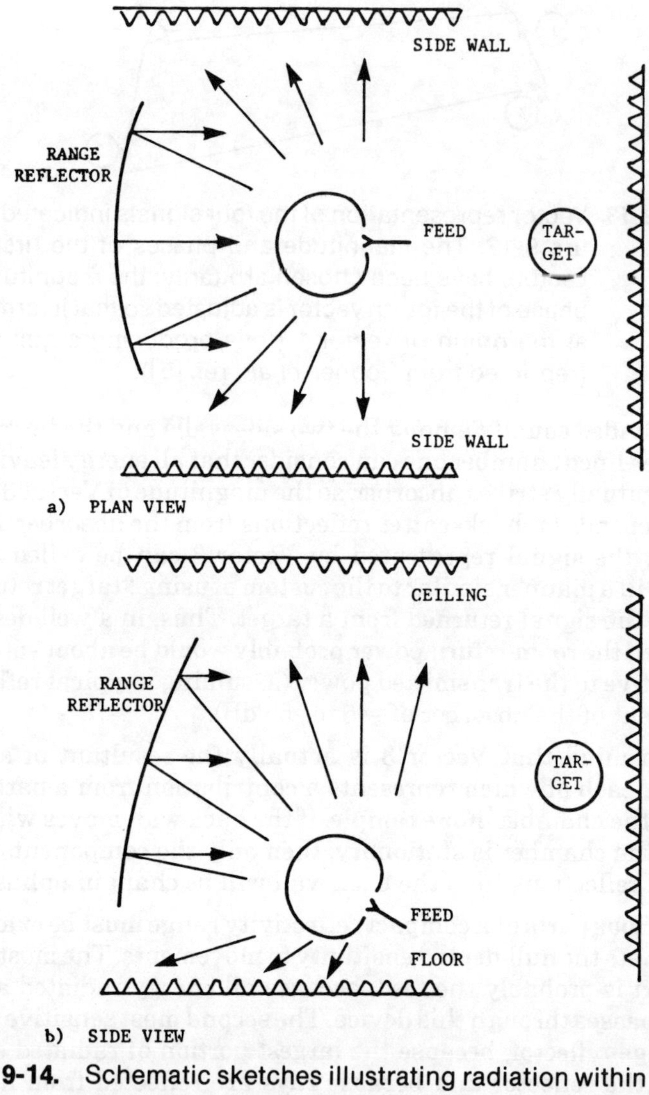


**Figure 9-14.** Schematic sketches illustrating radiation within a compact range chamber (reprinted from Bodnar, *et al.*, ref. [5]).

changing phase at the same rate. Most of the energy is radiated along the feed axis, so axial movement should cause the largest phase changes in Vector 3.

To simplify the problem, consider that the feed is vibrating (moving) axially and that all energy is radiated and received axially. In typical compact rànge feed horns, about 80% of the energy is radiated and received within a ±45° cone; this represents a level only 1 dB below the total radiated power, so the assumption that all energy is radiated and received axially is reasonable.

Next assume that the feed moves axially a distance $\Delta \ell$. Then the total round-trip (transmission and reception) path length change is $2\,\Delta \ell$, and the phase change of Vector 3 is

$$\Delta \xi = \frac{2\pi}{\lambda}(2\,\Delta \ell) = 4\pi \frac{\Delta \ell}{\lambda} \tag{9-12}$$

Let $E_3$ be the field magnitude of Vector 3. If under stationary conditions the received signals have been nulled out, and Vector 3 changes phase by $\Delta\,\xi$ the null will be replaced by a signal of field magnitude $E$ as illustrated in Figure 9-15. The value of $\Delta E$ is

$$\Delta E = E_3 \Delta \xi = 4\pi \frac{\Delta \ell}{\lambda} E_3 \tag{9-13}$$

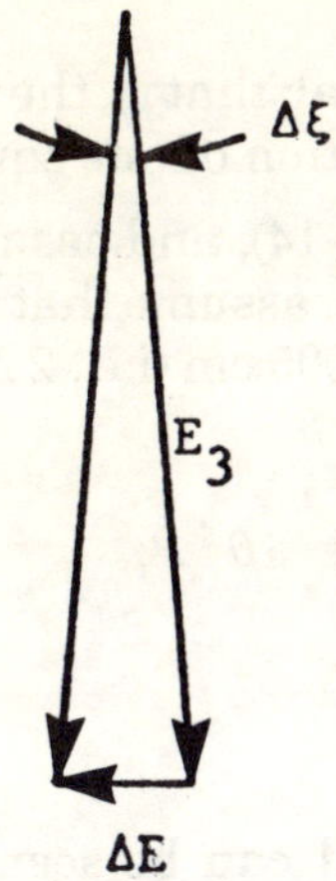


**Figure 9-15.** Vector diagram illustrating a change of phase in a room-return signal and the generation of a disturbing signal which replaces the null (reprinted from Bodnar, *et al.*, ref. [5]).

This is also the peak electric field magnitude of the disturbing signal for axial vibrations of the feed having peak-to-peak excursions of $2\,\Delta \ell$. The peak instantaneous power level in the disturbing signal is related to the power level in the room return by

$$\Delta P = (4\pi)^2 \left(\frac{\Delta \ell}{\lambda}\right)^2 P_3 \tag{9-14}$$

where $P_3$ is the power represented by Vector 3. This signal should be referenced to the signal returned from a target having a radar cross section of one square meter; the latter can be calculated from Equation (9-6).

It is instructive to perform some simple calculations. Assume that the compact range has a 12-foot range reflector (height and width) and a 12-foot focal length, then the following values would be typical for X-band operation:

$G = 8$

$\lambda = 3\text{ cm}$

$R = 4\,m$

then from Equation (9-6), for a 1 $m^2$ target,

$$P_r = \left( \frac{(8)^2 (3 \times 10^{-2})^2}{(4\pi)^3 (4)^4} \right) P_t$$

$$P_r = 1.1 \times 10^{-7} P_t \tag{9-15}$$

This is the reference signal; that is, the power returned from a target having a radar cross section of one square meter.

Now refer to Equation (9-14), and assume the room-return power is -60 dB relative to $P_t$. Also, assume that the peak-to-peak axial vibration of the feed is only 0.005 cm (i.e., $2\Delta\ell = 0.005$ cm). Substituting in (9-14)

$$\Delta P = (4\pi)^2 \left( \frac{.0025}{3} \right)^2 10^{-6} P_t$$

or

$$\Delta P = 1.1 \times 10^{-10} P_t \tag{9-16}$$

From (9-15) and (9-16), it can be seen that the peak value of the disturbing signal is about -30 dBsm for the assumed conditions, and this would be the limit of the achievable null depth. Similar calculations were made for other vibration magnitudes and for other values of room return, and the results are presented in Figure 9-16.

Note that the achievable null depth is quite sensitive to feed movements or vibrations. For example, if the room return is -55 dB relative to transmitted power, the peak-to-peak feed vibrations would have to be limited to about 0.0001 inch to achieve a -50 dBsm null at X-band frequencies. From Equation (9-14), it can be seen that the sensivity to vibrations would be less at lower frequencies. If one

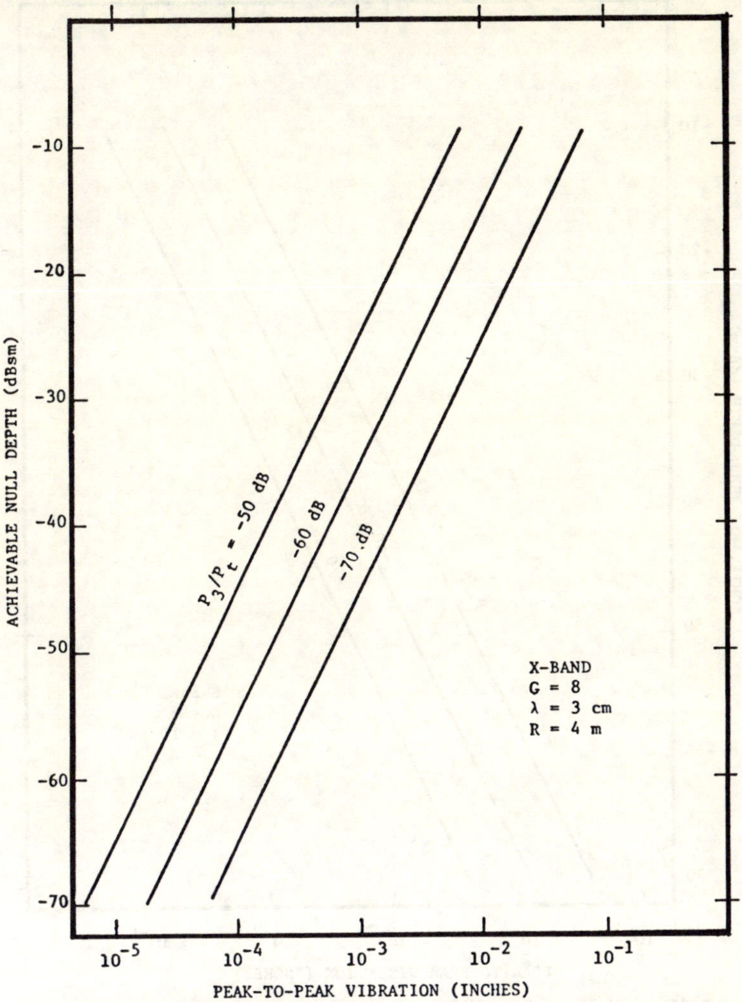


**Figure 9-16.** Achievable null depth versus peak-to-peak vibrations of the feed for various ratios of return to transmitted power at X-band (reprinted from Bodnar, *et al*., ref. [5]).

assumes a λ of 10 cm (i.e., S-band frequency), then the curves are as presented in Figure 9-17. Although the null depth is sensitive to very small movement of range components, good range performance can be achieved in practice with reasonable care.

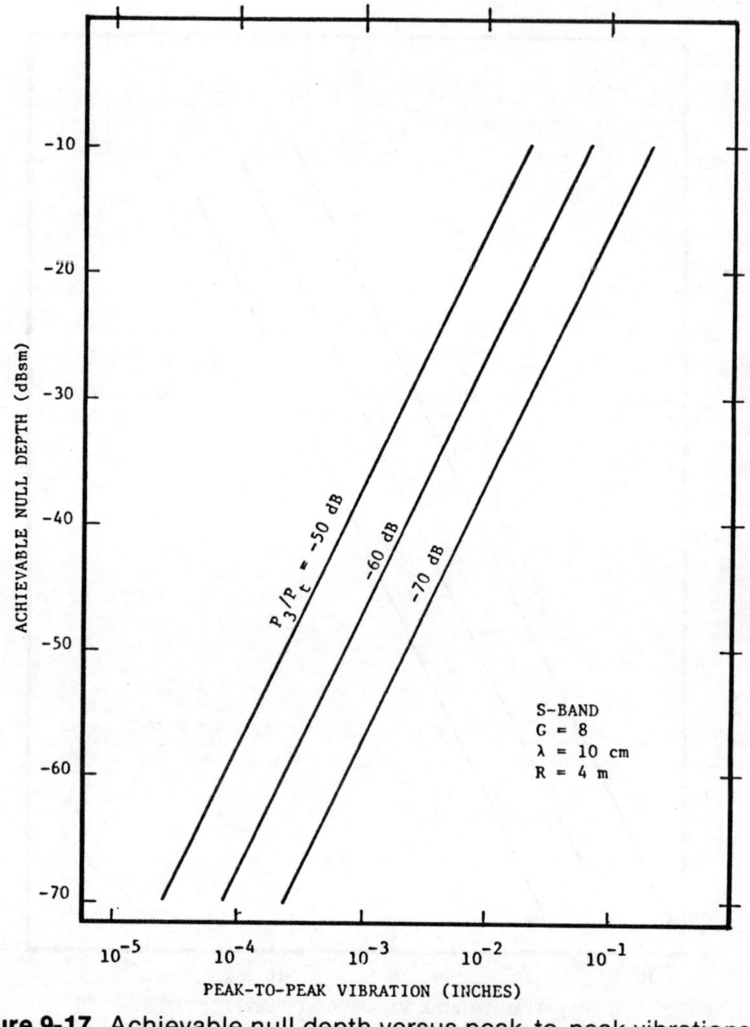


**Figure 9-17.** Achievable null depth versus peak-to-peak vibrations of the feed for various ratios of return to transmitted power at S-band (reprinted from Bodnar, *et al.*, ref. [5]).

*Environmental Changes*

By environmental changes, it is meant changes in parameters such as temperature and humidity. Of these, the most sensitive parameter usually will be temperature. As an example, refer to Figure 9-12 and assume that the waveguide circuitry is separated from the feed by 20 feet. Next estimate what the effects would be if the temperature of the 20-foot section of waveguide (between the feed and the circulator) changed 1°C.

At X-band frequencies, 20 feet of waveguide is about 150 guide wavelengths. Assume that the coefficient of thermal expansion (linear) for the waveguide is about 18 x $10^{-6}$/°C. Then, the change in length of the waveguide for a 1°C change in temperature would be

$$\Delta \ell = 150\,\lambda\,(\,18\ \times\ 10^{-6}\,) \qquad (9\text{-}17)$$

$$= 2.7\ \times\ 10^{-3}\ \lambda$$

Such a change in length would affect the phase of the signal reflected from the feed (Signal 2 in Figure 9-12) the same as that calculated in Equation (9-12). Assume that the VSWR of the feed is 1.1; then, the field reflection coefficient, Γ, would be about 0.05. The field magnitude of the disturbing signal can be calculated from an equation similar to (9-13), and the power level can be calculated from an equation similar to (9-14).

$$\Delta P = (4\,\pi)^2\,(\,\Delta\ell/\lambda\,)^2\,\Gamma^2\,P_t \qquad (9\text{-}18)$$

$$\Delta P = (4\,\pi)^2\,(\,2.7 \times 10^{-3}\,)^2\,(\,0.05\,)^2\,P_t \qquad (9\text{-}19)$$

$$= 2.9 \times 10^{-6}\,P_t$$

The disturbing signal would be about −55 dB relative to the transmitted power. Recall from (9-15) that the return signal from a 1 $m^2$ target at X-band is about −70 dB relative to the transmitted power; thus, the disturbing signal would be about 15 dB higher than the return from a 1 $m^2$ target.

As an alternative situation, assume that the waveguide length is 2 feet, the VSWR of the feed is 1.01, and the temperature change is only 0.1°C. Then, the disturbing signal would be 60 dB below the above example, or about −45 dB relative to the return from a 1 $m^2$ target.

This discussion points out that temperature effects can be critical on CW reflectivity ranges; however, they can be controlled by keeping VSWRs low and line lengths short in critical waveguides, and by working in a temperature controlled room.

## REFERENCES

1. Mentzer, J. R., "The Use of Dielectric Lenses in Reflection Measurements," *Proc. IRE*, Vol. 41, February 1953, pp. 252-256.
2. Johnson, R. C., H. A. Ecker, and R. A. Moore, "Compact Range

Techniques and Measurements," *IEEE Transactions on Antennas and Propagation*, Vol. AP-17, September 1969, pp. 568-576.

3. Knott, E. F., "Lens Feasibility and Design Study," Interim Report on Boeing Company Purchase Order 64-132568-0750N, Georgia Institute of Technology, Atlanta, Georgia, December 1981.

4. Olver, A. D., and A. A. Saleeb, "Lens-Type Compact Range," *Electronics Letters*, Vol. 15, July 5, 1979, pp. 409-410.

5. Bodnar, D. G., C. P. Burns, R. C. Johnson, and J. W. Cofer, "Compact Range Study for RATSCAT," Final Technical Report on Contract F29601-77-C-0051, Georgia Institute of Technology, Atlanta, Georgia, January 1978.

6. Kouyoumjian, R. G., and L. Peters, Jr., "Range Requirements in Radar Cross Section Measurement," *Proc. IEEE*, Vol. 53, August 1965, pp. 920-928.

7. Hey, *et al.*, "The Scattering of Electromagnetic Waves by Conducting Spheres and Discs," *Proc. Phys. Soc.* (London), 1969, pp. 1038-1049.

# CHAPTER 10
# RCS MEASUREMENTS ON SCALE MODELS

**H.A. Corriher, Jr.**

**Engineering Experiment Station
Georgia Institute of Technology**



## 10.1 INTRODUCTION

Scale models of physical entities are widely used in making engineering measurements. Well-known examples are aircraft models in wind tunnels, ship models in water tunnels, and three-dimensional models of waterways. These all depend on the principle of similitude which allows some known relationship to be established between model and full-scale behavior. For example, in fluid mechanics a common modeling parameter is the Reynolds Number.

Electromagnetic modeling has been extensively developed during and since World War II,[1] while the modeling of antenna behavior dates back to 1919.[2] The earliest known work on model measurements of RCS was at the MIT Radiation Laboratory in 1942.[3] Despite a number of problems in preparing the models themselves and in making the measurements, scale models provide an important means either of determining (or at least estimating) the radar cross section (RCS) of targets for which no other method is practical, or of providing independent verification of values found in other ways. For example, a full-scale target may not be available for measurement, or the effect on RCS of a design variation may need to be determined before the final design is chosen.

This chapter gives an overview of model measurements using three different techniques: noncoherent optical, ultrasonic (sonar), and radio frequency (RF). The reader interested in more details should refer to other sources (e.g., references [1], [4-8]) and the references cited therein.

## 10.2 SCATTERING BY A COMPLEX TARGET

In considering a scale model or any other type of radar scatterer, it is worthwhile to point out that we really must deal with the reflection of electromagnetic fields. The targets that we are interested in are frequently complex in the sense that they have more than one scattering center (or scattering element). For a single polarization, the electric field $\overline{E}_r$ reflected from the target may be expressed as the phasor summation (i.e., resultant) of the fields from the $N$ separate scattering elements which constitute the target ensemble. If interaction among scatterers is ignored, this may be written in terms of the incident field and the complex scattering coefficients of the individual elements as

$$\overline{E}_r \propto E_o \sum_{i=1}^{N} A_i \, e^{j\phi_i} \tag{10-1}$$

where

$\overline{E}_r$ = electric field reflected by the target as a whole

$E_o$ = amplitude of the field incident on the target

$A_i$ = scattering amplitude for the $i^{th}$ element

$\phi_i$ = relative phase of the field scattered by the $i^{th}$ element

It is recognized that the phase and amplitude characteristics of each and every one of the $N$ scatterers cannot actually be obtained for use in Equation (10-1). Thus, mathematical modeling of all instantaneous effects cannot be expected. However, the identification and description of major scattering sources usually enables scattering peaks to be well approximated, if not the nulls. The point here is that radar scattering is a physical process in which a collection of scatterers is illuminated by a transmitter of electromagnetic energy and each scatterer re-radiates some of this energy in the direction of a receiver. The individually re-radiated (scattered) fields have phase and amplitude, and combine to produce a resultant field. This resultant field may be calculated or it may be measured for the real target or a scale model thereof.

Equation (10-1) is expressed in terms of electric fields rather than radar cross section, but the two are related by

$$\sigma \propto |\bar{E}_r|^2 \qquad (10\text{-}2)$$

where

$\sigma$ = radar cross section of the overall target
$|\bar{E}_r|$ = magnitude of the electric field reflected by the target as a whole

In Figure 10-1 we may take the reference phase as 0 for element 1 and then we have a certain amplitude and relative phase for each of the other elements: 2, 3, 4, 5, 6, etc. They add up to give a resultant that is the scattered field from that collection of scattering centers. Since "nature" is coherent on a sample-to-sample basis, each element that is being illuminated produces a field component with a certain amplitude and phase that comes back to be added up as discussed above.

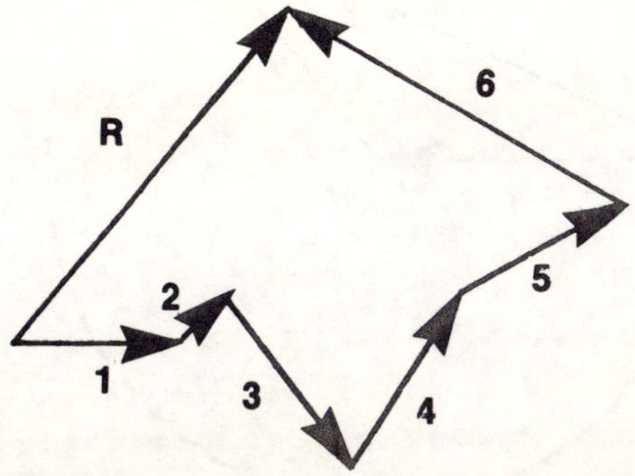


**Figure 10-1.** Addition of phasors.

Actually this is only part of the total picture, since this is for only one polarization component. The orthogonal polarization and two cross polarization terms are required to represent the complete electromagnetic situation. The individual polarization components are phasors which, along with the vectors for the polarization, constitute a vector-phasor relationship. We will not be concerned with such details, but only with some of the relationships that enter into the production of a resultant scattered field, and the fact that radar cross section is proportional to the square of the magnitude of that scattered field.

Several phasor diagrams are shown in Figure 10-2. In Figures 10-2(a) and 10-2(b) there are two phasors (1 and 2) which could represent either two scattering centers or the resultants of two collections of scatterers that produce phasor 1 and phasor 2. Notice in Figure 10-2(a) that, if the phasors have the same amplitude and if the relative phase of one can vary through 360°, it is possible to get a zero resultant if phasor 2 is anti-parallel to phasor 1, or to get twice the field if the two are parallel. So, for two nearly equal scatterers it is possible to get either an approximately zero return field if the scatterers interfere *destructively*, canceling each other, or twice the field of one scatterer if they add together, interfering with each other *constructively*. Since RCS is proportional to the square of the field, the RCS for these two interfering equal scatterers can vary between zero and four times the RCS of one of them.

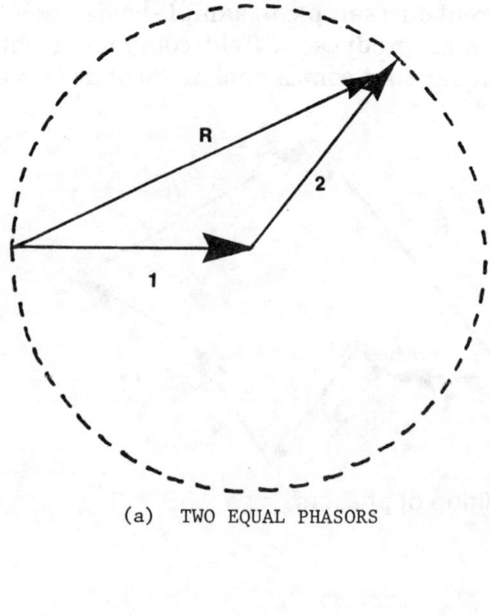


(a) TWO EQUAL PHASORS

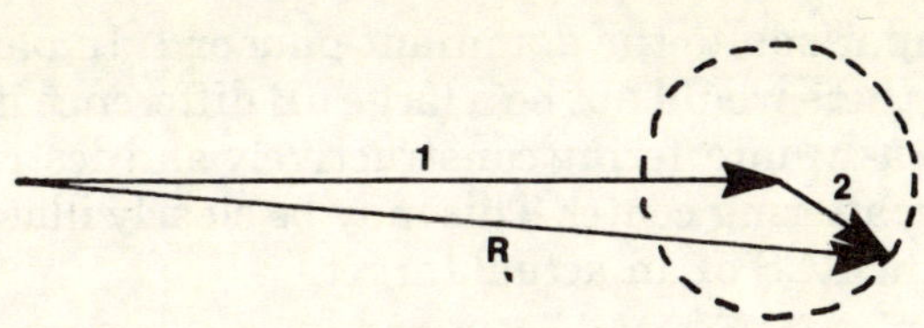


(b) ONE DOMINANT, ONE SMALL, PHASOR

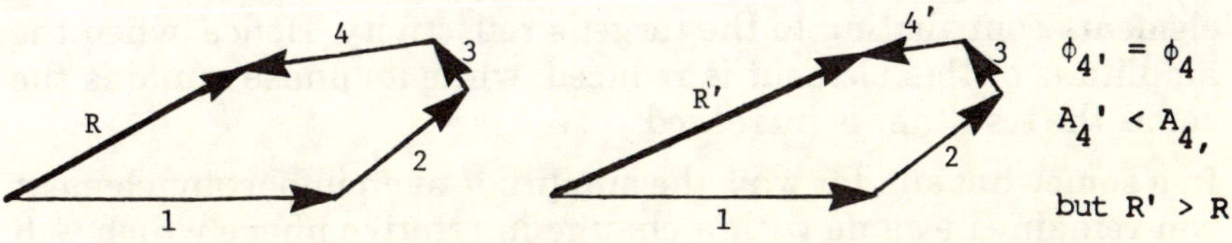


(c) AMPLITUDE CHANGE WITHOUT PHASE CHANGE OF LARGE PHASOR

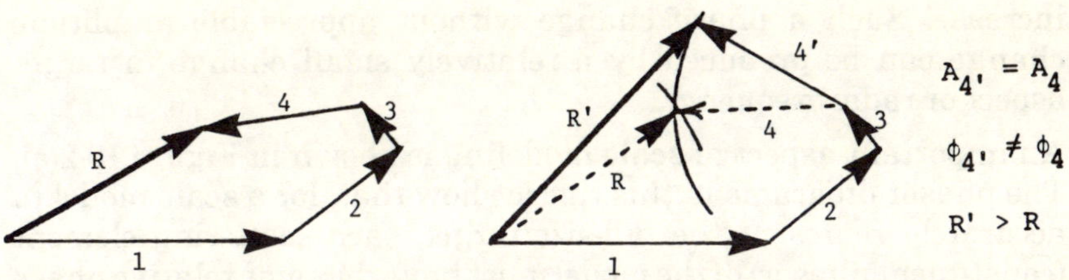


(d) PHASE CHANGE WITHOUT AMPLITUDE CHANGE OF LARGE PHASOR

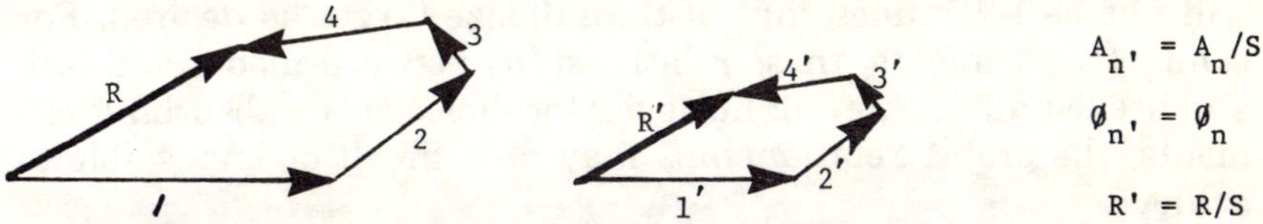


(e) SCALING OF PHASORS IN MODELING

**Figure 10-2.** Phasor addition for five situations.

Another situation of interest is shown in Figure 10-2(b), where phasor 1 represents a dominant scatterer and there is another small scatterer (phasor 2) which has variable phase (or phasor 2 could be the resultant of a collection of scatterers, none of which is large). In this case, there would be a variation in the resultant which would be small in comparison to the dominant phasor 1. In particular, on a decibel scale there would not be a large dB difference if there were a lot of little "trash" interfering constructively and destructively with a dominant scattering center. This may be clearly illustrated by the behavior of the RCS of an actual target.

The phasor diagram of Figure 10-2(c) illustrates that reduction of the return from a single large scatterer on a complex target can actually increase the overall return. In this illustration, element 4 has a relative phase which is generally opposite to that of the other elements contributing to the target's reflectivity. Hence, when the amplitude of this element is reduced, while its phase remains the same, the resultant is increased.

In a somewhat similar way, the amplitude of an important element can remain the same with a change in relative phase which will profoundly affect the resultant. This is shown in Figure 10-2(d) in which element 4 retains its amplitude but has a change in phase which (for the situation illustrated) causes the overall return to increase. Such a phase change without appreciable amplitude change can be produced by a relatively small change in target aspect or radar frequency.

An important aspect of scale modeling is shown in Figure 10-2(e). The phasor diagrams in this figure show that, for a scale model to accurately represent the full-size target, each scattering element (constituent phasor) of the model must have the same relative phase and an amplitude of $1/S$ times that of the real target. (As explained in section 10.4 below, $S$ is the scale factor of the model.) If these relationships are not maintained, the measured RCS of the model will not be $1/S^2$ times that of the full-size target as desired. For conductive elements, these relationships between model and full-size are usually achieved; however, for dielectric or absorbing elements, the proper relationships may be difficult or impossible to achieve.

Figure 10-3 is a classic picture (Figure 6-32 of reference [9]) of a B-26 on the ground. The radar has a wavelength of 10 $cm$, and there is substantial variation in the RCS. Why? Because there are many scattering centers that are contributing to this overall pattern and they interfere with each other in constructive and destructive manners. Both the phases and the amplitudes of the fields from the individual scattering centers interact with each other to produce this rather complex picture. A 10° median curve has been added to show the regularity of the pattern, as well as a curve of 10° peaks.

Note in Figure 10-3 that the overall pattern exhibits fairly clean peaks and "noisy" minima — this matches the general behavior illustrated in Figures 10-2(a) and 10-2(b). For an airplane, ship, or tank, at the target's broadside, ends, or top (also, in the case of an airplane, the bottom) there usually are relative peaks. In between the peaks there will be more or less variation of the cross section, depending on the wavelength. Wavelength enters because the relative phases of these scattering centers depend on the number of wavelengths separating them (i.e., $\phi \propto 4\pi d/\lambda$, where $\phi$ is relative phase and $d$ is separation). When we talk about a large radar target or a small target, what we might say more precisely is "a target that is large with respect to wavelength" or "a target that is small with respect to wavelength." In this chapter we will primarily be considering targets that are large with respect to wavelength, but in some

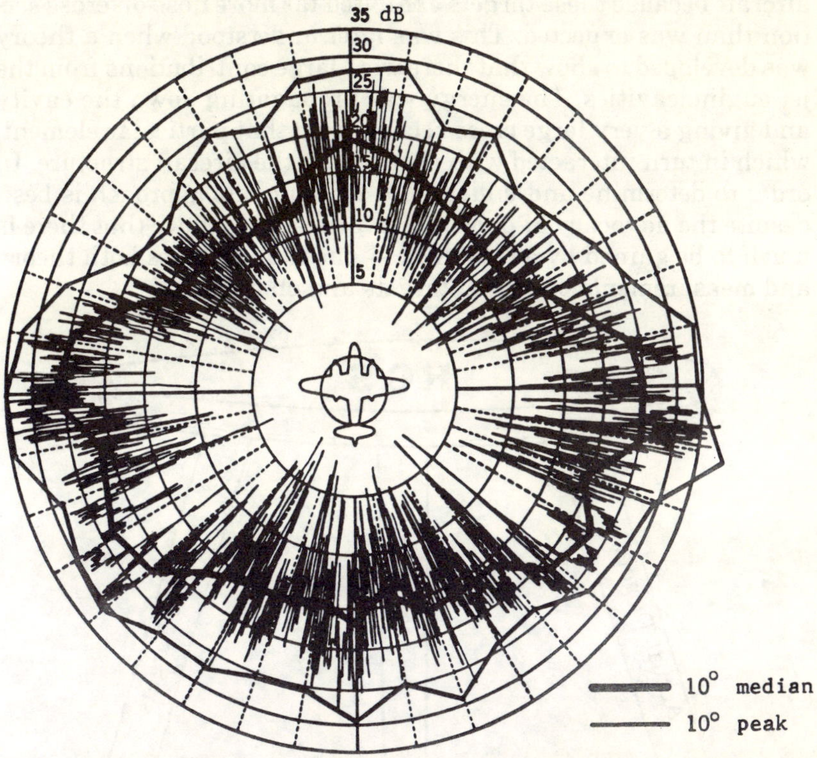


**Figure 10-3.** Return power vs. azimuth for B-26 at 10 cm with 10° median and 10° peak curves added (adapted from Kerr, ref. [9]; © 1951 McGraw-Hill Book Company).

cases there will not be many wavelengths separating the dominant scatterers, a condition which gives a relatively slowly varying, not very complex RCS pattern. If there are many wavelengths between the scatterers, the result is a complex pattern such as the one of Figure 10-3.

In order to better understand radar cross section we can look at it from three different standpoints: theory, actual targets, and scale models. The concept of a three-legged base of understanding is cartooned in Figure 10-4. Actual targets and scale models are complementary in terms of the kinds of information that can be obtained from measurements, as will be explained later. Theory is necessary because it provides a way to understand *why* a particular target configuration behaves in a particular manner.

For example, there was a degree of surprise in the early days of jet aircraft because these targets exhibited far more nose-on cross section than was expected. This was later understood when a theory was developed to show that there were large contributions from the jet engine cavities. The energy was propagating down the cavity and giving a very large cross section from that particular element, which in turn interacted with elements of the aircraft structure. In order to determine and understand RCS, no one approach is best, despite the advocacy of personal favorites. We believe that there is much to be gained by a total look — one that includes both theory and measurements on actual targets and on models.

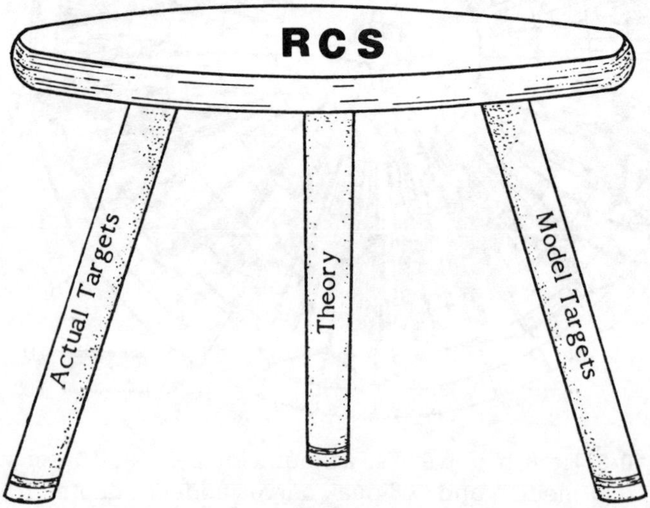


**Figure 10-4.** The three legs supporting an understanding of RCS.

The RCS plot of Figure 10-3 is just one cut through what is actually a three-dimensional target-centered pattern; in this case an azimuth cut was made in the plane of the wings. It is useful to think about the RCS of a target as being a "kind of lumpy" three-dimensional shape. There may be rounded lumps, or there may be very sharp spikes depending on the target shape and, especially, its size in terms of wavelength. Great circle or conical cuts through this pattern produce plots such as that of Figure 10-3.

In addition to RCS amplitude, the return from a target also exhibits variations in phase (*see* Figures 10-15 - 10-17 below). The phase behavior is responsible for target glint which is a part of the angular error of a tracking radar (e.g., monopulse). The radar attempts to point normal to the phase front and this can result in a pointing direction which is outside the physical limits of the target. (*See* Chapter 14 for a discussion of radar angle measurements.) Relative phase could also be considered as a three-dimensional pattern.

## 10.3 TYPES OF MODEL MEASUREMENTS

Most of this chapter is concerned with RF modeling, but we will also briefly discuss optical and ultrasonic modeling.

### 10.3.1 Noncoherent Optical Models

If a polished model of a target is illuminated by a collimated beam of noncoherent light, regions of high return and regions of low return may be observed. The high return regions are the specular (mirror-like) regions, and they reflect the light just like a mirror. Figure 10-5 is an illustration from *Electronics*[10] and shows that as the aircraft model changes aspect the pattern of the specular reflections changes.

Of primary interest to us is the question: Can you measure RCS using an optical model? The honest answer is probably "sort of."

Optical modeling was first done during World War II,[11] and was looked at very favorably several years ago.[12] The technique is now more or less out of favor. One of the reasons is that it is difficult to make and handle an accurate optical model.* For example, the magnitude of the reflection is affected by scratches on the model. Many targets have trihedral elements which give very large returns only if the three faces of the trihedral are precisely orthogonal. This configuration is very hard to model optically (although precise opti-

* The same cautions apply when laser "light" is used for coherent measurements on small, detailed models.

cal corner reflectors are made for a number of applications). Also, there are many, many optical wavelengths across the target, so the equivalent radar phase relations are not modeled, and quantitative measurement of the reflected light amounts to a noncoherent estimate of radar reflectivity rather than a coherent estimate (i.e., power is added rather than the phasor fields).

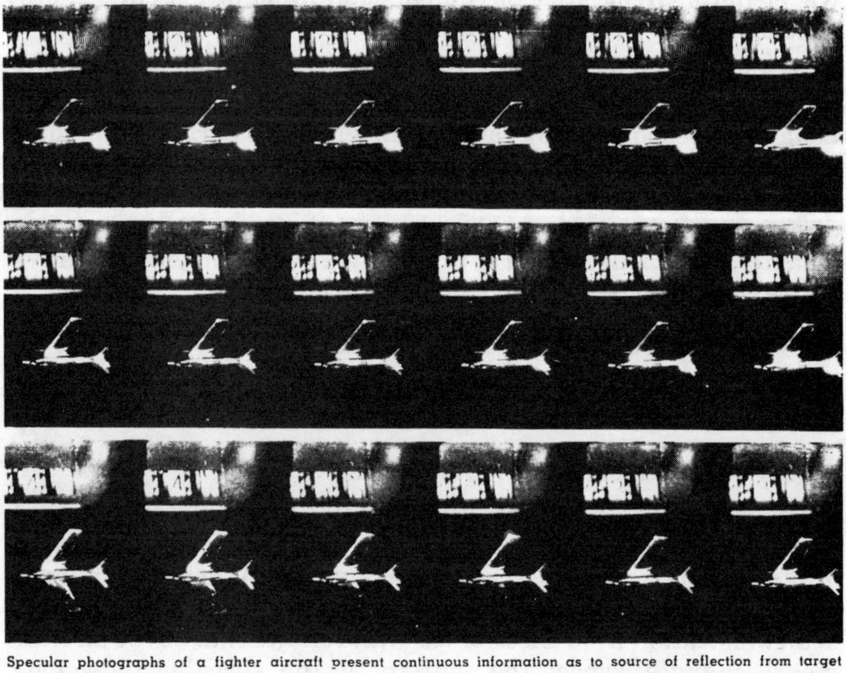


**Figure 10-5.** Variation with aspect of optical highlights of aircraft model (reprinted from Cole, ref. [10]; © 1955 *Electronics*).

What should be emphasized is that optical modeling gives a pictorial view of target reflectivity so that you can get an idea of where some of the specular reflections are. Of course, you can probably look at the target and tell that too, but having a picture can be useful. Figure 10-6 shows optical reflections at several aspects from another aircraft, a crude model of a B-57. The pictures on the left side of Figure 10-6 have lower exposure than those on the right and this picks out the higher amplitude scatterers. The right side, however, shows more of the shape of the aircraft. (RF data on this model at an equivalent frequency of 130 MHz are given in Figures 10-16 and 10-17 below.)

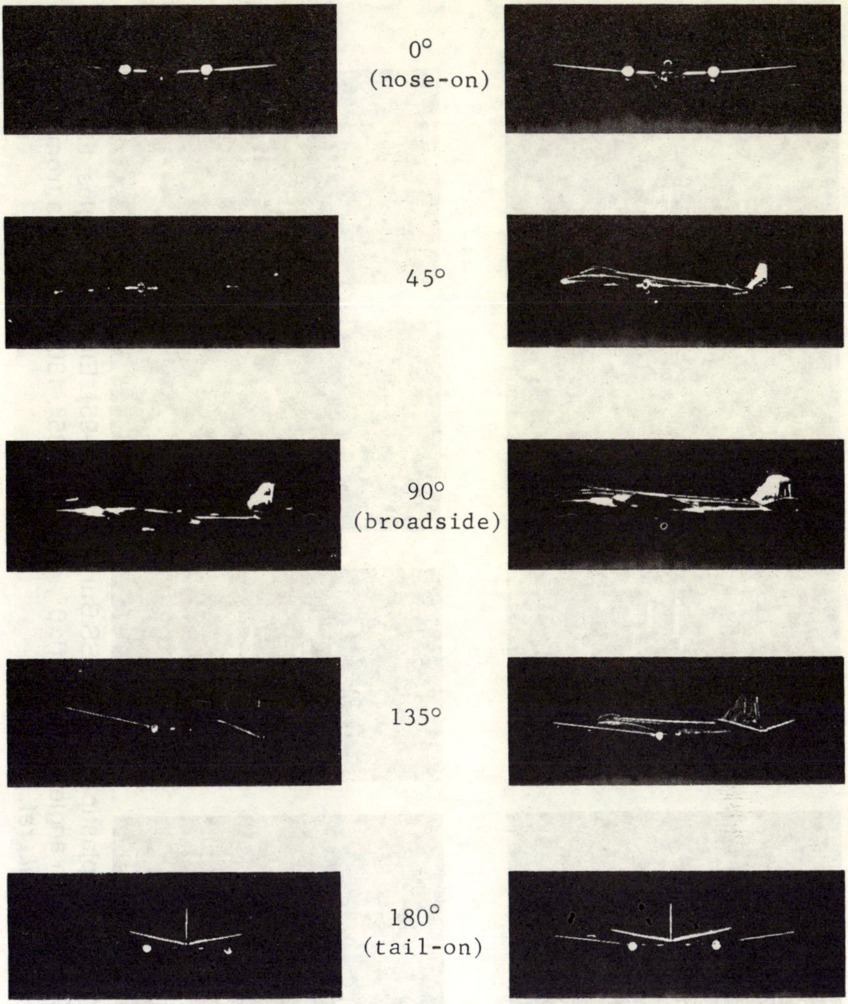


**Figure 10-6.** Patterns of optical highlights of polished model of B-57 aircraft (left-hand column has lower exposure) (reprinted from Corriher, ref.[6]).

Figure 10-7 illustrates optical reflectivity of a ship model.[13] This model is arranged over a reflecting surface so that at broadside you see both the ship and its image in the reflecting surface. This is not polished metal but simply a Revell "hobby" model sprayed with glossy black paint. However, it has a smooth surface and by shining collimated light on it you can get this pattern of reflection.

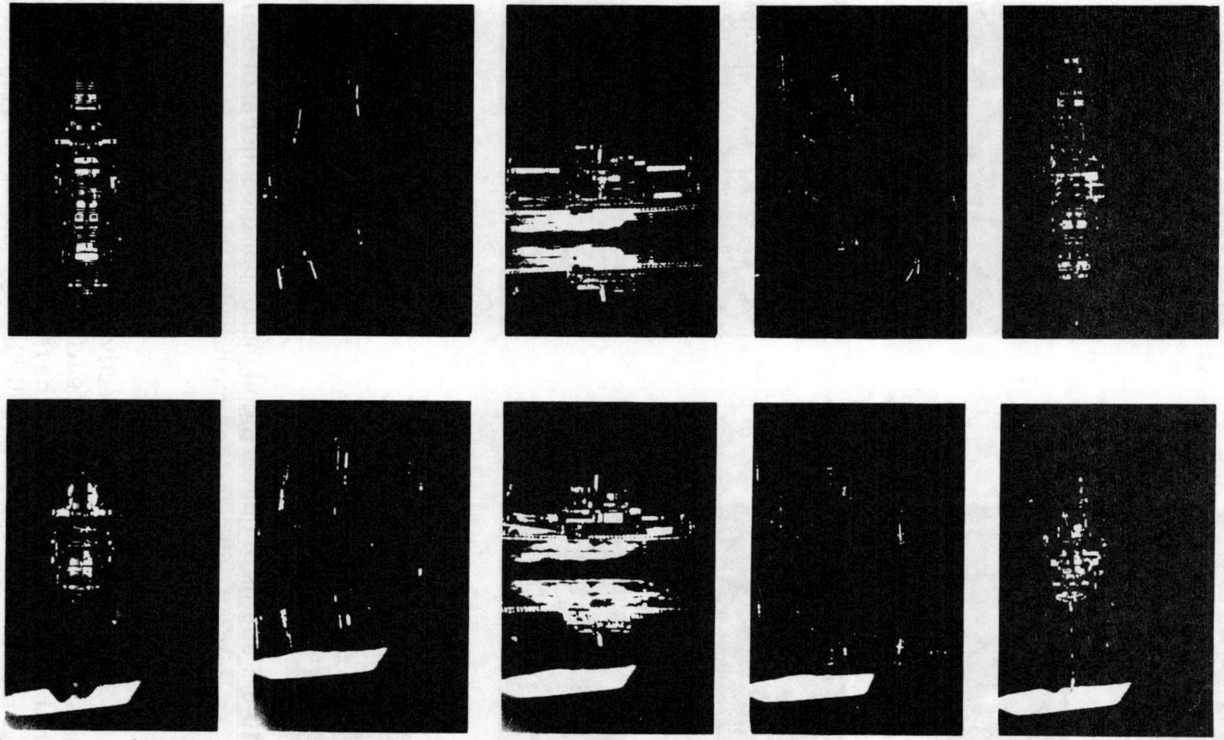

**Figure 10-7.** Optical highlights of plastic model of USS Burleigh (APA-95). Elevation angles: top 45°; bottom 20°. Azimuth angles (from left): 0°, 45°, 90°, 135°, 180° (reprinted from Corriher and Pyron, *et al.*, ref. [13] ).

There have been numerous calibrated optical measurements made in attempts to obtain equivalent RCS data. Figure 10-8 reproduces results from the Naval Missile Center on a ¼ scale jet aircraft model.[14] The microwave data were smoothed and can be compared to the optical simulation. There is a general pattern of similarities, but the curves differ by several dB at many aspects.

In making RCS measurements, we should not expect all methods to yield identical results. Even if you make the same types of measurements from one time to another using a different setup, or a different model, or a different actual target, the results will not be identical.* Whether the differences matter or not depends on the intended use of the RCS data. Therefore, it is worthwhile to know the probable sources of high reflectivity on the real target and how they are simulated on a model.[15] Optical models can assist in understanding mathematical or physical modeling, but it is not surprising that numerical RCS values obtained by noncoherent optical means do not precisely match RF data. In particular, the traveling wave contribution which is frequently important for low RCS bodies is not present in noncoherent optical modeling.

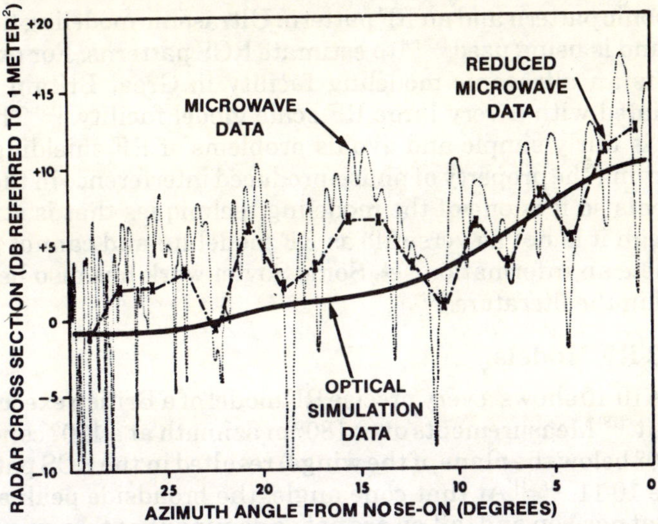


**Figure 10-8.** Comparison of jet aircraft RCS data from noncoherent optical simulation and microwave measurements (reprinted from Rozendal, ref. [15];© 1965 IEEE).

* For "confidence" in a set of numbers, do not replicate the measurement! (*See* also Figure 10-24.)

### 10.3.2 Ultrasonic Models

Another technique for model measurements is to use ultrasonic waves (sonar). If the model is built to reflect sound waves, immersed in a liquid, and illuminated with a sonar, the reflectivity pattern is very similar to the RF reflectivity pattern. There are, however, some important differences. There is no polarization information because sound waves are scalar waves and are not polarized. Also, traveling wave contributions to RCS are not modeled. However, there are relative phase relations which produce interference of the sound waves reflecting from the different parts of the target. Therefore, there is a complex reflectivity pattern which scales with the ratio of the sound wavelength to the model dimensions in the same manner as would the RF wavelengths with respect to the real target. The velocity of acoustic propagation in fresh water compares to the velocity of electromagnetic propagation in air approximately as 1:210,000. An ultrasonic source at 1.4 MHz will produce waves of 1 mm wavelength, which is a useful length for scale model measurements.

Figure 10-9 shows some Japanese results [16] which compare an ultrasonic pattern and an RF pattern. Ultrasonic modeling has been used and is being used [17-19] to estimate RCS patterns. For example, there is an ultrasonic modeling facility in Great Britain that is associated with a very large RF scale model facility. [19] The technique is fairly simple and avoids problems of RF shielding while preserving the property of phase-produced interference in the measurements so it is one of the modeling techniques that is still used although it is not as versatile as RF modeling and cannot give as complete an information set. Some Soviet work has also been described in the literature.[20]

### 10.3.3 RF Models

Figure 10-10 shows a very precise RF model of a British executive jet aircraft. [18] Measurements over 180° of azimuth at a 120° cone angle (i.e., 30° below the plane of the wings) resulted in the RCS pattern of Figure 10-11. [18,21] At that cone angle, the broadside peak and the peaks at nose-on and tail-on are not as dominant as they would be in the plane of the wings. These results are for an equivalent wavelength of 10 cm (3 GHz) and horizontal polarization. If data for another polarization or wavelength were desired, the measurement parameters could be changed and another pattern run.

RF models are discussed further below, but it may be pointed out

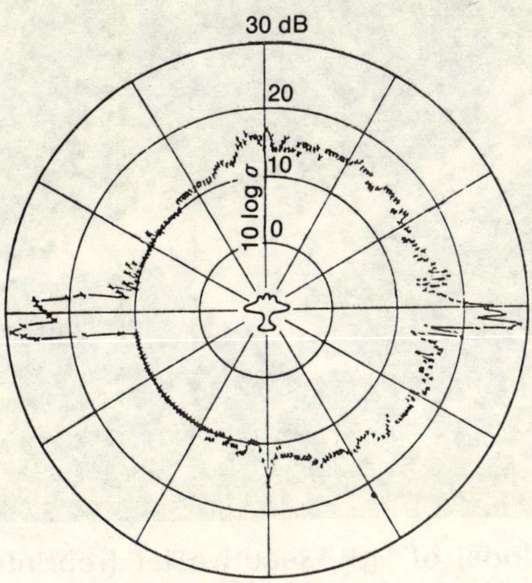


**Ultrasonic cross section of an AT-11 aircraft model.**

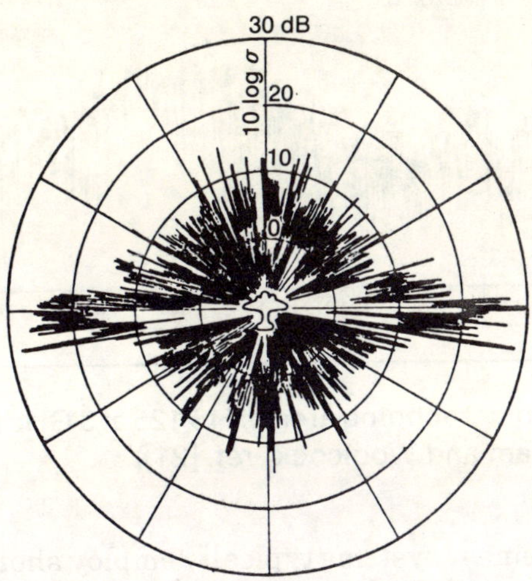


**Radar cross section of an AT-11 aircraft.**

**Figure 10-9.** Comparison of RCS patterns from an ultrasonic model and an actual aircraft (reprinted from Suzuki, ref.[16]; © 1964 IEEE).

**Figure 10-10.** Model of HS executive jet (reprinted from Cram, ref. [18]).

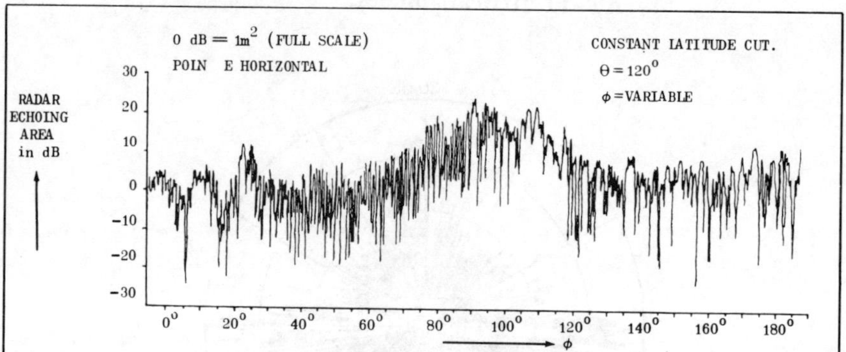


**Figure 10-11.** Typical echoing area of HS 125 at 3 GHz (reprinted from Cram and Woolcock, ref. [21]).

that the measurement systems typically employ short wavelengths in order to preserve the ratio of wavelength to target dimensions that would exist for the actual radar and target. For this reason, millimeter wave systems are common. [19,22,23]

Bahret, at the Air Force Wright Aeronautical Laboratories, has been concerned with RCS data for many years and his seven criteria for good measurements are summarized in Table 10-1. [24] While he

**TABLE 10-1**
**Bahret's Seven Characteristics of Good RCS Measurements**
(*Adapted from Bahret, ref.* [*24*].)

1. Must be able to correlate RCS data with target aspect and range.
2. Must record at a high enough data rate to preserve scintillation information.
3. Must be able to collect data at all frequencies and polarizations of interest.
4. Must have adequate S/N for lowest RCS levels desired.
5. Must have adequate dynamic range to preserve amplitude information.
6. Must have real-time monitoring capability to judge the quality of data.
7. Must record data to facilitate processing.

was primarily concerned with dynamic cross section measurements, these seven points can be used to discuss some of the advantages and disadvantages of model measurements.

As stated in Point 1, you must be able to correlate RCS data with target aspect and, for dynamic measurements, with the target range. Scale models are unexcelled in meeting this need because their measurement aspect can be precisely controlled in angle, and the range is fixed. Measurements can be made at any aspect because the model is a small replica of the real target and can be mounted on a precision positioner. For glint measurements, the interest is in very rapid phase changes. Angular scintillation (glint) can be related to phase shifts occurring with aspect changes on the order of hundredths of a degree. Teledyne Micronetics in San Diego, for example, typically produces data that has an angular resolution of 0.01°. The British make measurements at as fine an angular resolution as 0.005°. With models one has control over the aspect, and then accurate aspect information is available. For an aircraft in the air, a ship at sea, or a tank bouncing over the ground it is very difficult to control aspect, or even to know precisely what the aspect was at the time of a measurement. With a model, one has a body-fixed coordinate system that is very precisely controlled.

The RCS presented to a radar can vary widely and rapidly under dynamic conditions. There must then be a high enough data rate to preserve the amplitude scintillation information (Point 2). Suppose that the real target is an airplane which is yawing, pitching, and rolling. The effects of these motions will be mixed together and one may or may not have sufficient instrumentation in the aircraft to even determine aspect (Point 1), but certainly one could expect rather high scintillation rates in RCS and phase in most cases. A CW measurement system would produce essentially continuous data but with poor range resolution. A pulse system would require a *prf* high enough to be able to detect these fluctuations in RCS and phase. Measurements on a model range of cross section variations or phase variations can be accomplished within a certain bandwidth limit by rotating the model at an appropriate angular rate.

There are many tradeoffs that can be made in model measurements with respect to aspect. For example, if one has an interest in the RCS of an aircraft under missile attack from above, it can be difficult to measure the proper aspects from a ground radar looking up. Sometimes, in fact, in order to measure the cross section of the upper hemisphere of an aircraft, the aircraft is flown upside down. There are other target aspects that are difficult to present to a field measurement radar such as the plane of the wings for an aircraft. Modeling is more flexible in that it provides good aspect control and virtually any data rate desired.

Point 3 is concerned with collecting data at various frequencies and polarizations of interest. It is usually easier to instrument a wide range of these parameters in the laboratory than in the field. The British make model measurements from around 35 GHz up to about 2 THz, using laser radars at the higher frequencies. [19,21,25] Soviet use of laser radars has also been reported. [26,27] Also, polarization control is more readily available at low power than at the higher powers required for dynamic measurements on full-scale targets.

In order to present certain aspects, a real target must be a long distance from the radar. For a fairly low cross section target (or for a fairly low cross section region of a large target) there may be a signal-to-noise problem (Point 4). Low S/N is not normally a problem for model radar reflection ranges, although there may be environmental noise problems (interference) in model measurements, and frequently absorbers must be used to obtain good data. The British also use a very short pulse length and range gate the return so that they only get signals from the vicinity of the target and not the reflections from the room environment.

Some targets have regions which can differ in RCS by factors of $10^5$

or more (Point 5). If the dynamic range of the measurement equipment for models is not this great, attenuators can be used to make multiple patterns which can be matched together because aspect angle can be controlled and repeated from run to run. This is virtually impossible for dynamic full-scale measurements.

Maintaining measurement data quality is relatively easy in the case of modeling because virtually everything is under control (Point 6). If there is a sudden rain storm in the vicinity of the target in a real world case, you may have to repeat measurements. But, how did you get the actual target? You usually begged, borrowed, or stole it from the operational forces for a limited period of time! Then, during the three days that you were going to make measurements, there is a blizzard. Once again, the quality of data is under better control on a model range.

Being on a range does not, however, automatically ensure the quality of the data. For example, if the model is poor, one will be collecting a lot of beautiful numbers that do not mean a thing. In the case of model measurements there is better potential ability to control the environment and therefore the data-taking itself, and ultimately, the data.

Bahret's Point 7 states that RCS measurement data should be recorded. Modern instrumentation allows good data recording in the field, but there is even better ability to provide instrumentation in a model facility. There would be differences in the instrumentation required for dynamic or static measurements and between models and full-scale targets. An NRL facility[28] makes dynamic measurements on actual targets (i.e., real world measurements) whereas the RATSCAT facility[29] is a static measurement range which can use either models or real aircraft, missiles, and other targets. There are actually four combinations of static or dynamic measurements on models or actual targets and, occasionally, dynamic measurements are made on models (e.g., in the manner described by Blore and Musal[30] ).

## 10.4 RF SCALE-MODEL MEASUREMENTS

### 10.4.1 Basis of Electromagnetic Modeling

The theory of electromagnetic modeling has been developed by Sinclair[31] and others[32,33] and is briefly summarized by Blacksmith, *et al.*[1] The most useful approach for RCS measurements is that of geometric scaling whereby the dimensions of what is essentially a perfectly conducting model are chosen to be the dimensions of the actual target divided by some scale factor, $S$. Relations between quantities in the actual (full-scale) and model systems are shown in Table 10-2.[1,31] As shown in this table, the wavelength used for measurements is also scaled by $S$ so that the scale model target has the same dimensions in scale wavelengths that the actual target would have in actual wavelengths. Depending on the measurement wavelengths available, the model can be built at a size appropriate to simulate the desired actual wavelength while still having physical dimensions that can be handled on the measurement range. The model can be 1/2 scale, 1/100 scale, 16/87 scale, or even 3/2 scale. The physical size of model, however, must be large enough that it can be fabricated with reasonable precision.

**TABLE 10-2**
**Geometric Scaling Relations**
(*Adapted from Blacksmith, ref.* [*1*];© *1965 IEEE.*)

| QUANTITY | FULL-SCALE SYSTEM | MODEL SYSTEM |
|---|---|---|
| Length | $\ell$ | $\ell' = \ell/S$ |
| Time | $t$ | $t' = t/S$ |
| Frequency | $f$ | $f' = fS$ |
| Wavelength | $\lambda$ | $\lambda' = \lambda/S$ |
| Conductivity | $\sigma_c$ | $\sigma'_c = S\sigma_c$ |
| Resistance | $R$ | $R' = R$* |
| Inductive Capacity (Permittivity) | $\epsilon$ | $\epsilon' = \epsilon$* |
| Permeability | $\mu$ | $\mu' = \mu$* |
| Antenna Gain | $g$ | $g' = g$ |
| RCS | $\sigma$ | $\sigma' = \sigma/S^2$ |

The aircraft target of Figure 10-6 is approximately 1/72 scale so $S$ is 72. RCS measurements were made at X-band which scales to about 130 MHz, a relatively low radar frequency. As shown later, the RCS patterns at this low frequency did not show the high angular frequency fluctuations of the B-26 in Figure 10-3.

In another case, measurements were needed on an 81 mm mortar round at S-band but only an X-band measurement radar was available. In order to make 3 *cm* measurements equivalent to 5 *cm* measurements, a 0.6 scale model was used ($S = 1.67$). This was not because

*These are the *desired* relationships, but the characteristics of the materials themselves may vary with frequency.

the real target was either unavailable or was too large to handle but because the behavior was needed at a different radar wavelength. (One can also build something larger than actual in order to look at the behavior at a scaled wavelength that is shorter.)

Time scales just like length. Therefore, if you wanted to look at something with the equivalent of a short pulse, say 2 nanoseconds, and the scale factor was 10 (a model object 1/10 the size of the real object), the measurement would require a 0.2 nanosecond pulse length to scale properly. Instead of wavelengths laid along the target, now it would be pulse lengths.

To scale according to the relations in Table 10-2, the model should have $S$ times the conductivity of the real target. If the model is metal and has virtually infinite conductivity (and most metals do at the wavelengths we are generally interested in) then there is no concern about scaling the conductivity. If the model is of wood or fiberglass and sprayed with silver paint, there must be enough skin depths of silver paint or the conductivity will not be close to that of the real metallic object. Resistance stays the same between target and model, but resistivity scales because it has dimensions of resistance per unit length. However, RCS, which has the dimensions of length squared, is scaled as the square of the scale factor.

There are numerous problems associated with modeling non-conductive surfaces which may be encountered in reduced RCS configurations. Note in Table 10-2 that the needed relationship is for $\mu$, $\epsilon$, and $R$ to be the same for the model and the full-scale target, but most non-conductive materials have properties which vary with frequency. Thomas[34] has discussed some of these aspects in modeling.

Figure 10-12 shows the relationships among model measurement frequency, model scale factor, and the equivalent full-scale frequency. For example, if 30 GHz were available as a measurement frequency, and you wanted to simulate a 10 GHz radar, then a 1/3 scale model would be used ($S$ would be 3). Generally, the model scale factor is chosen to allow an available measurement frequency to simulate the desired full-scale radar frequency.

### 10.4.2 Model Accuracy

The degree of detail required in the model depends upon how accurately you need the RCS information, and what the contributions of individual elements are to the overall radar cross section. In addition, surface tolerance can be an important factor. For very low

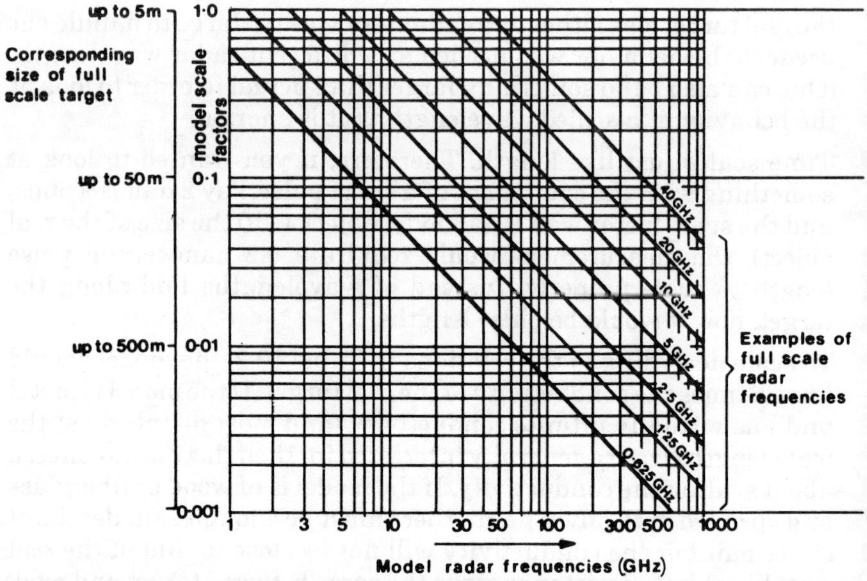


**Figure 10-12.** Chart of radar modeling relationships (reprinted from Cram and Woolcock, ref. [21]).

cross sections, it is important to have a very smooth surface if, in fact, the real target has a smooth surface in terms of roughness with respect to wavelength. Senior has reported measurements made on rough spheres. [35] In Figure 10-13, $d$ is the depth of the bumps on the sphere and $w$ is the width of the bumps. This 10-inch sphere was observed with four different frequencies. When the sphere was not very rough with respect to wavelength (i.e., $d/\lambda$ and $w/\lambda$ small) there was virtually no difference in RCS as the rough sphere was rotated. As the roughness became larger with respect to wavelength, then there was a variation up to 2 - 4 dB in sphere cross section at different angles. Senior [35] concludes that a surface smoothness of about $10^{-3}\,\lambda$ will permit data good to a fraction of a dB at peak values and about one dB at minima.

If the target has a dominant scatterer, it will make little difference if a small scatterer is rough or not. However, for a low cross section shape such as a re-entry vehicle, its roughness can matter very much. Also important can be other details of the construction. For example, in the RCS of a cone sphere a major contribution comes from the tip, and another major contribution is from the creeping wave that goes around the sphere portion and is then re-radiated by the tip. If the join between the cone and the sphere on a model differs

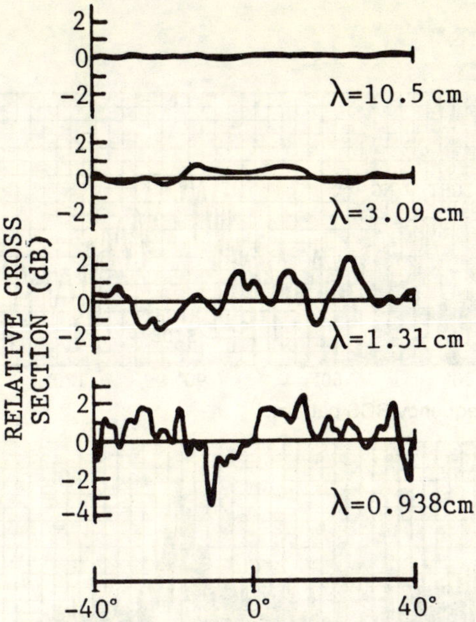


Relative back scattering cross section of 10-inch diameter sphere.

SPHERE PARAMETERS

| Freq. (Gc/s) | λ(cm) | *ka* | *d*/λ | *w*/λ |
|---|---|---|---|---|
| 2.87 | 10.5 | 7.73 | $7 \times 10^{-3}$ | $3 \times 10^{-2}$ |
| 9.70 | 3.09 | 26.1 | $2 \times 10^{-2}$ | $10^{-1}$ |
| 22.97 | 1.31 | 62.0 | $6 \times 10^{-2}$ | $2 \times 10^{-1}$ |
| 31.97 | 0.938 | 86.2 | $8 \times 10^{-2}$ | $3 \times 10^{-1}$ |

**Figure 10-13.** Effect of surface roughness on RCS of a sphere (reprinted from Senior, ref. [35]; © 1965 IEEE).

from that of the real target, there will be a different creeping wave contribution to RCS. For a low RCS target, you would then have to be concerned with the conductivity, the surface finish, and the joins, as well as the shape.

## 10.5 DATA FROM RF SCALE MODELS

### 10.5.1 Monostatic Examples

The plots in Figure 10-14 contrast low-frequency RCS behavior of a model aircraft target with high-frequency behavior[36]; there are

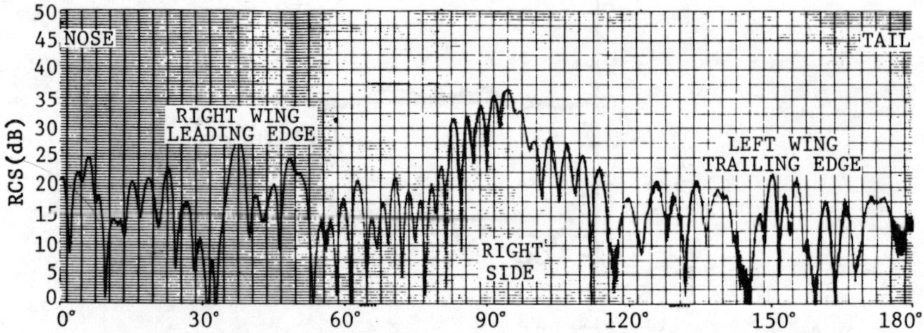


(a) Typical low frequency RCS pattern.

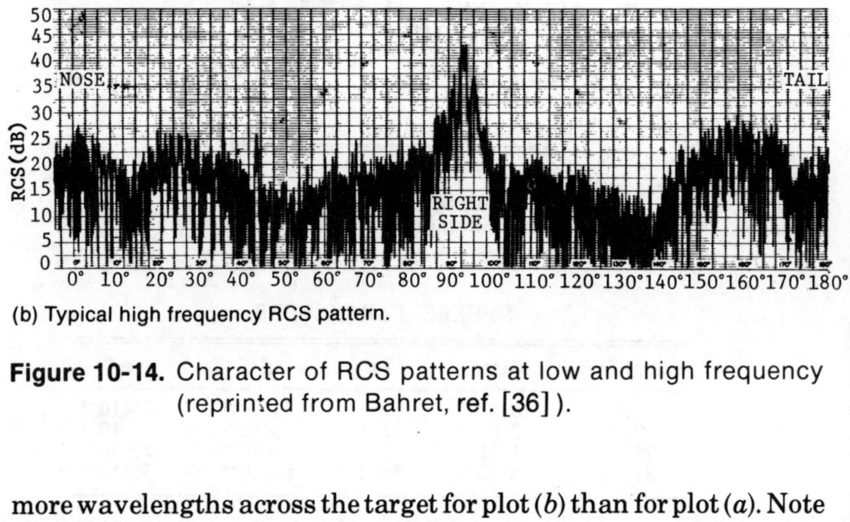


(b) Typical high frequency RCS pattern.

**Figure 10-14.** Character of RCS patterns at low and high frequency (reprinted from Bahret, ref. [36]).

more wavelengths across the target for plot (*b*) than for plot (*a*). Note that the left side and the right side curves of plot (*a*) are different — the right side is smeared. This resulted from having the wind blow the model while the right-side portion of the pattern was being taken. This meant that some of the relative phase relations that were fixed on the left were being varied by the motion of the target and were thereby varying the RCS while the right-side data were being taken.

At the higher frequency shown in plot (*b*), the overall noisiness of the pattern would obscure such rapid variations. If the wind blew, there could be erroneous data, and one would not even know it (but it is doubtful that one would be interested in that much detail). The general behavior of RCS or the statistics of the RCS for a particular aspect region are usually of interest.

Considering the concept of a lumpy three-dimensional surface, what RCS does the radar see? The radar sees the value where the line-of-sight intersects the three-dimensional surface. If there is a large RCS lobe, and the radar sees the lobe fly by, it will see the effect of a target RCS that slowly rises and then falls. If the surface is intersected in a region of very fine spikes, then the radar will see this rapid variation of RCS with aspect converted into a rapid time variation in return power to the radar.

Figure 10-15 presents data on an aircraft model. [37] The top plot is of radar echoing area (the British term for radar cross section). Although the vertical dB scale does not show numerical values, there

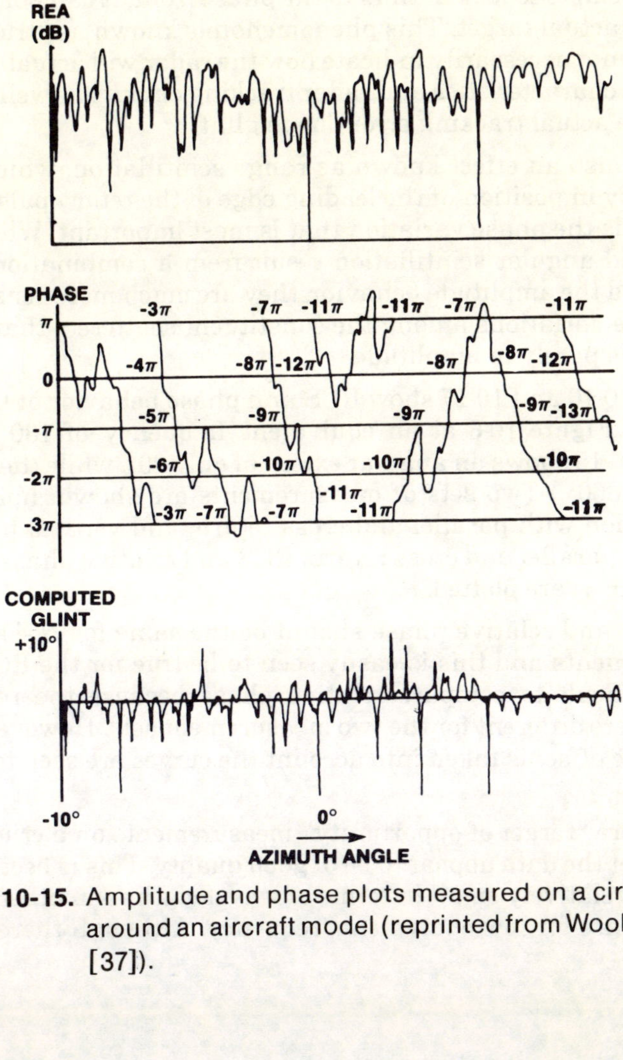


**Figure 10-15.** Amplitude and phase plots measured on a circular path around an aircraft model (reprinted from Woolcock, ref. [37]).

are some very deep nulls. The center plot is of relative phase over a range of $4\pi$ radians ($+\pi$ to $-3\pi$). The bottom plot shows glint (angular scintillation) computed from the variation of the relative phase. Notice that in regions where the phase changes rapidly there are minima in the RCS pattern. For each angle where there is a rapid change in phase there is also a deep RCS null and a glint spike. However, for each RCS null there is not always a rapid change in phase or a glint spike.

From such model measurements, relative phase behavior can be determined. A tracking radar attempts to point normal to the phase front, and when there is a rapid angular change in phase, and the radar attempts to look normal to the phase front, it can point away from the actual target. This phenomenon is known as "true" glint and does not necessarily indicate how the radar will actually track. Antenna characteristics and radar tracking circuits as well as S/N affect the actual tracking error due to glint.

There is also an effect known as range scintillation, which is an instability in position of the leading edge of the return pulse. Once again, it is the phase variation that is most important. While both range and angular scintillation result from a combination of the phase and the amplitude behavior, they are much more sensitive to the phase variations among the constituent scatterers than to the variations purely in amplitude.

Figures 10-16 and 10-17 show RCS and phase behavior of the B-57 model of Figure 10-6 at an equivalent frequency of 130 MHz.[6] (Figure 10-15 shows an angular extent of only 20°, while these plots are over 360°.) Two sets of measurements are shown: horizontal polarization with parallel and cross return, and vertical polarization with parallel and cross return. RCS and relative phase (over a range of $\pm\,\pi$) are plotted.

The RCS and relative phase should be the same for *HV* and *VH* measurements and this is easily seen to be true for the RCS. It is more difficult to compare relative phase because the reference values were different for the two measurement sets. However, when the phase offset is taken into account the curves are seen to match closely.

These were "target-of-opportunity" measurements on a crude model, and yet the data appear to be of good quality. This is because the target is relatively small in terms of wavelength — in other words, the data are for low equivalent frequency. Although there are no glint data, the deep RCS nulls correlate well with rapid changes in the relative phase.

The data of Figures 10-16 and 10-17 would be very difficult to take on a real target. One would need a radar with dual-polarized transmission as well as dual-polarized reception to be able to get the *HH*, *HV*, *VV*, and *VH* returns over many aspects. However, knowing that this is the kind of behavior expected from the target, then you can collect real world data and see how it compares in a general sense (i.e., in an aspect-cell sense rather than point by point).

The data of Figure 10-18 are from a 1/72 scale model of a B-25 over about a 45° aspect variation [38] and show what happens when measurement frequency is varied. Notice that the overall pattern of returns changes over the 45° aspect region. While the scattering amplitudes of the individual scattering centers are probably not very frequency sensitive over the approximately 5% frequency range, relative phase can change appreciably. In a sense, the relative phases among scattering centers are being changed electrically by changing the frequency rather than physically by changing aspect. A frequency-agile radar changes the relative phases among scattering centers to average out nulls, *et cetera*. Phase relations change in the case of a real aircraft flying toward you because it is yawing, pitching, and rolling, as well as changing in gross aspect due to translation. Although elemental returns will add somewhat differently at different instants of time, Figure 10-18 shows the kind of general behavior expected when a target is viewed with different frequencies.

We have previously discussed the detail required for modeling and also the fact that with jet engines you see into the cavity and get a large return. Not only does the radar see into the cavity and get a high return, it also sees the rotating compressor blades (in fact, several stages of the compressor). These many scatterers have a range of relative phases which leads to a complex return. Figure 10-19 shows the *VV* and *HH* RCS (REA as labeled by the British) of a jet engine that has been modeled for the Royal Signal and Radar Establishment at Malvern.[37] The aspect range is ±90° from nose-on.

The modeling of an engine is difficult, and if the RCS were desired for the engine alone, it might be easier to use an actual engine even with the difficulty of getting the engine, setting it up on the turntable, *et cetera*. Although the model will not look exactly like a real

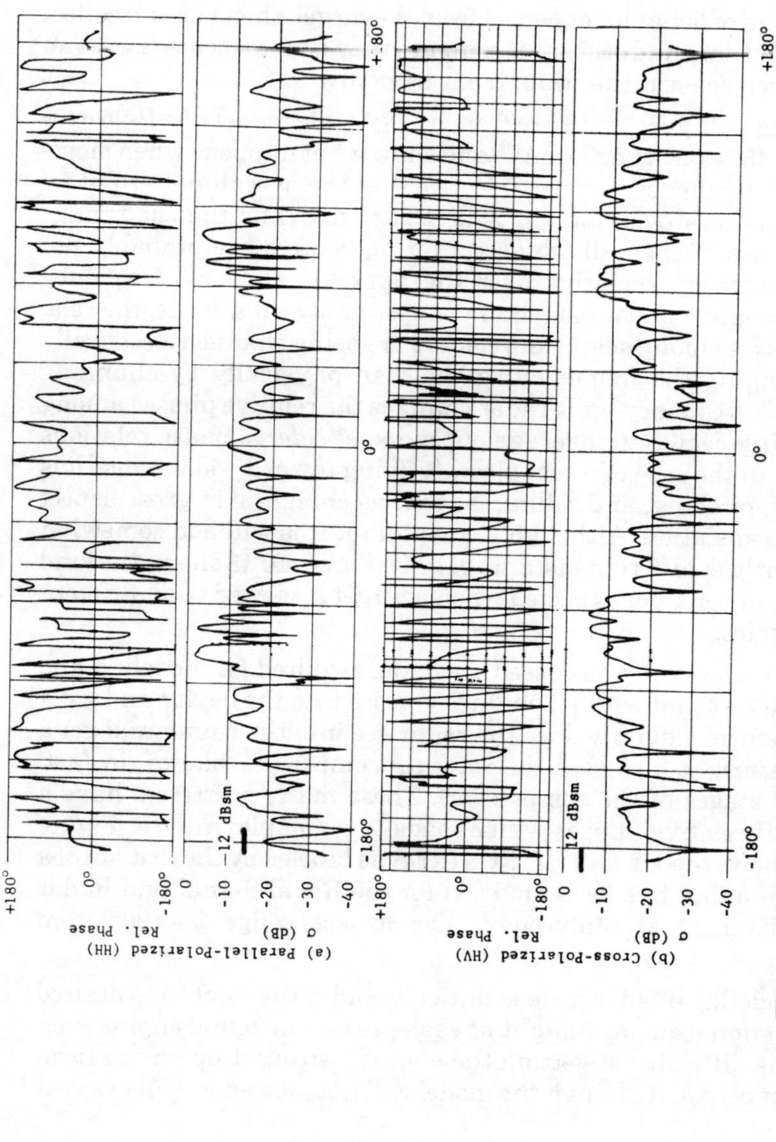


**Figure 10-16.** Parallel and cross polarized relative phase and RCS for 1/72-scale B-57 at 0° elevation (horizontal transmission at simulated frequency of 130 MHz) (reprinted from Corriher, ref. [6]).

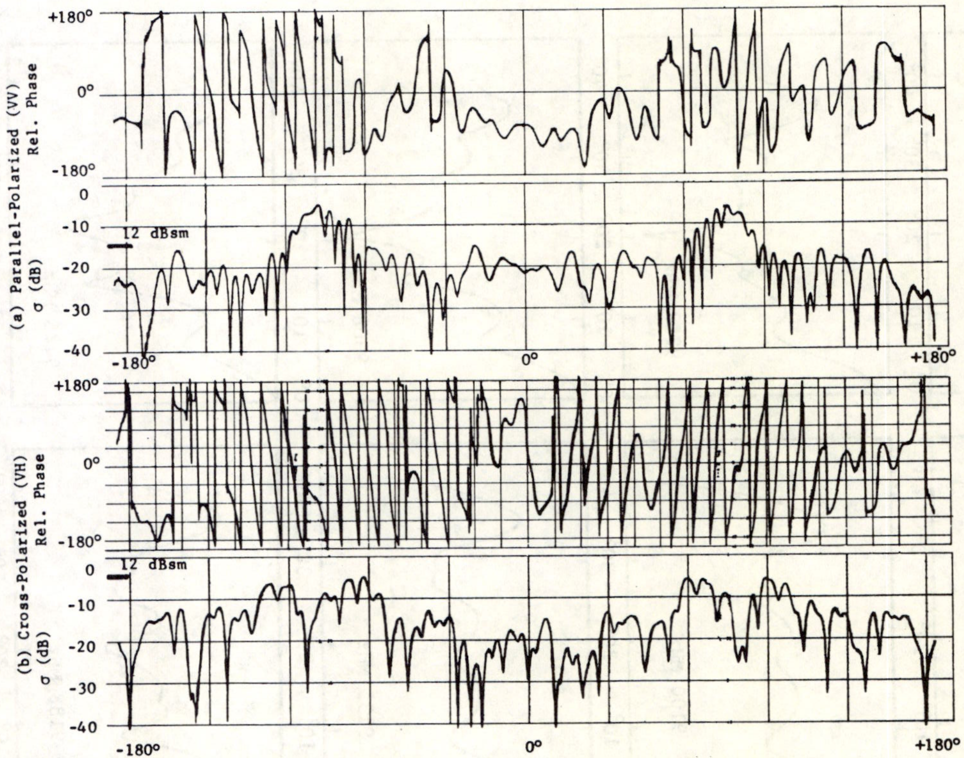


**Figure 10-17.** Parallel and cross polarized relative phase and RCS for 1/72-scale B-57 at 0° elevation (vertical transmission at simulated frequency of 130 MHz) (reprinted from Corriher, ref. [6]).

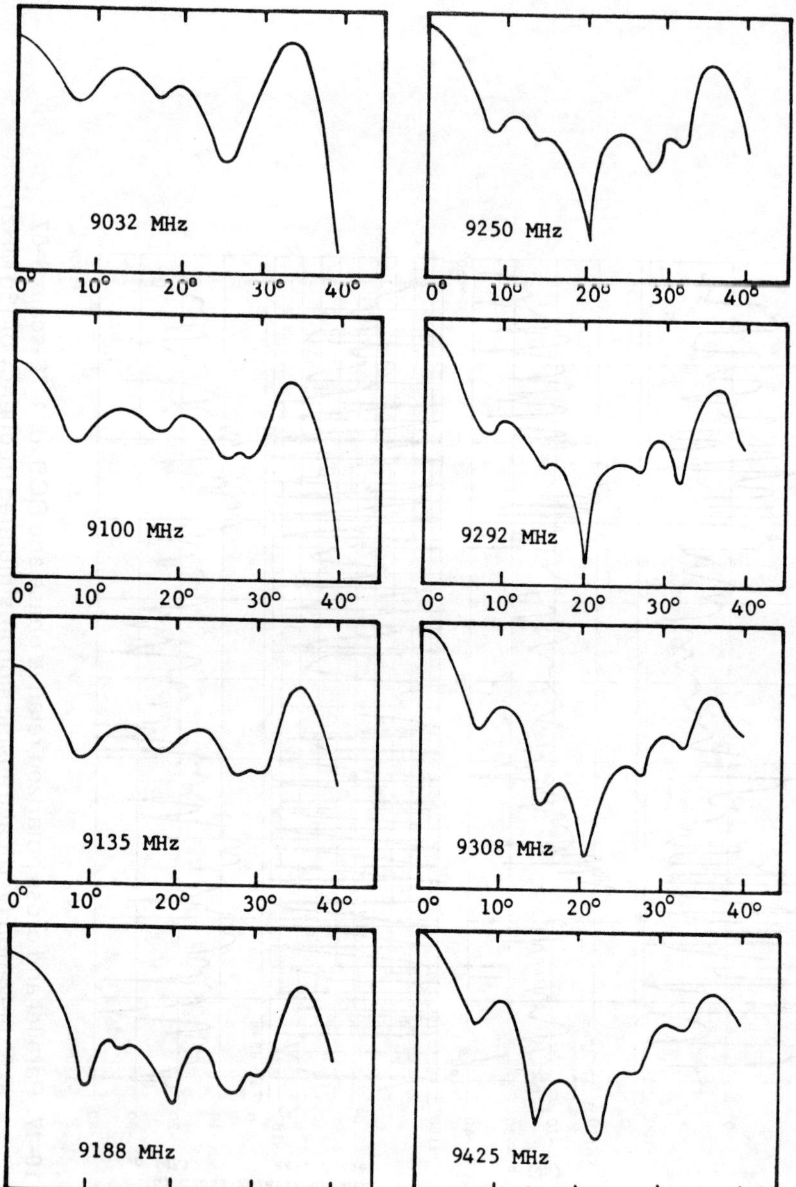


**Figure 10-18.** Wavelength dependence of return from 1/72-scale B-25 at measurement frequencies near 9200 MHz (130 MHz equivalent frequency) (reprinted from Meeks, *et al.*, ref. [38] ).

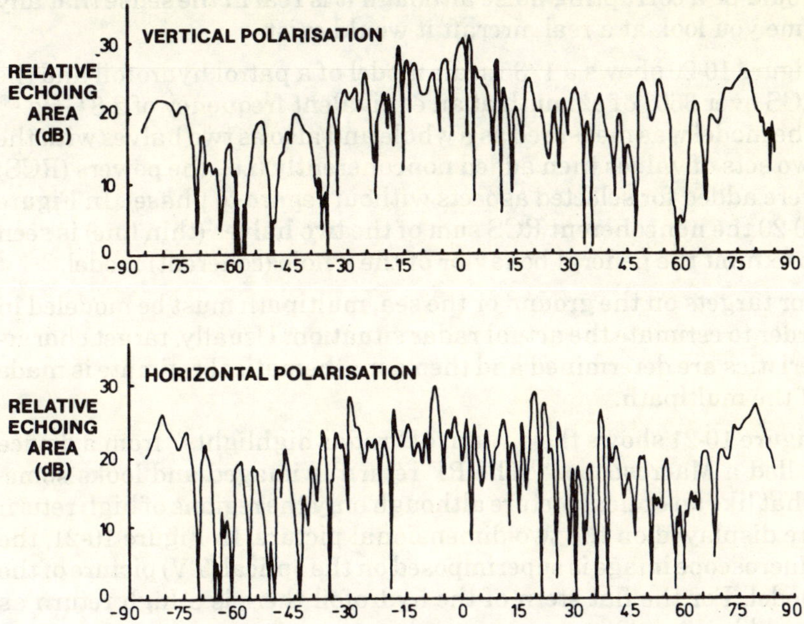


**Figure 10-19.** X-band equivalent RCS of aircraft engine model (reprinted from Woolcock, ref. [37]).

engine, one real engine and another real engine may not look exactly the same either. (The engines are probably going to be closer to identical than would the overall aircraft.) The value of having a model engine is in being able to better model the entire aircraft. A real aircraft will have jet engine modulation on the return power and, if the radar *prf* is high enough, Doppler components can be observed at the engine modulation frequencies.

In addition to engines, other features on a real aircraft are difficult to model. These include the cockpit, the radar compartment behind the radome, the radome itself (in some cases), and radar/ECM/ESM antennas. In other words, all complex "cavities" cause difficulties.

It would be difficult to model the dynamic behavior of an aircraft on a model range since a real aircraft — just like a real bird — "flaps" its wings. If the wings are major contributors interfering with other scatterers, their motion will introduce a temporal modulation into the RCS. For an actual aircraft, the modulation could be either temporal or spatial depending on the aspect of the radar. If the interest is in the aspect variability of aircraft RCS, such modulation would be a corrupting noise although it is real in the sense that any time you look at a real aircraft it would exist.

Figure 10-20 shows a 1/35 scale model of a patrol hydrofoil and the RCS over 360° of azimuth at an equivalent frequency of 2.8 GHz.[39] The model was measured as a whole and also as two halves with the two sets of values then added noncoherently (i.e., the powers (RCS) were added for selected aspects without regard to phase). In Figure 10-20 the noncoherent RCS sum of the two halves (thin line) is seen to exhibit the general behavior of the whole (coherent) model.

For targets on the ground or the sea, multipath must be modeled in order to estimate the actual radar situation. Usually, target characteristics are determined and then a mathematical estimate is made of the multipath.

Figure 10-21 shows the pattern of "radar highlights" from a device called a Macroscope.[39] The RF return is imaged and looks somewhat like an optical picture although only the regions of high return are displayed on the two-dimensional picture. In Figure 10-21, the Macroscope image is superimposed on the optical (TV) picture of the model. For the flat stern of the hydrofoil there is a high return as would be expected.

As shown in Table 10-2, the resistance should not change between the full-scale and model targets. However, length scales, so resistance per unit length would also scale. In Figure 10-22,[40] on the left are RCS data for *conducting* tetrahedra — four equal-length rods formed into a tetrahedron. On the right, the rods are carbon (i.e., *resistive*). The purpose here was to examine aerospace materials that now include carbon fiber laminates and to establish what happens to the model cross sections of such bodies. RCS scales as the square of the scale factor, which was 7.5 for these measurements (full-scale frequency of 400 MHz from a model frequency of 3000 MHz). For the carbon model, the median RCS is –27.9 dBsm, which may be scaled and compared to –12.2 dBsm measured for the full-scale tetrahedron. The scale correction for the model is $(7.5)^2$ (in dB, $10 \log (7.5)^2 = 20 \log 7.5 = 17.5$ dB). Adding $-27.9 + 17.5 = -12.4$ dBsm, which is very close to the measured full-scale value of –12.2 dBsm.

### 10.5.2 Bistatic Example

Scale models are very useful for bistatic measurements since angles can be controlled well. Plots of RCS for a model C-47 target at bistatic angles of 70° and 110° are shown in Figure 10-23.[41] For

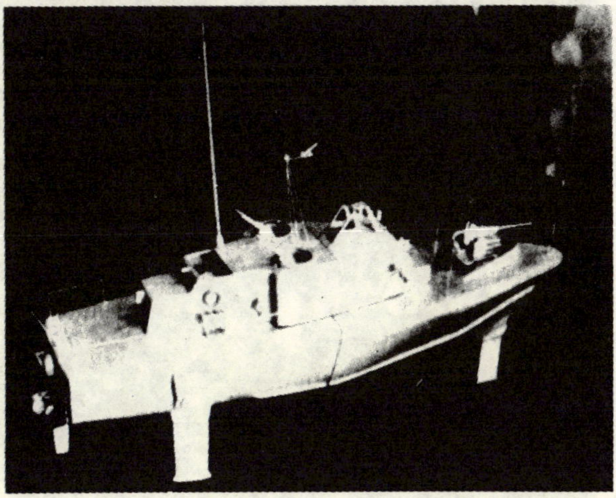

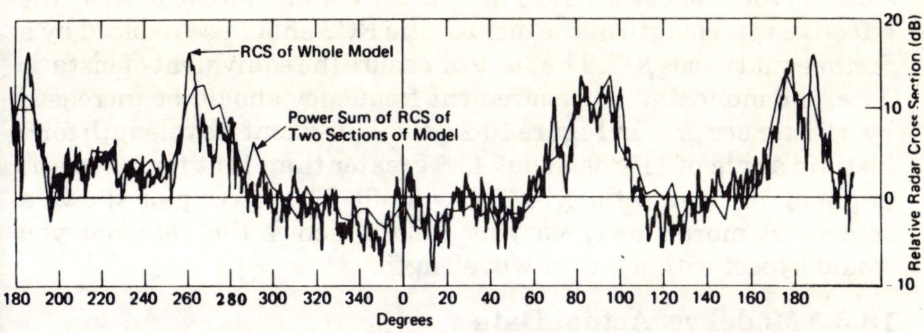


**Figure 10-20.** RCS of 1/35-scale PGH-2 at equivalent frequency of 2.8 GHz (RCS of whole model compared to two halves summed together) (reprinted from Paddison, *et al.*, ref. [39]; © 1978 IEEE).

**Figure 10-21.** TV display of stern view of PGH-2 hydrofoil ship on conducting ground plane with radar bright spots superimposed (reprinted from Paddison, *et al.*, ref. [39]; © 1978 IEEE).

small bistatic angles, bistatic RCS for a given aspect is about the same as the monostatic RCS at half the bistatic angle, $\beta$. Also, the effective wavelength of the monostatic RCS should be reduced by a factor equal to cos $\beta/2$. That is, to measure the equivalent of bistatic data, the monostatic measurement frequency should be increased by a factor sec $\beta/2$. In Figure 10-23, the equivalent wavelength for a bistatic angle of 110° is about 43% greater than that for a bistatic angle of 70° (the ratio 1.74/1.22 = 1.43). The lower plot shows a somewhat more slowly varying RCS which is the behavior you would expect with a longer wavelength.

### 10.5.3 Model vs. Actual Data

While Figure 10-24 is not a comparison of model data with full-scale target data, it shows measurements of the same model on two different model ranges, [42] and thereby indicates the fact that there will be some variations even in "identical" measurements. Notice that the two sets of curves are actually very, very close to each other. Although the author is disappointed in their similarity, it should be noted that these are very low RCS values (the largest differences are at values of about 0.01 $m^2$ and the differences are small with respect

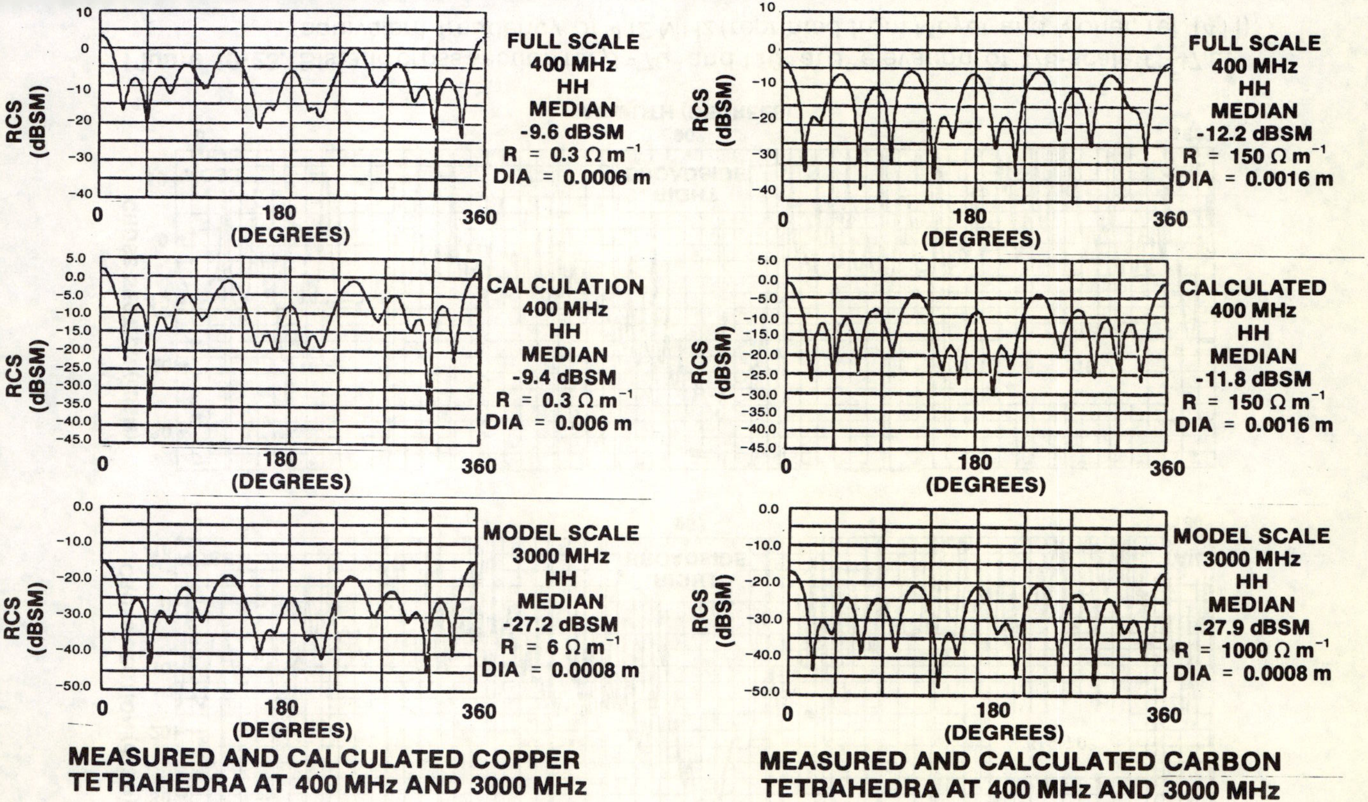


**Figure 10-22.** RCS of conductive and resistive tetrahedra (reprinted from Elphick, *et al.*, ref. [40]; © 1977 IEEE).

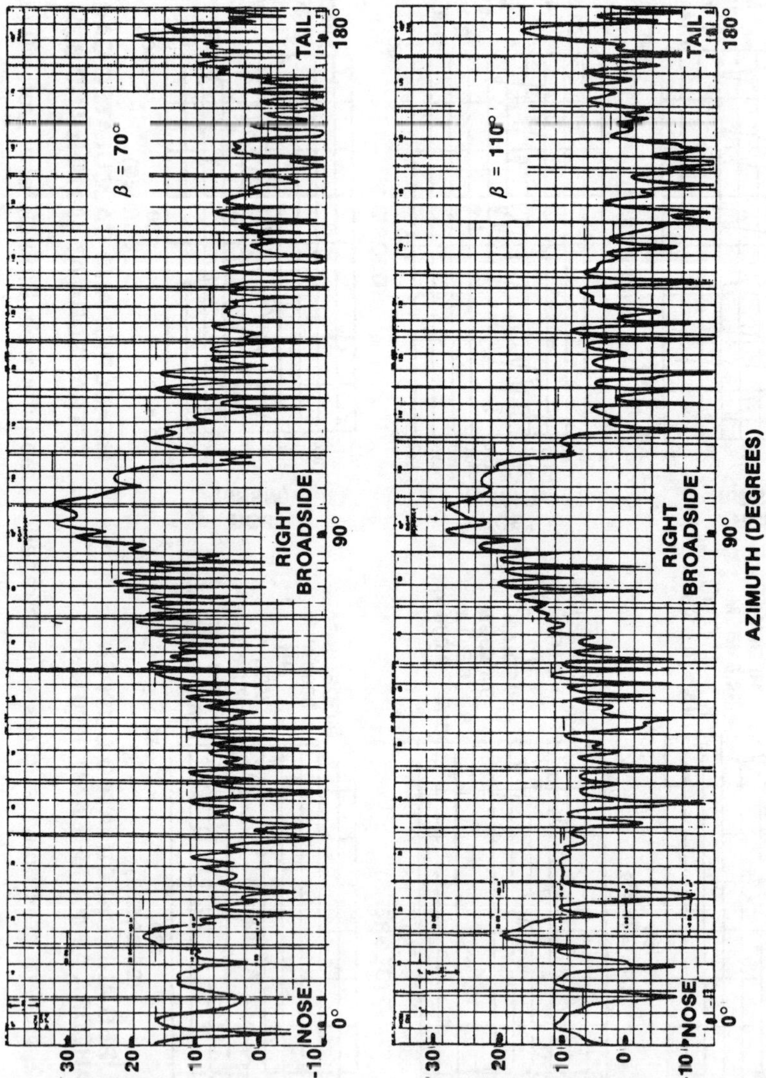


**Figure 10-23.** Bistatic cross section for $\beta = 70°$ and 110° at 0° elevation of 1/8-scale C-47 at equivalent frequency of 412 MHz (reprinted from Meyer and Moner, ref. [41]).

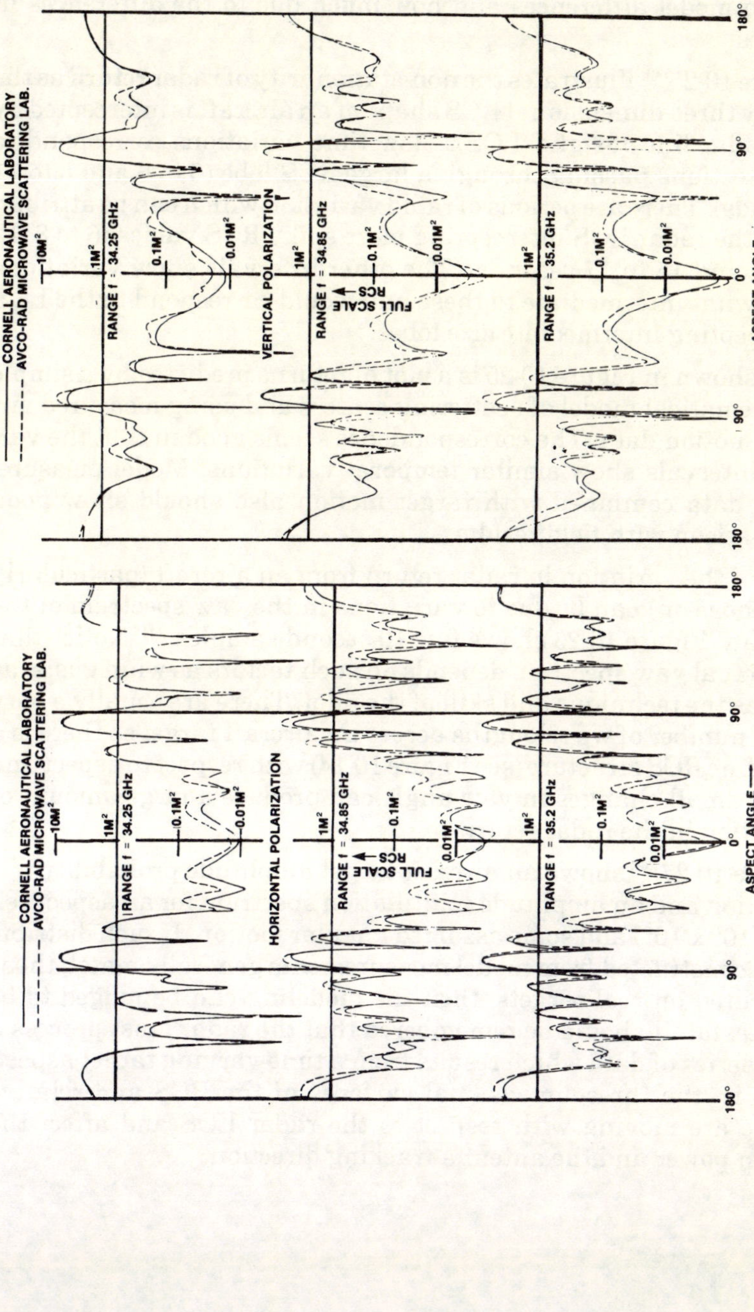


**Figure 10-24.** Comparison of low RCS measurements on two model ranges (reprinted from Kell and Pedersen, ref. [42]; ©1965 IEEE).

to a dB). A more interesting picture would have had each of the two facilities make a model and then have the two models measured on both ranges. This would have shown how much of the variation was due to model differences and how much due to the differences in ranges.

Figure 10-25[43] illustrates the non-stationarity of radar return as the lumpy three-dimensional RCS shape of an aircraft is intercepted by the radar line-of-sight (LOS). Slow time variations correspond to intersections passing through a large RCS lobe (0 to 6 and 8 to 11 seconds). There are periods of rapid variation which can be attributed to the radar LOS intercepting more spiky RCS values (6 to 8, 12 to 13, and 15 to 17 seconds). The other intervals show variations somewhat intermediate to these and would correspond to the LOS intercepting intermediate size lobes.

Also shown in Figure 10-25 is a plot of returns predicted by a simple mathematical model of scattering centers and using measured aircraft motion data. The correspondence seems good in that the various intervals show similar temporal variations. Model measurement data combined with target motion also should show good comparison with field results.

Part of the variation in radar return from an aircraft (particularly near nose-on) can be due to variations in the yaw spectrum of the aircraft. Figure 10-26 shows five 10-second samples. [44] Notice that the actual yaw spectrum depends on such factors as wind gusts, as well as the technique and skill of the pilot. There are usually a very large number of wavelengths across an aircraft target so there is a very fine RCS structure (see Figure 10-14) with respect to aspect and fairly small changes in yaw angle can produce a large amount of variation in the radar return.

Figure 10-27[36] shows an aircraft target amplitude probability distribution and an amplitude scintillation spectrum for an aspect cell (e.g., 10° x 10°) and some assumed angular motion. If such distributions constructed from model measurements generally match those measured on real targets, then the modeling can be judged to be successful. It should be remembered that the radar must process a time series of data which results from a time-varying target aspect. That is, the three-dimensional surfaces of the RCS and relative phase are moving with respect to the radar LOS and affect the return power and the antenna tracking direction.

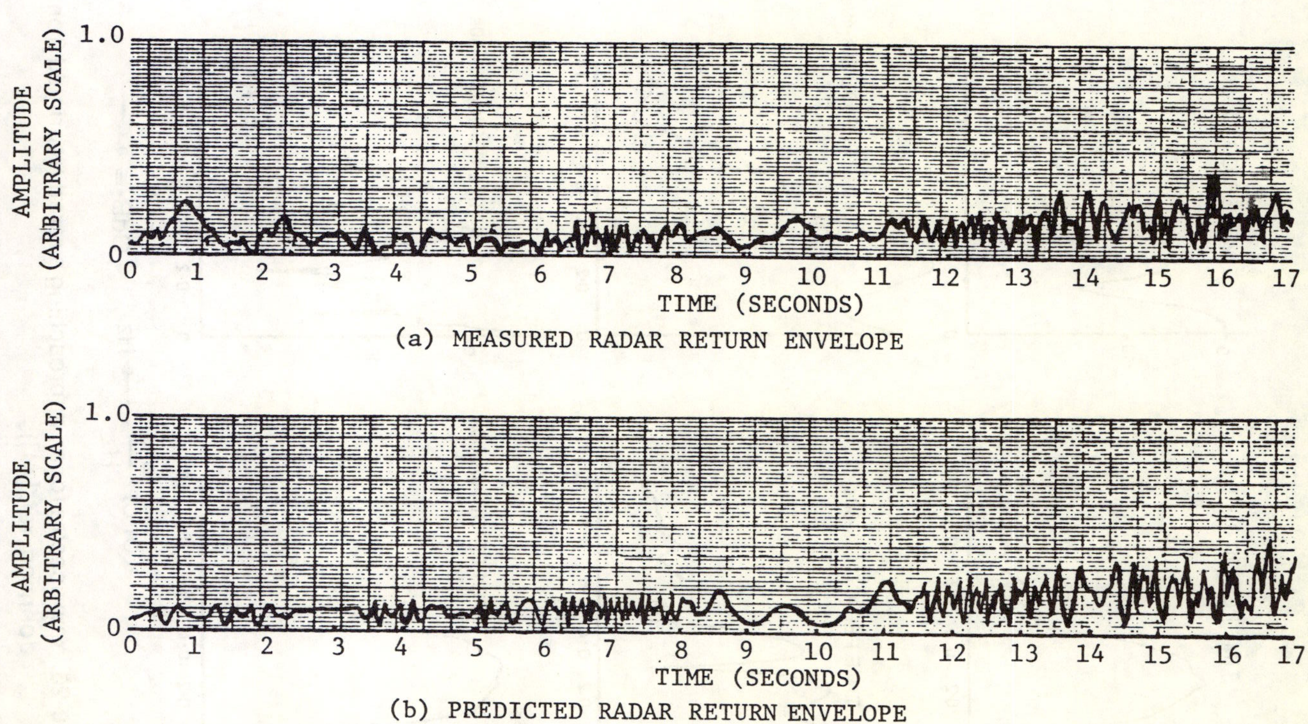


**Figure 10-25.** Measured and predicted time histories of X-band return with twin-jet aircraft showing non-stationarity (reprinted from Moll, ref. [43];© 1967 IEEE).

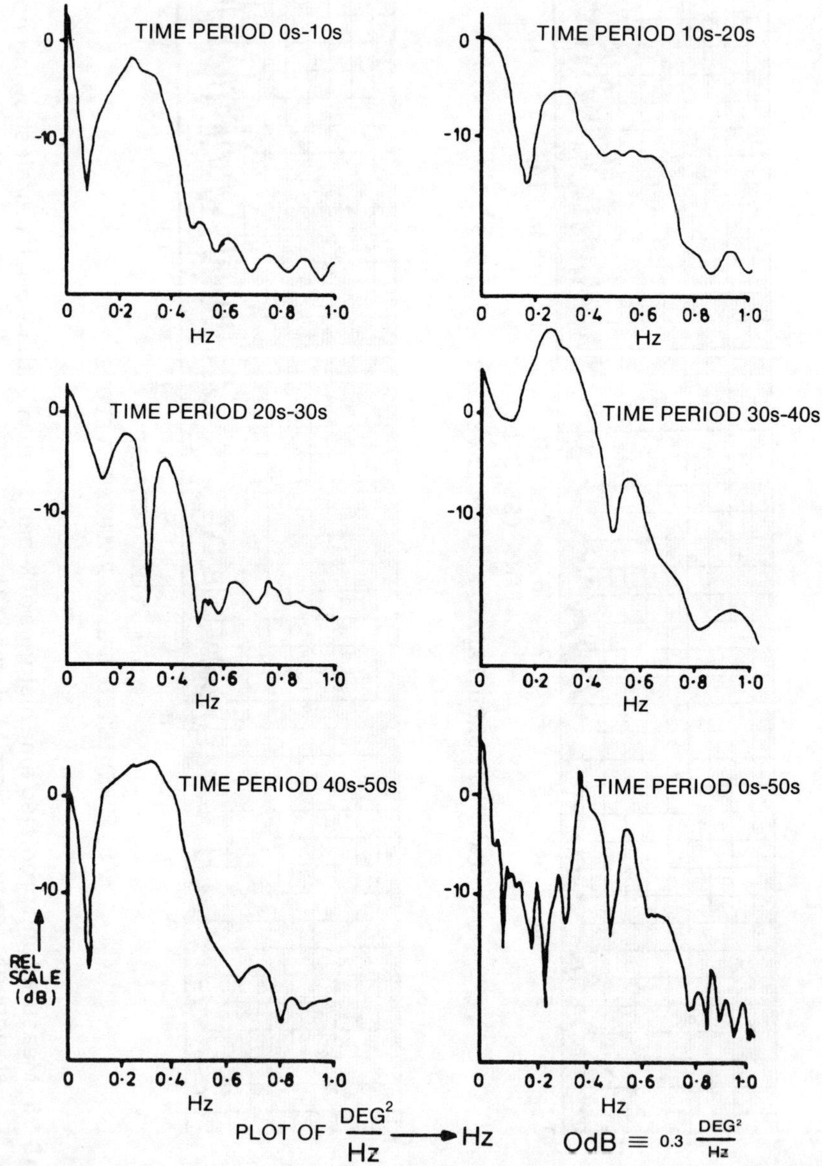


**Figure 10-26.** Yaw spectra for an approaching aircraft (reprinted from Johnson, ref. [44]).

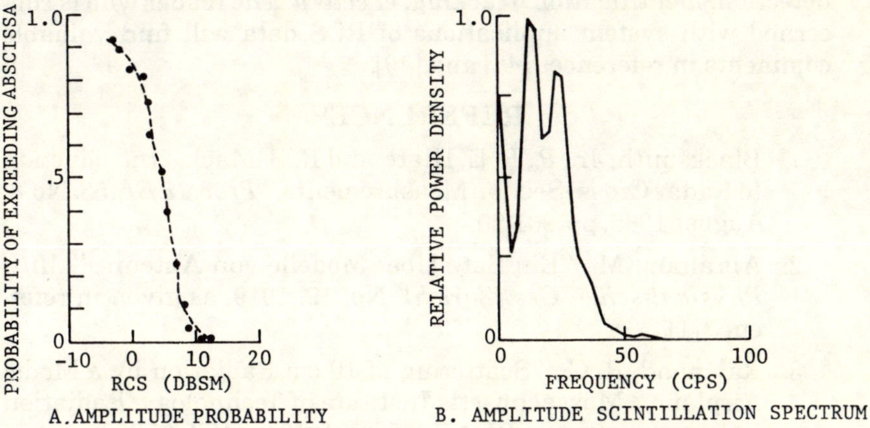


**Figure 10-27.** Typical RCS data for an aspect cell (reprinted from Bahret, ref. [36]).

## 10.6 CONCLUDING REMARKS

This chapter has discussed the RCS and relative phase behavior of radar targets, with emphasis on the use of scale models to make measurements employing optical, ultrasonic, and RF methods. Characteristics of each of these methods have been summarized, and several examples of scale-model data have been presented.

Model measurements are usually less expensive and can be controlled better than field measurements, but there are advantages and disadvantages in each of the various ways to determine target characteristics. Use of more sophisticated measurement equipment at higher frequencies[45,46] will require greater attention to construction details of physical models and of measurement setups. Of particular current interest is the measurement of low-RCS objects. Insofar as such objects use absorbers which have properties which do not scale with frequency, model measurements will be difficult to use. However, modeling can be used to assess the effects of shaping which is an important RCS reduction technique. With any very low-RCS measurement, the RCS contribution of model supports can severely degrade the results. Some adaptation of Peter Fritsch's "spin-drop" method[47] may hold promise in such situations.

The objective of any determination of radar target characteristics is to allow prediction of radar behavior in some realistic situation — detection, identification, tracking, *et cetera*. The reader who is concerned with system applications of RCS data will find valuable comments in references [48] and [49].

## REFERENCES

1. Blacksmith, Jr., P., R. E. Hiatt, and R. B. Mack, "Introduction to Radar Cross-Section Measurements," *Proc. IEEE,* 53, No. 8, August 1965, pp. 901-20

2. Abraham, M., "Ein Satz Uber Modelle von Antenne," *Mitt. Physikalischen Ges.(Zurich)*, No. 19, 1919, as given in reference [1].

3. Raymond, R. C., "Scattering of 10 cm Radiation by a Model Airplane," Massachusetts Institute of Technology, Radiation Laboratory, Report RL-156, May 21, 1942, ATI 45 262.

4. "Radar Reflectivity Measurements Symposium," RADC-TDR-64-25, Vols. I and II, April 1964, AD 601 364 and AD 603 365.

5. Bachman, C. G., "Some Recent Developments in RCS Measurement Techniques," *Proc. IEEE,* 53, No. 8, August 1965, pp. 962-72.

6. Corriher, Jr., H. A., "Radar Reflectivity of Aircraft," Georgia Institute of Technology, Atlanta, Georgia, Final Report on Project A-1025, Contract N00014-67-A-0159-003, January 14, 1970, AD 882 842.

7. *Determination and Use of Radar Scattering Characteristics*, AGARD Lecture Series No. 59 (AGARD-LS-59), September 1973 (available from NTIS as AD 769 277).

8. Walden, J., *et al.*, "Submillimeter Modeling of Millimeter Radar Systems," *SPIE Vol. 259 - Millimeter Optics*, 1980, pp. 152-57.

9. Kerr, D. E., ed., *Propagation of Short Radio Waves*, MIT Radiation Laboratory Series, Volume 12, McGraw-Hill Book Company, New York, 1951.

10. Cole, Jr., E. B. "Airplane Models Reveal How to Avoid Radar," *Electronics,* 28, January 1955, pp. 122-25.

11. Goudsmit, S. A., and R. R. Weiss, "Possible Measurement of Radar Echoes by Use of Model Targets," Massachusetts Institute of Technology, Radiation Laboratory, Report RL-196, March 4, 1943.

12. Sletten, C. J., and F. S. Holt, "Antenna Laboratory Reports to the Steering Committee," Air Force Cambridge Research Laboratories, Report TR-15, Section 3-F, October 1950.

13. Corriher, Jr., H. A., and B. O. Pyron, *et al.*, "Radar Reflectivity of Sea Targets," Georgia Institute of Technology, Technical Report No. 1, Vol. I, on Contract NOnr-991(12), December 15, 1966, AD 829 538.

14. Rozendal, J. K., "Comparison of Radar Cross-Section by Microwave and Optical Simulation Techniques," *Proc. IEEE,* 53, No. 8, August 1965, p. 1093.

15. Bartle, R. C., C. I. Beard, J. E. Burke, and V. Twersky, "Optical Scattering Data and Microwave Approximations," *Applied Optics,* 4, No. 10, October 1965, pp. 1253-66.

16. Suzuki, C., "Model Measurements of the Radar Cross-Section by Ultrasonic Waves," *Electronics and Comm. in Japan,* 47, May 1964, pp. 12-18 (translation of *J. Inst. of Elec. Comm. Engrs. of Japan*).

17. Meyer, E., and P. Schnupp, "Model Experiments on Electromagnetic Reflected Beam Direction Finding (Radar) Using Ultrasonic Waves (Sonar)," *Acustica,* 10, No. 1, January 1960, pp. 1-13.

18. Cram, L. A., "Modeling Methods of Determining Radar Echo Characteristics," paper 1(e) in reference [7].

19. Cram, L. A., and J. R. Staveley, "Recent Developments in Scale Modeling of Radar Reflections by Radar and Sonar Methods," Paper in *Radar-77,* IEE Conference Publication No. 155, International Conference, London, October 25-28, 1977, pp. 473-77.

20. Basalov, F. A., and V. Ya. Kucheryavyi, "Ultrasonic Modeling of Radar Bearing," *Izv. VUZ Radioelektronika,* 15, No. 11, November 1972, pp. 1388-90.

21. Cram, L. A., and S. C. Woolcock, "Review of Two Decades of Experience Between 30 GHz and 900 GHz in the Development of Model Radar Systems," 16 pp., paper 1.6 presented at meeting below but not originally in pre-print.
*Millimeter and Submillimeter Wave Propagation and Circuits,* AGARD Conference Pre-Print No. 245 (AGARD-CPP-245), papers presented at the 25th Meeting of the AGARD Electromagnetic Wave Propagation Panel, Munich, September 4-8, 1978.

22. Gabsdil, W., and W. Jacob, "Model Simulation of Target Characteristics and Engagement Situations Employing Millimeter Wave Radar Systems," 17 pp., paper presented at meeting of reference [21] but not originally in pre-print.

23. Bird, D., M. W. Plaster, and C. G. C. Wilson, "Measurement of the Depolarization of Radar Scattering Using Scale Modelling Techniques," *Conference Proceedings: Military Microwaves '82*, London, October 20-22, 1982, pp. 429-34.

24. Bahret, W. F., "Dynamic Full Scale Measurement of RCS," paper 1(d) in reference [7].

25. Power, B. E., and H. Herman, "Scale Model Measurement Facilities and Their Use for Obtaining Scattering Data Pertinent to Millimetric-Wave Radars," *Conference Proceedings: Military Microwaves '82*, London, October 20-22, 1982, pp. 414-22.

26. Mirovitskiy, D. I., V. A. Torgovanov, and G.P. Cherkunova, "Optical Simulation of Reflection and Scatter of Microwaves," *Radiotekhnika i Elektronika,* 15, No. 1, January 1970, pp. 38-50.

27. Vasil'ev, V. N., "Optical Nonscale Modeling," *Izv. VUZ Radiofizika*, 20 No. 8, August 1977, pp. 1146-52.

28. Olin, I. D., and F. D. Queen, "Dynamic Measurement of Radar Cross Sections," *Proc. IEEE,* 53, No. 8, August 1965, pp. 954-61.

29. Marlow, H. C., D. C. Watson, *et al.*, "The RATSCAT Cross-Section Facility," *Proc. IEEE,* 53, No. 8, August 1965, pp. 946-54.

30. Blore, W. E., and H. M. Musal, Jr., "The Radar Cross Section of Plasma-Coated Metal Bodies," *Proc. IEEE,* 53, No. 8, August 1965, pp. 1122-23.

31. Sinclair, G., "Theory of Models of Electromagnetic Systems," *Proc. IRE,* 36, No. 11, November 1948, pp. 1364-70.

32. Stratton, J. A., *Electromagnetic Theory*, pp. 488-90, McGraw-Hill Book Company, New York, 1941.

33. King, R. W. P., *Fundamental Electromagnetic Theory*, pp. 316-20, Dover Publications, 1963 (originally published as *Electromagnetic Engineering*, McGraw-Hill Book Company, New York, 1945).

34. Thomas, A. S., "Modeling of RAM-Covered Complex Bodies," Vol. I of reference [4], pp. 542-48.

35. Senior, T. B. A., "Surface Roughness and Tolerances in Model Scattering Experiment," *IEEE Trans. Ant. and Prop.*, AP-13, No. 4, July 1965, pp. 629-36.

36. Bahret, W. F., "The Use of RCS Data," paper 3(c) in reference [7].

37. Woolcock, S. C., "Target Characteristics," paper 2(a) in reference [7].

38. Private communication from F. S. Holt, Air Force Cambridge Research Center as reproduced (pp. 186-87) in M. L. Meeks, C. H. Wilcox, and N. A. Logan, "Reflection Properties of Radar Targets," Georgia Institute of Technology, Atlanta, Georgia, Contract NOrd-11886, August 1954, AD 60 221.

39. Paddison, F. C., C. A. Shipley, A. L. Maffett, and M. H. Dawson, "Radar Cross Section of Ships," *IEEE Transactions on Aerospace and Elec. Systs.,* AES-14, No. 1, January 1978, pp. 27-34.

40. Elphick, B. L., M. J. Chappell, and R. Batty, "The Measured and Calculated RCS of a Resistive Target Model," paper in Conference Publication of reference [19], pp. 120-24.

41. Meyer, E. G., and C. L. Moher, "Lightweight Radar Alarm System," Radiation, Inc., Final Report A-04773 (RADC-TR-65-59-Vol. I-Part I), Contract AF 30(602)-3554, March 1966, AD 482 920.

42. Kell, R. E., and N. E. Pedersen, "Comparison of Experimental Radar Cross-Section Measurements," *Proc. IEEE,* 53, No. 8, August 1965, pp. 1092-93.

43. Moll, J. W., "On the Radar Echo from Aircraft," *IEEE Trans. on Aerospace and Elec. Systs.,* AES-3, May 1967, pp. 574-77.

44. Johnson, R. H., "The Effect of Target Characteristics on Radar System Performance," Royal Radar Establishment, Malvern, as referenced in "Use of Radio Modeling Data," S. C. Woolcock, paper 3(b) in reference [7].

45. Brown, E. G., M. W. Plaster, and S. C. Woolcock, "Millimeter-Wave Measurement Radars for 140 and 280 GHz," *Conference Proceedings: Military Microwaves '80,* London, October 22-24, 1980.

46. Burke, H. S., and W. P. Hansen, Jr., "Millimeter Radar Reflective Range System," *SPIE Vol. 337 — Millimeter Wave Technology,* 1982, pp. 116-28.

47. Fritsch, P. C., "A Free-Space Method of Measuring Radar Cross Section in the Laboratory," *Proc. IEEE,* 51, No. 9, September 1963, pp. 1271-72.

48. Cram, L. A., "State-of-the-Art and Future Prospects in Radar Scattering," paper 3(e) in reference [7].

49. Smith, A. G., and D. J. Bryant, "Accuracy Requirements in Radar Cross Section Measurements from a Systems Viewpoint," *Proc. IEEE,* 53, No. 8, August 1965, pp. 1159-60.

# CHAPTER 11

# ORGANIZATION OF A TEST PROGRAM

**Nicholas C. Currie**
**Engineering Experiment Station**
**Georgia Institute of Technology**

## 11.1 OVERVIEW

The measurement of the radar reflectivity of targets is a deceptive process. On the surface the basic concepts are straightforward, such as the very simple form of the radar equation. On the other hand the attainment of accurate data of a suitable type to meet the objectives of a test program can be extremely difficult to achieve. The key to obtaining accurate, useful test data is the proper planning of the experiment prior to the execution of the measurement process.

The experimental plan needs to consider the test objectives, the cost/performance tradeoffs, the required data set size, the measurement sensor characteristics, environmental characterization, the calibration process, and the desired data characteristics and format. In this chapter the considerations which enter into the acquisition of radar reflectivity data will be discussed with the goal of assisting the reader in the proper planning and execution of radar reflectivity experiments.

## 11.2 REASONS FOR REFLECTIVITY MEASUREMENTS

In general, radar reflectivity measurements are designed to determine the radar "size" of a given target of interest. The basic unit of measure of the radar "size" is its radar cross section $\sigma$ (RCS) which, as was discussed in Chapter 2, is defined to be

$$\sigma = 4\pi \frac{\textit{Power unit solid angle reflected toward the receiver}}{\textit{Power density incident on target}} \qquad (11\text{-}1)$$

and has units of area. In general, RCS is a function of a number of variables such as frequency, target type and size, and the aspect angle from which the target is viewed. To give an idea of target magnitudes commonly encountered, at 95 GHz, a perfectly conducting 6-inch sphere will have an RCS of 0.007 $m^2$, and a 6-inch triangular corner reflector will have a peak RCS of 227 $m^2$. A normalized value, called radar cross section per unit area ($\sigma°$) or per unit volume ($\eta$) is used when describing area extensive or volume extensive clutter. (*See* Chapter 2 for a more detailed discussion of radar cross section.)

The radar cross section is, in general, a statistical parameter and is thus best described by statistical measures such as mean, median, standard deviation, correlation time, frequency spectrum, and amplitude distribution. These various parameters, once known, can be utilized to calculate the system design parameters for a proposed radar system, for computer modeling of a proposed radar system to determine its performance before committing to the expense of building a prototype system, for evaluation of signal processing techniques, and for the determination of the performance of an actual system.

The use of the radar range equation along with known radar reflectivity and environmental parameters allows the calculation of the required radar power and receiver sensitivity parameters to achieve a given detection range. Once radar parameters are determined, then the radar performance can be modeled by calculating the receiver operating curves in which the probability of detection (based on the target plus noise or clutter parameters) is plotted *versus* the probability of false alarm (based on noise or clutter parameters) as a function of signal-to-noise (signal-to-clutter) ratio. Signal processing techniques can be evaluated by determining the effects of the signal processing on the receiver operating curves. Finally, a careful characterization of the reflectivity of targets and clutter during a system test allows extrapolation of the performance to other environments. The end use of reflectivity data obtained during a measurement program strongly influences the experiment and must be considered in the planning process since it is virtually impossible to remove all of the experimental dependencies on geometry, environment, radar measurement system, and data recording system. Thus, as far as is practical, the experimental parameters must be carefully selected to yield RCS data which apply to the intended use. For example, the beamwidth, frequency, and pulse length should be very close to the intended radar system for which the data are to be applied.

## 11.3 EXPERIMENTAL PLANNING

### 11.3.1 Definition of Objectives

The objectives of a program will vary depending on the application. This may include obtaining data to increase the data base of a given type or a specific application involving a specific target, environment, or signal processing technique. Since time and money for such endeavors are normally finite, it is important to have the objectives clearly defined so that these parameters are optimized. The objectives should include a precise description of the questions to be answered, the data needed to answer the questions, the measurements required to obtain the needed data, the specific site and equipment parameters needed, and the type and format of data to result from the test program.

The most important factor in the development of objectives is the setting of priorities. Priorities may include the order in which different types of data will be taken or in which different targets will be measured. (One never knows when a major component will fail in the measurement or recording system thus preventing the collection of any more data in the remaining available time). Priorities may also involve a time limit to obtain the data, or else, specify a limited amount of available funds which necessitate taking short cuts in the measurement program. Once the priorities are determined, a test plan can be generated which describes how the prioritized objectives will be attained.

### 11.3.2 Development of a Test Plan

Traditionally, a test plan has been a telephone book sized document in which the combined author's knowledge about radar is intertwined with the discussion of the experiment. All that is really needed is a brief summary of the objectives along with a concise description of how the objectives are to be attained. This should include a list of priorities, the effects of various constraints on the program (such as lack of money, measurement system inadequacies, and site deficiencies), and a summary of the expected results. When a formal test plan is prepared, informational descriptions of the test site, measurement equipment, test procedures, data format, support activities, and a timetable should be included to provide suitable background for someone unfamiliar with the experiment. Table 11-1 gives an example of the Table of Contents for a good test plan. In the example test plan, Section 1 lists the overall program objectives and defines the priorities. Section 2 summarizes the measurement and recording equipment to be utilized in the experiment. Section 3 describes the test sites to be used including geometries, facilities, and available support. Section 4 delineates the responsibilities of each participant in the tests. (In the example there is a contractor, a sponsor, and a subcontractor providing site support involved in the program.) This helps to avoid misunderstandings between participants and prevents details from "falling into the cracks" due to lack of communication between participants. Section 5 describes the actual test procedures which are to be followed for each test site including system checkout, calibration, data collection, and ground truth. Finally, Section 6 discusses the analysis to be performed on the resulting data including both "quick-look" analysis in the field to verify data validity and more detailed analysis after the data collection. The expected results based on previous

**TABLE 11-1**
**Test Plan Table of Contents**

work in the area are included to serve as a base line for comparison with the newly acquired data.

The most important part of the test plan (other than setting priorities) is the development of a test matrix. This is the road map that will be followed during the tests, and thus, must define the measurements that will be performed in order to meet the objectives in the order determined by the priorities. Sometimes it is not easy to develop a suitable test matrix. For example, Table 11-2 shows an experimental matrix for a test program in which it is desired to obtain radar reflectivity measurements for three targets, five aspect angles, three antenna aperture sizes, seven elevation angles, six transmit modes, seven pulse lengths, three types of clutter background, four types of environmental conditions, and six modes of radar frequency agility. Obviously, the goal of the tests is to determine the effects of the resolution cell size and frequency agility step size on the RCS of several military vehicles for several clutter backgrounds. At first glance this seems to be a quite reasonable matrix but if one calculates the number of measurements required to achieve all of the combinations, one finds that over 1,000,000 target runs are required and over 50,000 clutter runs are needed to complete the test matrix. Thus, both the data acquisition time and the analysis time would be prohibitive for this test matrix. Table 11-3 gives a reduced test matrix which is more reasonable in size, but allows at least a sample of the effects of varying all of the parameters. The number of targets is reduced to one (the most important target), the depression angle is allowed to vary for only one aperture size and aspect angle, and the same is true for the aspect angle dependence measurement. The resulting runs require a total of only 372 target runs and 244 clutter tests which are much more reasonable numbers. However, in spite of the reduction in the total number of data runs, some measure of the dependence of the target RCS on each of the various parameters is still obtained from the tests while the cost of the program is kept at a practical level. Once the measurement program is begun the test matrix may have to be modified again due to other circumstances such as equipment failures, target availability, weather, etc. Whenever test matrix changes are contemplated the priorities of the test program must again be considered.

### 11.3.3 Selecting the Experimental Parameters

In order to perform meaningful RCS measurements, both the system characteristics and data set size must be commensurate with

**TABLE 11-2**
**Complete Test Matrix**

| Target | Aspect Angle | Aperture | Elevation Angle | Transmit Mode | Pulse-width ($\mu$sec) | Clutter | Environment | Frequency Agility |
|---|---|---|---|---|---|---|---|---|
| JEEP | 0° | 1.0° | 5° | *H* | 0.01 | TREES | CLEAR WINTER | Coherent (non-agile) |
| TRUCK | 180° | 4.5° | 7° | *V* | 0.02 | WEEDS | CLEAR SUMMER | 1MHz |
| TANK | 90° | 9.0° | 10° | *R* | 0.03 | GRASS | FOG | 3MHz |
| | 45° | | 13° | *L* | 0.05 | | RAIN | 5MHz |
| | 135° | | 19° | AGILE (*H*,*V*) | 0.10 | | | 10MHz |
| | | | 25° | AGILE (*R*,*L*) | 0.20 | | | 50MHz |
| | | | 35° | | | | | |

3 x 5 x 3 x 7 x 6 x 6 x 3 x 4 x 6 = 1,116,480 *target tests*

1 x 1 x 3 x 7 x 6 x 6 x 3 x 4 x 6 = 54,432 *clutter tests*

**TABLE 11-3**
**Practical Test Matrix**

| Target | Aspect Angle | Aperture | Depression Angle | Transmit Mode | Pulse-Width | Clutter | Environment | Frequency Agility |
|---|---|---|---|---|---|---|---|---|
| TANK | 180° | 1.0° | 5° | *(H-V)* | .01 | WEEDS | CLEAR WINTER | Coherent |
| | 45° | | 10° | *(R-L)* | .20 | | CLEAR SUMMER | 10MHz |
| | | | 13° | | | | | |
| | | | 19° | | | | | |
| | | | 25° | | | | | |
| TANK | 0° | 1.0° | 10° | *(H-V)* | .01 | TREE LINE | CLEAR WINTER | Coherent |
| | 180° | | | *(R-L)* | .20 | | CLEAR SUMMER | 10MHz |
| | 45° | | | | | | | |
| | 135° | | | | | | | |
| TANK | 135° | 1.0° | 10° | *(H-V)* | .01 | TREE LINE | CLEAR WINTER | Coherent |
| | | 4.5° | | *(R-L)* | .03 | | CLEAR SUMMER | 10MHz |
| | | | | | .05 | | | |
| | | | | | .20 | | | |
| TANK | 135° | 1.0° | 10° | *(H-V)* | .01 | TREE LINE | CLEAR WINTER | Coherent |
| | | | | *(R-L)* | .20 | | CLEAR SUMMER | 10MHz |
| | | | | | | | FOG/RAIN | |

| | | | | | | | | |
|---|---|---|---|---|---|---|---|---|
| TANK | 135° | 1.0° | 10° | (*H-V*)<br>(*R-L*) | .01<br>.03 | TREE LINE | CLEAR WINTER<br>CLEAR SUMMER | Coherent<br>1MHz<br>3MHz<br>10MHz<br>50MHz |

| | | | | |
|---|---|---|---|---|
| 2 x 5 x 2 x 2 x 2 x 2 | = | 160 | → | *80* |
| 4 x 2 x 2 x 2 x 2 | = | 64 | → | 16 |
| 2 x 2 x 4 x 2 x 2 | = | 64 | → | 64 |
| 2 x 2 x 3 x 2 | = | 24 | → | 24 |
| 2 x 3 x 2 x 5 | = | 60 | → | 60 |
| | | 372 *target tests* | | 244 *clutter tests* |

the desired application. In many cases, the major radar parameters such as:

Frequency

Antenna Beamwidth/Illuminated Area

Pulsewidth/Resolution

are determined by the characteristics of the system for which data are being gathered. Peak power and receiver noise figure should be selected such that there is an adequate signal-to-noise (S/N) ratio for accurate measurements to be performed; typically, a minimum of 15 to 20 dB S/N ratios are desirable.

Signals often fluctuate widely in amplitude, requiring substantial system dynamic range, and if the range to the target is changing, or multiple targets at different ranges must be accommodated, the dynamic range requirements may be increased. Normally the dynamic range is determined by the characteristics of amplifiers in the receiver chain or by quantization errors of the data acquisition system. To illustrate these points, consider an amplifier with a gain of 30 dB, maximum output power of 0 dBm, 10 dB noise figure, and 1 MHz bandwidth. The output noise power is then -114 dBm (thermal noise) +10 dB (noise figure) +30 dB (gain) = -74 dBm. If one assumes 15 dB S/N ratio is required, then the instantaneous dynamic range at the output is 59 dB. If an A/D converter is used for data acquisition, this may further limit dynamic range. For example, an eight-bit converter provides for 256 discrete levels; the smallest level is 1/256 of the largest, and if a linear receiver is used the dynamic range will be

$$DR = 20 \log_{10} 64 = 48.2\ dB \tag{11-2}$$

Of course, integration of a number of fluctuating returns may improve the apparent dynamic range. If a logarithmic receiver is used, the resolution becomes linear in dB which also represents a dynamic range limitation since the resolution accuracy becomes a function of the power level. (This is usually only a problem if the log data are to be "linearized" as a part of the analysis.)

Once the desired number of A/D bits has been selected, system bandwidth will determine the required delay stability or pulse jitter of the system. For wideband systems having high accuracy requirements with short pulse lengths, the requisite delay stability may be difficult to achieve due to the necessity of generating long range delays with jitters of less than a nanosecond.

Once the major system parameters have been selected, other important parameters to be determined are system sample rate, the number of samples per data run, and the number of runs for each target/environment condition. These central points are often ignored and may be difficult to quantify, but are essential to the success of any measurement program. A key point in the estimation of data set size is the accuracy of the estimate desired; often "*confidence intervals*" are used as a measure of the validity of such an estimate. If, for example, a 95% confidence interval is specified, then 95% of all identical RCS measurements would fall within the interval. When discussing error limits one must keep in mind that there are two types of error limits; the first are errors that occur during a given data run while the second are errors associated with changes from run-to-run for supposedly identical target conditions.

### 11.3.4 Sizing the Experimental Data Set

In many cases the radar return from a target or clutter will fluctuate in a manner which is not completely understood. In such cases it is desirable to be able to estimate the mean or the variance of the return within some accuracy bounds, and the number of data samples required to achieve this aim must be determined. The concept of the *confidence interval* is often used to perform this calculation. Simply stated, the *confidence interval* is defined as an interval within which a given statistical parameter (such as the mean) is estimated to lie with respect to some characteristic (such as the number of samples). Thus, a 90% confidence interval for the mean value of a random process implies that the computed interval will contain the true mean value of the random process 90% of the time.

There are two frequently encountered situations where confidence intervals arise; in the first situation an estimate of the standard deviation is known, a desired confidence interval is given, and the required number of measurements must be determined. If the process is assumed to be normal or Gaussian, then one procedure is:[1]

1. Choose the desired confidence interval (or confidence level), $1 - \alpha$ (often 90 or 95%)
2. Look up $Z = Z_{1-\alpha/2}$ in Table 11-4, where $Z$ is the so called *standard normal variable*. ($Z = [(x-m)/\sigma]$, where $x$ is the random variable, $m$ is the true mean, and $\sigma$ is the true standard deviation)
3. Compute the number of samples from the equations:

$$X_U = \overline{X} + Z \frac{\sigma}{\sqrt{n}} \tag{11-3}$$

## TABLE 11-4
### Cumulative Normal Distribution — Values of $z_p$
(Reprinted from Natrella, ref. [1]; p. T-3.)

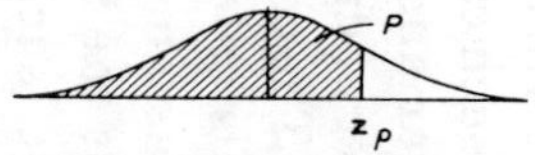


**Values of $z_P$ corresponding to $P$ for the normal curve.**
**$z$ is the standard normal variable**

| P | .00 | .01 | .02 | .03 | .04 | .05 | .06 | .07 | 08 | .09 |
|---|---|---|---|---|---|---|---|---|---|---|
| .00 | — | −2.33 | −2.05 | −1.88 | −1.75 | −1.64 | −1.55 | −1.48 | −1.41 | −1.34 |
| .10 | −1.28 | −1.23 | −1.18 | −1.13 | −1.08 | −1.04 | −0.99 | −0.95 | −0.92 | −0.88 |
| .20 | −0.84 | −0.81 | −0.77 | −0.74 | −0.71 | −0.67 | −0.64 | −0.61 | −0.58 | −0.55 |
| .30 | −0.52 | −0.50 | −0.47 | −0.44 | −0.41 | −0.39 | −0.36 | −0.33 | −0.31 | −0.28 |
| .40 | −0.25 | −0.23 | −0.20 | −0.18 | −0.15 | −0.13 | −0.10 | −0.08 | −0.05 | −0.03 |
| .50 | 0.00 | 0.03 | 0.05 | 0.08 | 0.10 | 0.13 | 0.15 | 0.18 | 0.20 | 0.23 |
| .60 | 0.25 | 0.28 | 0.31 | 0.33 | 0.36 | 0.39 | 0.41 | 0.44 | 0.47 | 0.50 |
| .70 | 0.52 | 0.55 | 0.58 | 0.61 | 0.64 | 0.67 | 0.71 | 0.74 | 0.77 | 0.81 |
| .80 | 0.84 | 0.88 | 0.92 | 0.95 | 0.99 | 1.04 | 1.08 | 1.13 | 1.18 | 1.23 |
| .90 | 1.28 | 1.34 | 1.41 | 1.48 | 1.55 | 1.64 | 1.75 | 1.88 | 2.05 | 2.33 |

**Special Values**

| P | .001 | .005 | .010 | .025 | .050 | .100 |
|---|---|---|---|---|---|---|
| $z_P$ | −3.090 | −2.576 | −2.326 | −1.960 | −1.645 | −1.282 |

| P | .999 | .995 | .990 | .975 | .950 | .900 |
|---|---|---|---|---|---|---|
| $z_P$ | 3.090 | 2.576 | 2.326 | 1.960 | 1.645 | 1.282 |

$$X_L = \overline{X} - Z \frac{\sigma}{\sqrt{n}} \tag{11-4}$$

where

$X_U$ = upper confidence limit
$X_L$ = lower confidence limit
$\overline{X}$ = geometric mean of $n$ samples $= 1/n \left( \sum_{i=1}^{n} X_i \right)$
= known standard deviation

Thus,

$$X_L - X_U = (1 - \alpha) \tag{11-5}$$

$$X_L - X_U = 2Z \frac{\sigma}{\sqrt{n}} \tag{11-6}$$

$$n = \left( \frac{2 Z \sigma}{1 - \alpha} \right)^2 \tag{11-7}$$

The second case of interest is where one is given a set of $n$ measurements and must determine the associated confidence limits. For this case the true standard deviation is unknown, so the following applies:

1. Choose the desired confidence level $1 - \alpha$
2. Compute $\overline{X}$ (the geometric mean, see above)
3. Compute $S$ from

$$S = \sqrt{\frac{n \sum_{i=1}^{n} X_i^2 - (\sum_{i=1}^{n} X)^2}{n(n-1)}} \tag{11-8}$$

4. Look up $t = t_{1-\alpha/2}$ for $n$–1 independent samples in Table 11-5
5. Compute

$$X_U = \overline{X} + t\, S/\sqrt{n} \tag{11-9}$$

$$X_L = \overline{X} + t\, S/\sqrt{n} \tag{11-10}$$

6. The confidence interval = $X_U - X_L$ (11-11)

A third case which sometimes arises is the determination of the number of samples required to estimate the standard deviation to a given accuracy within a desired confidence limit. Figure 11-1 is useful for such an estimation. The curve relates the confidence level 1-$\alpha$ to the accuracy of the standard deviation estimate $P$ in percent and the number of independent samples (degrees of freedom).

**TABLE 11-5**
**Percentiles of the Z Distribution**
(Reprinted from Natrella, ref. [1]; p. T-5.)

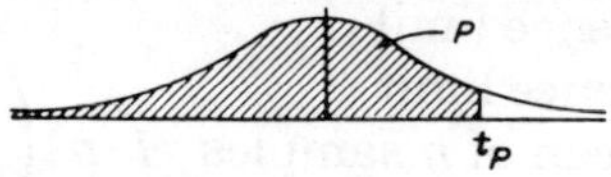


| df | $t_{.60}$ | $t_{.70}$ | $t_{.80}$ | $t_{.90}$ | $t_{.95}$ | $t_{.975}$ | $t_{.99}$ | $t_{.995}$ |
|---|---|---|---|---|---|---|---|---|
| 1 | .325 | .727 | 1.376 | 3.078 | 6.314 | 12.706 | 31.821 | 63.657 |
| 2 | .289 | .617 | 1.061 | 1.886 | 2.920 | 4.303 | 6.965 | 9.925 |
| 3 | .277 | .584 | .978 | 1.638 | 2.353 | 3.182 | 4.541 | 5.841 |
| 4 | .271 | .569 | .941 | 1.533 | 2.132 | 2.776 | 3.747 | 4.604 |
| 5 | .267 | .559 | .920 | 1.476 | 2.015 | 2.571 | 3.365 | 4.032 |
| 6 | .265 | .553 | .906 | 1.440 | 1.943 | 2.447 | 3.143 | 3.707 |
| 7 | .263 | .549 | .896 | 1.415 | 1.895 | 2.365 | 2.998 | 3.499 |
| 8 | .262 | .546 | .889 | 1.397 | 1.860 | 2.306 | 2.896 | 3.355 |
| 9 | .261 | .543 | .883 | 1.383 | 1.833 | 2.262 | 2.821 | 3.250 |
| 10 | .260 | .542 | .879 | 1.372 | 1.812 | 2.228 | 2.764 | 3.169 |
| 11 | .260 | .540 | .876 | 1.363 | 1.796 | 2.201 | 2.718 | 3.106 |
| 12 | .259 | .539 | .873 | 1.356 | 1.782 | 2.179 | 2.681 | 3.055 |
| 13 | .259 | .538 | .870 | 1.350 | 1.771 | 2.160 | 2.650 | 3.012 |
| 14 | .258 | .537 | .868 | 1.345 | 1.761 | 2.145 | 2.624 | 2.977 |
| 15 | .258 | .536 | .866 | 1.341 | 1.753 | 2.131 | 2.602 | 2.947 |
| 16 | .258 | .535 | .865 | 1.337 | 1.746 | 2.120 | 2.583 | 2.921 |
| 17 | .257 | .534 | .863 | 1.333 | 1.740 | 2.110 | 2.567 | 2.898 |
| 18 | .257 | .534 | .862 | 1.330 | 1.734 | 2.101 | 2.552 | 2.878 |
| 19 | .257 | .533 | .861 | 1.328 | 1.729 | 2.093 | 2.539 | 2.861 |
| 20 | .257 | .533 | .860 | 1.325 | 1.725 | 2.086 | 2.528 | 2.845 |
| 21 | .257 | .532 | .859 | 1.323 | 1.721 | 2.080 | 2.518 | 2.831 |
| 22 | .256 | .532 | .858 | 1.321 | 1.717 | 2.074 | 2.508 | 2.819 |
| 23 | .256 | .532 | .858 | 1.319 | 1.714 | 2.069 | 2.500 | 2.807 |
| 24 | .256 | .531 | .857 | 1.318 | 1.711 | 2.064 | 2.492 | 2.797 |
| 25 | .256 | .531 | .856 | 1.316 | 1.708 | 2.060 | 2.485 | 2.787 |
| 26 | .256 | .531 | .856 | 1.315 | 1.706 | 2.056 | 2.479 | 2.779 |
| 27 | .256 | .531 | .855 | 1.314 | 1.703 | 2.052 | 2.473 | 2.771 |
| 28 | .256 | .530 | .855 | 1.313 | 1.701 | 2.048 | 2.467 | 2.763 |
| 29 | .256 | .530 | .854 | 1.311 | 1.699 | 2.045 | 2.462 | 2.756 |
| 30 | .256 | .530 | .854 | 1.310 | 1.697 | 2.042 | 2.457 | 2.750 |
| 40 | .255 | .529 | .851 | 1.303 | 1.684 | 2.021 | 2.423 | 2.704 |
| 60 | .254 | .527 | .848 | 1.296 | 1.671 | 2.000 | 2.390 | 2.660 |
| 120 | .254 | .526 | .845 | 1.289 | 1.658 | 1.980 | 2.358 | 2.617 |
| ∞ | .253 | .524 | .842 | 1.282 | 1.645 | 1.960 | 2.326 | 2.576 |

In order to illustrate some of these points, consider an application where it is necessary to perform a series of sea clutter measurements using a radar with a logarithmic receiver, requiring a 2 dB, 90% confidence interval. It is known from the literature and earlier experiments that the sea return may be considered approximately

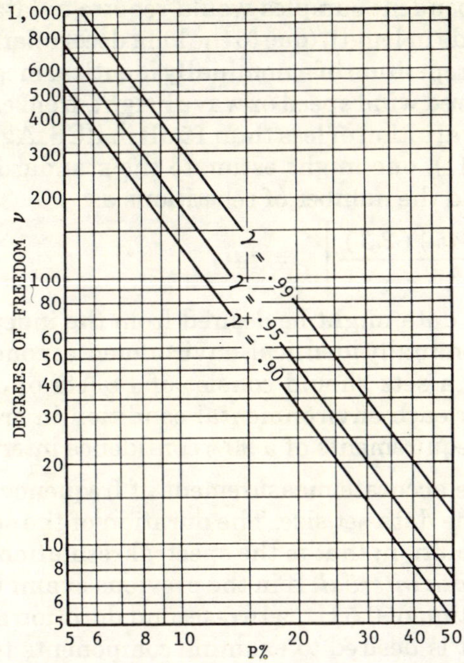


**Figure 11-1.** Curve which relates standard deviation to number of samples, confidence level, and accuracy (reprinted from Natrella, ref. [1]).

lognormally distributed, permitting the assumption of normally distributed video voltage out of the log detector to be used. The logarithmic standard deviation from sea clutter generally ranges from 3 dB to greater than 7 dB. The decorrelation time for sea clutter is less than 0.1 seconds. As was noted in Chapter 3, the maximum data rate for which independent samples can be obtained from a target is inversely proportional to the decorrelation time. Thus, only about 10 independent samples per second can be obtained for sea clutter regardless of the actual sample rate. Following the procedures given above, one can estimate the number of independent samples by first observing that from Equation (11-7)

$$n = \left(\frac{2\,Z\,\sigma}{1 - \alpha}\right)^2 \tag{11-12}$$

and for $\sigma = 7$ dB, $Z = 1.645$ (for $X_U = 95\%$), and $1 - \alpha = 2$ dB,

$$n = \left[\frac{2\,(1.645)\,(7)}{2}\right]^2 = 133 \tag{11-13}$$

These 133 independent samples would require a data set approximately 14 seconds in length (due to the long decorrelation time of 0.1 seconds). Upon repetition of a nominally identical measurement for the same measured wind speed or wave height, from experience one might expect a variation of less than 15 dB in RCS. Again returning to Equation (7-11), one might estimate the standard deviation of $\sigma = 15/7 = 2.3$ and the number of repetitions as:

$$n = \left[ \frac{(2)(1.645)(2.3)}{2} \right]^2 = 14 \qquad (11\text{-}14)$$

Thus, 14 sets of data might be desired from the independently determined, but nominally identical, environmental conditions. Thus, the required data sets should consist of 14 sets of data, each 14 seconds long for each environmental condition in order to satisfy RCS accuracy requirements of a 90% confidence interval of 2 dB.

The need for the accurate measurement of frequency spectra may also influence the data set size. The duration of the set determines the spectral resolution; that is the spectral resolution in Hz is proportional to 1/*data interval*. If in the previous example it is desired to have a resolution of 0.5 Hz, a two-second duration sample would be required. If it is desired to examine components to 100 Hz (ten times the major components), a data rate of at least 200 samples per second would be required. In practice, several times this value might be used — perhaps 500 or 1000 PPS might be appropriate. It should be noted that without weighting, the contamination from adjacent frequencies falls off as sin $X/X$ which might require quite long samples to observe small components close to a large carrier signal.

In the previous paragraphs, an attempt was made to show how the accuracies and resolution of the required data descriptions influence the size and duration of the data sets, and estimation of these sizes is an essential part of the experiment planning process. For more detailed information on the sizing of experimental measurements, *see* reference [1].

## 11.4 RADAR INSTRUMENTATION SYSTEM CONSIDERATIONS

Typically, it is desirable that the measurement accuracy of test equipment be ten times as great as the allowable error in the measurement being made. For similar reasons, it is desirable that the integrity of the measurement radar system be considerably better than the typical operational system whose design will follow from the analysis of the data being collected. The integrity of the instrumentation system is measured for various parameters such as stability of the amplitude of a noncoherent system, the stability of the frequency and phase for a coherent system, and the polarization isolation for the dual polarized system, to name a few. It is sometimes difficult to achieve a factor of ten improvement in integrity beyond that of an operational system, particularly when the operational system is using state-of-the-art technology. However, it is at least desirable to have the instrumentation system be as good as, if not better than, the radar system for which data are being collected and analyzed.

### 11.4.1 Sensor Calibration

In addition to the integrity of an instrumentation system, one element which sets it apart from an operational system is the calibration capability built into the radar system. Typically, in a measurements program the instrumentation system is operated for a variety of tests at various times of the day and night. The data are typically recorded on either analog instrumentation recorders or digital recorders. Due to the long-term drift characteristics of the instrumentation and recording systems, periodic calibration procedures must be performed. This results in a set of calibration signals being recorded just prior to and/or directly following the actual measurement. For a noncoherent system, the amplitude calibration procedure involves one of two types as discussed in Chapter 4. The first type is the so-called indirect calibration and involves the injection of either a signal generator signal into the antenna connection port or a directional coupler into the receiver waveguide. The signal generator level is varied over the dynamic range of the receiver at fixed increments in signal level. This procedure allows a record of the receiver output voltage to be recorded as a function of input power level. The other procedure used is called the mixed calibration method and involves boresighting the radar antenna to a standard calibrated radar reflector at a known distance and inserting attenuation in the receiver section until the reflector is attenuated down to the receiver noise level. This procedure provides a record of the receiver output voltage as a function of radar cross section measured by the receiver at a given range. Both of the above described procedures should be used in calibrating the radar set whenever possible. The signal generator calibration is used in conjunction with the radar range equation to obtain agreement with the radar reflector calibration. Agreement between these two independent calibration procedures indicates a high integrity and confidence factor in the operation of the instrumentation system. A dual-polarized receiver in the radar requires that the calibration procedure be performed for each polarization.

If the radar receiver has a phase channel, or has a coherent receive capability or a psuedo-coherent detection capability, then the phase must be calibrated as well as the amplitude. The phase is calibrated by incrementally inserting different phase shifts in the receiver channel and recording the output of the phase detector. This provides a record of the phase detector output voltage as a function of received phase. As with the amplitude calibration, when using a multichannel receiver, the phase must be calibrated for each channel.

Accurate calibration of radar measurements is always a difficult task, and is more of a problem in the millimeter wave region than at lower frequencies. Millimeter system components are not as stable as at lower frequencies so that changes in transmitted power or receiver drift are more likely. The only solution to this type of problem is to calibrate often. Of course, this slows down the measurement process. Another problem encountered is that absolute calibration reference targets may have measured radar cross sections significantly lower than their theoretical cross sections due to being non-ideal, particularly at the higher frequencies. The solution to this problem is to make calibration targets as accurate as possible, and then to calibrate them independently before beginning field tests. This requires a great deal of attention to calibration targets and their mounts since the pointing accuracy of calibration targets, such as diplanes and trihedral reflectors, becomes much more important as the frequency increases. One other minor nuisance encountered is that since millimeter calibration targets tend to be smaller than those used at microwave frequencies, they are harder to see optically, making pointing of the radar antennas for calibration more difficult.

One helpful property of millimeter wave calibration is that multipath tends to be less of a problem since antenna beamwidths tend to be smaller, and the terrain along the propagation path appears rougher, so that the effect of multipath lobing on calibration is reduced. (*See* Section 11.5 for a discussion of multipath effects on low angle measurements.)

### 11.4.2 Adaptability

Another element of an instrumentation radar system is its adaptability. It would be expected that a typical instrumentation radar would be used for a variety of measurement programs, and that each of these programs would have a unique application. Furthermore, the application for the data collected will determine the operational configuration of the instrumentation sensor. Therefore, a variety of operating modes must be available for an instrumentation system to be useful for more than one measurement program. Operating modes which are typically incorporated in instrumentation radars include a variety of frequency agility modes, polarization modes, pulse lengths, pulse repetition frequencies, and a number of signal processing capabilities which may be performed prior to recording the data. The pulse lengths typically of interest in developing modern operational systems are from 10 nanoseconds up to about 1 microsecond. Pulse repetition frequencies of interest typically are from 1 kHz up to about 1 MHz. Frequency agility modes of interest are intrapulse frequency agility or chirp modes, pulse-to-pulse frequency agility (which may also be coherent), and FM-CW. Polarization modes of interest are linear, including horizontal or vertical on transmit, dual-polarized horizontal and vertical on receive, and alternate left/right circular polarizations on transmit and receive. The use of polarization agility is sometimes of interest in developing modern system concepts. It is sometimes required that measurements be made for a given program under a variety of radar operating mode conditions. In this case, it is convenient to have operator control of these parameters. The state of the art in component technology up through at least 95 GHz is such that most of these operating parameters can be implemented in instrumentation radar systems. Proper design of the timing and control circuitry allows for remote control of these parameters.

### 11.4.3 Component Effects on Radar Performance

Instrumentation grade radar systems are generally not developed around state of the art components since calibration, parameter flexibility, and accuracy are more important than maximum detection range, minimum sensitivity, or physical size. For higher frequencies, such as the millimeter region, however, the current level of component technology and availability has necessitated the use of state of the art components in the development of many instrumentation radars in order to obtain acceptable measurement performance, thus creating problems in component stability and reliability. Parameters such as transmitter peak power, and power/frequency stability, receiver noise figure, and dynamic range strongly affect the performance of a measurement system. Currently, at frequencies up through $K_u$-band, components are available which provide large peak power (megawatts incoherent and kilowatts coherent power) and adequate stabilities for most RCS measurement applications. However, at 35 GHz and above the situation is not quite as good.

Typical transmitter peak powers of 150 kW (35 GHz), 30 kW (70 GHz), and 2 kW (95 GHz) are available from magnetron oscillators which may also be frequency agile with tuning bandwidths of 500 MHz (35 GHz, 70 GHz). [2] The magnetrons are capable of fixed pulsewidth operation over the range from 20 to 500 nanoseconds with some tubes capable of short pulse, 7-nanosecond (95 GHz) operation. Tube amplifiers are available at 95 GHz with 2 kW peak power, 30 dB gain, and 300 MHz bandwidth from extended interaction amplifiers (EIAs), and 1 kW peak power, 40 dB gain, and 600 MHz bandwidth for traveling wave tubes (TWTs) at 95 GHz. Solid state transmitters using pulsed Impatt or Gunn devices are available which deliver 10 watts peak (35 GHz), 1 watt (70 GHz), and 5 watts (95 GHz), with a chirp bandwidth of approximately 500 MHz. Pulsewidths are typically 10 to 100 nanoseconds, and may be variable over a limited range. Gunn diode amplifiers are also currently available which can deliver up to 3 watts peak power at 35 GHz, for variable pulsewidths from 10 to 250 nanoseconds and bandwidths of 1 GHz.

Millimeter receiver components are now being developed with specifications similar to their lower frequency counterparts. [2] Integrated mixer-preamplifiers with double sideband noise figures of 4.0 dB (35 GHz), 6.0 dB (70 GHz) and 8.0 dB (95 GHz) are available, along with low-noise Gunn diode local oscillators. Ferrite circulators are available with isolations as high as 40 dB (35 GHz) and 20 dB (70 GHz and 95 GHz), and insertion losses of 0.1 dB (35 GHz) and 0.7 dB (70 GHz and 95 GHz). Duplexers and conventional receiver protectors are not readily available, although TR limiters at 35 GHz are available for 100 kW power with typical insertion losses of 2 dB. Ferrite switches are commonly used in the receiver line to provide 35 dB isolation during the transmission period, but these have high insertion losses of 1 to 2 dB. Where high-speed recovery times are required (short range operation), pin diode switches are available which produce isolations of 30 dB (35 GHz) and 15 dB (95 GHz), with insertion losses of 1 dB (35 GHz) and 2 to 3 dB (95 GHz). Taking into account the devices currently available and the resultant system losses, practical instrumentation radar sensitivities for 20 MHz IF bandwidths are -90 dBm (35 GHz), -84 dBm (70 GHz), and -82 dBm (95 GHz).

Although the system sensitivities are worse at millimeter waves than at lower frequencies, the increase in antenna gain for a given aperture size results in approximately the same range performance as at lower frequencies for a given transmitted power and target cross section, if atmospheric propagation effects are ignored.

## 11.5 ENVIRONMENTAL CHARACTERIZATION

One area that is quite often overlooked in the design of an experimental test program is the inclusion of appropriate procedures for characterizing the environment. For a carefully executed measurement program, cumulative errors in calibration are typically less than 3 to 4 dB, as discussed in Chapter 4. However, variation in the data measured for targets and clutter can vary 10 dB or more. This variation usually occurs because some physical property of the target or clutter which strongly affects the reflectivity properties was not properly characterized. This is particularly true for low angle measurements and for millimeter wavelengths. It has been determined that reflectivity from rain depends as strongly on drop-size distribution as rain rate,[3] that snow backscatter is very strongly affected by the surface roughness and the free water content,[4] that the backscatter from foliage depends on the surface moisture present,[5] and that sea return is strongly affected by wind generated droplets and small facets on the ocean surface[6] as well as atmospheric propagation conditions for low angles.

### 11.5.1 Environmental Measurements

Thus, environmental parameters must be accurately quantified if the physical interaction mechanisms between electromagnetic waves and the scattering objects are to be understood. Unfortunately, the instrumentation and techniques for such measurements are in most cases not yet fully developed. A number of rain drop-size spectrometers are now available including both optical and hydrophonic types. However, they are expensive and measure the drop-size for only a very small area. Snow characterization is still a process of collecting snow samples, freezing them, and then weighing them for determination of the free water content, and the use of pictures to estimate surface roughness. Sea surface characterization is still confined to wave gauges or to stereophotography. However, neither technique accurately characterizes airborne droplets which may have strong effects on the backscattering properties of the sea. Finally, no practical technique has been developed for sensing the surface moisture present on foliage or on the ground.

Infrared sensors can give relative ground moisture readings, but are hard to calibrate since the readings depend on soil type as well as moisture content. Thus, it can be seen that environmental characterization is as difficult as it is important. However, it is extremely important that such measurements be performed as a part of the test program in spite of the difficulties. (*See* Chapter 12 for a more detailed description of ground truth measurement techniques.)

### 11.5.2 Quantifying Propagation Effects

*Millimeter Wave Measurements*

At millimeter frequencies, radiation propagating through the atmosphere can be strongly affected by meteorological conditions. For clear air, appreciable attenuation occurs for electromagnetic energy traveling through the medium even in the so-called transmission windows near 35 GHz, 95 GHz, 140 GHz, and 300 GHz, as is shown by Table 11-6, which gives clear air attenuation values for 10 GHz through 95 GHz in the radar "windows." However, for the ranges which are typically encountered for high angle measurements, clear air attenuation can be neglected. This is not true for the precipitable atmosphere where heavy fog, rain, snow, and hail can have significant effects on propagating millimeter wave energy through both attenuation and scattering mechanisms. Table 11-7 summarizes attenuation losses due to propagation through falling rain at 10, 35, 70, and 95 GHz.[3] When measurements are made under these conditions, these atmospheric effects must be taken into account. Also, anomalous propagation is present at millimeter frequencies as well as at lower frequencies. Thus, ground truth measurements of the precipitating atmosphere must be included in any tests involving adverse weather.

**TABLE 11-6**
**Typical Values for Clear Air Attenuation**

| Frequency | Relative Humidity | Attenuation Constant $\alpha$ (dB/Km) |
|---|---|---|
| 10 GHz | 25% | 0.010 |
| | 100% | 0.015 |
| 16 GHz | 25% | 0.020 |
| | 100% | 0.040 |
| 35 GHz | 25% | 0.035 |
| | 100% | 0.100 |
| 95 GHz | 25% | 0.100 |
| | 100% | 0.230 |

One method of accounting for atmospheric effects on a radar signal is to periodically inject a known power level into the receiver while monitoring a calibration target continuously. The power from the radar calibration target is related to the injected power by the radar equation using the known system parameters. Then any changes in power from the reflector become immediately obvious and can be taken into account. In addition, measurement of meteorological parameters such as rain rate or drop size distribution can confirm the meteorological effects detected by the above method since previous experiments have related such parameters to attenuation and backscatter.[3]

**TABLE 11-7**
**Attenuation Due To Rain**
(from Currie, *et al.*, ref. [3].)

| $f$ = 10 GHz | 35 GHz | 70 GHz | 95 GHz |
|---|---|---|---|
| $\lambda$ = 3.2 cm | 0.86 cm | 0.43 cm | 0.32 cm |
| $A$ = .00919 | $A$ = .273 | $A$ = .634 | $A$ = 1.6 |
| $B$ = 1.1600 | $B$ = .985 | $B$ = .868 | $B$ = .64 |

| R (*mm/hr*) | $\alpha$ (*dB/Km*) = $AR^B$ | | | |
|---|---|---|---|---|
| | 10 GHz | 35 GHz | 70 GHz | 90 GHz |
| 0 | 0 | 0 | 0 | 0 |
| 5 | 0.0509 | 1.33 | 2.56 | 4.48 |
| 10 | 0.133 | 2.64 | 4.68 | 6.98 |
| 15 | 0.213 | 3.93 | 6.65 | 9.05 |
| 20 | 0.297 | 5.22 | 8.54 | 10.88 |
| 25 | 0.385 | 6.50 | 10.36 | 12.55 |
| 30 | 0.475 | 7.78 | 12.14 | 14.11 |
| 35 | 0.568 | 9.06 | 13.88 | 15.57 |
| 40 | 0.663 | 10.33 | 15.18 | 16.96 |
| 45 | 0.760 | 11.60 | 17.26 | 18.29 |
| 50 | 0.859 | 12.87 | 18.91 | 19.56 |
| 55 | 0.960 | 14.14 | 20.55 | 20.79 |
| 60 | 1.062 | 15.40 | 22.16 | 21.99 |
| 65 | 1.165 | 16.67 | 23.75 | 23.14 |
| 70 | 1.270 | 17.93 | 25.32 | 24.27 |
| 75 | 1.375 | 19.19 | 26.89 | 25.36 |
| 80 | 1.482 | 20.45 | 28.44 | 26.43 |
| 85 | 1.590 | 21.71 | 29.98 | 27.48 |
| 90 | 1.700 | 22.97 | 31.50 | 28.50 |
| 95 | 1.809 | 24.22 | 33.02 | 29.50 |
| 100 | 1.920 | 25.48 | 34.52 | 30.49 |

A second method for determining the atmospheric effects on the measurements involves transmitting a known power level from the target site to the radar site at a slightly different frequency from the radar. (A second receiver would be required.) Variations in the received power indicate changes in propagation conditions.

*Low Angle Measurements*

For low angle measurements the propagation factor $F$ becomes important in its effects on reflectivity measurements. The three most important parameters are refraction, multipath, and ducting or trapping of the energy. Refraction results in targets being visible beyond the normal radar horizon, multipath causes large fluctuations in the apparent target return with variations in range or height, and ducting can result in an enhancement of the reflected energy over that expected in free space. Normal refraction effects can be minimized by operating above the free space radar horizon. Multipath effects can be predicted analytically, or can be measured by raising the height of a small target above the surface and recording the return. Ducting conditions can be predicted by recording the temperature, pressure, and humidity as a function of height, and calculating the refractive index of the air *versus* height. Alternatively, for sea return the range dependence of the sea backscatter

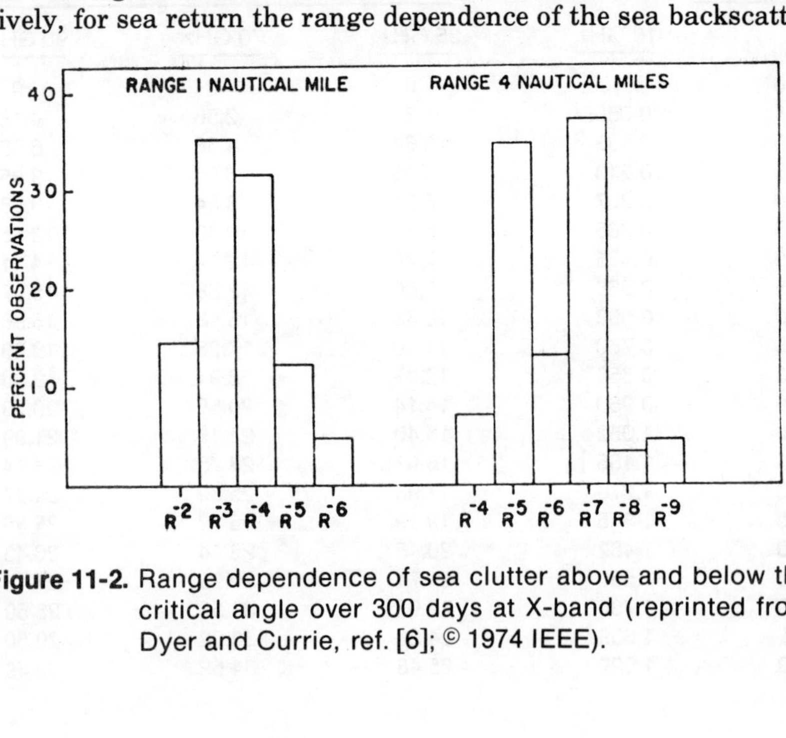


**Figure 11-2.** Range dependence of sea clutter above and below the critical angle over 300 days at X-band (reprinted from Dyer and Currie, ref. [6]; © 1974 IEEE).

can be measured as shown in Figure 11-2. If the range dependence is not $R^{-3}$ above the critical angle and $R^{-7}$ below the critical angle (which would be expected for an area scatterer; *see* Chapter 2), then abnormal propagation conditions are present.

The effects of multipath can be accounted for by inclusion of the propagation factor $F$ in the radar range equation, where $F$ may be written as:

$$F = 1 + \rho \, D \exp(-j\,\delta) \qquad (11\text{-}15)$$

where

$\rho$ = reflection coefficient of the earth's surface

D = divergence factor to account for the curved earth

$\delta$ = phase difference between the direct and reflected waves of the target

When the angle between the direct and reflected (from the earth's surface) rays is small, the radar range equation takes the following form:

$$P_r = \frac{P_t\, G^2\, \sigma \lambda^2\, F^4\, L^2}{(4\pi)^3\, R^4} \qquad (11\text{-}16)$$

where

$P_r$ = received power from the target

$P_t$ = transmitter power

$G^2$ = antenna gain

$\sigma$ = RCS of target

$\lambda$ = transmitted wavelength

$F$ = propagation factor given above

$L$ = waveguide and atmospheric losses ($L$ is less than 1)

$R$ = slant range from radar to target

To examine the characteristics of multipath effects more easily, initially consider a simplified flat earth geometry ($D = 1$) where the reflection coefficient $P$ is -1. Under these conditions (11-3) reduces to

$$P_r = P_{rfs}\, F^4 = P_{rfs}\,(2 - 2\cos K\,\Delta R)^2 \qquad (11\text{-}17)$$

where

$$\Delta R = \frac{2h_1\, h_2}{R} \qquad (11\text{-}18)$$

where

$P_{rfs}$ = received power from the target in free space

$h_1$ = radar antenna height

$h_2$ = target height

$R$ = range

For a fixed range and antenna height, the received power varies as $(1 - \cos\psi)$ where $\psi$ linearly increases with target height. Therefore, the received power varies from a minimum of 0 to a maximum of 16 $P_{rfs}$ as a function of target height. The maximum occurs where

$$\Delta R = \frac{2h_1 h_2}{R} = \frac{(2n + 1)\lambda}{2}, n = 0, 1, 2, \ldots \tag{11-19}$$

and the minimum where

$$\Delta R = \frac{2h_1 h_2}{R} = n\lambda, n = 0, 1, 2, \ldots \tag{11-20}$$

Figure 11-3 is a plot of the propagation factor, $|F|^4$, in dB as a function of $\Delta R$. The effect of including earth curvature and actual reflection coefficients may alter the exact position and depths of the nulls of the curve in Figure 11-3, but does not alter the shape and characteristics of the curve.

Consider again equations (11-3) and (11-4), and the flat earth geometry where the reflection coefficient is -1. If instead of varying the target height the antenna and target heights are fixed, and range is permitted to vary, then $\Delta R$ will increase as range decreases. Therefore, a lobing structure in the field will also be observed with variations in range. However, the nature of the field variations with range is quite different from the variations with receiver height. Since $\Delta R$ monotonically decreases with increasing R, there is a range beyond which the received power also decreases monotonically. At this point

$$\Delta R = \frac{\lambda}{2} = \frac{2h_1 h_2}{R} \tag{11-21}$$

or

$$R = \frac{4h_1 h_2}{\lambda} \tag{11-22}$$

To examine variations in field strength with antenna height, it is only necessary to consider $F^4$, since range is by definition constant. To examine field variations with range, $F^4/4$ must be examined. To

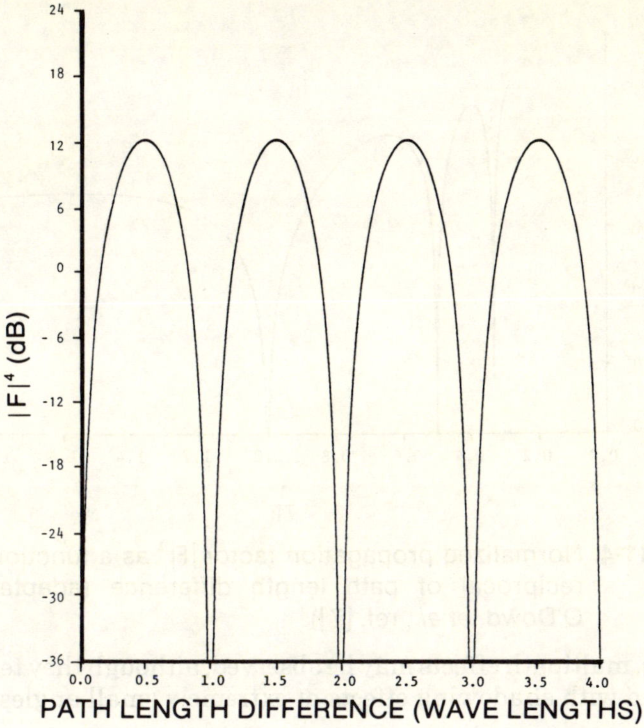


**Figure 11-3.** Plot of propagation factor $|F|^4$ *versus* path length difference for unity reflection coefficient (adapted from O'Dowd, *et al.*, ref. [7]).

**remove a specific range and geometry from the immediate problem, (11-12) can be written as**

$$P_R = K\,\Delta\, r^4\,(1 - \cos r\Delta r) \qquad (11\text{-}23)$$

$$= K\,\Delta\, r^4\, F^4 \qquad (11\text{-}24)$$

**where $K$ is the appropriate constant. Figure 11-4 shows $F^4\,r^4$ in dB as a function of $1/R$ for the ideal flat earth case. Since $1/R$ is linearly related to $R$, the abscissa in Figure 11-4 is directly proportional to range.**

**Thus, it can be seen that the return from any small target changing in either range or height will fluctuate due to the multipath effects. Therefore, the multipath field must be mapped whenever direct calibrations are performed, and multipath considerations must be taken into account for low angle target measurements. For clutter measurements, if the beam area is large, the multipath effects tend to be averaged out. However, for small beam sizes or short pulse**

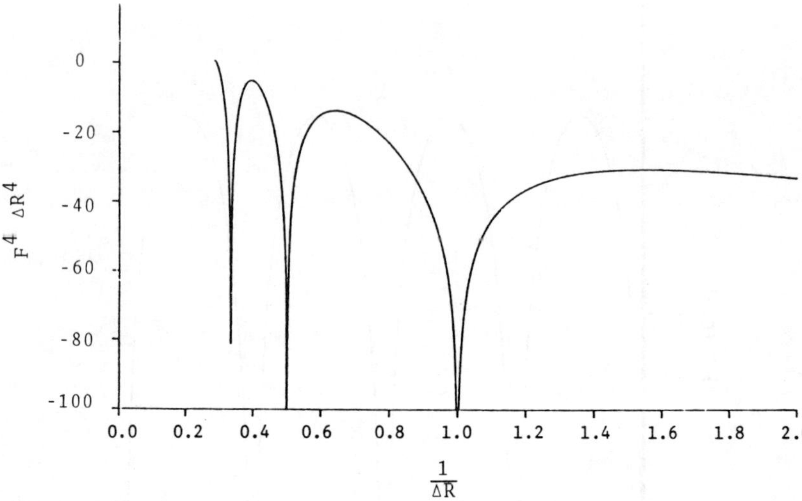


**Figure 11-4.** Normalized propagation factor $|F|^4$ as a function of the reciprocal of path length difference (adapted from O'Dowd, *et al.*, ref. [7]).

lengths, multipath effects may be observed, although they tend to be confused with shadowing effects at extremely small angles.

## 11.6 MEASUREMENT TECHNIQUES

Measurement techniques for the acquisition of RCS data are typically dependent upon the type of system used and the type of target to be observed. However, to illustrate the typical procedures that might be followed, consider the use of a pulse radar system to observe ground clutter. Such a measurement system has a number of features which are necessary if accurate RCS measurements are to be performed. Among these are:

A means of accurately pointing the antenna

Provision for measuring transmitted power

A point to inject test signals into the receiver

A receiver having adequate dynamic range to encompass the range of amplitude fluctuation of the received signal

A data acquisition and/or processing system which can deal with data on a pulse-by-pulse basis

Before such a measurement system is taken to the field, careful measurements should be made of radar parameters such as antenna gain, operating frequency, transmitted signal characteristics, and receiver performance. Once in the field, the procedure on each test day might be as follows:

Measure transmitted power and minimum detectable signal

Inject a calibrated test signal into the receiver front end and step the power level in known increments through the receiver dynamic range while recording/processing the receiver output to obtain an indirect calibration

Acquire data from a known RCS standard (absolute calibration)

Compare the results of the two calibration techniques (close the calibration)

Perform the measurements called for in the test matrix

Repeat one of the calibration procedures (either absolute or indirect) and compare with the calibration run performed prior to the data runs.

During the measurements the transmitter power should be monitored at the directional coupler on the antenna input. A number of well-established techniques may be used, and the recent availability of reasonably accurate thermistor mounts up through 95 GHz has simplified the monitoring problem somewhat. This same directional coupler may be used to inject test signals into the receiver front end for the indirect calibration. The small beamwidths often encountered in modern radar systems make the accuracy of pointing the antenna quite important; boresighted reflex cameras, T.V. cameras, or precision optics are often used to aid in beam pointing.

When closing the calibration, initial discrepancies of 10 dB or more are often discovered when the results of the two calibration methods are compared. Cause for such discrepancies include: large waveguide losses in the antenna run (due to moisture or flexguide breaks), multipath effects on the RCS calibration standard, or transmitter/receiver spectral mismatch, to name a few. In general, a match of 3 dB between these two methods is considered to be quite good.

The procedure outlined above helps to ensure reasonable, calibrated data; if commerically available power meters and signal generators are used, gathering data can be a relatively straightforward operation. However, as the frequency increases above 35 GHz, less and less of such equipment is available, while calibration procedures become somewhat less convenient and certainly more expensive. For this reason, different procedures are often used at these higher frequencies. One commonly used alternative requires that a precision attenuator be inserted in the receiver chain and the radar be used to observe an extremely large standard target. The attenuator is then varied in discrete steps to provide the desired calibration information. Care must be used with such a procedure, since no independent check on calibration is readily available, and factors such as improper aiming of the RCS standard and the presence of multipath can contribute to calibration errors. Furthermore, the calibration standard must reflect enough power to the radar to ensure receiver saturation when the attenuator is set to zero. Otherwise, the posibility that target returns might be stronger than the highest level calibration point exists, making the returns difficult to calibrate.

## 11.7 DATA ANALYSIS CONSIDERATIONS

The objective of any measurement program lies in the application of data to some real world problem; the key to this aspect of the program lies in the analysis and presentation of results. Results might include presentation of average quantities such as radar cross section, radar cross section per unit area, and attenuation coefficient. These parameters can be coupled with meteorological, temporal, or environmental parameters in the final presentations of results. Fluctuation characteristics such as amplitude distributions, frequency spectra, or correlation functions are also important for many applications. However, considerable care is needed in the interpretation of such data. Finally, there are data which are related to specific signal processing techniques, such as special polarization or frequency processing, which will require their own specific presentations.

It is important to note that a measure of target or clutter reflectivity goes beyond the need to merely measure the one-time reflectivity figure or radar cross section associated with the particular target. The measurement, to be useful, needs to include the amplitude statistics, and the spectral and correlation properties.

Figures 11-5 through 11-7 give examples of some of the types of data analysis results obtained from measurement programs.[3,4,5] Figure 11-5 shows the average radar cross section values per unit area ($\sigma°$) for a patch of smooth snow as a function of time of day for the parallel and orthogonally polarized received components (vertical transmitted polarization) at 35 GHz. This figure illustrates the large variations with time of backscatter which occur when the frozen surface crust melts. Figure 11-6 shows the incoherent frequency spectra roll-off characteristics for deciduous trees for two ranges of wind speed at 95 GHz. The curves shown represent averages of

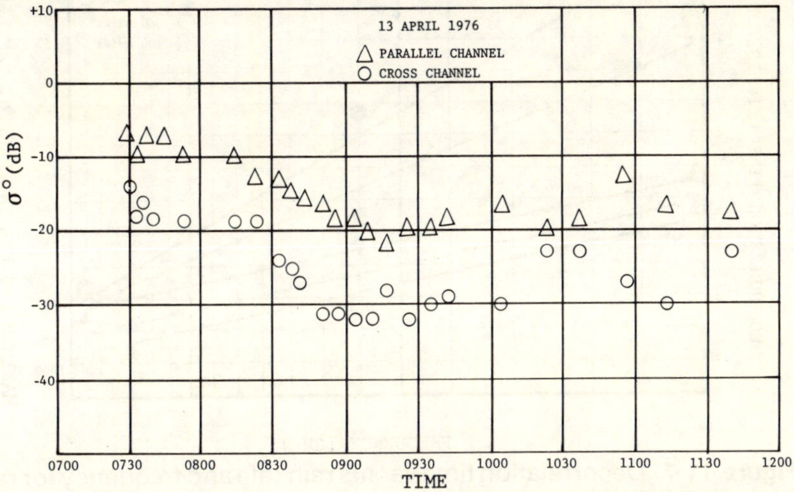


**Figure 11-5.** Comparison of the parellel and cross polarized components of the return from a snow-covered field for vertically (*VV*) polarized transmission; 35 GHz (reprinted from Currie, *et al.*, ref. [4]).

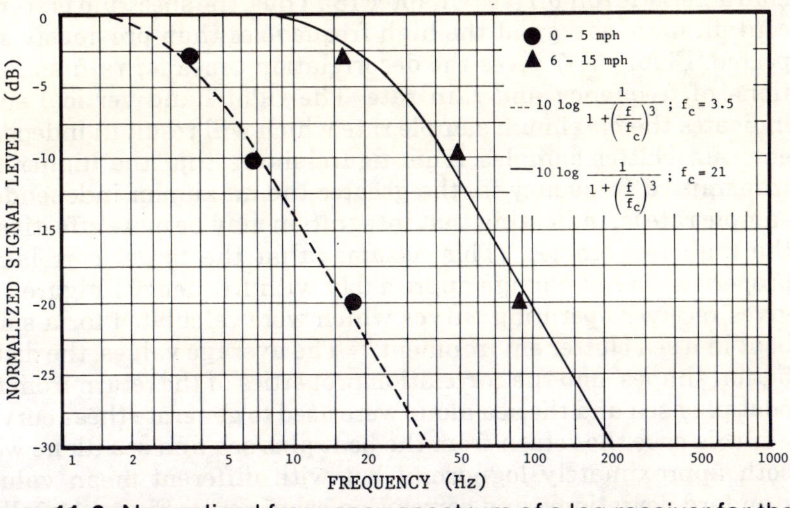


**Figure 11-6.** Normalized frequency spectrum of a log receiver for the return from deciduous trees for two ranges of wind speed; 95 GHz, vertical (*VV*) polarization (reprinted from Currie, *et al.*, ref. [5]).

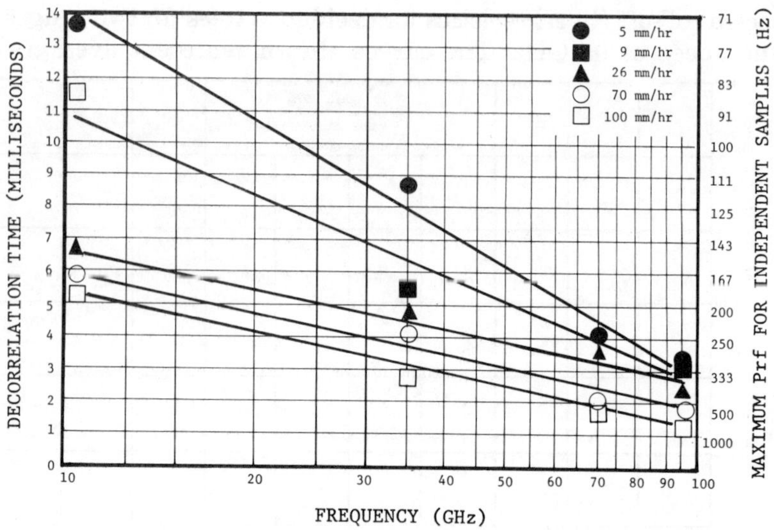


**Figure 11-7.** Decorrelation time versus rain rate and frequency for rain backscatter (reprinted from Currie, *et al.*, ref. [3]).

about 50 data runs of approximately 10 seconds duration. These curves show that the spectra are not Gaussian in nature as has been assumed by many system designers, but rather, have a $1/(1+(f/f_c)^2)$ characteristic roll-off (*see* Chapter 13). Thus, the spectrum is seen to contain more energy at the high frequencies than previously suspected. Figure 11-7 gives the decorrelation times for rain as functions of frequency and rain rate. The right hand vertical scale indicates the maximum sample rate which will result in independent rain clutter samples. This figure shows that the higher the transmitted frequency is, the greater the maximum independent sample rate is, indicating that integration may be more effective at the high frequencies. (This assumes that the target correlation properties do not change appreciably with frequency.) Figure 11-8 gives receiver operating curves which were calculated for a small boat in a sea clutter environment.[18] The average values, the distribution shapes, and the correlation properties of the return from the boat-plus-sea and the sea alone were used to generate these curves. For this case the return from the boat-plus-sea and sea alone were both approximately lognormal but with different mean values, standard deviations, and correlation coefficients. Figure 11-9 illustrates how the clutter spectral characteristics can be used to determine Doppler radar performance. In the figure, the clutter characteristics of 4/mm/hr rain are plotted for 9.4 GHz, 16.5 GHz, and

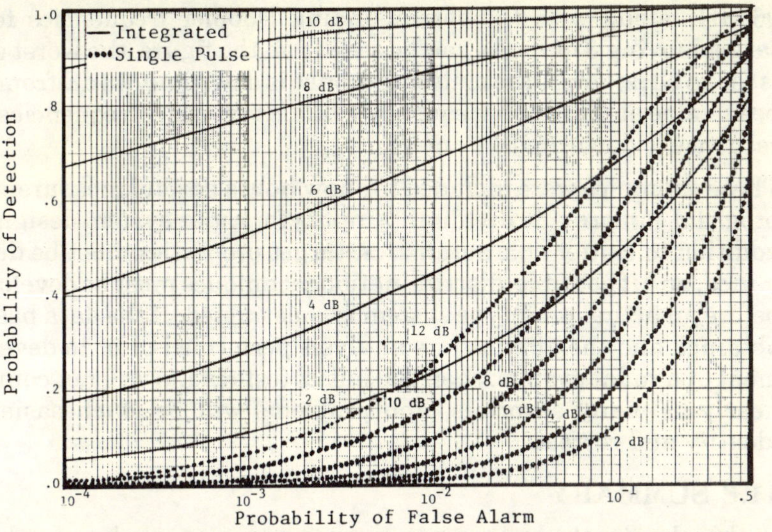


**Figure 11-8.** Comparison of integrated and single-pulse probability of detection versus probability of false alarm as a function of single to background ratio (reprinted from Currie, *et al.*, ref. [8]).

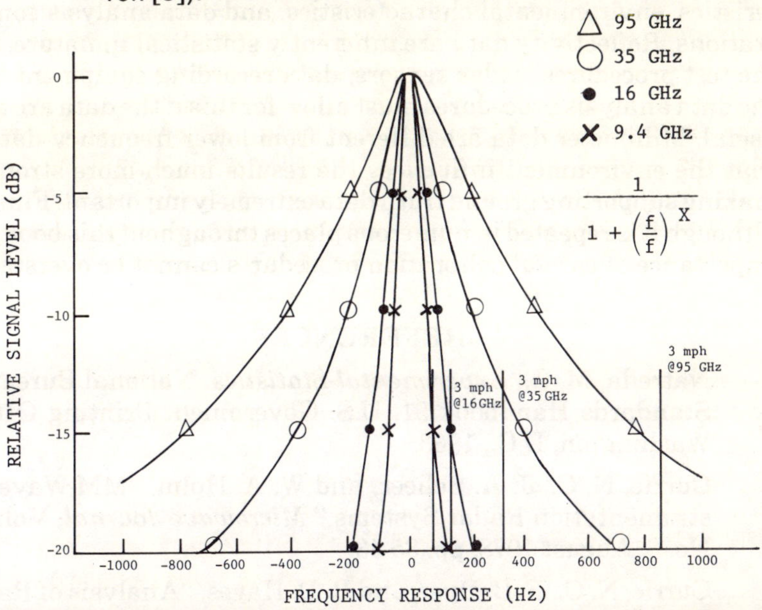


**Figure 11-9.** Noncoherent frequency spectra from rain, 4 mm/hr (courtesy of R. D. Hayes).

96 GHz along with the corresponding Doppler frequencies for a target moving 3 mph. As can be seen from the figure, the worst case is for 35 GHz, where the clutter response is down only 13 dB from the peak value at the 3 mph frequency while, at the other frequencies the response is down 15 dB or more.

These examples were included to show how a careful measurement program can provide invaluable inputs for radar system design. In addition to clutter characteristics, target properties must be determined, and if possible, discrimination techniques must be developed before an actual prototype system is constructed. If the data base is obtained and the appropriate analysis performed prior to development of a radar system, then the initial test results will not contain nearly as many surprises, and problems will be much easier to identify and correct.

## 11.8 SUMMARY

In this chapter the key components of experimental planning have been discussed, including: the definition of the test objectives, the setting of priorities, the development of a test plan, and the primary considerations for radar experiments such as radar sensor characteristics, environmental characteristics, and data analysis considerations. Reflectivity data are inherently statistical in nature, and the test procedures, radar sensors, data recording equipment, and the data analysis procedures must allow for this if the data are to be useful. Millimeter data are different from lower frequency data in that the environment influences the results much more strongly, making supporting ground truth data extremely important. Finally, although it is repeated in numerous places throughout this book, the importance of careful calibration procedures cannot be overstated.

## REFERENCES

1. Natrella, M. G., *Experimental Statistics*, National Bureau of Standards Handbook 91, U.S. Government Printing Office, Washington, D.C., 1963.

2. Currie, N. C., J. A. Scheer, and W. A. Holm, "MM-Wave Instrumentation Radar Systems," *Microwave Journal*, Vol. 21, No. 8, August 1978, pp. 35-43.

3. Currie, N. C., F. B. Dyer, and R. D. Hayes, "Analysis of Radar Rain Return at Frequencies of 9.375, 35, 70, and 95 GHz," Technical Report No. 2 on Contract DAAA25-73-C-0256, Geor-

gia Institute of Technology, Atlanta, Georgia, February 1975, ADA007254.

4. Currie, N. C., F. B. Dyer, and G. W. Ewell, "Millimeter Backscatter Measurements From Snow," Technical Report No. AFATL-TR-77-4 on Contract F08635-76-C-0221, Georgia Institute of Technology, Atlanta, Georgia, June 1976.

5. Currie, N. C., *et al.*, "Radar Millimeter Wave Measurements: Volume I, Snow and Vegetation," Report No. AFATL-TR-77-92 on Contract F08635-76-C-0221, Georgia Institute of Technology, Atlanta, Georgia, July 1977.

6. Dyer, F. B., and N. C. Currie, "Some Comments on the Characteristics of Radar Sea Clutter," *1974 IEEE AP-S International Symposium Digest*, June 1974.

7. O'Dowd, W. M., Jr., F. B. Dyer, and M.T. Tuley, "Effects of the Ocean Surface on the Forward Scatter of Radar Signals at Low Incidence Angles," Final Report on Contract N00039-76-C-0368, Georgia Institute of Technology, Atlanta, Georgia, September 1978.

8. Currie, N. C., M. M. Horst, and J. C. Butterworth, "Shore Tests on the AN/APS-127 Airborne Radar System," Final Test Data Package - Phase I on Contract N62269-76-C-0232, Georgia Institute of Technology, Atlanta, Georgia, February 1977.

# CHAPTER 12
# GROUND TRUTH MEASUREMENTS

**Nicholas C. Currie**

**Engineering Experiment Station**
**Georgia Institute of Technology**



## 12.1 OVERVIEW

The radar reflectivity of a target is dependent on the physical properties of the target as well as the characteristics of the incident electromagnetic radiation. Thus, knowledge of these physical properties is necessary to understand the reflectivity characteristics. This knowledge is normally used in one of two ways. First, once the relationship between physical parameters and radar reflectivity is known, the reflectivity properties can be used to measure the physical properties; for example, the measurement of the amount of free water in a snow pack by remote sensing from an airborne platform or satellite. Second, the knowledge of the relationship between physical and reflectivity properties can be used to extend the measurements of the performance of a radar in one environment to different environments, thus allowing the prediction of radar performance under widely varying conditions.

Unfortunately, physical measurement techniques have not kept up with the advances in reflectivity techniques. However, some studies have been made in recent years, and increased attention is now being given to the measurement of physical parameters. In this chapter, the physical parameters currently considered to be of importance to reflectivity properties will be summarized, and the state of the art in the measurement of these parameters will be described along with current problem areas.

At this time physical measurements can be divided into several areas including: meteorological parameters, surface snow and ice characteristics, soil and vegetation parameters, and sea surface measurements. These four areas will be discussed below.

## 12.2 METEOROLOGICAL PARAMETERS

Rain is the atmospheric parameter having the most significant effect on microwaves; therefore, the measurement and recording of rain characteristics are the predominant criteria used to characterize scattering from the atmosphere. Characteristics of rain and fog having the greatest effect on microwaves are the total amount of water, the mean drop size, and the shape of the drop size distribution. Ideally, these characteristics should be measured and recorded to define precipitation over the complete propagation path between the measurement system and target. Precipitation spectrometers provide the most complete measurement of rain characteristics; however, their cost prohibits their use for defining precipitation characteristics over the complete propagation path. Precipitation spectrometers are normally used to accurately define precipitation characteristics at one point along the propagation path, preferably near the vicinity of the target. Precipitation characteristics along the rest of the propagation path are measured by more conventional and cost effective means; i.e., tipping bucket rain gauges.

Other physical measurements are also considered important to microwave reflectivity, including: wind speed and direction, air temperature, barometric pressure, path loss, relative humidity, and optical visibility.

The basic parameters that are normally measured, and the instrumentation required to perform the measurements, are summarized in Table 12-1. The required accuracy for many of the measurements is unknown because the relationships between the physical parameters and backscatter properties have not been established. The establishment of these relationships is one objective of combined microwave and physical measurements.

The most significant meterological effects on electromagnetic energy are attenuation, backscatter, and depolarization due to rain. Since these effects are dependent on the size of raindrops, it is necessary to measure not only the rainfall rate, but also the raindrop size distribution.

Attenuation due to fog is not normally expected to be large, except at high frequencies, but it should be measurable at times, with magnitudes of approximately one or two dB one-way on a 2-km path in extreme conditions at 100 GHz. A measure of the fog water content made by a transmissometer usually provides sufficient documentation for effects on radar scattering up through the millimeter wave region.

**TABLE 12-1**
**Meteorological Parameters and Instrumentation**

| PARAMETER | INSTRUMENTATION |
|---|---|
| Rain Rate | Tipping Bucket Rain Guage |
| Raindrop Size | Precipitation Spectrometer |
| Atmospheric Attenuation | Transmissometer |
| Visibility | Forward Scatter Meter |
| Temperature | Weather Station |
| Pressure | Weather Station |
| Humidity | Weather Station |
| Wind Velocity | Weather Station |
| Sky Temperature | Calibration Radiometer |
| Target Background Temperature | Calibration Radiometer |

The measurement of optical visibility provides a useful adjunct to attenuation measurements made by the transmissometers. Since optical extinction is proportional to the second moment of the drop-size distribution, correlation of visibility measurements with transmissometer attenuation measurements permits an estimate of mean drop size. The visibility measurement also provides a measure of weather severity in fog or rain; this measure can be related to people's common experience and can provide a semiquantitative measure of meteorological effects on microwave scattering.

Because of the strong dependence of attenuation on temperature, the measurement of temperature is essential. Although the standard parameters (i.e., humidity, pressure, wind velocity) are of somewhat lesser importance, they are valuable for the complete documentation of millimeter wave effects due to adverse weather conditions and may be essential for documentation of tests in clear weather when effects of refractive index fluctuations may be measurable. Some of the primary instruments will be described in more detail below.

### 12.2.1 Tipping Bucket Rain Gauges

Rain accumulation is generally measured by a tipping bucket rain gauge or some other instrument that gives an output indication when a prescribed quantity of water has been accumulated. Rain rate can then be determined by measuring the water accumulated over a period of time. A typical example of a tipping bucket rain gauge is the Model P501-1 manufactured by Weather Measure Corporation® which is shown in Figure 12-1 with wind vanes designed to reduce cross wind effects.[1]

The standard configuration of the Model P501-1 rain gauge measures rain accumulation of 0.01 inch (0.25 mm) with a nominal accuracy of 1.0 percent for rain rates below about 60 mm/hr; at rain rates above 60 mm/hr, an increasing amount of water is lost as the bucket tips. This loss is illustrated by Figure 12-2 which gives the number of tips per minute and the percent error due to water lost during a tip as a function of rain rate. This loss can be compensated for in subsequent data analysis if the total rainfall is accumulated in the gauge and measured at the end of the observation period or at the end of the rain storm.

At rain rates below about 10 mm/hr the Model P501-1 rain gauge has the additional disadvantages of (1) being slow to respond to the onset of rain due to the required initial wetting of the intake funnel, and (2) providing no information about the small-scale temporal variations, due to the long interval (90 seconds or more) between successive tips (*see* Figure 12-2). The latter can be eliminated by expanding the intake. Funnels with diameters of 10 and 100 times that of the standard intake have been used successfully to provide rainfall data at intervals down to 9 seconds at 1 mm/hr rainfall rates. For testing that involves measurements over a wide range of rainfall rates, the most effective configuration of rain gauges would be a set of gauges having standard, 10X, and 100X intakes. However, if one gauge is to be used, then a 3X expander would be best for the range of rainfall rates of greatest interest (3 mm/hr to 20 mm/hr).

**Figure 12-1.** Tipping bucket rain gauge with wind vanes.

The optimum spacing of rain gauges may be determined from consideration of response times and available data on the spatial variability of precipitation. Measurements cited by Hogg and Chu[2] suggest that for propagation distances less than a few kilometers, a single measurement of rain fall rate is sufficient to characterize the propagation conditions. Because these results were based on year-long measurements, however, it is uncertain as to whether the quantitative conclusion can be applied to the dynamic testing environment.

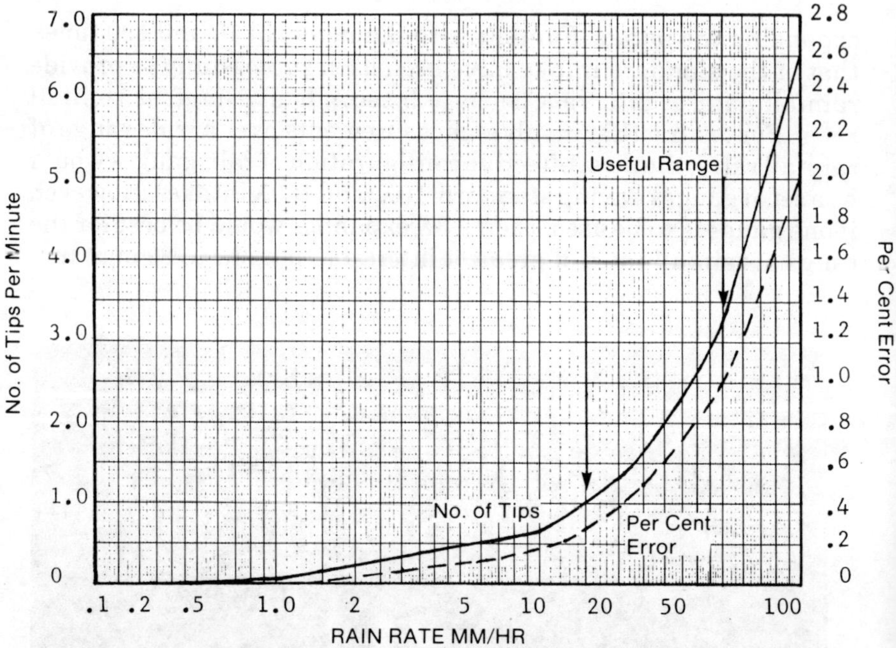


**Figure 12-2.** Number of tips per minute and per cent error *versus* rain rate for a typical tipping bucket rain sensor.

The measurement of rainfall rate by increments of rainfall implies a minimum time for each measurement, and this implies a spatial scale due to air movement. For times of 10 to 20 seconds between tips of the gauge, corresponding to rainfall rates of 45 to 90 mm/hr (very heavy to intense), a translational speed of 10 mm/hr implies movement of 100 to 200 meters between tips. Tips at these rates would be registered by a 10X expanded gauge for rainfall of 4.5 to 9.0 mm/hr (moderate) and by a 100X expanded gauge for rainfall of 0.45 to 0.9 mm/hr (very light). In the latter case, movement of the weather system might not exceed 5 mm/s; this would imply a translation of only 50 to 100 meters during the measurement. In such light rainfalls, however, spatial variability is much less, and a horizontal spacing of 100 to 200 meters between gauges would be sufficient.

### 12.2.2 Precipitation Spectrometers

Raindrop size distribution is the most important measurable parameter affecting electromagnetic radiation, particularly in the millimeter wave region; attenuation and backscatter are both dependent on raindrop size and the distribution of the sizes.

Typical instruments for measuring raindrop size include optical array precipitation spectrometers, momentum spectrometers (distrometer), and electrostatic raindrop spectrometers. The optical array precipitation spectrometers appear to be the most useful of the types. The Model GBPP-100V manufactured by Particle Measuring Systems, Inc.® appears currently to be the best of the lot.

The Model GBPP-100V precipitation spectrometer counts raindrops in diameter increments of 200 micrometers over the range of 200 micrometers to 12.4 millimeters. The largest size measurable is well beyond the observed size of raindrops. The smallest observable size is 142 micrometers in diameter, which, with a fall speed of 0.7 m/s is near the lower size limit of what is generally considered rain precipitation. Raindrop size is measured by the occluding of one or more elements of a photo-diode array. Derivation of the rain drop size distribution can be accomplished either in real time or in post-operation analysis by using appropriate data processing equipment. The computation involves the fall speed of the drops and the effective aperture of the instrument as functions of drop size. The drop size distribution can then be used to compute water content and rainfall rate for comparison with measurements by the transmissometers and microwave reflectivity. The rain rate can be expressed as follows:[1]

$$R\,(mm/hr) = 3600\,H/T \qquad (12\text{-}1)$$

where $H$ is the height of accumulated water over the sensor head in mm and $T$ is the time in seconds.

$H$ is equivalent to the volume of the water divided by the sensor head area. The volume is given by:

$$V = \frac{\pi}{6} \sum_{i=1}^{n} A_i D_i^3 \qquad (12\text{-}2)$$

where:

$A_i$ = the number of drops of a given diameter $D_i$ (±0.05 mm) which strike the sensor head, and

$D_n$ = the maximum drop diameter

Combining the above factors yields:

$$R = \frac{1885}{A \cdot T} \sum_{i=1}^{n} A_i D_i^3 \qquad (12\text{-}3)$$

The volumetric distribution (number of drops in a cylindrical column above the sensor head with volume of one cubic meter) is related to the measured distribution over some time interval by:

$$N(vol) = \frac{N_i D_i H}{T V_t} \tag{12-4}$$

where:

$N_i$ = the number of drops striking the sensor in time $T$ with diameter $D_i$

H = the height of a cylindrical column with the same diameter as the sensor head and a volume of 1 $m^3$

$V_t$ = the terminal velocity of a drop diameter $D_i$

The terminal velocities used to calculate the volumetric distributions can be obtained from Gunn and Kinzer.[3]

Figure 12-3 gives typical raindrop size distributions which were measured for approximately a 10 mm/hr rain rate illustrating the large differences that can occur over short intervals of time.

### 12.2.3 Forward Scatter Meter

Visual range can be measured by a forward scatter meter that senses the scattering of light in a small torus-shaped sample volume between a light projector and a light-sensitive receiver to measure the visual range. A typical instrument for application in adverse weather is the Model 207 forward scatter meter manufactured by EG&G International, Inc., Environmental Equipment Division®. This meter uses an optical system in the projector assembly to provide the desired beam pattern of high intensity light and an identical optical system in the receiver assembly to collect the light scattered in the torus volume onto a light sensitive detector. The projector light is modulated and the detector synchronously demodulated to exclude interference from external lights.

### 12.2.4 Weather Station

Standard meteorological parameters are readily measured by a variety of weather stations including standard sensors for temperature, humidity, wind velocity, and barometric pressure. A typical weather station, manufactured by Meteorology Research, Inc., provides calibrated outputs of these parameters for recording purposes. Typical accuracies of these parameters are given in Table 12-2.

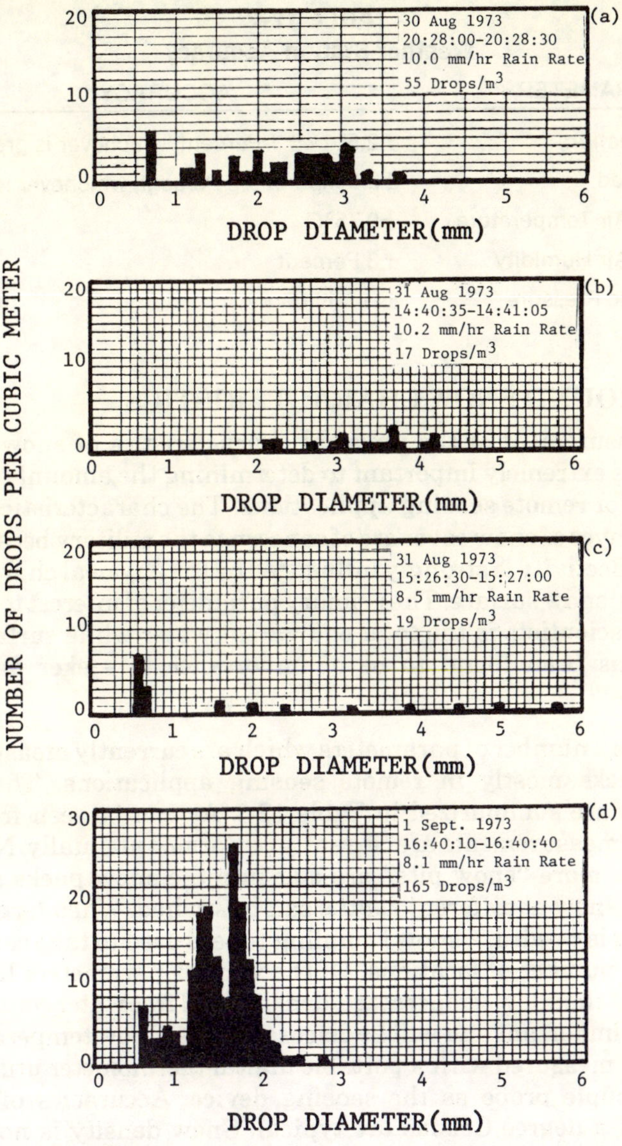


**Figure 12-3.** Four different drop size distributions which were measured for similar rain rates (reprinted from Currie, *et al.*, ref. [1]).

**TABLE 12-2**
**Weather Station Accuracy**

| PARAMETER | ACCURACY |
|---|---|
| Wind Direction | ±2.5° or ±1 Percent, whichever is greater |
| Wind Speed | ±2.5 mph or ±2 Percent, whichever is greater |
| Ambient Air Temperature | ±0.75°C |
| Ambient Air Humidity | ±3 Percent |
| Barometric Pressure | ±.1 Millibars |

## 12.3 GROUND SNOW CHARACTERISTICS

The measurement of the physical characteristics of snow on the ground is extremely important to determining the amount of water present for remote sensing applications. The characteristics of the surface layer of snow cover are of concern to the military because the radar reflectivity is strongly influenced by the physical characteristics of the snow surface. The primary parameter of interest to remote sensing scientists is the total equivalent water while surface free water has been found to be of importance to seeker detection problems.[4]

There are a number of parameters which are currently measured on snow packs mostly in remote sensing applications. These parameters are summarized in Table 12-3. As can be seen from the table, all measurements are currently performed manually. Normally, one or more "snow pits" are dug into the snow packs so that layers of snow at different depths can be studied. When large scale sampling is needed, a snow sampling tube is used to take core samples at a number of locations. Of the various parameters listed in Table 12-3 the snow temperature, density, and free water content are the most important and will be discussed here. Snow temperature is typically measured with a portable digital thermometer utilizing a thermocouple probe as the sensing device. Accuracies of a few tenths of a degree Celsius are typical. Snow density is normally measured with a snow density kit such as is manufactured by

Hydro-Tech®. The kit consists of a snow sampler of a known volume and a scale calibrated in *gm/cc,* which is small enough and rugged enough to be carried around a snow pack. The measured densities are then manually recorded. Figure 12-4 shows a typical snow density measurement kit.

**TABLE 12-3**
**Snow Ground Truth Parameters**

| NO. | PARAMETER | CURRENTLY ACCEPTED PROCEDURES |
|---|---|---|
| 1 | Depth | Manual Measurement with Scale |
| 2 | Density Profile | Manual Measurement of Layers with Density Kit |
| 3 | Free Water | Manual Calorimetric Technique |
| 4 | Temperature Profile | Themocouple Probes |
| 5 | Total Water Equivalent | Calculated From Density and Depth |
| 6 | Crystal Structure | Visual Observation/Microscope Photograph |
| 7 | Surface Roughness | Photograph of Grid Inserted into Snow |
| 8 | Solar Loading | Photometer |
| 9 | Snow Resistance (Mechanical) | RAM Penetrometer (Manual) |

There are a number of techniques for measuring snow free water content including the hot water calorimeter, cold calorimeter, centrifuge, capacitance, and dielectric constant techniques. However, to date only the cold calorimeter technique has been found to be both repeatable and insensitive to snow structure. Thus, a description of this technique adapted from Stiles and Ulaby will be given here.[5]

The equipment used for the cold calorimeter snow free water measurement is shown in Figure 12-5. The calorimeter is an insulating container with provisions for measuring temperature. Typically, a thermos bottle with terminals for an internal thermocouple thermometer is used as the calorimeter. A known amount of toluene (cooling agent) is allowed to reach equilibrium inside the calorimeter; then the wet snow is added, and the solution again allowed to reach equilibrium. Weight and temperature measurements at the two equilibrium temperatures allow calculation of the wetness.

If the calorimeter is assumed lossless, heat will be conserved between initial and final states:

**Figure 12-4.** View of a typical snow density measurement kit.

$$H_i = H_f \tag{12-5}$$

where

$H_i$ = initial heat content of all constituents

$H_f$ = final heat content of solution

The heat content of the final solution of toluene and ice is

$$H_f = T_f(W_i + E_c)C_{tf} + T_f W_s C_{sf} \tag{12-6}$$

where

$T_f$ = final equilibrium temperature
$W_i$ = weight of toluene
$E_c$ = calorimeter constant
$W_s$ = total weight of snow
$C_{tf}$ = specific heat of toluene at $T_f$
$C_{sf}$ = specific heat of ice at $T_f$

The two terms are the heat content of the toluene and calorimeter, and the heat content of the ice. The heat content of the initial constituents (toluene, snow, free water) is

$$H_i = T_i ( W_i + E_c ) C_{ti} + L_f W_f + T_s W_d C_s + T_s W_f C_w \quad (12\text{-}7)$$

where

$T_i$ = initial temperature of toluene
$T_s$ = snow temperature
$W_f$ = weight of free water in snow
$W_d$ = weight of dry snow, $S = D + F$
$C_{ti}$ = specific heat of toluene at $T_i$
$C_s$ = specific heat of snow at $T_s$
$C_w$ = specific heat of water at $T_s$
$L_f$ = heat of fusion of water = 79.7 *cal/g*

The first term of Equation (12-7) is the initial heat content of the toluene and calorimeter, and the second term is the heat required to change the state of the free water to ice. The third term is the heat content of the ice component of the snow, and the last term is the heat content of the free water in the snow. If the temperature of the snow ( $T_s$ ) is less than 0°C, then the last term of Equation (12-6) should be changed to $T_s W_f C_s$, since there should be no free water.

Using Equations (12-5), (12-6), and (12-7) and solving for the fraction of free water yields:

$$\frac{m_w}{100} = \frac{W_f}{W_s} = \frac{( W_i + E_c ) ( T_f C_{tf} - T_i C_{ti} )}{W_s ( T_s C_w + L_f )} + \frac{( T_f W_s C_{sf} - T_s W_d C_s )}{W_s ( T_s C_w + L_f )} \quad (12\text{-}8)$$

The volumetric free water ratio is then

$$m_v = \rho m_w \quad (12\text{-}9)$$

where $\rho$ is the snow density.

The procedure used for this measurement, repeated from Stiles, reference [5] is as follows:

1. Weigh the empty calorimeter, measure its temperature, and record the results.
2. Add 300 *ml* of toluene, shake gently 4 to 5 minutes, or until the temperature stabilizes. The toluene must be well below 0°C.
3. Weigh and measure the temperature of the calorimeter and cold toluene.
4. Remove the cap and add the snow sample. The snow volume should be approximately 500 *ml*. Replace the cap and shake gently for 3 minutes or until the temperature stabilizes.
5. Weigh and measure the final temperature of the calorimeter, toluene, and snow sample.
6. Clean the calorimeter.

The total time to perform this measurement and to clean up is about 45 minutes. For monitoring the wetness change in the surface layer on a warm sunny day, this time resolution was not as small as desired. This time interval can be reduced by using smaller sample

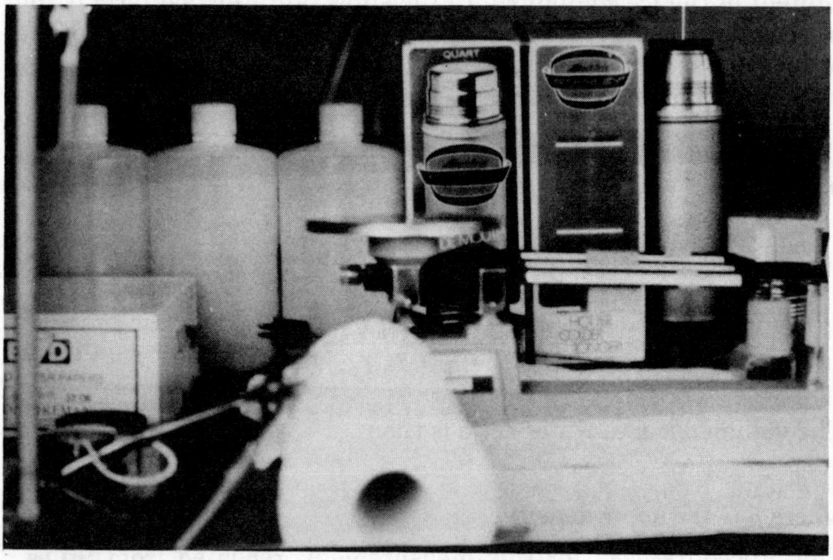

**Figure 12-5.** Equipment needed for performing the cold calorimeter measurement of snow free water.

sizes and by using several calorimeters. Typically, the equation for free water is solved by hand or with a calculator, although programmable calculators are starting to be used to speed up the process. Figure 12-6 compares the reflectivity of snow at 35 GHz with the free water content and surface temperature measured at Verona, New York in 1978.[6] It can be seen that there is a strong correlation between these two physical parameters and the reflectivity, indicating the importance of snow ground truth for radar applications.

## 12.4 SOIL AND CROP CHARACTERISTICS

Soil characteristics are of primary interest to remote sensing applications in order to assist in the remote assessment of crop yields, *et cetera*. However, ground reflectivity at millimeter waves has been found to be influenced by surface moisture on the ground or foliage, and thus, soil ground truth has recently been of interest for military applications.

Table 12-4 summarizes the most common soil parameters currently measured while Table 12-5 summarizes the primary measured crop parameters. Many of the parameters measured for soil are related to those for snow, and as in the case of snow, nearly all of the measurements are performed manually. The parameter of most interest for remote sensing applications is the field capacity of soil. The field capacity is defined as the per cent moisture remaining in the soil three days after a saturating rain. The field capacity can also be expressed as:

$$FC = A + B \cdot \textit{Percent Sand} + C \cdot \textit{Percent Clay} \qquad (12\text{-}10)$$

where

$FC$ = the field capacity

$A$, $B$, and $C$ = measured constants

Practically, the field capacity is calculated from the equation:

$$M_{FC} = 100 \times M_v / M_t \qquad (12\text{-}11)$$

where

$M_{FC}$ = percent field capacity

$M_v$ = volumetric soil moisture

$M_t$ = soil moisture at a tension of 1/3 mm of mercury

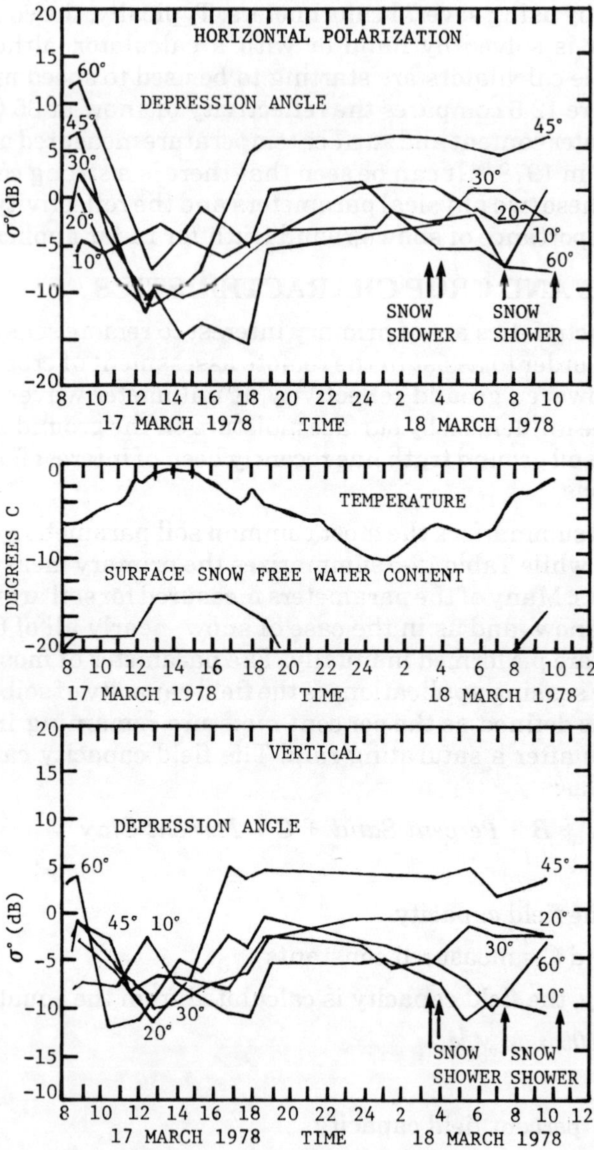


**Figure 12-6.** Comparison of snow reflectivity at 35 GHz with free water and air temperature (reprinted from Hayes, *et al.*, ref. [6]; ©1980 IEEE).

**TABLE 12-4**
**Soil Ground Truth Parameters**

| NO. | PARAMETER | CURRENTLY ACCEPTED PROCEDURE |
|---|---|---|
| 1 | Gravimetric Soil Moisture | Manual Technique Comparing Wet and Dried Soil by Weight |
| 2 | Soil Bulk Density | Measured Average of a Number of Soil Samples |
| 3 | Volumetric Soil Moisture | Calculated From Gravimetric Moisture and Density |
| 4 | Soil Tension | Mercury Manometer (Pressure Chamber) |
| 5 | Soil Roughness | Macro Roughness — Visual Observation<br>Micro Roughness — Photograph of Grid Driven into Soil |
| 6 | Field Capacity | Calculated from Volumetric Moisture and Soil Tension |

**TABLE 12-5**
**Vegetation Ground Truth Parameters**

| NO. | PARAMETER | CURRENTLY ACCEPTED PROCEDURE |
|---|---|---|
| 1 | Leaf Area Index | Manual Count of Leaves on Crops |
| 2 | Plant Wet and Dry Weight | Manual Measurement |
| 3 | Plant Density | Manual Measurement |
| 4 | Plant Growth Shape | Manual Observation |
| 5 | Leaf Water Potential | Mercury Manometer |
| 6 | Spectral Reflectance | Photometer |

The volumetric soil moisture is calculated from the gravimetric soil moisture and the soil density (assumed previously known). The tension is defined as the ability of the soil to hold water; it is determined by measuring the amount of water forced out of saturated soil as pressure above one atmosphere is applied.

Vegetation parameters such as leaf area index and plant moisture are determined by physically counting leaves or weighing plant material. To date, no feasible solution for separating surface and internal moisture has been devised.

## 12.5 SEA PARAMETERS

The backscattering properties of the sea are much more dynamic and dependent on weather parameters than land properties because the sea surface is not fixed but moves in response to wind and other atmospheric conditions. As a result, ground truth is an absolute necessity to the understanding of the backscattering properties of the sea. In general, the sea backscatter has been found to depend on the wind speed and direction, the wave height and direction, the presence of rain, and the presence of so-called anomalous propagation conditions. The measurement of these parameters is discussed below.

### 12.5.1 Atmospheric Parameters

Atmospheric parameters are measured by weather stations in a manner similar to that described in section 12.2, except that normally one likes to have one weather station near the ocean surface and another at 50 feet or higher. Large differences in the measured temperature and relative humidity may indicate the presence of anomalous propagation conditions which may result in "trapping" of radar waves in the vertical dimensions, thus resulting in the received power from the sea having a range dependence different from $R^{-3}$ as would normally be expected. The measurement of the slope of the return from the sea is a much better way of determining the presence of such conditions, as discussed in Chapter 11. For determining the wind speed and direction effects on sea backscatter, the weather station near the ocean surface should normally be used.

### 12.5.2 Wave Height

The wave height is most commonly measured for fixed installations by a wave gauge although other instruments such as pressure transducers have been used with limited success. The standard techniques for constructing a wave gauge involve the use of a vertical length of resistive wire which extends from several feet above the mean sea level to several feet below the mean sea level. When the ocean waves pass over the vertical wire, the salt water shorts out varying lengths of the wire, thus changing its resistance. This changing resistance is used to vary the frequency of an audio CW oscillator, thus producing an FM modulated waveform whose frequency is proportional to the water height passing the vertical wire.

**Figure 12-7.** View of the top portion of a typical resistive wire wave gauge (courtesy of the Naval Coastal Systems Center, Panama City, Florida).

This signal can then be converted to a slowly varying ac signal through the use of an FM discriminator. Once the frequency of the oscillator *versus* water height is calibrated, then the wave gauge gives a very accurate picture of the sea surface. Figure 12-7 gives a picture of the top of a typical wave gauge.

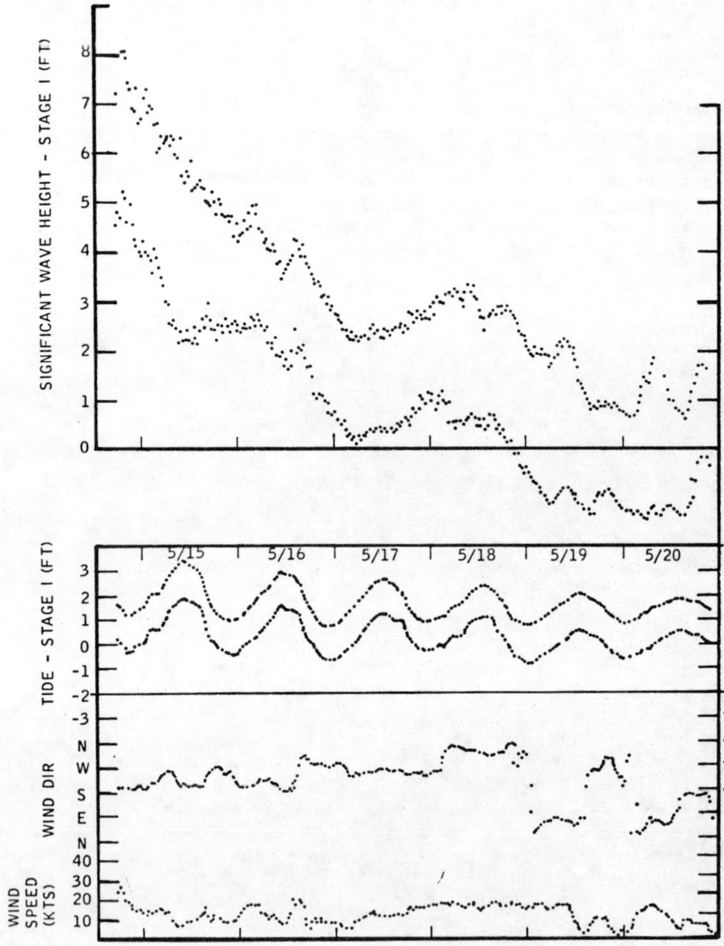


**Figure 12-8.** Plot of statistical summaries of wave data for the Gulf of Mexico (courtesy of the Naval Coastal Systems Center, Panama City, Florida).

The signal from the wave gauge can be processed via computer to produce a number of statistical parameters including *rms* wave height, significant wave height, standard deviation of the wave height, tide levels, and wave spectral content. Figure 12-8 gives a plot of typical parameters which were recorded on the Stage I tower at the Naval Coastal Systems Center at Panama City, Florida. If an array of four or more wave gauges is used, then a directional spectrum can be computed, as illustrated in Figure 12-9.

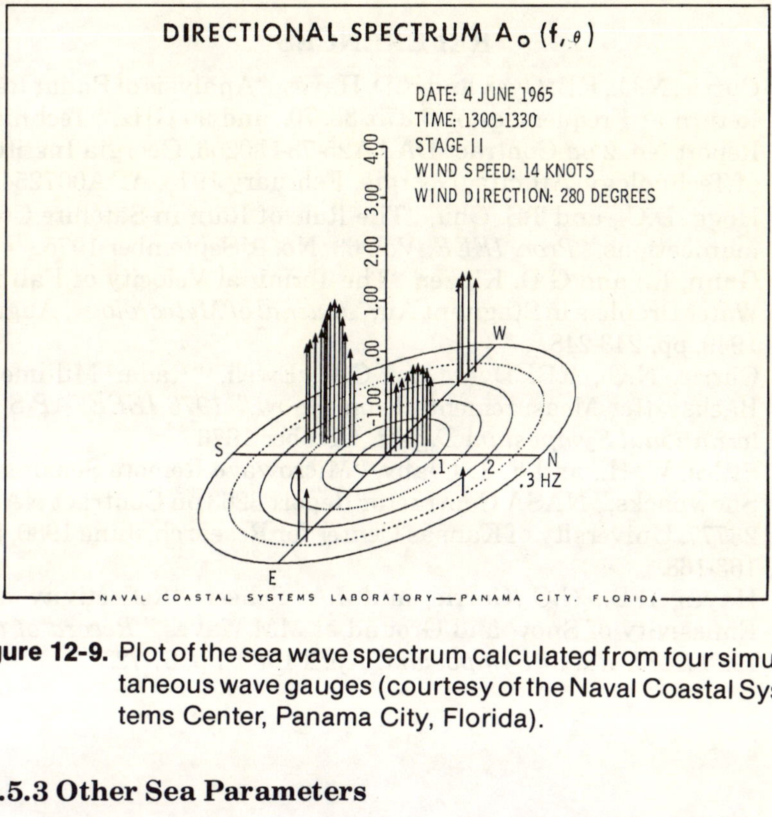


**Figure 12-9.** Plot of the sea wave spectrum calculated from four simultaneous wave gauges (courtesy of the Naval Coastal Systems Center, Panama City, Florida).

### 12.5.3 Other Sea Parameters

Other sea ground truth parameters are often collected in addition to those discussed above. These include wave pressure, both near the surface and on the bottom, water temperature, water conductivity, and current speeds at various depths. In general, these parameters have less correlation with radar backscatter than those discussed above.

## 12.6 FUTURE WORK

Due to increased interest in ground truth in recent years some work is being conducted to improve ground truth techniques. The thrust of the work is aimed at developing electrical, microwave, or optical techniques which allow rapid measurement of physical parameters without disturbing the medium and will decrease the amount of physical labor needed to achieve the desired results. This an area in which new and novel ideas are needed to improve the methods of data collection.

## REFERENCES

1. Currie, N.C., F.B. Dyer, and R.D. Hayes, "Analysis of Radar Rain Return at Frequencies of 9.375, 35, 70, and 95 GHz," Technical Report No. 2 on Contract DAAA25-73-C-0256, Georgia Institute of Technology, Atlanta, Georgia, February 1975, ADA007254.
2. Hogg, D.C., and T.S. Chu, "The Rule of Rain in Satellite Communications," *Proc. IEEE,* Vol. 63, No. 9, September 1975.
3. Gunn, R., and G.D. Kinzer, "The Terminal Velocity of Fall for Water Droplets in Stagnant Air," *Journal of Meteorology,* August 1949, pp. 243-248.
4. Currie, N.C., F.B. Dyer, and G.W. Ewell, "Radar Millimeter Backscatter Measurements From Snow," *1976 IEEE/AP-S International Symposium Digest,* October 1976.
5. Stiles, W. H., and F. T. Ulaby, "Microwave Remote Sensing of Snowpacks," NASA Contractor Report 3263 on Contract NA55-23777, University of Kansas Center for Research, June 1980, pp. 163-168.
6. Hayes, R.D., N.C. Currie, and J.A. Scheer, "Reflectivity and Emissivity of Snow and Ground at MM Waves," *Record of the 1980 IEEE Radar Symposium,* April 1980, pp. 67-72.

# CHAPTER 13
# STATISTICAL PROPERTIES OF DATA

**Maurice W. Long**
**Private Consultant**



## 13.1 OVERVIEW

This chapter includes discussions of the statistical properties of radar echoes. Specifically, material is included on the causes of echo fluctuations on the basic statistical principles used in radar echo analyses. And samples of actual radar clutter data that illustrate statistical variations in amplitude (short-term amplitude statistics), correlation functions, and frequency spectra are also included. Finally, a brief discussion is given on the statistical models that are ordinarily used for radar targets.

## 13.2 CAUSES AND RATES OF FLUCTUATION

Figure 13-1 shows short-term variations in the echo strength of a pulse, noncoherent, C-band (5 cm), dual polarization radar. Vertical polarization was transmitted, and signals were simultaneously received for both vertical (*V*) and horizontal (*H*) polarizations. The antenna was held stationary so that the antenna beam was searchlighted on a number of pine trees. From Figure 13-1, it can be seen that

a. the amplitudes of the signals seem to fluctuate in a more or less random manner as a function of time;
b. amplitude variations in excess of 10 dB occur frequently;
c. signal variations occur for both polarizations with little or no cross-correlation;
d. the rates of fluctuation of the radar echoes are proportional to windspeed; and
e. there are only a few amplitude reversals per second, i. e., the predominance of echo power is within the frequency range of zero and a few Hertz.

The causes for echo fluctuations will now be discussed. From Figure 13-1, since the rates of fluctuation of tree echo are proportional to windspeed, it is apparent that echo fluctuation is due to the relative motion of the scatterers (leaves, limbs, etc.) that are moved by the wind and are within a radar resolution cell. To help visualize this phenomenon, recall that tree element displacements caused by the wind may be comparable to one-half the radar wavelength or larger. Thus, if one thinks of the vectorial summation of the contributions from the numerous small elements (scatterers), wherein the phases are varying randomly due to the variations in path lengths between the radar and the tree elements, it is seen that relatively small displacements can cause dramatic amplitude changes in the echo. A list of factors that can cause echo fluctuations includes:

a. changes in multipath effects due to changes in the radar/target geometry due, for example, to an aircraft flight pattern;

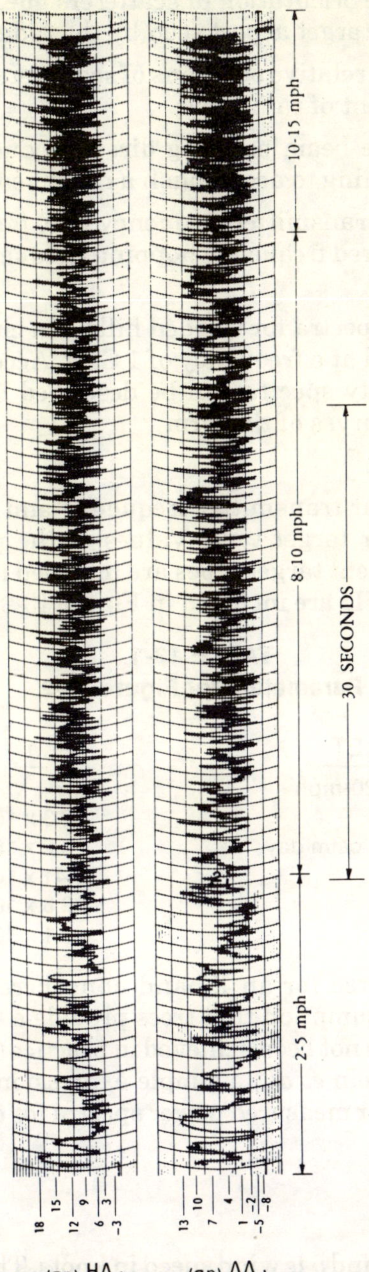


**Figure 13-1.** Pine tree echo for VV and VH polarizations (windspeed 2 - 15 mph) (reprinted from Hayes, *et al.*, ref. [1]).

b. changes in the orientation of scatterers due, for example, to a change in the target aspect seen by the radar;

c. changes in the relative positions of scatterers due, for example, to the movement of raindrops;

d. changes in the beam pointing direction due, for example, to antenna scanning to accomplish a radar search function; and

e. variations in transmitter frequency due, for example, to performing a desired frequency hopping (frequency agility) radar function.

Barlow[2] reported spectra for wooded hills, sea clutter, rain clouds, and chaff measured at a frequency of 1 GHz. According to Barlow, clutter power density spectra can be described by exponential or Gaussian-shaped curves of the form

$$P(f) = \exp\left[-a(f/f_o)^2\right] \tag{13-1}$$

where $f_o$ is the radar transmitter frequency and $a$ is a parameter dependent on radar target type. Values of the parameter $a$ that correspond to different target types are included in Table 13-1. Several spectra for 1 GHz are included in Figure 13-2.

**TABLE 13-1**
**Parameters for Figure 13-2**

| TARGET | $a$ | Curve in Figure 13-2 |
|---|---|---|
| Heavily wooded hills, 20-mph wind blowing | $2.3 \times 10^{17}$ | 1 |
| Sparsely wooded hills, calm day | $3.9 \times 10^{19}$ | 2 |
| Sea echo, windy day | $1.41 \times 10^{16}$ | 3 |
| Rain clouds | $2.8 \times 10^{15}$ | 4 |
| Window "jamming" | $1 \times 10^{16}$ | 5 |

Performance predicted for an X-band noncoherent pulse-Doppler radar, under the assumption that trees provide a Gaussian-shaped power spectrum, has not been achieved in practice (Fishbein *et al.*[3]). According to Fishbein *et al.*, a simple expression that gives good agreement with their measured power spectra for deciduous foliage is

$$P(f) = \frac{1}{1 + (f/f_c)^3} \tag{13-2}$$

where $f_c = 1.33e^{0.1356v}$ and $v$ is wind speed in knots. The measurements were made for horizontal polarization.

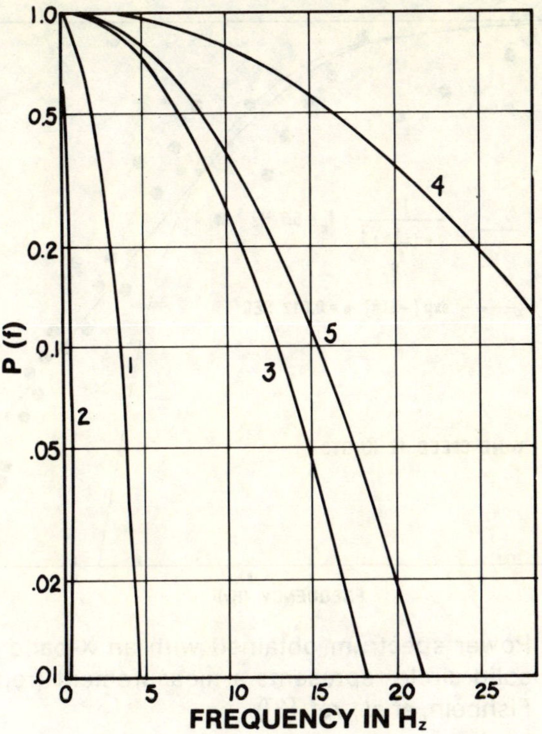


**Figure 13-2.** Frequency spectra of various types of complex targets at 1 GHz. See Table 13-2 to identify curves (reprinted from Barlow, ref. [2];© 1949 IEEE).

Figure 13-3 gives measurement results reported by Fishbein *et al.*[3] for trees in a 12-knot wind at X-band. The figure also includes a Gaussian-shaped spectra and one calculated with Equation (13-2). It is clear that the Gaussian curve drops off with frequency more rapidly at the higher frequencies than do the measured results. Currie *et al.*[4] observed that the spectral shapes of tree echo at 9.5, 16, 35, and 95 GHz all decreased slowly (consistent with Fishbein *et al.*) with increases in Doppler frequency.

For discussion on clutter data with phase coherent radar, *see* Blake,[5] Long,[6] and Nathanson.[7] The fluctuation spectra are roughly Gaussian in shape; but the center frequency can be either positive or negative depending on the translational velocities between transmitter, echo sources, and receiver.

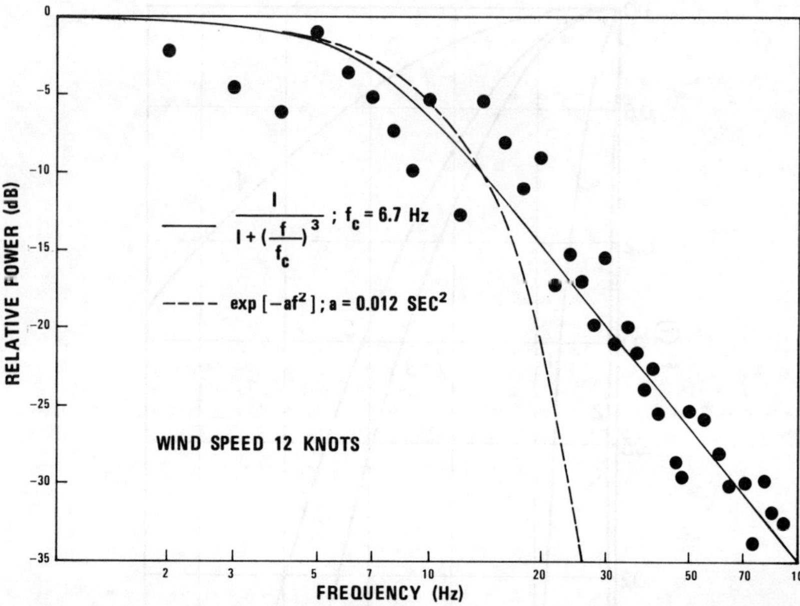


**Figure 13-3.** Power spectrum obtained with an X-band radar. Each solid circle represents a measurement (reprinted from Fishbein, *et al.*, ref. [3]).

## 13.3 BASICS OF STATISTICS

*Probability* is the expected fractional frequency of occurrence of a statistical event, and therefore probability is always positive and never exceeds unity. For example, assume that the probability that a variable $X$ lies between $X_1$ and $X_2$ is 0.2. Then,

$$Pr\{X_1 \leq X \leq X_2\} = 0.2$$

and $Pr\{X \text{ is any value}\} = \{\infty \geq X \geq -\infty\} = 1$.

The *probability density function* $W(X)$ is defined so that

$$W(X) = Pr\{X \text{ is between } X \text{ and } X + dX\}$$

Then

$$Pr\{X_1 \leq X \leq X_2\} = \int_{X_1}^{X_2} W(X)\,dX \tag{13-3}$$

and

$$\int_{-\infty}^{\infty} W(X)\,dX = 1 \tag{13-4}$$

The *mean* or, equivalently, the *average* of a variable $X$ is denoted by an overbar or brackets as $\overline{X}$ or $\langle X \rangle$ and can be obtained through use of $W(X)$ as follows:

$$\overline{X} = \langle X \rangle = \int_{-\infty}^{\infty} X\,W(X)\,dX \tag{13-5}$$

If $X$ is a time dependent function, a time average or running average over a time interval $\tau$ can of course be obtained as follows:

$$\overline{X} = \frac{1}{\tau} \int_{o}^{\tau} X(t)\,dt \tag{13-6}$$

The *variance* $S^2$ is a mathematical quantity used to provide a "feeling" for the range of fluctuations from the average or mean. Specifically $S^2$ is the average of the deviations (from the mean) squared and can be expressed as

$$S^2 = \langle (X-\overline{X})^2 \rangle = \int_{-\infty}^{\infty} (X - \overline{X})^2\,W(X)\,dX \tag{13-7}$$

The *standard deviation* $S$ is another mathematical quantity used to express deviation from the mean and is expressed as

$$S = \sqrt{variance}$$

An advantage of using $S$ instead of $S^2$ is that its dimensional unit is that of the variable. For example, if $X$ is in watts, $S$ is in watts but $S^2$ is in $(\text{watts})^2$.

The *median* $X_m$ of a variable $X$ is the value of $X$ for which $X$ is less than or is exceeded one-half of the time. Depending on the probability density function $W(X)$, the median $X_m$ can be less than, equal to or greater than the average $\overline{X}$. Then, the following relationships exist:

$$Pr\{X \geq X_m\} = Pr\{X \leq X_m\} = 0.5 = \int_{o}^{X_m} W(X)\,dX \tag{13-8}$$

The statistical properties of RCS data are frequently graphed as cumulative distributions either as a "less than" probability,

$$Pr\{\sigma > \sigma_1\} = Pr\{\sigma_1 \geq \sigma \geq 0\} = \int_{0}^{\sigma_1} W(\sigma)\,d\sigma \tag{13-9}$$

or as a "greater than" probability,

$$Pr\{\sigma > \sigma_1\} = Pr\{\infty \geq \sigma \geq \sigma_1\} = \int_{\sigma_1}^{\infty} W(\sigma)\,d\sigma \tag{13-10}$$

Probability of false alarm, PFA, is often used in radar to express the probability that an undesired signal exceeds a given level; therefore PFA due to clutter can usually be obtained from a "greater than" cumulative probability distribution. "Less than" and "greater than" distributions are interrelated as follows:

$$Pr\{\sigma < \sigma_1\} = 1 - Pr\{\sigma > \sigma_1\} \tag{13-11}$$

Figures 13-4, 13-5, and 13-6 are "less than" cumulative distributions, each with the statistical variable expressed in decibels and graphed

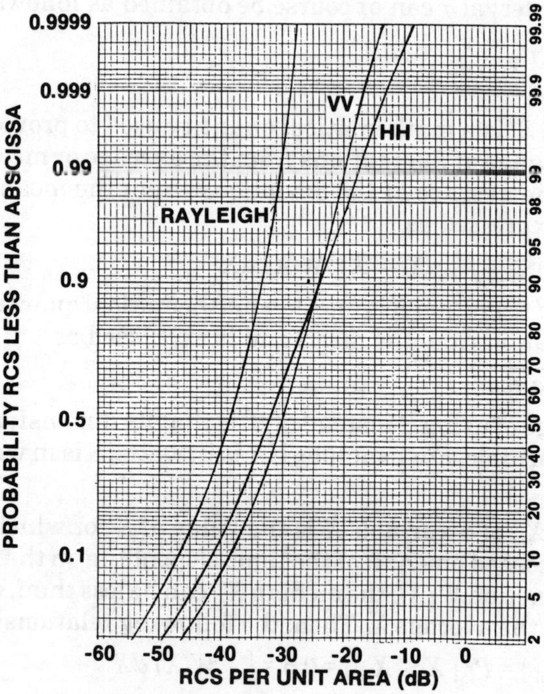


**Figure 13-4.** Cumulative distribution for X-band sea echo (reprinted from Rivers, ref. [8]).

along the abscissa. Figure 13-7 is a "greater than" cumulative distribution.

## 13.4 PROBABILITY DENSITY FUNCTIONS AND DISTRIBUTIONS

The well known bell-shaped probability curve is, specifically, the Gaussian or normal probability density function. This density function is used in many scientific disciplines and is the one ordinarily used to describe experimental errors. The four most commonly used distributions for describing RCS fluctuations are the Rayleigh, Ricean, lognormal, and the Weibull. Properties of the above-mentioned distributions are given later.

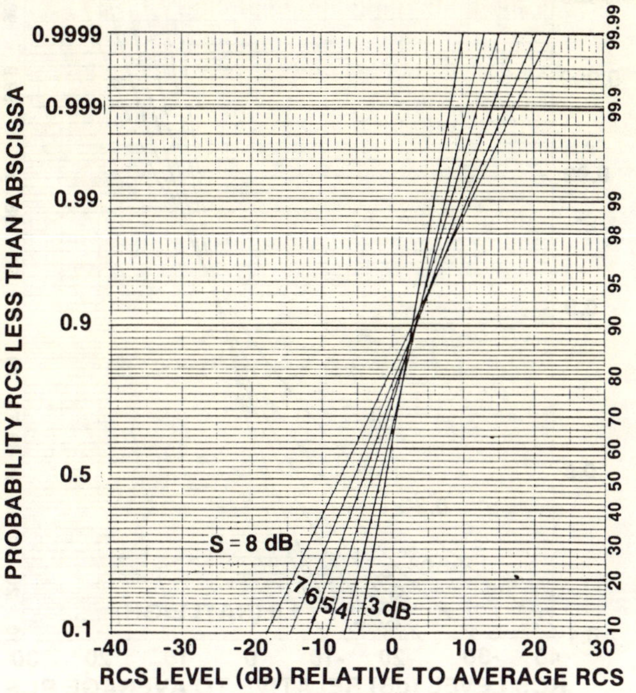


**Figure 13-5.** Lognormal cumulative distributions (reprinted from Rivers, ref. [8]).

### 13.4.1 Normal or Gaussian Distribution

The normal or Gaussian probability density function is

$$W(X) = \frac{1}{S\sqrt{2\pi}} \exp\left[-\frac{1}{2}\left(\frac{X-\overline{X}}{S}\right)^2\right] \qquad (13\text{-}12)$$

For a given $W(X)$, $S$ and $\overline{X}$ are constants, $S > 0$, and $X = \overline{X}_m$. Figure 13-8 shows a universal Gaussian probability density function with $S \cdot W(X)$ plotted *versus* $X/S$, when $\overline{X} = 0$. The Gaussian function has the special property that $< X^2 >$ equals $S^2$ if $\overline{X} = 0$.

For the Gaussian function, 68% of the data points are within plus or minus one standard deviation of the median, and 95% of the points are within plus or minus two standard deviations of the median.

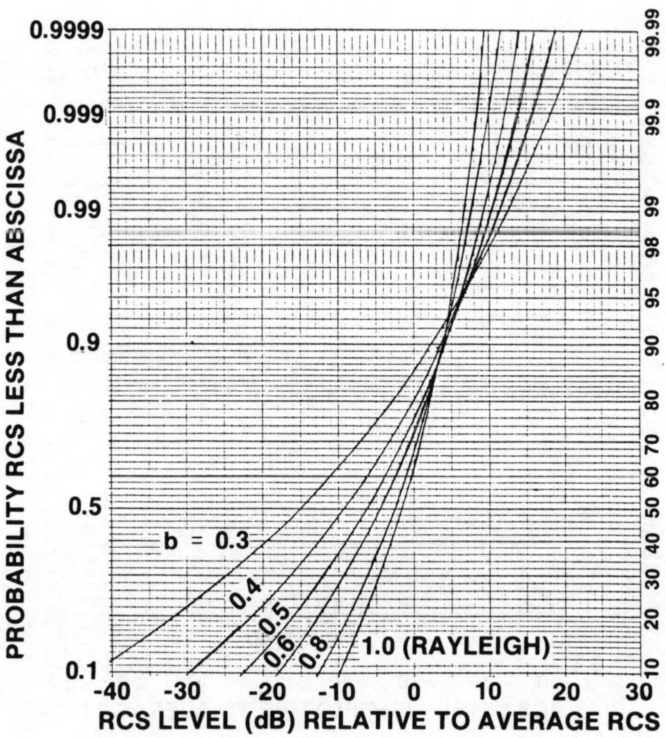


**Figure 13-6.** Weibull comulative distributions (reprinted from Rivers, ref. [8]).

### 13.4.2 Rayleigh Distribution

The Rayleigh density function is

$$W(X) = \frac{2X}{\alpha} \exp\left(-\frac{X^2}{\alpha}\right) \tag{13-13}$$

where $X \geq 0$, $\alpha > 0$, and $\alpha$ is a constant for a given $W(X)$. The Rayleigh density function describes the amplitude (magnitude of voltage) due to noise from a linear network or due to "Rayleigh" clutter, as described below.

The exponential or Rayleigh power density function characterizes power fluctuations and is different than the Rayleigh density function that characterizes amplitude fluctuations. Let $\sigma$ be proportional to power or RCS and therefore let $\sigma$ be proportional to $X^2$ of Equation (13-13). Then, it can be shown that

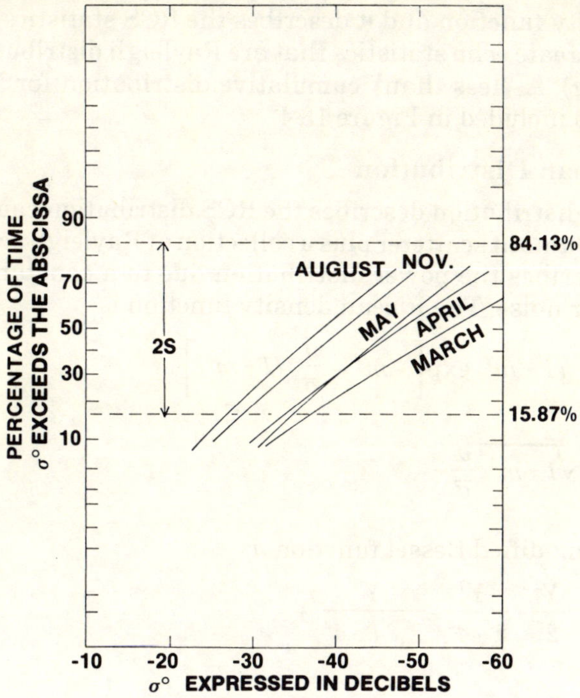


**Figure 13-7.** $\sigma^\circ$ for cultivated terrain at different times of the year, depression angle 1.25° (reprinted from Linell, ref. [1]).

$$W(\sigma) = \frac{1}{\overline{\sigma}} \exp\left(-\frac{\sigma}{\overline{\sigma}}\right) \tag{13-14}$$

where $\sigma > 0$ and, for a given $W(\sigma)$, $\overline{\sigma}$ is a positive constant and $\sigma_m = 0.693\,\overline{\sigma}$. A useful relationship between the average and the median for the exponential or Rayleigh power density function is

$$\overline{\sigma}(\text{in } dB) = \sigma_m(\text{in } dB) + 1.6\ dB \tag{13-15}$$

In the radar field, $W(\sigma)$ of Equation (13-14) is frequently called (although erroneously) a Rayleigh density function when, in fact, Equation (13-13) is the Rayleigh density function. This terminology stems from the fact that Equations (13-13) and (13-14) describe the amplitude (e.g., voltage) and power (or radar cross section), respectively, obtained from "Rayleigh" clutter — a large number (ensemble) of equal (or nearly equal) but randomly located scatterers.

Therefore, strictly speaking $W(\sigma)$ of Equation (13-14) is the Rayleigh *power* density function and it describes the RCS statistics for scatterers that create echo statistics that are Rayleigh distributed (Rayleigh clutter). A "less than" cumulative distribution for Rayleigh scatterers is included in Figure 13-4.

### 13.4.3 Ricean Distribution

The Ricean distribution describes the RCS distribution caused by a constant dominant scatterer plus a collection of Rayleigh scatterers. Also, it describes the power distribution due to a constant signal plus receiver noise. The Ricean density function is

$$W(\sigma) = \frac{1}{\bar{\sigma}} (1 + m^2) \exp\left[ - m^2 - \frac{\sigma}{\bar{\sigma}} (1 + m^2) \right] \tag{13-16}$$

$$\times I_o \quad 2 m \sqrt{1 + m^2 \frac{\sigma}{\bar{\sigma}}}$$

where $I_o$, a modified Bessel function, is

$$I_o(Y) = 1 + \frac{Y^2}{2^2} + \frac{Y^4}{2^2 \cdot 4^2} + \frac{Y^6}{2^2 \cdot 4^2 \cdot 4^2} + \ldots \tag{13-17}$$

and

$m^2$ = ratio of RCS of the constant echo to the average RCS of the random echo

In the absence of a constant echo, $m^2$ equals zero and Equation (13-16) reduces to the expression for the Rayleigh scatterers [Equation (13-14)]. For the limit of very large values of $m^2$ when the constant component completely dominates the echo, $W(\sigma)$ becomes sharply peaked and its average and median values are equal to the RCS corresponding to the constant component. Because of the complexity of Equation (13-16), calculations involving $W(\sigma)$ for intermediate values of $m^2$ are difficult to accomplish; fortunately, however, extensive numerical tables have been published on the cumulative distributions for the Ricean function (Norton, *et al.*[9]). The dependence on $m^2$ for both experimental data and calculated values is illustrated in Figure 13-9.

### 13.4.4 Lognormal Distribution

Lognormal describes a distribution of a variable whose logarithm has a density function that is normally distributed. In other words, if the frequency (probability) scale of a normal density function is maintained and its linear amplitude scale is exchanged for a logarithmic scale, the normal density function is replaced by the lognormal density function which follows:

$$W(\sigma) = \frac{1}{\sigma S\sqrt{2\pi}} \exp\left[\frac{-(\ln\sigma - \mu)^2}{2S^2}\right] \qquad (13\text{-}18)$$

where

$\mu = \overline{\ln \sigma}$

and

$S^2$ = variance of $\ln \sigma$

Other useful relationships for lognormal statistics include[6]

$$\bar{\sigma} = e^{\mu + S^2/2} \qquad (13\text{-}19)$$

$$\sigma_m = e^{\mu} \qquad (13\text{-}20)$$

$$\frac{\bar{\sigma}}{\sigma_m} = e^{S^2/2} \qquad (13\text{-}21)$$

$$\frac{\bar{\sigma}}{\sigma_m}\text{(in dB)} = 0.115\left[S\text{(in dB)}\right]^2 \qquad (13\text{-}22)$$

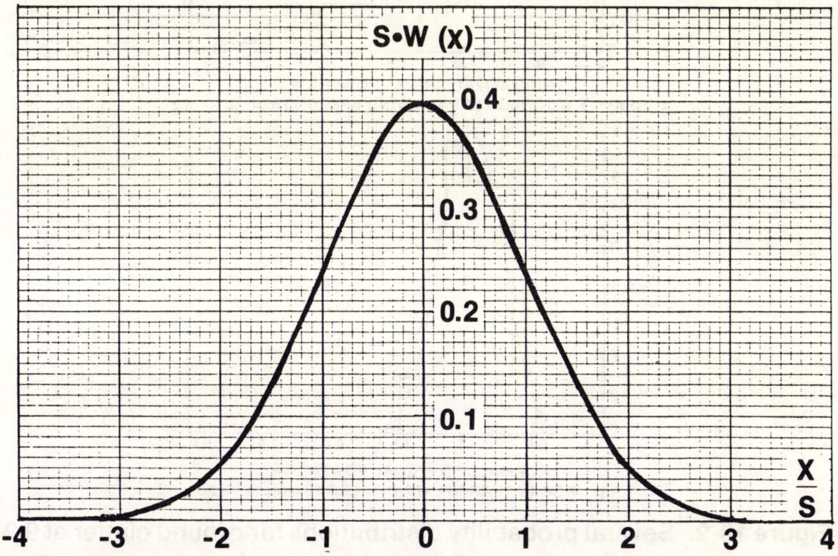


**Figure 13-8.** The Gaussian probability density function W(X). The curve has been made independent of S by plotting S·W (X) against X/S.

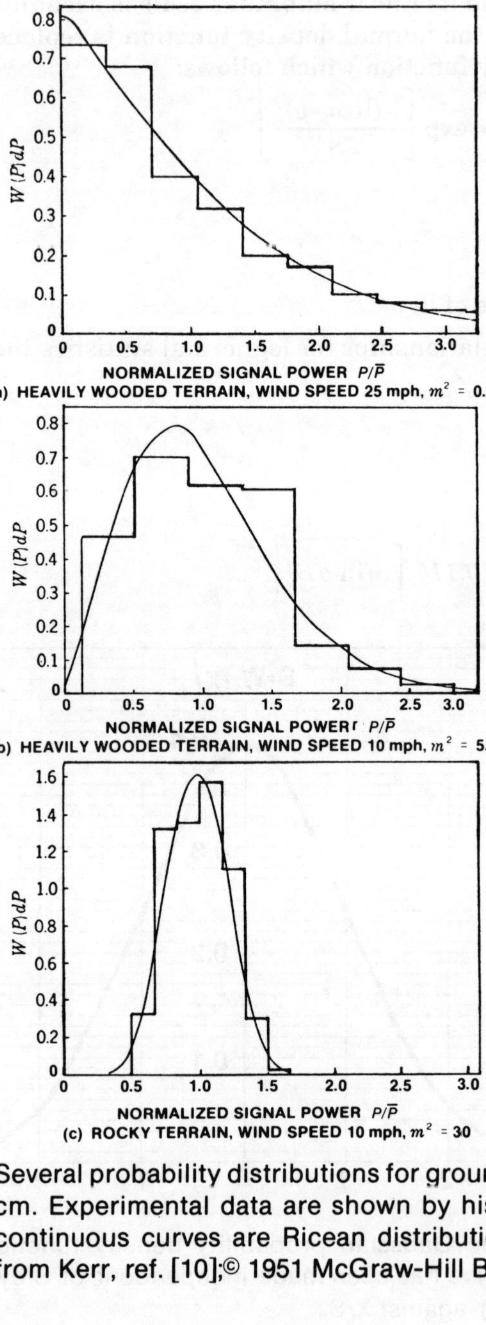


**Figure 13-9.** Several probability distributions for ground clutter at 9.2 cm. Experimental data are shown by histograms. The continuous curves are Ricean distributions (reprinted from Kerr, ref. [10];© 1951 McGraw-Hill Book Company).

Figure 13-5 includes cumulative distributions, drawn relative to average value, for lognormal density functions with standard deviation $S$ of 3, 4, 5, 6, 7, and 8 dB. The curves were reproduced from a research report by Rivers.[8] It should be noted that each cumulative distribution of Figure 13-5 is a straight line, consistent with the fact that normal probability graph paper is used and the amplitude scale (abscissa) is logarithmic instead of linear because the units are decibels.

The lognormal distribution is sometimes useful for describing clutter of the type that attains very large RCS values, e.g., 10 or 20 dB above its median, for much higher percentages of time than does clutter that obeys either Rayleigh or Ricean statistics. Two experimental sea clutter curves, marked *HH* and *VV*, and a Rayleigh cumulative distribution are included in Figure 13-4. It is apparent that the small RCS values of the experimental curves are best represented by the Rayleigh shape; however, the large RCS values are better represented by straight lines or lognormal distributions.

Figure 13-7 shows measured distributions *versus* time of year reported by Linell[11] for cultivated terrain. The data were obtained by receiving at various radar ranges and scanning the antenna beam in azimuth. Standard deviation *S(dB)* for the data can be obtained by using the fact that each dashed line is separated from the median by one standard deviation. Therefore, the *S(dB)* values of Figure 13-7 range from about 10 dB up to approximately 17 dB for the March data. These *S(dB)* values are considerably larger than are obtained when the time-dependent fluctuations of clutter, from a given radar resolution cell, are measured.

### 13.4.5 Weibull Distribution

The Weibull distribution has properties that lie between those of the Rayleigh and the lognormal distributions. Also, measured data (*see*, e.g., Figure 13-4) often have properties that lie between the two distributions. It is therefore not surprising that the Weibull distribution has been used, in recent years, with increasing frequency for describing clutter.

Robert R. Boothe was, it seems, the first person to use the Weibull function to describe radar clutter statistics. Boothe[12] noted that the curves in Figure 13-7, for example, can be described with Weibull statistics as well over the range of measured amplitudes as was previously done by Linell[11] with lognormal statistics.

The Weibull density function is

$$W(\sigma) = \frac{1}{\overline{\sigma}}\, b\sigma^{(b-1)} \exp\, \frac{-\sigma^b}{\overline{\sigma}} \tag{13-23}$$

It is of interest to notice that for $b$ equals unity, Equation (13-23) reduces to Equation (13-14). In other words, the Rayleigh function is a member of the Weibull family.

For the Weibull function, the ratio of average RCS $\overline{\sigma}$ to median RCS $\sigma_m$ is given (Schleher[10]) by

$$\frac{\overline{\sigma}}{\sigma_m} = \frac{\Gamma(1+1/b)}{(\ln 2)^{1/b}} \tag{13-24}$$

where

$\Gamma(1 + 1/b)$ = the Gamma function of the argument $(1 + 1/b)$

$\ln 2$ = the natural logarithm of 2

Figure 13-6 includes several Weibull cumulative distributions for various values of the parameter $b$ of Equation (13-23). It should be noted that each of these distributions is graphed with respect to its average or mean. These curves were taken from a research report by Rivers.[8]

To help illustrate certain differences between the lognormal and the Weibull functions, the reader should compare the $S$ = 8 dB curve of Figure 13-5 with the $b$ = 0.6 curve of Figure 13-6. First, recall that each curve is normalized to 0 dB average value. Then, notice that for

(a) probability less than 0.9 and probability greater than 0.1, i.e., for 80% of the time, the ranges of RCS values for the two curves are approximately equal; and yet

(b) the median for the Weibull is approximately 3 dB greater than the median for the lognormal curve.

It is also apparent that for the case of the two curves with equal averages and nearly equal ranges for 80% of the time, large RCS values are far more likely with a lognormal than with a Weibull distribution. For example, the $b$ = 0.6 curve exceeds the average by 14 dB with probability $10^{-4}$ (ordinate = 0.9999) but the $S$ = 8 dB curve exceeds the average by 14 dB with probability $4 \times 10^{-3}$ (ordinate equals 0.996). In other words, for this specific example the lognormal exceeds a given large value 40 times as often as does the Weibull distribution and yet the median of the Weibull exceeds that of the lognormal by 3 dB.

## 13.5 COMPARISONS OF AVERAGES WITH MEDIAN AND 0.9 CUMULATIVE PROBABILITY VALUES

Radar cross section data are usually reported in terms of an average value or a median value, but experimenters rarely report both values. However, radar designers often desire both values and, unfortunately, the median and the average values of a statistically varying parameter are interrelated through the sometimes quite unpredictable probability density function. Invariably, it seems, there are two prevailing motivating influences: experimenters prefer to measure medians because system non-linearities tend to have little effect on the median (the value which is exceeded half the time), but clutter analysts usually prefer averages because median RCS when normalized to illuminated surface area depends on surface area but average, normalized RCS for homogeneous clutter is independent of surface area. Therefore, after considering the various requirements for measurement accuracies and the end uses for the data, it is usually found that one must obtain detailed statistical data (distributions) on the echo amplitude variations. Then, an average value can be obtained by appropriately integrating the measured distribution.

The large range of values that can exist for the ratio of average to median RCS will now be illustrated. For large radar cells of resolution, the temporal variations of clutter are typically Rayleigh. Then, from Equation (13-15) the ratio $\overline{\sigma}/\sigma_m$ will be 1.6 dB. If a strong fixed echo were present, the distribution would be Ricean and $\overline{\sigma}/\sigma_m$ would be less than 1.6 dB but greater than zero dB. For short pulse marine radar, sea clutter frequently is observed to be approximately lognormal with a standard deviation of about 7 dB. Then, from Equation (13-22), $\overline{\sigma}/\sigma_m$ will be 5.5 dB. Even higher values of $\overline{\sigma}/\sigma_m$ can occur. Recall that the distributions in Figure 13-7 are approximately straight lines and therefore those curves appear to approximate lognormal distributions. The standard deviations are about 10 dB or larger and one of the distributions (for March) has a standard deviation of 17 dB. If that distribution were in fact lognormal with a standard deviation of 17 dB, $\overline{\sigma}/\sigma_m$ would equal 33 dB. For a 10 dB standard deviation, Equation (13-22) provides a value of 11.5 dB for $\overline{\sigma}/\sigma_m$. Therefore, even for a 10 dB standard deviation and a lognormal distribution, an average obtained from a measured median value with an erroneous assumption that the distribution were Rayleigh would be in error by a factor of 10.

Thus far it has been shown that the ratio of the average to the median value of a random variable is a quite sensitive function of the probability density function. Therefore, the ratio of average to median, instead of always being near unity as is sometimes assumed, can in fact be a sensitive indicator of the statistical properties of measured data.

Rivers[8] observed a very interesting and powerful relationship between the average RCS and the RCS at the 0.9 cumulative probability point for clutter, which is

$$\overline{\sigma} = (\sigma_{0.9} - 3.5\ dB) \pm 0.5\ dB \tag{13-25}$$

where

$$Pr\ \{\ \sigma < \sigma_{0.9}\ \} = 0.9 \tag{13-26}$$

Rivers found that this relationship held not only for the wide range of clutter characteristics that were available to him, but also held for theoretical Rayleigh, Weibull, and lognormal distributions having standard deviations ordinarily encountered with clutter data. Thus, a sometimes useful way of estimating the average value for clutter is to place it 3.5 dB below the 0.9 cumulative probability point. The reader can test and verify this relationship by reviewing the distributions included in Figures 13-7 and 13-9.

## 13.6 COMPARISONS OF SPATIAL AND TEMPORAL STATISTICS

Clutter is composed, generally, of distributed clutter and discrete clutter. Examples of discrete clutter sources include water towers, buildings, tree trunks, floating debris, and sea-wave prominences. Depending on circumstances, discrete clutter can of course be either undesirable clutter or targets of interest. Distributed clutter is due, for example, to tree foliage, grass, small waves, wind ripples, and rain.

Generally speaking, discrete clutter sources are "sprinkled" randomly throughout the distributed clutter. For the outer Boston environment, McEvoy[14,15] noted that at S-band sources of discrete clutter with RCS exceeding $10^3\ m^2$ were too numerous to count, that there were progressively fewer discretes for larger RCS values, and that there were actually a number of objects that provide RCS values above $10^6\ m^2$. These large RCS values are due to buildings, water towers, etc. The McEvoy studies suggest a ten-fold reduction in the large RCS density for rural terrain.

Clutter fluctuates both with spatial (range, azimuth, etc.) changes and with time in a given range-azimuth cell. Spatial and temporal distributions are generally different. For land, temporal distributions (fixed range-azimuth resolution cells) are usually Ricean, which includes the Rayleigh as one limit and a non-fluctuating echo as the other limit. However, for X-band frequencies (9 GHz) and higher, Currie *et al.*[4] have reported lognormal shapes and Hayes[16] reported that at 95 GHz the temporal distributions, although usually lognormal, are sometimes Weibull in shape. Spatial distributions for land usually range from the Rayleigh distribution to large standard deviation distributions that have been described both by the lognormal and the Weibull.

Since the sea surface varies temporally within a given range-azimuth cell and sea waves propagate through range-azimuth cells, temporal and spatial sea statistics are closely related. The temporal statistics are usually reported as Rayleigh for depression angles above five degrees and for large cell areas. For smaller angles and pulse lengths of 0.5 microsecond or less, the temporal statistics are reported with different shapes falling between the Rayleigh and the lognormal. Obviously, conditions that cause temporal sea echo variations with large standard deviations will also cause spatial sea echo variations with large standard deviations. Therefore, spatial sea echo distributions are expected to also lie between Rayleigh and lognormal.

## 13.7 AUTOCORRELATION FUNCTION AND POWER DENSITY SPECTRUM

There are two general descriptors for specifying the continuum in fluctuation frequency or rate that constitutes a radar signal. One of the descriptors is the well-known power density spectrum, i.e., a curve showing power level per unit bandwidth or power density *versus* frequency like Figure 13-2 or 13-3. The second and the lesser known of the descriptors is the autocorrelation function. This function is a curve that depicts the degree of the correlation in the strength of a signal with the strength of the same signal, when delayed, *versus* amount of delay.

Either of the two descriptors can be obtained mathematically from the other one through a Fourier transformation. Therefore, in a mathematical sense only one of the two descriptors is required for describing signal fluctuations. However, in practice, because of available instrumentation, processing means, or planned data use, one descriptor may be preferred over the other. For example, although the power density spectrum is used more often, the autocorrelation function has proved itself as a useful device for recognizing the existence of two or more independent scattering mechanisms. In fact, by means of an autocorrelator, separate scattering mechanisms that cause slow and fast fluctuation spectra have been observed for both land and sea clutter (Long[6]).

The autocorrelation function $R(\tau)$ of a stationary (that is, temporally homogeneous) function $X(t)$ is defined as

$$R(\tau) = \lim_{T \to \infty} \frac{1}{2T} \int_{-T}^{T} X(t) \cdot X(t+\tau)\, dt \tag{13-27}$$

In physical terms, $R(\tau)$ expresses the similarity between the echo signal $X(t)$ at time $t$ and at some later time $t+\tau$. After some reflection by the reader, it will be apparent that

a. for a slowly varying signal with predominantly low frequency components, $R(\tau)$ decreases slowly with $\tau$; and
b. for a rapidly varying signal with predominantly higher frequency components, $R(\tau)$ decreases very rapidly with $\tau$.

The power density spectrum

$$P(f) = \lim_{T \to \infty} \frac{1}{2T} \left| \int_{-T}^{T} X(t)\, e^{-j2\pi f\tau} dt \right|^2 \tag{13-28}$$

is the Fourier transform of the correlation function; that is,

$$P(f) = \int_{-\infty}^{\infty} R(\tau) \cdot e^{-j2\pi f\tau}\, d\tau \tag{13-29}$$

Hence, by inverse transformation,

$$P(f) = \int_{-\infty}^{\infty} P(f) \cdot e^{j2\pi f\tau}\, df \tag{13-30}$$

Therefore, the total power contained in the spectrum is equal to the autocorrelation coefficient at $\tau = 0$:

$$R(0) = \int_{-\infty}^{\infty} P(f)\, df \tag{13-31}$$

The inverse or reciprocal relationship between frequency and correlation is exhibited in a useful reciprocal relationship between half-power bandwidth and decorrelation time. First, let half-power bandwidth $f_{1/2}$ be the frequency at which the power density $P(f)$ is one-half its value at zero frequency. Now, let decorrelation time be the time delay $\tau_0$ at which the correlation function $R(\tau)$ is one-half its value for zero time delay. Then, following these definitions

$$P(f_{1/2}) = \frac{1}{2} P(0) \tag{13-32}$$

and

$$R(\tau_o) = \frac{1}{2} R(0) \qquad (13\text{-}33)$$

Since $P(f)$ and $R(\tau)$ are related through the Fourier transform, the exact relationship between $f_{1/2}$ and $\tau_o$ depends on the detailed shapes of $P(f)$ and $R(\tau)$. However, for the Rayleigh case of many equal scatterers, $P(f)$ and $R(\tau)$ are each Gaussian in shape, and it has been shown (Long,[6] p.41) that

$$f_{1/2} = \frac{0.22}{\tau_o} \qquad (13\text{-}34)$$

Figure 13-10 shows a correlation function drawn for the purpose of providing a curve that was "typical" of a large collection of sea clutter data. The actual data were receiver output signals recorded on magnetic tape. Let it be assumed that $X(t)$ of Equation (13-25) represents magnitude of voltage. Then, for $\tau$ equals zero,

$$R(0) = \langle X(t)^2 \rangle \qquad (13\text{-}35)$$

and therefore $R(0)$ is proportional to total (mean) power. Thus, since $R(\tau)$ for a very long time delay, i.e., $R(\infty)$, should approach zero providing no periodic components are present, the measurement

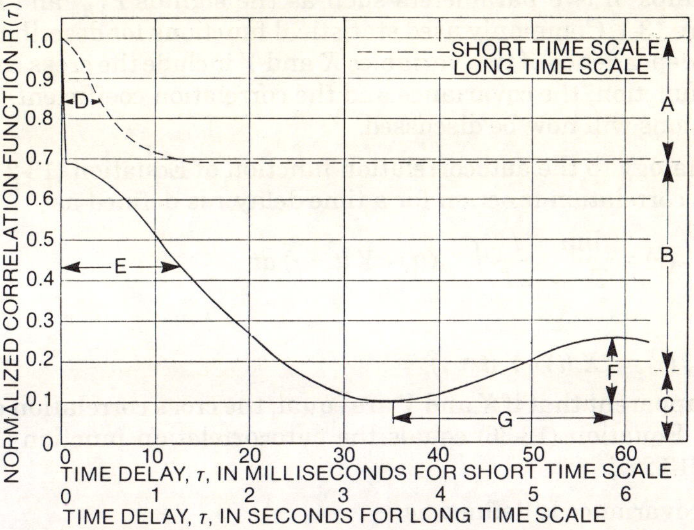


**Figure 13-10.** Illustration of the characteristics measured on correlation functions (reprinted from Long, ref. [17]; © 1974 IEEE).

$R(0)$ minus $R(\infty)$ is proportional to the randomly fluctuating component of power.

The following observations can now be made regarding the correlation function (curve) of Figure 13-10.

a. The curve has a constant offset of amplitude $C$.
b. The curve contains a periodic component with amplitude related to measurement $F$ and period related to measurement $G$.
c. The curve contains data on a random, high-frequency fluctuation with amplitude proportional to measurement $A$ and decorrelation time corresponding to measurement $D$.
d. The curve contains data on a random low-frequency fluctuation with amplitude proportional to measurement $B$ and decorrelation time corresponding to measurement $E$.

It is therefore apparent that the correlation function can serve as a useful device for recognizing the existence of two or more independent scattering mechanisms.

## 13.8 INTERDEPENDENCE OF TWO VARIABLES

It is sometimes desirable to express mathematically the interrelationships of two parameters such as the signals $P_{VH}$ and $P_{VV}$ of Figure 13-1. Commonly used statistical functions for describing the interdependence of two variables $X$ and $Y$ include the cross correlation function, the covariance and the correlation coefficient. These functions will now be discussed.

In analogy to the autocorrelation function of Equation (13-27), the cross correlation function for a time delay $\tau$ is defined as

$$R_{XY}(\tau) = \lim_{T \to \infty} \frac{1}{2T} \int_{-T}^{T} X(t) \cdot Y(t+\tau)\, dt$$

or (13-36)

$$R_{XY}(\tau) = \langle X(t) \cdot Y(t+\tau) \rangle$$

It is apparent that if $X$ and $Y$ are equal, the cross correlation coefficient Equation (13-36) equals the autocorrelation function [Equation (13-27)].

The covariance is defined as:

$$\text{cov}(X, Y) = \langle XY \rangle - \langle X \rangle \langle Y \rangle \qquad (13\text{-}37)$$

If $X$ is not a function of $Y$, i.e., if $X$ and $Y$ are independent of one another, the covariance of $X$ and $Y$ is zero. If the covariance is positive, $X$ and $Y$ vary in harmony; if the covariance is negative, $X$

and $Y$ vary in opposition. The reader is cautioned that although the covariance is zero if $X$ and $Y$ are independent, the covariance can in fact be zero if $X$ and $Y$ are dependent. For example, if $X$ equals sin $\Theta$ and $Y$ equals cos $\Theta$, $X$ and $Y$ are dependent and yet the covariance of $X$ and $Y$ is zero because $< \sin\Theta >$, $< \cos\Theta >$ and $< \sin\Theta \cos\Theta >$ are each zero.

A relationship between the covariance and the cross correlation function will now be addressed by returning to Equation (13-36). If $X$ and $Y$ are completely unrelated, i.e., statistically independent, $R_{XY}$ equals $<X><Y>$. Such as condition of independence is expected for two variables that have been displaced in time by a large value of $\tau$. Thus, except for situations where $X$ and $Y$ contain periodicities, $R_{xy}$ in the limit of $\tau$ approaching infinity is expected to be $<X,Y>$. Now notice from Equation (13-36) that $R_{XY}$ equals $<XY>$ for $\tau$ equals zero. Therefore, it is now apparent from Equation (13-37) that the covariance equals the difference between the cross correlation function for $\tau$ of zero and for $\tau$ approaching infinity.

The correlation coefficient is defined as

$$\rho_{XY} = \frac{<XY> - <X><Y>}{\sqrt{(<X^2> - <X>^2)(<Y^2> - <Y>^2)}} \tag{13-38}$$

It is related to the covariance by a normalization function that forces its magnitude to be unity if $X$ and $Y$ are linearly related. Specifically, the correlation function of $X$ and $Y$ equals the covariance of $X$ and $Y$ divided by the standard deviation of $X$ times the standard deviation of $Y$. In summary,

(a) $\rho_{XY}$ equals zero if $X$ and $Y$ are independent,
(b) $X$ and $Y$ vary in harmony if $\rho_{XY}$ is positive,
(c) $X$ and $Y$ vary in opposition if $\rho_{XY}$ is negative, and
(d) $X$ and $Y$ are linearly dependent if the magnitude of $\rho_{XY}$ is unity.

Finally, a cautionary note on correlation that has already been underscored for covariance, i.e., it is not necessarily true that $X$ and $Y$ are statistically independent if $X$ and $Y$ are uncorrelated. However, if $X$ and $Y$ are independent, the variables $X$ and $Y$ are uncorrelated.

## 13.9 TARGET MODELS

Statistical models are also used for describing the fluctuations in target RCS and, as with clutter, target statistics are difficult or impossible to describe with certainty. For a comprehensive introduction to the statistics of target fluctuations, the reader should see Blake[5], pp. 69-78.

Curves for determining detection probability and probability of false alarm versus the ratio of average target power to average noise power have been available for some years for non-fluctuating targets and for the four best known target models which are called Swerling models, cases 1 through 4 (*see*, e.g., Blake,[5] Nathanson,[7] Skolnik.[18] Two of the Swerling models are for "slow" fluctuations and two are for "fast" fluctuations. The definitions of slow and fast relate to radar pulse repetition frequency and antenna scan rate. If the target echo varies from scan-to-scan but is essentially constant from pulse-to-pulse over the time required for the antenna to scan across the target, the target fluctuation is called slow. On the other hand, if the target echo fluctuates so that there is little correlation in the signal from pulse-to-pulse, the target fluctuation is called fast. If target echo is constant from scan-to-scan, as is true for a sphere, the echo is sometimes called Swerling zero.

The most frequently used Swerling models are cases 1 and 3, which are for slow fluctuations. For cases 1 and 2 the target RCS statistics are assumed to be Rayleigh [Equation (13-34)] and the probability density function for cases 3 and 4 is

$$W(\sigma) = \frac{4}{\overline{\sigma}^2} \quad \exp\left( - \frac{2}{\overline{\sigma}} \right) \tag{13-39}$$

According to L. W. Brooks (Blake,[5] p. 76), the Swerling cases 3 and 4 [Equation (13-39)] behave like a Ricean for which the coherent-to-random power ratio is approximately 3 dB.

The statistical properties of some targets are more spiky, i.e., large RCS values occur with higher probability, than are predicted by the Swerling models. Examples of target shapes that can cause spiky echo include flat plates, cylinders, and cones. Radar antennas when seen as targets are also known to create spiky echoes. For simplicity, as in the case of clutter, the lognormal distribution is sometimes used for approximating the RCS of spiky targets.

The chi-square family of statistics has been widely used in target and clutter modeling.[19] Actually, the Swerling models and a target set called the Weinstock models are both members of the chi-square family. Weibull statistics have also been used for target modeling.

## 13.10 SUMMARY AND DISCUSSION

Radar echo is caused by the vectorial summation of electromagnetic fields that are caused by re-radiation from objects, i.e., scatterers, that are illuminated by the radar. Scatterers that are either large or small compared to the radar wavelength can contribute significantly to the total echo at a given time or from a given area in space. Changes in echo, i.e., fluctuations, can be caused by changes in the positions or in the orientations of the scatterers or by changes that cause different scatterers to be viewed by the radar.

Clutter sources are composed of a continuum of scatterer types, from point sources, i.e., discrete clutter, at one extreme to diffuse scatterers, i.e., distributed clutter, at the other extreme. Examples of discrete clutter sources include water towers, buildings, tree trunks, floating debris, sea-wave prominences. Distributed clutter is caused by a large number of separate scatterers such as tree canopy, grass, small sea waves, wind ripples, rain. Depending on the radar application, an echo source can of course be either undesirable clutter or a target of interest.

The echo strength of a target or clutter usually varies over ranges and at rates that are wide and unpredictable. The rates of amplitude change are usually described by frequency spectra which have shapes that depend on the particular collection of scatterers being viewed. For a large number of moving scatterers of nearly equal size, the theoretical spectrum in terms of power density has the shape of a Gaussian curve. Experience has shown that echoes due to trees generate spectral shapes that do not drop off with increased Doppler frequency as fast as the Gaussian shape does. For a coherent Doppler spectrum, the spectrum center depends on the velocities between transmitter, echo sources, and receiver.

Because of advantages in measurement techniques or in convenience of analysis, rates of fluctuation are sometimes described by the autocorrelation function of a signal instead of the frequency spectrum. The information available is, in principle, equivalent because the autocorrelation function and the frequency spectrum are interrelated by the Fourier transform.

The range of amplitudes, i.e., the percentages of occurrences of radar echo at various amplitude levels, is usually described by a cumulative statistical distribution. Distributions that are commonly used to describe the amplitude variations of echo from targets or clutter are the Rayleigh, Ricean, lognormal, and Weibull. The Rayleigh function describes the voltage variations caused by thermal noise or by changes in the relative positions of a large number of scatterers of nearly equal magnitude. The power distribution, i.e., a distribution proportional to the distribution of voltage squared, caused by fluctuations that obey Rayleigh statistics has an exponential density function. This function is sometimes called the Rayleigh power density function. The Ricean distribution is a well-known statistic for describing a voltage that fluctuates about a constant value; it is caused by the vector sum of a constant signal plus a Rayleigh fluctuation (e.g., noise). Lognormal and Weibull functions are both useful for describing a target or clutter when relatively large signals occur with a much higher probability of occurrence than results from a Rayleigh fluctuation.

Clutter, as seen from a particular fixed position, varies both in time from a given range-azimuth cell and spatially. Spatial and temporal distributions are generally different. For land, temporal distributions (observed spatial grid or footprint fixed) are generally considered to be Ricean, which includes the Rayleigh as one limit and a non-fluctuating echo as the other limit. However, for X-band frequencies (9 GHz) and higher, lognormal shapes have been reported and at 95 GHz the temporal distributions, although usually lognormal, have been reported as sometimes being Weibull in shape. Spatial distributions for land usually range from the Rayleigh to large standard deviation distributions that have been described both by lognormal and Weibull statistics.

Since sea waves propagate through a given spatial grid, temporal and spatial radar sea statistics are closely related. As in the case of land, the data on temporal sea statistics are extensive. Temporal variations are usually reported as Rayleigh for depression angles of five degrees or more, especially if the radar cell areas are large. For smaller angles and cell sizes, the temporal distributions have shapes that lie between the Rayleigh and the lognormal. Obviously, conditions that cause temporal sea echo variations that have large standard deviations will also cause spatial sea echo variations that have large standard deviations. Therefore, depending on sea surface details, spatial sea echo distributions are expected to lie between Rayleigh and lognormal. As for ground clutter, another set of useful statistics for sea clutter is the Weibull family.

A number of clutter descriptors, including the average RCS value, are generally needed for systems analyses. However, because of system nonlinearities and other operational features, the average value of RCS is not always readily available. An interesting and useful relationship has been noted from sea clutter observations: ordinarily, the average value of RCS for clutter is 3.5 dB ±0.5 dB below the 0.9 probability point.

There are perhaps as many statistical models available from the literature for targets as there are for clutter. In fact, Rayleigh, Ricean, lognormal, and Weibull statistics have each proved to be useful for describing the fluctuations in signal amplitude due to both targets and clutter.

## REFERENCES

1. Hayes, R. D., *et al.*, "Study of Polarization Characteristics of Radar Targets,"Final Report on Contract DA-36-039SC-64713, Georgia Institute of Technology, Atlanta, Georgia, October 1958.
2. Barlow, E. J., "Doppler Radar," *Proc. IRE*, vol. 37, April 1949, pp. 340-355.
3. Fishbein, W., S. W. Graveline, and O. E. Rittenbach, "Clutter Attention Analysis," U.S. Army Electronics Command, Technical Report ECOM-2808, March 1967.
4. Currie, N. C., F. B. Dyer, and R. D. Hayes, "Radar Land Clutter Measurements at Frequencies of 9.5, 16, 35, and 95 GHz," Technical Report No. 3, Contract DAAA25-73-C-0256, Georgia Institute of Technology, Atlanta, Georgia, April 1975.
5. Blake, L. V., *Radar Range-Performance Analysis*, D. C. Heath and Company, Lexington, Massachusetts, 1980.
6. Long, M. W., *Radar Reflectivity of Land and Sea*, Artech House, Dedham, Massachusetts, 1983.
7. Nathanson, F. E., *Radar Design Principles*, McGraw-Hill Book Company, New York, 1969.
8. Rivers, W. K., "Low-Angle Radar Sea Return at 3-mm Wavelength," Final Report on Contract N62269-70-C-0489, Georgia Institute of Technology, Atlanta, Georgia, November 1970.
9. Norton, K. A., *et al.*, "The Probability Distribution of the Amplitude of a Constant Vector Plus a Rayleigh - Distributed Vector," *Proc. IRE*, vol. 43, October 1955, pp. 1354 - 1361.
10. Kerr, D. E., *Propagation of Short Radio Waves*, McGraw-Hill Book Company, New York, 1951.
11. Linell, T., "An Experimental Investigation of the Amplitude Distribution of Radar Terrain Return," 6th Conference of the Swedish National Committee on Scientific Radio, March 13, 1963, English translation dated October 1966.

12. Boothe, R. R., "The Weibull Distribution Applied to the Ground Clutter Backscatter Coefficient," Technical Report RE-69-15, U.S. Army Missile Command, Redstone Arsenal, Alabama, June 1969, AD 691109.

13. Schleher, D. C., "Radar Detection in Weibull Clutter," *IEEE Transactions on Aerospace and Electronic Systems*, November 1976, pp. 736-743; July 1977, p. 435.

14. McEvoy, W. J., "Discrete Clutter Measurements Program: Operations in Western Massachusetts," Mitre Report MTR-2084, March 1972, AD 742297.

15. McEvoy, W. J., "Discrete Clutter Measurements Program: Operations in the Metropolitan Boston Area," Mitre Report MTR-2085, March 1972, AD 742298.

16. Hayes, R. D., "95 GHz Pulsed Radar Return from Trees," *EASCON '79 Conference Record*, October 1979, pp. 353-356.

17. Long, M. W., "On a Two-Scatterer Theory of Sea Echo," *IEEE Transactions on Antennas and Propagation*, September 1974, pp. 667-672.

18. Skolnik, M. I., *Introduction to Radar Systems*, McGraw-Hill Book Company, New York, 1980.

19. Meyer, D. P. and H. A. Mayer, *Radar Target Detection*, pp. 64-68, Academic Press, New York, 1973.

# CHAPTER 14
# RADAR ANGLE MEASUREMENTS

**Joseph. A. Bruder**

**Engineering Experiment Station**
**Georgia Institute of Technology**

## 14.1 OVERVIEW

In the measurement of target and terrain cross section it is frequently important to measure the angular positions of the scatterers in the beam. This is especially important in the case of complex targets to enable the determination of the angle errors (called glint) of a target and to be able to calculate amplitude scintillation associated with the complex target. In the measurement of terrain backscatter, measurement of the angular position of the scatterer can enable true mapping of nonhomogeneous terrain features in the antenna beam. The measurement of angular errors themselves are important, and may be associated with RCS measurement programs as well. In the case of dynamic cross section measurements it may be necessary to angle track the target to keep the antenna boresight fixed on the target.

This chapter describes the basic types of angle-sensing radars which can be used to perform angle measurements, and includes their limitations and applications. The contaminants that can affect the measurement of true angular information with respect to the target or terrain are also discussed. The possible error sources in the angle measurement process are described, and some of the effects and possible solutions are outlined. Thus, a summary of angle tracking techniques and phenomena is given for those persons interested in performing angle measurements of targets and clutter.

## 14.2 PROPERTIES OF TARGET ANGLE RETURNS

The function of an angle sensing radar is to interpret the received signals at the radar aperture and provide an error voltage which defines the apparent direction of the target from antenna boresight and the magnitude of its angular distance from boresight. The ability of a radar to accurately perform this function is strongly affected by the nature of the target, the propagation environment, and the radar characteristics.

Unfortunately, many actual radar targets such as aircraft consist of multiple scatterers separated in angle, and as a result, the scatterers add in and out of phase causing amplitude and angular scintillation (glint). Because of the complex nature of real targets, the measured cross section of the target can vary as much as 40 dB with only a small change in aspect angle. Glint caused by the complex scatterers can cause the apparent centroid of the target returns to wander about the total extent of the target, and at times will even be outside the target boundaries. An example of glint measured by J. H. Dunn[1] is shown in Figure 14-1. If one considers a simple two point scatterer, when the returns from two nearly equal scatterers approach 180° in-phase, the resultant amplitude return is minimized while the apparent angular measurement error due to glint is maximized, as shown by D. D. Howard[2] in Figure 14-2. The correlation between amplitude scintillation and glint becomes complicated for a complex target, and as a result the glint becomes erratic in nature.

The propagation environment also affects the ability of the radar to accurately measure the angular direction to the target. These propagation effects include path loss, multipath, refraction, and ionospheric effects. The most important of these is multipath for most radar installations, particularly those situated in an over the sea environment, or those radars required to track low angle targets. Multipath can result in biases which cause the boresight to deviate drastically from the target location, or may reduce the S/N ratio. The loss of signal amplitude can introduce angle errors, since the target signal-to-noise ratio is a strong factor in the radar's ability to accurately track the target.

## 14.3 ANGLE SENSING METHODS

The determination of angle to a target within the antenna beam can be performed by a number of radar measurement techniques. Operational radars are often required to furnish target angle bearing data, but this information is limited to a knowledge that the target is within the beam. In order to provide information to cause the antenna servo motors to drive the antenna boresight to the target centroid, the radar must provide a sense of where the target is within the antenna beam: left, right, up, or down from the antenna boresight. Also for servo stability requirements, it is necessary to provide angle error signals which are approximately linearly related to the angle of the target from boresight.

A number of angle sensing techniques have been developed, starting with sequential lobe type radars, and evolving to a conical-scan radar (SCR-584) (developed during World War II by the MIT Radiation Laboratory). This led eventually to the development of monopulse angle sensing. Monopulse radars have the ability to provide angle sensing on a single pulse basis, which eliminates the deleterious effects of amplitude scintillation on conical scan and sequential lobe radars. Another method of providing angle sensing is performed by measuring the Doppler component of the return as a

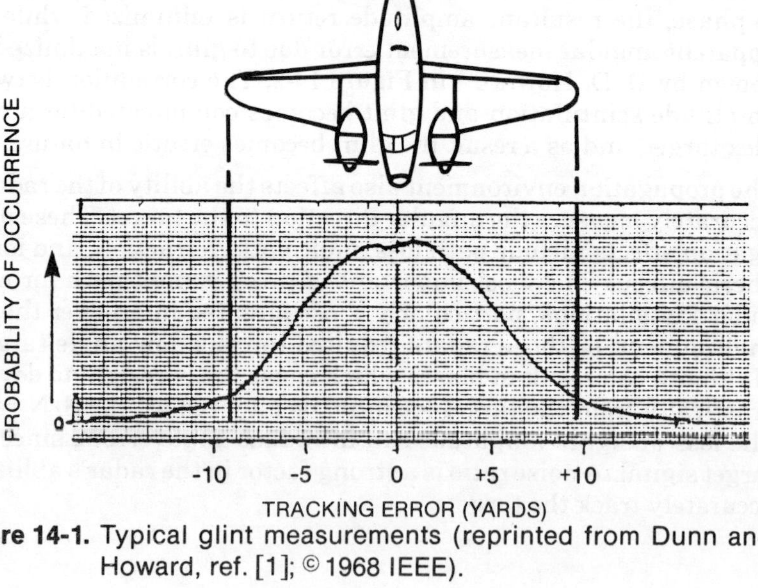


**Figure 14-1.** Typical glint measurements (reprinted from Dunn and Howard, ref. [1]; © 1968 IEEE).

function of vehicle motion (Synthetic Aperature Radar techniques) or of rotation of the observed target ($d\phi/d\psi$). Unlike the other methods which require multiple beam positions in space, Doppler techniques require only a fixed pencil beam and a rotating target.

### 14.3.1 Conical Scan

Conical scan provides angle sensing by rotating an offset pencil beam about the antenna boresight. The conical scan radar shown in the block diagram of Figure 14-3 is basically the simplest type of angle sensing radar. The feed of the conical scan antenna is offset so that the pencil beam is squinted off boresight. By rotating or nutating the feed, the pencil beam will rotate the squinted pencil beam around the antenna boresight. Alternatively, some conical scan radars rotate a subreflector that is tilted with respect to the plane of the antenna. Regardless of the actual implementation, a mechanical means is used to rotate the pencil beam about the boresight of the antenna. If a steady target is located on the antenna boresight, no modulation will be present on the received video output. However, if a steady target is located off boresight, a sinusoidal modulation of the received video will be given by the equation.[3]

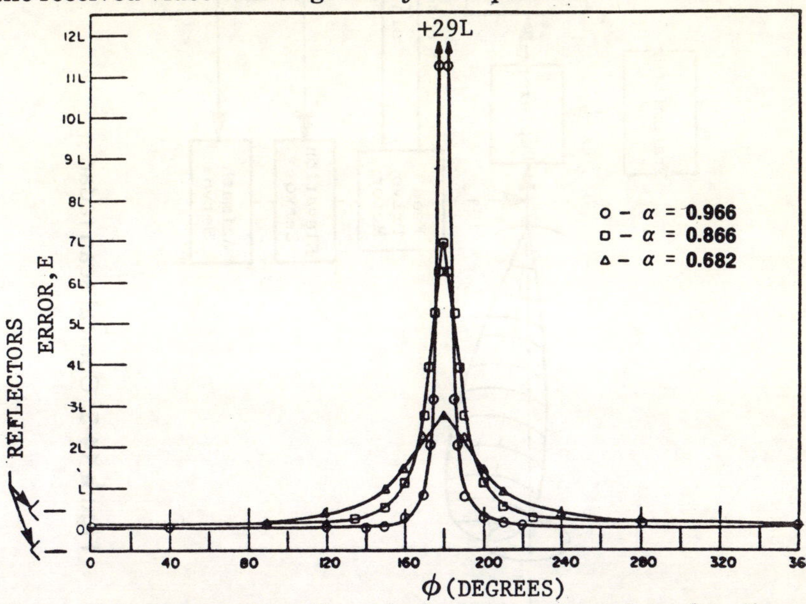


**Figure 14-2.** Apparent location of a dual source target as a function of relative phase for three values of relative amplitude, *a*, where the reflector separation length is *L* (reprinted from Howard, ref. [2]).

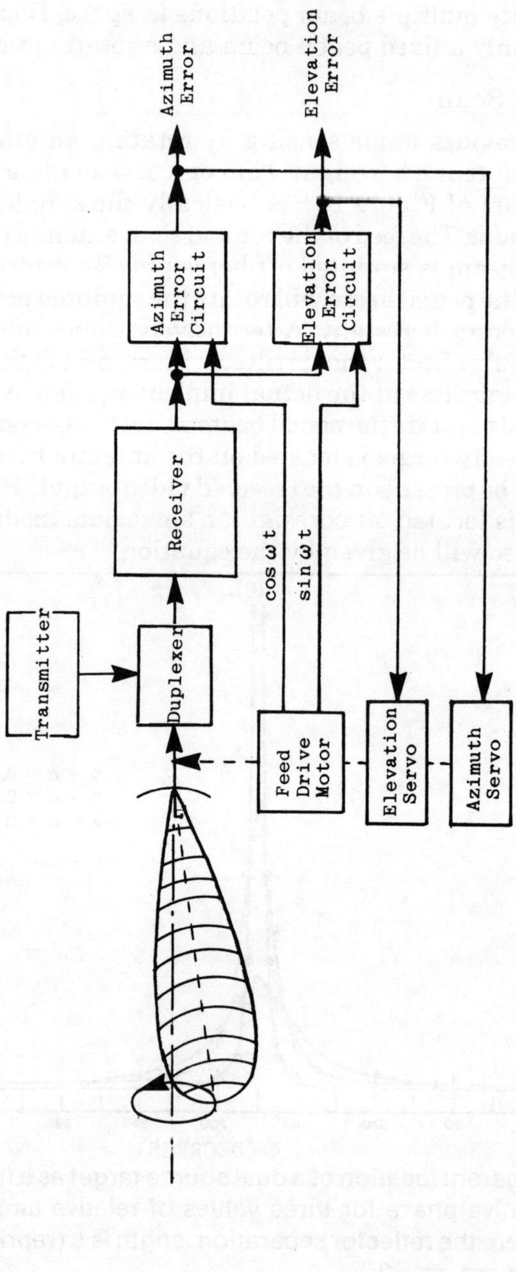


**Figure 14-3.** Conical scan radar.

$$E_t = E_o [ 1 + K_s \epsilon \cos ( \omega_s t + \phi ) ] \quad (14\text{-}1)$$

where

$E_o$ = the average magnitude of the received signal

$\epsilon$ = the angular distance of the target from boresight

$K_s$ = the error slope (determined by the antenna beam squint angle from boresight)

$\omega_s$ = the antenna scan frequency

$\phi$ = phase angle of the return modulation with respect to the scan frequency (determined by the direction of the target off boresight)

This equation can be rewritten in the form[3]

$$E_t = E_o ( 1 + K_s \epsilon_t \cos \omega_s t + K_s \epsilon_e \sin \omega_s t ) \quad (14\text{-}2)$$

where

$\epsilon_t$ = the transverse error component

$\epsilon_e$ = the elevation error component

The transverse and elevation error components are then coupled to the transverse and elevation motors which drive the antenna boresight to the target location.

### 14.3.2 Sequential Lobe

Sequential lobing radars provide angle sensing by sequencing between multiple off set beam locations. Sequential lobing is similar to conical scan except that it employs multiple fixed-squinted beams (normally four for azimuth and elevation sensing) and sequentially switches between the beams. The beam switching is normally accomplished by electronic means enabling greater switching speed and avoiding problems associated with mechanical scanners

For a four beam sequential lobing radar, the beam patterns would normally be as shown in Figure 14-4. The receiver is sequentially switched to the four beams, and the elevation error is determined by comparing the output of the *A* and *B* beams, while the azimuth error is computed from the difference in the *C* and *D* beams.

### 14.3.3 Monopulse

Monopulse radars provide angle sensing using multiple simultaneous antenna beams. A considerable advancement in angle sensing was achieved with the development of monopulse. Monopulse angle sensing with its simultaneous lobes permits instantaneous angle

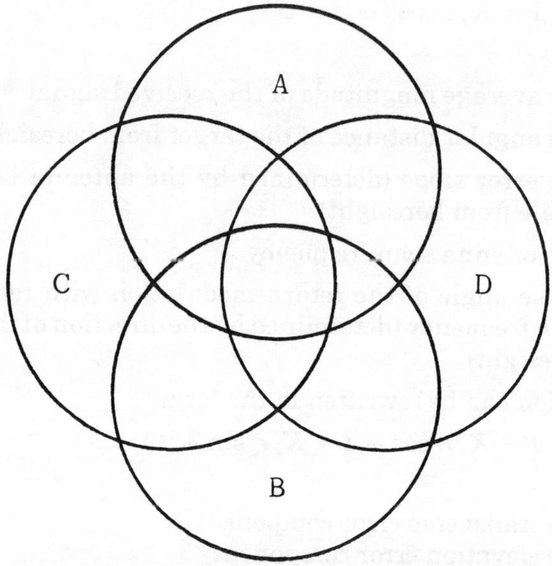


**Figure 14-4.** Sequential lobing radar beam pattern.

sensing on a single pulse basis. In comparison to the previous angle sensing techniques which required a number of pulses for angle measurement, monopulse is inherently immune to angle tracking errors caused by target amplitude fluctuations.

Monopulse radars are described as simultaneous lobe systems because parallel processing channels permit instantaneous angle determination on a single pulse basis. A single plane monopulse antenna uses two offset beams, while a dual plane (azimuth and elevation) monopulse radar uses four off set beams. The processing for a dual plane (four horn) monopulse radar is similar to that of a single plane (two horn) monopulse radar except the processing is performed in both planes.

*Single Plane Monopulse*

Single plane monopusle provides angle sensing in only one plane (either vertical or horizontal), and normally has a pencil beam in the other plane. An example of such a radar, as shown in Figure 14-5, is one in which the antenna is scanned in azimuth, and monopulse elevation patterns are then used to measure the elevation, and hence the altitude, of a target.

The two basic types of monopulse antennas are the amplitude monopulse antenna and the phase monopulse antenna. An amplitude monopulse antenna, such as that shown in Figure 14-6, has a single aperture with the *A* and *B* antenna beams squinted off the antenna boresight. For a target located above boresight, the amplitude return in the *A* beam from the target is greater than the return in the *B* beam. Thus, comparing the amplitude returns from the target between the two beams enables the determination of the target angle from boresight (amplitude sensing). A phase monopulse antenna, such as shown in Figure 14-7, has two antenna apertures with both beams parallel to each other. For a target located above boresight the amplitude return in the two apertures will be approximately equal in amplitude but the phase of the return in the *A* aperture will lead the phase of the return in the *B* aperture. Measuring the relative phase between the returns from the two antennas will indicate the angle of the target off the antenna boresight (phase sensing).

According to Rhodes,[4] amplitude monopulse antenna outputs can be used for amplitude sensing, or converted to phase sensing. Similarly, phase monopulse antenna outputs can be used for phase sensing or converted to amplitude sensing. The outputs from either antenna type can be converted to sum and difference (Σ and Δ) sensing.

Σ and Δ sensing is the most common implementation and is the one that will be described in this chapter. In practice, a monopulse comparator (such as magic "T") is used to convert the *A* and *B* signals at RF into Σ ($A + B$) and Δ ($A - B$) signals. If the target is located above boresight as shown in Figure 14-6, the signal in the *A* lobe is greater than the signal in the *B* lobe, so that the resulting Δ is

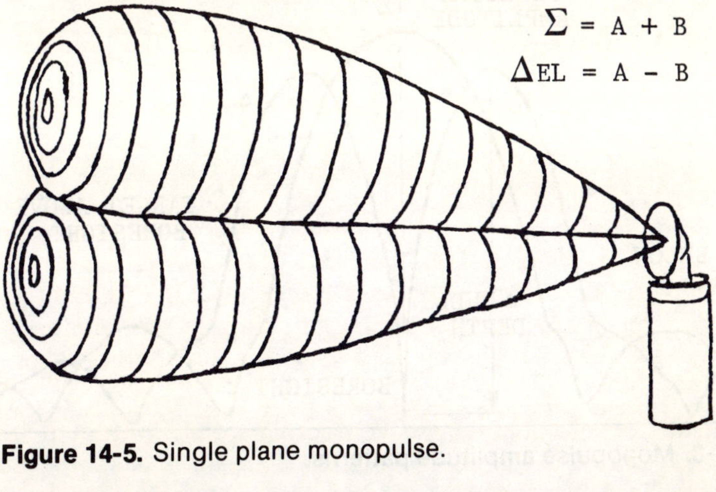


**Figure 14-5.** Single plane monopulse.

positive (in phase with the Σ signal). If the target is below boresight, the signal in the $B$ lobe is greater than the signal in the A lobe so that Δ is negative; that is, the resulting Δ signal is 180° out of phase with the Σ signal.

Another way of looking at the monopulse outputs is to view the antenna patterns for the Σ and Δ signals. The amplitude patterns for a point source in Figure 14-8 shows that the |Σ| signal peaks when

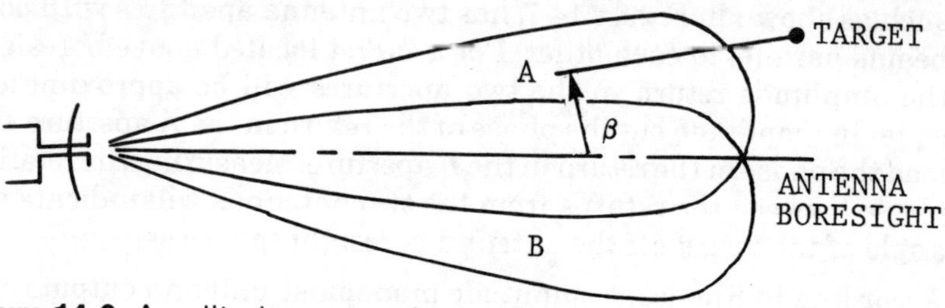


**Figure 14-6.** Amplitude monopulse antenna.

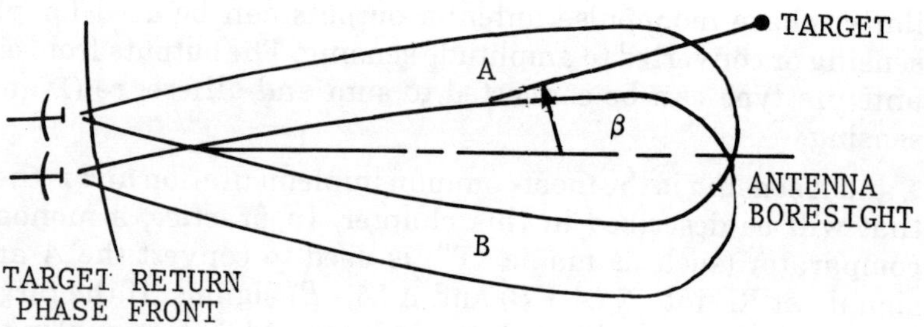


**Figure 14-7.** Phase monopulse antenna.

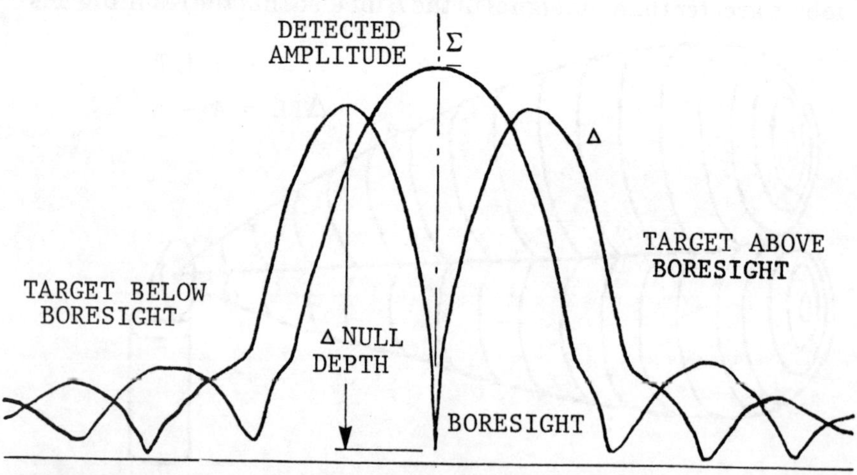


**Figure 14-8.** Monopulse amplitude patterns.

the target is on the antenna boresight. The $|\Delta|$ signal is nulled on the antenna boresight but increases rapidly as the target moves in either direction off boresight. The phase of the $\Delta$ signal shown in Figure 14-9 is zero when the target is above boresight and rapidly switches to 180° as the target goes below boresight.

Angle sensing is accomplished by using the function $\beta \cong [K(|\Delta|/|\Sigma|)\cos\phi]$ where $\phi$ is the phase angle between the $\Delta$ and $\Sigma$ signal, and $K$ is the error slope (a constant related to the antenna parameters). The function $\beta \cong [K(|\Delta|/|\Sigma|)\cos\phi]$ *versus* the target off boresight angle is plotted in Figure 14-10, and for a well-designed antenna, the function $\beta$ is approximately linear over the 3 dB beamwidth of the antenna $\Sigma$ pattern.

### *Dual Plane Monopulse*

Dual plane monopulse radars provide simultaneous angle sensing in both elevation and azimuth, as well as range, thus permitting the location of a target in three dimensional space. In most radar tracking applications angle sensing in both the elevation and azimuth plane is required to keep the antenna boresight directed at the target. There are many radar configurations, too numerous to describe in this chapter, that have been used to obtain monopulse angle sensing. However, there are certain generic types of radar configurations which form the basis for many radar systems. The monopulse antenna feed is normally a cluster of horns which can

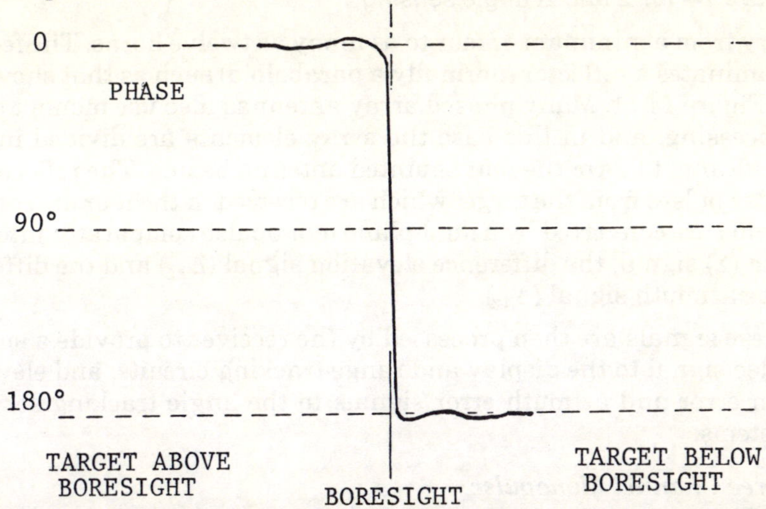


**Figure 14-9.** Monopulse antenna phase patterns ($\Delta$ phase relative to $\Sigma$).

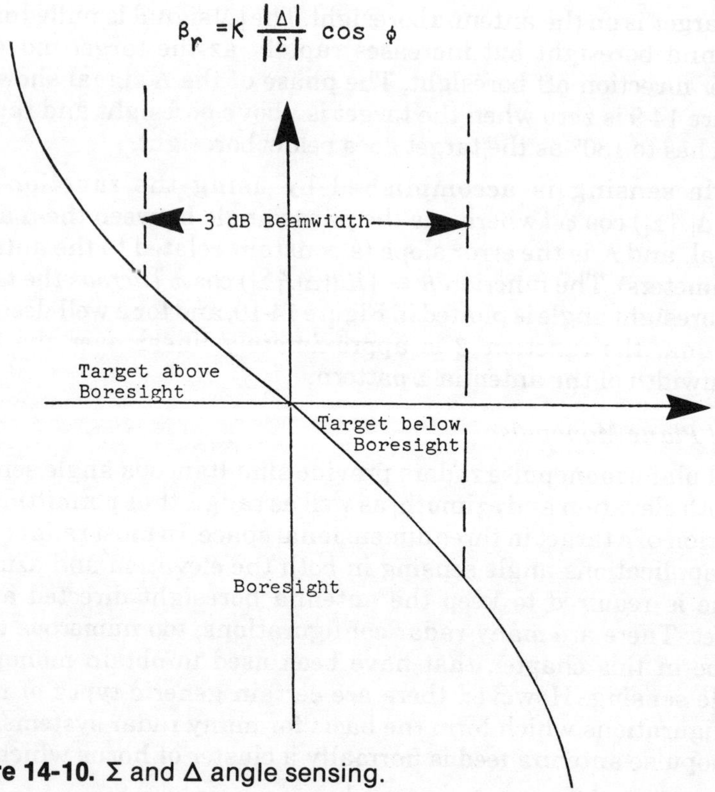


**Figure 14-10.** Σ and Δ angle sensing.

vary from a minimum of four to as many as twelve horns. The feed illuminates a reflector (normally a paraboloid) such as that shown in Figure 14-11. Many phased array antennas also use monopulse processing, and in this case the array elements are divided into quadrants to form the four squinted antenna beams. The reflected radar pulses from the target which are received in the four antenna beams are converted by a dual-plane monopulse comparator into a sum (Σ) signal, the difference elevation signal ($\Delta_{EL}$) and the difference azimuth signal ($\Delta_{AZ}$).

These signals are then processed by the receiver to provide a sum video signal to the display and range tracking circuits, and elevation error and azimuth error signals to the angle tracking servo systems.

*Three Channel Monopulse*

The most common configuration for a three channel monopulse

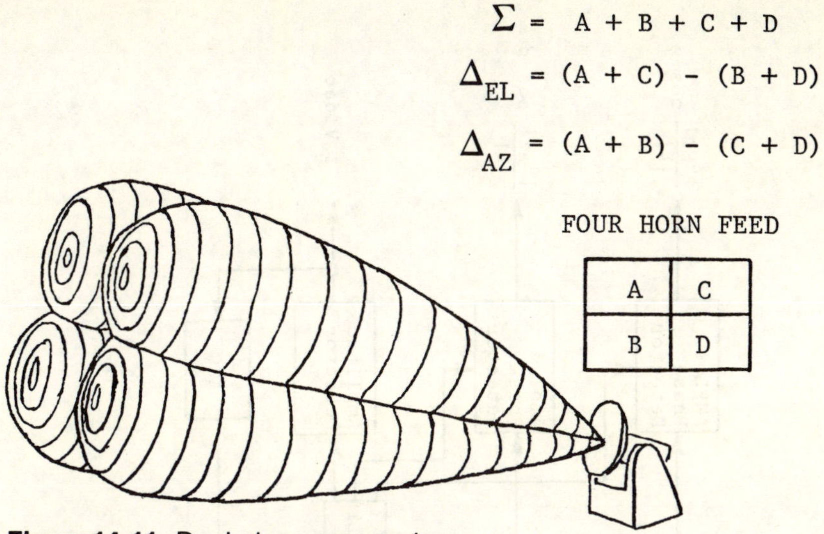


**Figure 14-11.** Dual plane monopulse.

radar is shown in Figure 14-12. The $\Sigma$, $\Delta_{EL}$ and $\Delta_{AZ}$ signals are converted from RF to IF by the mixer preamps whose outputs are connected to three gain and phase matched IF amplifiers, and the output of the $\Sigma$ IF amplifier is amplitude detected to provide a $\Sigma$ video signal. In addition to providing $\Sigma$ video to the displays and range tracking circuits, the $\Sigma$ video is also fed to the automatic gain control (AGC) circuits. The function of the AGC is to control the gain of the three IF amplifiers so that the detected sum signal at the target range is maintained at a fixed amplitude. Because the gains of the difference IF amplifiers are also controlled by the AGC, the signals out of the difference IF amplifiers are proportional to $\Delta/\Sigma$.

These signals are then detected in the amplitude sensitive phase detectors whose output is proportional to $|\Delta/\Sigma| \cos \phi$ where $\phi_1$ and $\phi_2$ are the phase angles of the $\Delta_{EL}$, $\Delta_{AZ}$ signals compared to those of the $\Sigma$ signal. The $|\Delta_{EL}|/|\Sigma| \cos \phi_1$ and $|\Delta_{AZ}|/|\Sigma| \cos \phi_2$ signals are proportional to the off boresight angles to the target in the elevation and azimuth planes.

### *Two Channel Monopulse*

Dual plane monopulse processing can be performed by a dual channel receiver such as shown in Figure 14-13. The $\Sigma$, $\Delta_{EL}$ and $\Delta_{AZ}$ signals are input to the RF signal sampler. The sample timing configures the RF signal sampler on one pulse repetition interval (PRI) so that the elevation $\Delta_{EL}$ signal is coupled to the sampler so

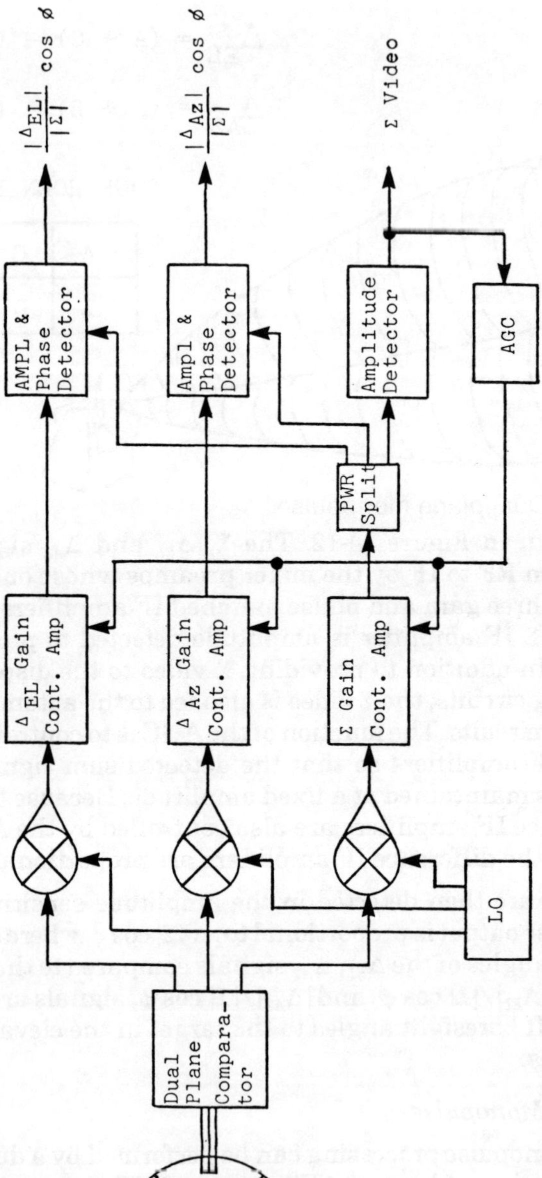


**Figure 14-12.** Three channel monopulse configuration.

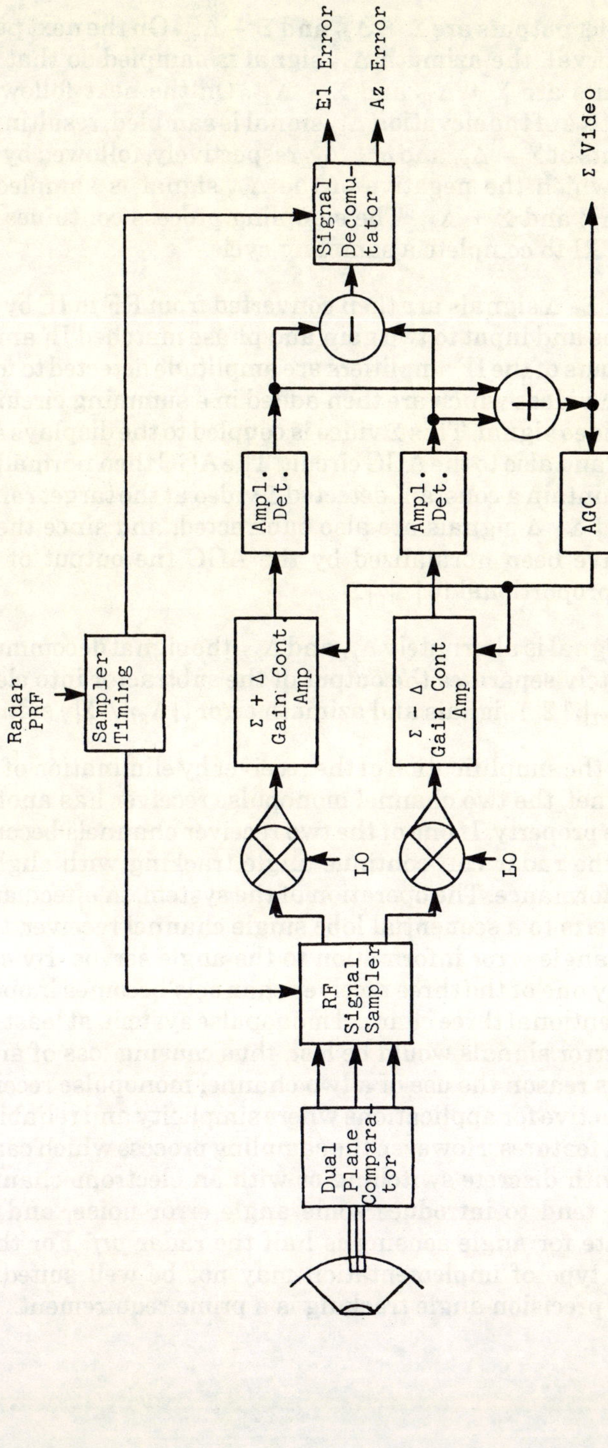


**Figure 14-13.** Two channel, dual plane monopulse configuration.

that the sampler outputs are $\Sigma + \Delta_{EL}$ and $\Sigma - \Delta_{EL}$. On the next pulse repetition interval, the azimuth $\Delta_{AZ}$ signal is sampled so that the sampler outputs are $\Sigma + \Delta_{AZ}$ and $\Sigma - \Delta_{AZ}$. On the next following PRI, the negative of the elevation $\Delta_{EL}$ signal is sampled, resulting in sampler outputs of $\Sigma - \Delta_{EL}$ and $\sigma + \Delta_{EL}$ respectively, followed by the next PRI in which the negative of the $\Delta_{AZ}$ signal is sampled to provide $\Sigma - \Delta_{AZ}$ and $\Sigma + \Delta_{AZ}$. The sampling process continues, requiring four PRI to complete a sampling cycle.

The $\Sigma + \Delta$ and $\Sigma - \Delta$ signals are then converted from RF to IF by the mixer preamps and input to two gain and phase matched IF amplifiers. The outputs of the IF amplifiers are amplitude detected to form $\Sigma + \Delta$ and $\Sigma - \Delta$ videos which are then added in a summing circuit to result in a $\Sigma$ video signal. This $\Sigma$ video is coupled to the displays and range tracker and also to the AGC circuit. The AGC then normalizes the gain to maintain a constant detected $\Sigma$ video at the target range. The $\Sigma + \Delta$ and $\Sigma - \Delta$ signals are also subtracted, and since the IF amplifiers have been normalized by the AGC the output of the subtractor is proportional to $|\Delta|/|\Sigma|$.

Since this $\Delta$ signal is alternately $\Delta_{EL}$ and $\Delta_{AZ}$, the signal decommutator appropriately separates the output of the subtractor into elevation error ($|\Delta_{EL}|/|\Sigma|$) signals and azimuth error ($|\Delta_{EL}|/|\Sigma|$) signals.

In addition to the simplification of the receiver by elimination of one receiver channel, the two channel monopulse receiver has another rather unique property. If one of the two receiver channels becomes inoperative, the radar can continue angle tracking with slightly degraded performance. The operation of the system, in effect, automatically reverts to a sequential lobe single channel receiver, thus maintaining angle error information to the angle servos. By comparison, if any one of the three receiver channels becomes inoperative in a conventional three channel monopulse system, at least one of the angle error signals would be lost, thus causing loss of angle track. For this reason the use of a two channel monopulse receiver becomes attractive for applications where simplicity and reliability are important features. However, the sampling process which can be done either with discrete switching or with an electromechanical resolver does tend to introduce some angle error noise, and the equivalent rate for angle sensing is half the radar *prf*. For these reasons, this type of implementation may not be well suited for radars where precision angle tracking is a prime requirement.

*Single Channel Monopulse*

A single channel dual-plane monopulse radar implementation is essentially the same as that for the two channel implementation, expcept that the second ($\Sigma$ - $\Delta$) receiver channel is eliminated. The RF sampler uses the same four step sequence as that for the two channel monopulse.

The elevation error is computed by subtracting the $\Sigma$ -$\Delta_{EL}$ in sequence step 3 with the $\Sigma + \Delta_{EL}$ of sequence step 1. The resultant $\Delta_{EL}$ error is then proportional to the angular distance of the target from boresight. In a similar manner, the azimuth error is computed by subtracting the $\Sigma - \Delta_{AZ}$ in sequence step 4 for the $\Sigma + \Delta_{AZ}$ in sequence step 2.

Actually, the term "single channel monopulse" is a misnomer in that the radar does not provide angle sensing on a single pulse basis, which is a prime feature of a monopulse system. In a sense, this single channel system is actually a form of conical scan radar which utilizes a minumum of four pulses to derive the azimuth and elevation errors. A prime feature of a single channel system is that all the signals are processed through a common receiver channel, thus eliminating the need for phase and gain matching in the receiver. This system does not require the mechanical scanning of the feed or reflector in a typical conical scan radar, and can be implemented to lobe on receive only.

### 14.3.4 Angle Sensing Using Doppler Methods

The development of Synthetic Aperture Radars (SAR) has demonstrated that a pencil beam antenna moved through space with a fixed relation to a vehicle can be used to provide precision angular measurements if the vehicle's flight path can be accurately located in space on a pulse to pulse basis. In this case the change in Doppler from the target on a pulse to pulse basis provides the information necessary to compute the pointing angle to the target with resolution which may be as much as 1000 times greater than that which would be attributed to the antenna beamwidth. The SAR technique was developed by the Environmental Research Institute of Michigan for high altitude terrain mapping. Quite extensive terrain sensing is being performed by the University of Kansas.[5] However, the technique is too complex to adequately describe in this chapter; for further information lthe reader is referenced to the book by Kovaly.[6]

Doppler sensing can also be used to provide the means to resolve individual scatterers from a fixed radar if the target rotates in a predictable or measureable manner. Referring to Figure 14-14, an object rotating at a radius (*d*) from a fixed point *C* will produce an angular rate of change of phase which is given by the equation.[7]

$$f_\Theta = \frac{1}{2\pi}\frac{d\phi}{d\Theta} = -\frac{2}{\lambda}\frac{dr}{d\Theta} = 2\left(\frac{d}{\lambda}\right)\cos\Theta = \frac{2x}{\lambda} \text{ cycles/radian} \tag{14-3}$$

where

$f_\Theta$ = the Doppler frequency component

$\phi$ = is the phase of the return received at the radar

$\Theta$ = the angular position of the target with respect to fixed point $C$ $(\Theta = \Omega t)$

$r$ = the distance of the object from the radar

$\lambda$ = the wavelength of the radar frequency

$x$ = the cross range distance

The cross range distance is equivalent to the product of the azimuth angle $\alpha$ and range $r$ for small values of $\alpha$. The Doppler component provides a direct measure of the lateral displacement of the reflecting object from the centroid of the rotation. Because of the specular nature of the reflector, several reflecting points at different distances from the centroid can be resolved by spectral analysis of the returns. The achievable cross range resolution is given by [7] where

$$\Delta x = \frac{\lambda}{2\Omega}\Delta f_t = \frac{\lambda}{2\Omega}\frac{1}{T} = \frac{\lambda}{2\Delta\Theta} \tag{14-4}$$

$T$ = is the processing time duration ( $\cong 1/\Delta f_t$ )

$\Delta\Theta$ = the angular rotation during processing time $T$

One direct application of the Doppler measurement technique is the signature characterization of targets mounted on a rotating pedestal or turntable. Using this technique the contributions of the cross range scatterers can be resolved into their individual contributions (for example, *see* Figure 14-15). The technique is also useful for imaging rotating objects such as missile bodies in space. For further detail the radar is referenced to the work by Mensa[7] and Richaczek.[8]

## 14.4 ANGLE MEASUREMENT CONSIDERATIONS

A number of factors related to the target, the angle sensing system,

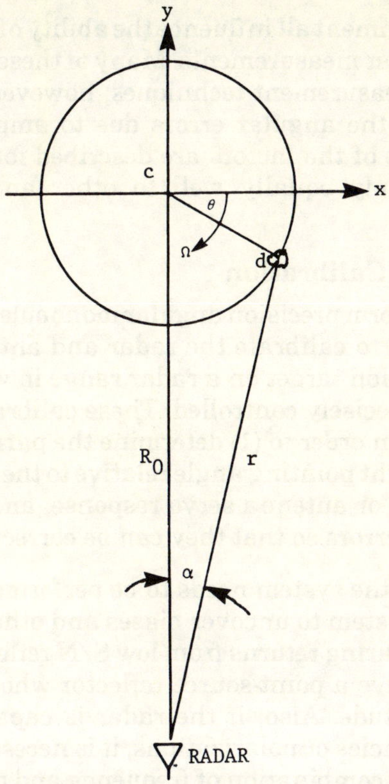


**Figure 14-14.** Doppler angle measurement.

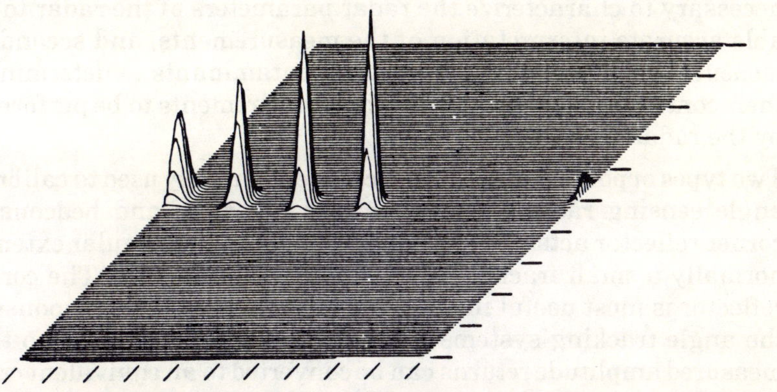

**Figure 14-15.** Doppler multiple target resolution (reprinted from Mensa, ref. [7]; ©1981 Artech House, Inc.).

and the environment all influence the ability of the radar to achieve accurate angular measurements. Many of these factors are common to all angle measurement techniques; however, monopulse radars can eliminate the angular errors due to amplitude scintillation. Although most of the factors are described for monopulse radars, the effects apply equally well to other angular measurement techniques.

### 14.4.1 Angle Calibration

In order to perform precision angular monopulse measurements it is first necessary to calibrate the radar and antenna using a point-source calibration target on a radar range in which the test conditions can be precisely controlled. These calibration measurements are necessary in order to (1) determine the parameters of the radar such as boresight pointing angle relative to the antenna axis, angular error slope, or antenna servo response; and (2) to uncover systematic radar errors so that they can be corrected.

Calibration of the system needs to be performed over the dynamic range of the system to uncover biases and other amplitude related effects of measuring returns from low S/N reflectors. To do this it is desirable to have a point-source reflector whose return is controllable in amplitude. Also, if the radar is capable of operating at several frequencies or polarizations, it is necessary to calibrate the system at each combination of frequency and polarization in order to fully characterize the radar for the expected operational modes. The purpose of the calibration of the radar is twofold. First, it is necessary to characterize the radar parameters of the radar to enable accurate interpretation of the measurements, and second, to measure the effects of radar system contaminants to determine if their contributions will degrade the measurements to be performed by the radar.

Two types of pointlike target sources are commonly used to calibrate angle sensing radars, namely corner reflectors and beacons. A corner reflector acts as a pointlike target since its angular extent is normally a small fraction of the antenna beamwidth. The corner reflector is most useful in checking the antenna servo response of the angle tracking systems and for calibrating the radar so that measured amplitude returns can be converted to an equivalent cross section. In order for the corner reflector to act as a point source, it is necessary to isolate it from other objects, including its supporting structure. This is often difficult to accomplish perfectly in practice, and as a result, reflections from the supporting structure can contribute angle biases to the boresighting process. The measurement of the antenna response from a corner reflector provides received data that are the result of the two-way antenna pattern effects (once on transmission through the antenna, and once on return to the antenna). Most receiver systems have limited dynamic range, and therefore, it is difficult to plot the sidelobe characteristics of the antennas using a corner reflector because of the two-way antenna gain reduction.

A beacon is an active signal source tuned to the transmitter frequency including an antenna which is directed at the monopulse antenna, and as such, acts as a point source target. The beacon source is inherently isolated from the supporting structure due to the time delay in the beacon response. Normally a variable delayed trigger pulse is sent to the beacon so that it can simulate the received signal from a target at a distant range. The use of a beacon results in receiving the one-way antenna pattern, thus reducing the dynanic range required by the receiver for recording the sidelobe levels of the antenna. The beacon thus provides a reliable means for measuring antenna amplitude and phase characteristics as well as the off boresight error function. The inclusion of a calibrated attenuator (preferably remotely controlled) on the output of the beacon source provides an easy means for calibrating the radar over its full dynamic range.

When using either a corner reflector or a beacon as a point source, it is desirable to locate the point source in the far field of the monopulse antenna (that is $2D^2/\lambda$, where $D$ is the diameter of the monopulse antenna and $\lambda$ is the wavelength of the transmit frequency). If the point source is located in the near field of the antenna, it will not affect the boresight pointing accuracy; however, the off boresight error function and the phase response may be affected.

Many precision tracking radars incorporate an optical device on the antenna pedestal which is co-boresighted to the tracking antenna. These optical devices are normally either a telescope with internal crosshairs or a closed circuit television camera fitted with a telephoto lens. The optical device is co-boresighted with the monopulse antenna boresight on a "clean" pointlike target, and can present a visual picture of where the monopulse boresight is actually tracking the target. In boresighting the optical device to the monopulse boresight, the alignment must account for parallax errors that result from the physical displacement of the optical device from the center of the monopulse antenna. This is normally accomplished in practice by off-setting the optical target by the appropriate distance from the point source target on the calibration tower.

Elimination of the multipath is a necessity for accurately calibrating and adjusting the monopulse system. Multipath will affect the boresight pointing accuracy, will result in phase differences between the sum and difference channel returns, will degrade the null depth of the monopulse system, and will result in non-symmetrical off boresight patterns. Multipath returns even as low as -30dB below the main lobe returns can affect the boresight pointing accuracy. Methods for eliminating multipath on a calibration range include the following:

a) Elevate the radar, the point source target, or both at high enough elevation so that when the antenna boresight is aligned on the point source target the multipath return is in low level sidelobes of the antenna.
b) Set up the point source target on a hill across a valley from the radar so that no multipath reflection is encountered.
c) Set up microwave fences (sheets of plywood covered with absorber) between the radar and the beacon or reflecting source to block off the multipath return.

Generally, microwave fences will not provide sufficient multipath reduction by themselves to permit precision angular calibration. Calibration range location and available tower heights may also preclude sufficient geometric configurations to eliminate multipath. Often a combination of the techniques a, b, and c must be used to provide a multipath-free environment for calibration. When applying these techniques, how do you know that you have a multipath free environment? You have a multipath free environment when:

1. Good difference null depths are obtained across the frequency range of the radar.
2. Low quadrature phase error is obtained across the beamwidth of the antenna and across the radar frequency range.
3. The off boresight antenna pattern will be symmetrical about boresight.
4. Radar boresight (on point target) will correspond to optical boresight.

### 14.4.2 Angle Tracking Contaminants

The ability to precisely measure the elevation and azimuth angle to a target is degraded by a number of possible contaminants to the angle measurement process. Depending upon the precision and accuracy required from the angle measurement data, a number of factors need to be considered in the angle measurement process.

*Angular Errors Due to Signal-to-Noise*

The angular error produced by thermal noise in an angle tracking system depends upon the radar configuration. For a monopulse tracker the signal in the reference channel sets the AGC gain of the reference and difference channel so that the noise in the difference channel is set by the reference channel signal-to-noise. Assuming that the S/N is well above unity, the angular error for a range gated error signal, which is averaged over the pulsewidth can be computed. The single pulse filtered angle error component present at the input to the angle servo is given by Barton[3] as

$$\sigma_1 = \frac{\Theta}{K_m \sqrt{2B\tau\,(S/N)}}, \qquad (B\tau \geq 1) \tag{14-5}$$

where

$\Theta$ = antenna beamwidth
$K_m$ = error slope constant
$B$ = IF bandwidth
$\tau$ = pulse duration
S/N = signal-to-noise ratio in the reference channel

Since the servo bandwidth is usually considerably less than the radar *prf*, the servo response effectively integrates the single pulse angular error component, thus reducing the angular error. The angular error component at the output of the angle servo is thus given by Barton[3] as

$$\sigma_t = \frac{\Theta}{K_m \sqrt{B\tau \left(\frac{S}{N}\right)\left(\frac{f_r}{\beta_n}\right)}} \tag{14-6}$$

where

$f_r$ = pulse repetition frequency
$\beta_n$ = noise bandwidth of the angle servo

*Radar Glint*

Radar glint is defined as the angular measurement error introduced by the complex nature of the observed target. Complex targets consist of multiple scatterers separated by a finite angle or range. The return from a complex target is made up of the ensemble of the returns from the various scattering elements. As the aspect angle to the complex target changes, the changes in the relative amplitude and phases of the scatterers result in amplitude scintillation and angular scintilaltion of the target returns. The angular scintillation is known as glint, and results in angle tracking errors when tracking a target. For many complex targets, the angular error appears to be noiselike, although it is generally not like white noise in nature.

One concept postulated for the existence of glint is that the apparent power centroid of the target returns moved about the target; however, this concept is invalid since the apparent angle to the target can often be outside the target extent. Howard[2] reported that the phenomenon was actually a warpage of the phase front. Since all angle measuring systems effectively determine the angle to the target by aligning the antenna with the apparent phase front of the target returns, a warpage in the phase front will cause the pointing angle of the antenna to be directed away from the centroid of the target.

Howard derives the equation for phase front warpage for two isolated targets as shown in Figure 14-16. The slope or tilt of the phase front with respect to the actual center between the two targets is given by Howard[2]

$$\delta' = \frac{L\cos\psi}{2}\left(\frac{1-a^2}{1+a^2+2a\cos(\phi+L\sin\psi)}\right) \tag{14-7}$$

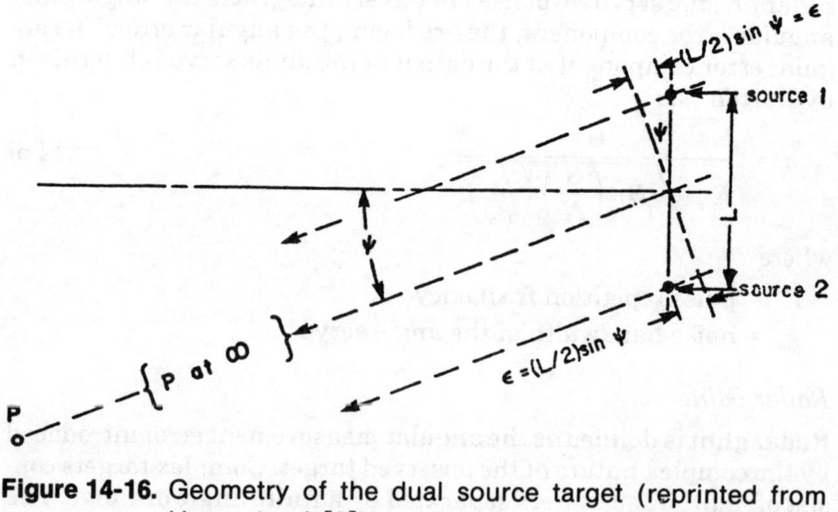


**Figure 14-16.** Geometry of the dual source target (reprinted from Howard, ref. [2]).

If we assume that $\psi = 0$ (the condition where the line between the two scatterers is perpendicularly bisected by the line to the radar) the equation reduces to

$$\delta'(\psi = 0) = \frac{L}{2}\left(\frac{1 - a^2}{1 + a^2 + 2a\cos\phi}\right) \tag{14-8}$$

This is identical to the equation derived by Meade[9] for two independent scatterers and has been proven experimentally as shown in Figure 14-2. For a complex target the equation for phase front tilt is given by Howard as

$$\gamma'_{\beta} = \frac{\sum_{i=1}^{n}\sum_{j=1}^{n} A_i A_j \Theta_i \cos(\omega_i - \omega_j)t}{\sum_{i=1}^{n}\sum_{j=1}^{n} A_i A_j \cos(\omega_i - \omega_j)t} \tag{14-9}$$

where

$A_i$ = the amplitude of the $i$th reflector

$\Theta_i$ = equal to $L_i \cos\beta$

$\beta$ = the angle between the perpendicular to the scattering array and the direction from the radar to the target

$\omega_i(t)$ and $\omega_j(t)$ = random angular frequencies

If a conical scan or beam sampling type of angle measuring radar is employed, then an additional component needs to be added to account for the effects of amplitude scintillation.

Dunn and Howard[1] also derived the Pointing vector for a distributed source target and showed that power flow can deviate from the radial direction and is normal to the phase-front contours. Thus, a search type radar will demonstrate the same angular pointing errors as an angle tracking radar.

This angular glint will cause the apparent indicated angle to deviate from the target centroid as illustrated in Figure 14-1, and at times will actually give an indicated angle outside the target extent. The spectral content of the glint for this target is shown in Figure 14-17.

Methods for minimizing the effects of glint generally use either aperture diversity (moving the antenna laterally or vertically) or frequency diversity (changing the radar transmit frequency). In order to be effective, the aperture or frequency diversity must be rapid with respect to the servo bandwidth and must be of sufficient extent to accomplish decorrelation of the scatterers on the target. The reduction of *rms* tracking error by using frequency agility has been computed by Loomis and is illustrated in Figure 14-18.

*Multipath*

Multipath is a phenomenon whereby part of the transmitted signal from the radar and the return signals from the target are reflected off a surface normally within the beamwidth of the radar. This return reflected off the surface combines with the direct reflected return from the target, and effectively warps the phase front of the target returns in a similar manner as radar glint. As a result, multipath can produce significant angle tracking errors, and on the average will bias the angular return toward the source of the multipath.

The effect of multipath depends upon a number of factors including surface roughness, target and multipath angular separation, and relative path length. The problems with low angle tracking has been investigated by Barton[11] along with an evaluation of techniques for

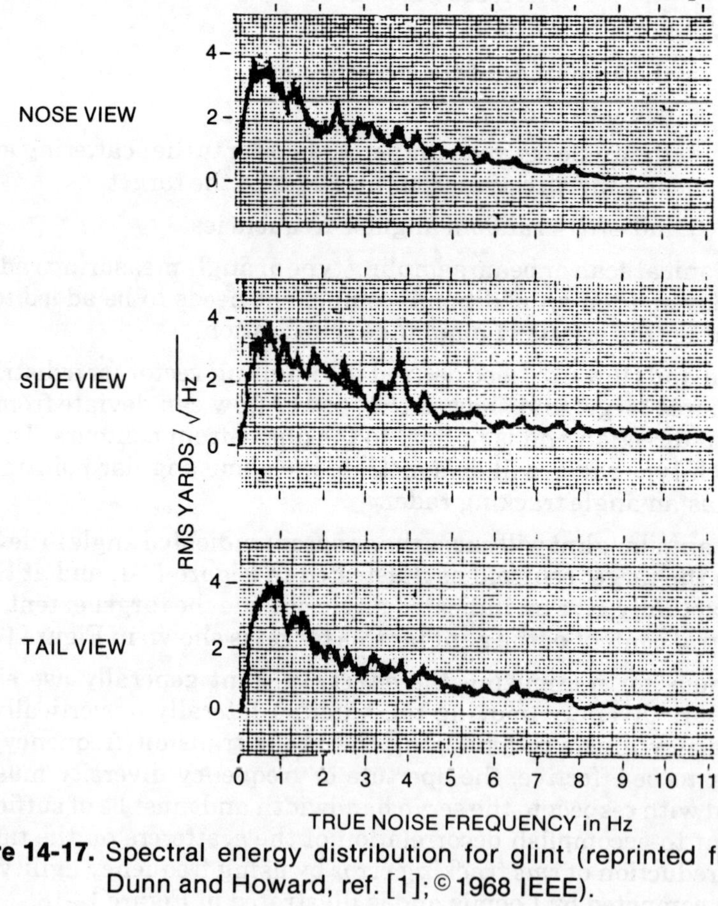


**Figure 14-17.** Spectral energy distribution for glint (reprinted from Dunn and Howard, ref. [1]; © 1968 IEEE).

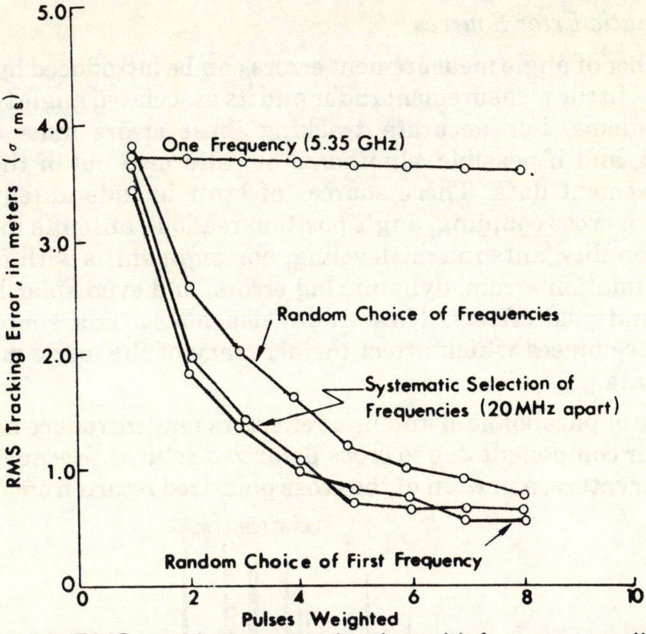


**Figure 14-18.** RMS tracking error reduction with frequency agility (reprinted from Loomis and Graf, ref. [10]; ©1974 IEEE).

the reduction of multipath effects. The problem is especially severe when tracking low altitude targets over the sea. This is illustrated in Figure 14-19 for the case of an aircraft being tracked over the Atlantic Ocean.[12]

Tracking a target over a sea with a glistening surface will result in diffuse scattering from the surface. Barton[11] shows the effects of this diffuse scattering and the angular error produced in Figure 14-20.

For a smooth sea the scattering is specular in nature providing an indicated angle half-way between the target and the multipath returns except at the nulls where the direct return is cancelled out by the multipath returns. For a medium sea, the angular error effects are more distrubuted, although they still provide large angular errors at low elevations. At sufficient altitude the effects of multipath are severely attenuated. For a rough sea, the angular error effects become less pronounced because of the diffuse effect of the multipath return.

Multipath can cause severe angle tracking errors, and its effects must be considered in the design of radar systems, especially those which will be situated in an over the water application.

*Systematic Error Sources*

A number of angle measurement errors can be introduced by imperfections in the measurement radar and its associated angle tracking mechanisms. For accurate tracking these errors must be considered, and if possible, eliminated or calibrated out of the angle measurement data. These sources of error include antenna and feedhorn cross coupling, angle position readout, antenna axis nonorthogonality, antenna misleveling, boresight shifts with frequency, collimation errors, dynamic lag errors, and even solar heating and wind gust effects. These radar dependent error sources can introduce biases which affect the accuracy of the angle measurement data.

The use of paraboloidal antenna reflectors may introduce an angular error component due to cross polarized returns. Because of the dish curvature, a portion of the cross polarized return from a target

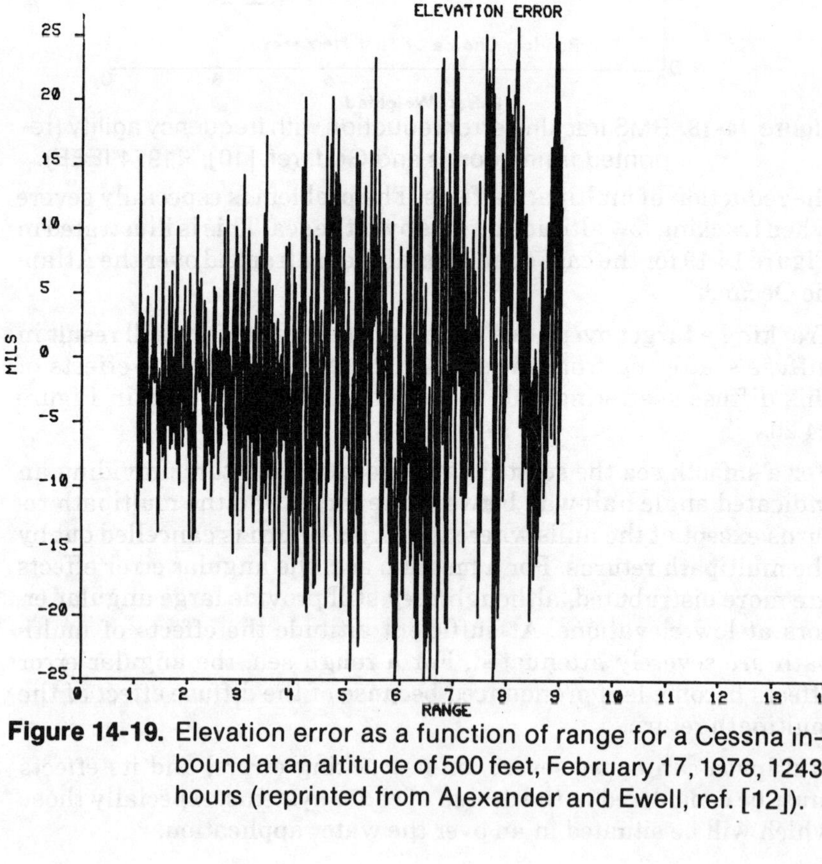


**Figure 14-19.** Elevation error as a function of range for a Cessna inbound at an altitude of 500 feet, February 17, 1978, 1243 hours (reprinted from Alexander and Ewell, ref. [12]).

off boresight in azimuth will be interpreted as an elevation error. Similarly, if there is cross coupling in the antenna feedhorn assembly, the cross coupling due to orthogonally polarized signals can introduce similar tracking errors that can even be more serious than those due to the antenna dish curvature. If this cross coupling is severe enough, it can actually cause the antenna to spiral off a target when significant energy is received in the orthogonal polarization. Another possible source of antenna error is that due to nonorthogonality of the elevation and azimuth planes of the antenna.

Collimation errors are those which cause a shift of the apparent boresight from the mechanical boresight. For a radar which must operate over a band of frequencies, the most significant source of collimation errors can be that due to frequency. A change in operating frequency can result in a shift of the boresight, which must then be calibrated out of precision measurement radar data. Other possible sources of collimation errors include receiver nonlinearity, LO tuning effects, signal amplitude, and temperature.

A number of possible measurement errors can be a result of the

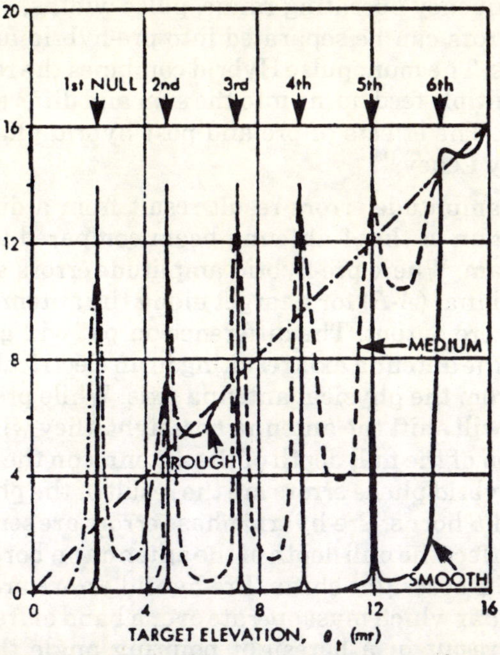


**Figure 14-20.** Multipath effects for various states of sea roughness (reprinted from Barton, ref. [11]; ©1974 IEEE).

antenna pedestal and associated mechanical and electrical interfaces. The antenna elevation and azimuth position data take offs are normally either precision potentiometers or digital encoders. Errors associated with data pick off include those associated with granularity and accuracy of the angle position data readouts. Dynamic lag errors can be introduced while tracking moving targets which are accelerating, or when the target presents a rapidly changing aspect angle with respect to the radar. Misleveling of the antenna pedestal can introduce angle errors in the measurement data with respect to absolute coordinates. Wobble in the elevation or azimuth pedestal bearings can also introduce tracking errors.

Solar heating of surfaces of the antenna and pedestal can introduce warpage which can affect the accuracy of angle measurements. Wind gusts can also result in biases which may result in dynamic measurement errors.

*Phase and Gain Track*

For a monopulse radar particular care must be exercised to provide good phase and gain track over the dynamic range of the system, over the frequency operating range, pulse widths, *et cetera.* Phase and gain errors can be separated into pre-hybrid and post-hybrid phase errors. The monopulse Hybrid combines the received signals from the multiple feed horns into the sum and difference signals for the receiver. The effects of pre and post hybrid phase errors were computed by Leney.[13]

Pre-hybrid amplitude errors result result from a difference in the amplitude gain of the *A* antenna beam compared to that of the *B* antenna beam. Where pre-hybrid amplitude errors are present, the difference signal (*A-B*) for a target along the antenna axis will not result in a zero output. The difference output will go to zero at an angle off of the antenna axis resulting in an electrical boresight that is skewed from the physical antenna axis. While pre-hybrid amplitude errors will shift the antenna boresight, they will not cause the deterioration of the null depth of the antenna on the electrical boresight. Pre-hybrid phase errors are the result of the phase imbalance in the *A* and *B* horns. Pre-hybrid phase errors are serious since they adversely affect the null depth of the antenna on boresight, and can also result in boresight shifts. Pre-hybrid errors are especially serious to a radar which must operate over a band of frequencies since they could result in a boresight pointing angle that shifts with frequency. Great care is normally exercised in designing and assembling the monopulse antennas, the feed structure and the RF components to minimize pre-hybrid errors. The monopulse comparator is normally mounted as close as physically possible to the antenna feed, and if possible it is right on the feed assembly itself. This is done to minimize the effects of differential phase and amplitude shifts with frequency.

Pre-hybrid amplitude errors can be measured in a number of ways. One method is to block off the *B* feed and run Σ patterns. Then remove the block from the *B* feed and block off the *A* feed and run Σ patterns on the *B* feed. The Σ patterns should be equal in amplitude. Another method is to mechanically align the face of the antenna so that it is perpendicular to a point-source calibration target. The amplitude error can then be determined by how far the antenna must be shifted to align the electrical boresight on the target. Normally, if care has been exercised in assembling the antenna and feed assembly, the shift will be samll. In general the only concern is whether the boresight shifts with frequency. It is generally sufficient to measure the boresight position over the desired frequency range of the radar, and then determine if the boresight shift is within acceptable bounds.

Pre-hybrid phase errors are usually checked by examining the depth of the difference null on boresight. In order to achieve a good null depth on boresight both the real and imaginary components of the difference signal must go to zero on boresight. This can only be achieved if the phase is matched in the *A* and *B* channels. It is also necessary to check the null depth over the frequency band of the radar.

Post-hybrid amplitude errors are the result of differences in gain between the Σ and the Δ channels. Post-hybrid gain errors will not affect the boresight pointing accuracy of the radar, but will result in changing the slope of the off boresight angle function (often called the angle error function in tracking radars). It is important to minimize the post-hybrid gain errors since the change in error slope will degrade the angle tracking performance, and will affect the angle sensing function used to measure the angular distance of a target off boresight.

Post-hybrid phase error will not affect either the boresight pointing accuracy or the null depth of the radar on boresight. The sensitivity of the angle error sensing function to phase depends upon the processing technique used. For example if the off boresight error function is $|\Delta|/|\Sigma| \cos \phi$, the phase angle $\phi$ is normally either 0 or 180 degrees depending upon which side of boresight the target is located. For this type of processing, phase errors of 10 to 20 degrees will not significantly affect the off boresight error function. However, if processing is used which depends upon the sine of the phase, then phase errors of more than 5 degrees will be significant.

Post-hybrid gain errors can best be measured by using a beacon source with a controlled signal level output. The error slope of the off boresight function can then be measured over the dynamic range of the receiver. The error slope *versus* received signal level should also be measured at several frequencies over the radar frequency range to assure that no frequency dependent effects are present. Ideally,the slope of the error function should remain constant over the dynamic range of the receiver and over the frequency range.

Post-hybrid phase errors can be measured if adjustable phase shifters are incorporated in the receiver IF channels. The phase shifter can be adjusted to provide maximum error voltage output at a given off boresight angle (this same phase shifter setting sould also correspond to a maximum negative error voltage output on the opposite side of boresight). If the radar incorporates a quadrature phase detector ($\sin\phi$), the phase error can be measured directly.

It is important that phase and gain track be maintained over the dynamic range of the system. This requires gain and phase matching of the receiver in the sum and difference channels. If the radar has selectable receiver bandwidths, the channels need to be matched at each of the bandwidths. The tolerance on phase and gain tracking depends upon the particular application, but for most angle tracking radars a gain track of ±1dB, and phase track of ±10° over the radar dynamic range will not seriously degrade the tracking performance.

*Near Field Effects*

In most cases, when collecting angular measurement data it is assumed that the target is in the far field of the radar, and that the target occupies a small fraction of the radar beamwidth. If the radar is calibrated using a point-source target such as a beacon which is located in the near field of the radar, the angle error measurements will be affected since the wave front arriving at the radar from the beacon is not planar but curves across the aperture. The effect of this curvature across the aperture will not deteriorate the boresight pointing accuracy, but will result in measuring a smaller error slope than if the target was in the far field.

The measurement of the angular component such as glint will also be affected if the target extent occupies an appreciable part of the antenna beamwidth. An analysis of the relationship between the angular extent of the target in the beam and the measurement of glint from a two target situation was analyzed by Ewell.[14] The analysis consisted of a computer simulation of the return from two point target sources with separations of fractional parts of a beamwidth. A sample of the results of this analysis is shown in Figure 14-21 in which $\delta$ is the square root of the ratio of the cross section of the two scatterers and $a$ is the separation of the two scatterers as a fraction of the antenna beamwidth. For the case where $a$ is equal to zero, the result is the classic two target return phenomena which shows a large angular deviation when the phase between the two returns approaches 180°. As the fraction at beamwidth distance $a$ between the two targets increases, the angular deviation decreases in the vicinity of 180° phase difference. However, for phase differences away from 180° the effect of glint can actually increase as the target extent in the beam increases. In the measurement of glint from an extended target it is important to be aware of the effects of target extent when relating the measured glint effects to another radar situation.

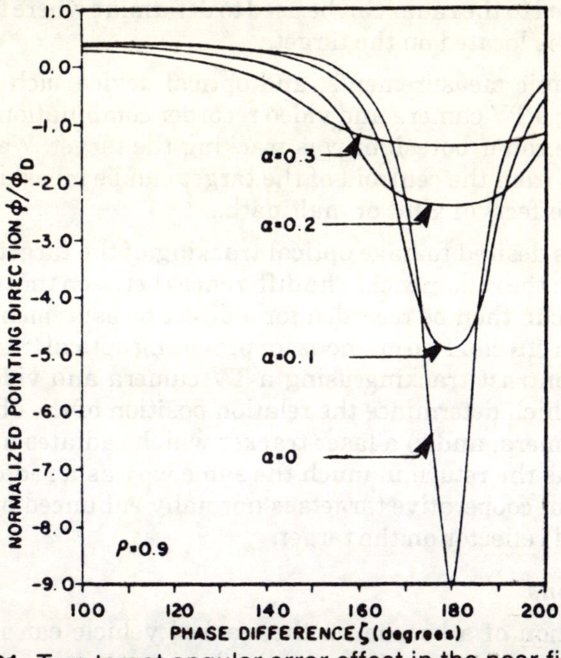


**Figure 14-21.** Two target angular error effect in the near field, $\rho = 0.9$ (reprinted from Ewell, ref. [14]; ©1971 IEEE).

### 14.4.3 Angle Location References

In order to evaluate the measurement errors produced by glint, a separate angle location reference must be provided which will independently indicate the true pointing angle of the antenna boresight. In using a location reference it is important to use one in which the location reference itself is not affected by the phenomenon to be measured.

*Optical*

The most common means for providing a location reference is to co-boresight an optical device to the radar antenna. This is normally accomplished by mounting the optical device either to the side or above the radar antenna, and boresighting the radar and the optical device on a common pointlike target. (Note: the optical target is normally offset from the radar point source target by the same distance as the optical device is located from the radar antenna centerline to account for parallax.)

A number of optical devices can be employed depending upon the particular application. For static measurements, a telescope co-boresighted to the radar can be used to determine where the antenna boresight is located on the target.

For dynamic measurements, and optical device such as a movie camera or a TV camera and video recorder combination will record where the radar boresight was tracking the target. Wander of the boresight from the centroid of the target can be measured to determine the effects of glint or multipath.

Often it is desired to have optical tracking of the target simultaneously with the radar track. The differences between the two tracking methods can then be recorded for a direct measurement of glint or multipath effects. Two methods for providing optical tracking are: 1) optical contrast tracking using a TV camera and video tracking system which determines the relation position of the object viewed by the camera, and 2) a laser tracker which radiates a laser beam and tracks the return in much the same way as a radar. The laser tracking of cooperative targets is normally enhanced by mounting an optical reflector on the target.

*RF Beacons*

The location of a beacon on the tracked vehicle can improve the tracking accuracy on a "cooperative" target. A beacon is an active device which responds to the received radar pulse by transmitting its own RF pulse either at the same frequency or at specific beacon frequency offset from the radar frequency. Also, the beacon normally incorporates a fixed time delay so that even if the beacon responds at the radar frequency it can still be separated in range. This technique is especially useful when tracking large complex objects, because the beacon acts as a point-source target and also aids in tracking vehicles in clutter environment. Another advantage in using a beacon is that the beacon signal only suffers a $1/R^2$ loss due to range as compared to $1/R^4$ loss for the reflected radar returns. Therefore, the use of a beacon usually results in increased signal-to-noise, which in itself increases target tracking accuracy.

However, beacons are not always the ultimate solution and care must be taken with their application. For example, if a beacon is linearly polarized and the target is free to rotate, the beacon polarization can become cross polarized with respect to a linearly polarized radar. Due to the inherent properties of paraboloidal reflectors, a cross polarized return with an elevation error can be interpreted as an azimuth error. Not only will an erroneous angle measurement result, but in certain cases the cross polarized return can actually cause the antenna to spiral off the target.

In general, circularly polarized antennas are used for beacons operating with linearly polarized radars. Thus, regardless of the aspect angle, reasonable signal levels should be retained. However, mounting locations for beacon antennas are often restricted, which frequently results in distorted illumination patterns. Phase front distortions resulting from the beacon antenna patterns can cause angular measurement errors, since the polarization patterns at certain aspect angles can be far from circular. Sometimes in order to achieve omnidirectional coverage on a vehicle, it is necessary to locate two or more beacon antennas on the target vehicle. However, when the radar is illuminated by two of the antennas at the same time, it can result in a situation equivalent to two target phenomenon which could result in a warped phase front possibly causing serious angle tracking errors. Special techniques need to be used in this situation to alleviate the tracking anomalies.

### *Target Attitude Determination*

To fully characterize the angle measurements on a target, it is necessary to record the target aspect angles. In this manner the target aspect angles can be correlated with the measured radar parameters. One method for the precise control of the target aspect angle is to mount the target on a pylon on a turntable base. The precise aspect angle can then be recorded along with the radar measurements.

For a dynamic target, the recording of target aspect angle is considerably more difficult. When tracking a cooperative aircraft target, a recorder aboard the aircraft could be used to record heading, altitude, pitch, roll, and a time sync. During post-mission data analysis the aircraft information and the radar data could be analyzed in the computer to provide the target aspect information.

If precise apsect angle information is not required, target aspect angles can be estimated from viewing of video or movie recordings from the flight.

### 14.4.4 Angular Presentations

Perhaps the simplest and most informative presentation for an angle tracking system is a numerical display of the antenna servo elevation and azimuth angle position, along with a readout of the tracked target range. These readouts define the position of the target in three-dimensional space, and for most applications provide the required information which can then be used for directing weapon systems, or recorded for later data analysis.

On boresight processing permits data to be presented on the display only when the target is within a small fraction of a beamwidth of boresight. As such, the on boresight processing provides an angle sharpening effect eliminating the blurring associated with viewing the target across the full beamwidth as the antennna scans by the target. For on boresight E-scan display, the antenna is scanned in elevation, and the range at which the boresight intersects the terrain is illuminated on the display. Thus, a terrain profile of elevation angular location to the targets is sharply defined as dots on the screen rather than a blur.

Above boresight processing provides display illumination only when the target or terrain is above the boresight of the radar. For an elevation scanning radar with an E-scan display, the terrain is sharply defined showing as a bright upper bound on the elevation angle/range display. This technique provides a brighter display since only the returns below boresight are discarded.

Off boresight processing can provide brighter displays since all the returns received in the radar beam are displayed, however the data processing is more complex. An off boresight E-scan display uses the elevation angle measured ($|\Delta_{EL}|/|\Sigma| \cos \phi$ for a monopulse processor) to determine the elevation angle on the display at each range increment. As such, the E-scan display can be generated without the need for elevation scanning.

A compensated PPI display can be implemented by using the azimuth angle ($\Theta$) to deflect the beam laterally as the trace sweeps out along the radial of the antenna pointing angle. The compensated display can be generated electronically using a conventional oscilloscope with vertical and horizontal deflection inputs. Referring to a target located as shown in Figure 14-22, the equation for the deflection axis for small angles of $\Theta$ are:

$$V_D = R \sin \psi + R\,\Theta \cos \psi \tag{14-10}$$

$$H_D = R \cos \psi - R\,\Theta \sin \psi \tag{14-11}$$

where

$V_D$ = the vertical deflection input to the oscilloscope

$H_D$ = the horizontal deflection input to the oscilloscope

$R$ = the range to the target

$\psi$ = the azimuth servo angle (azimuth boresight direction)

$\Theta$ = the measured off boresight angle to the target

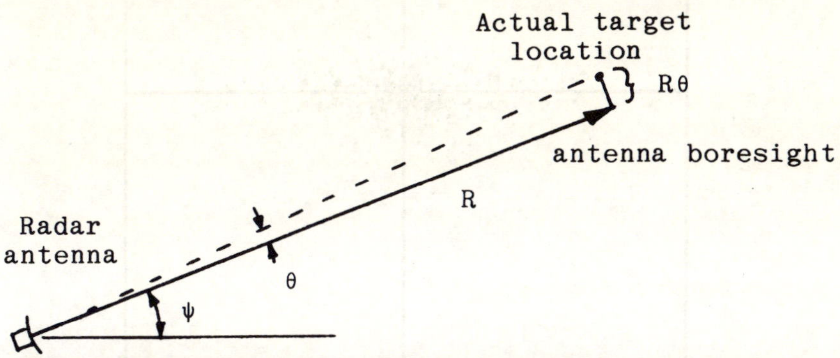


**Figure 14-22.** Off-boresight angle determination.

Another type of off boresight angle display is a transverse profilometer (C-scan) which displays the elevation *versus* azimuth angle cut at a fixed range to the terrain. Also angle profiles at several ranges can be presented on a common display as shown in Figure 14-23.

One further off boresight processing display is one which displays the real component $\beta_R = |\Delta_{EL}| / |\Sigma| \cos \Theta$ *versus* the orthogonal imaginary component $\beta_i = |\Delta_{EL}| / |\Sigma| \sin \phi$ as shown in Figure 14-24. The

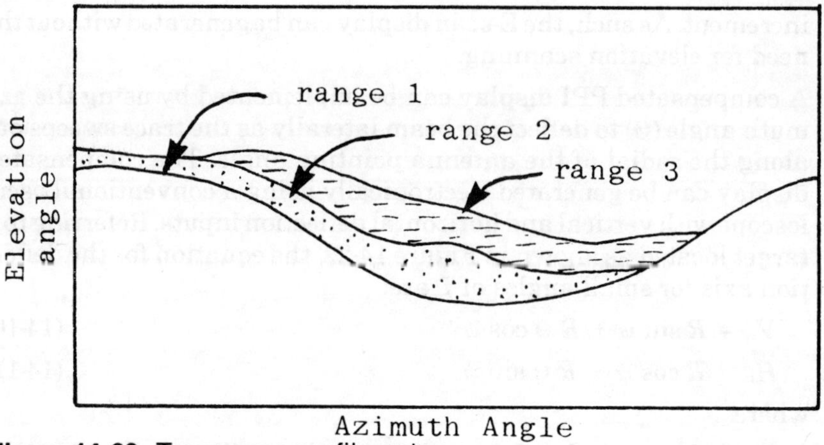


**Figure 14-23.** Transverse profilometer.

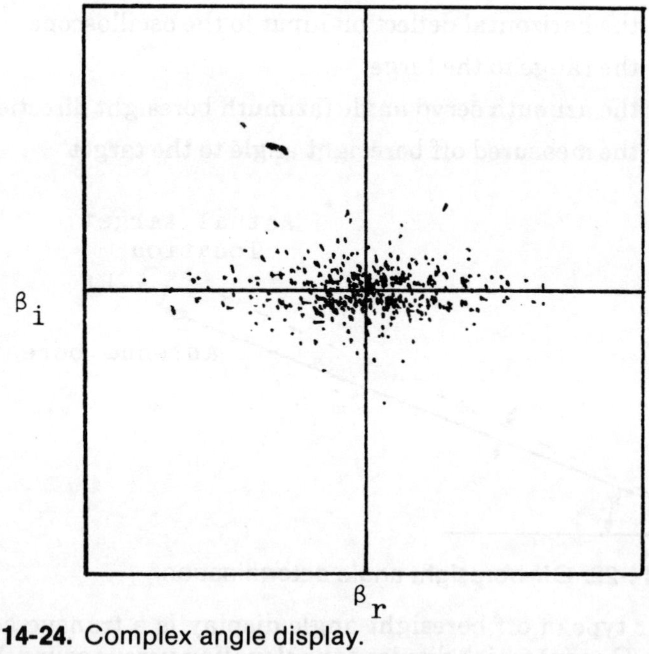


**Figure 14-24.** Complex angle display.

imaginary component is indicative of the angular spread of a complex target in the antenna beam.

### 14.4.5 Data Analysis

Processing of the measured angular data $\beta_r$ and $\Theta_r$ provides the ability to accurately determine the true elevation and azimuth angles normal to the received target return wavefront. For a dynamic target, lags in the antenna servo system can result in servo angle readouts which lag the actual position of the target in space. If the off boresight angle functins $\beta_r$ and $\Theta_r$ are calibrated as a function of volts *versus* degrees from boresight, then the measured $\beta_r$ $\Theta_r$ can be recorded and used to compensate for the lag out of the angle position data recordings.

The measurement of the true angular position of the target will be affected by the contaminants mentioned previously which will cause the target track to deviate from the true target trajectory. Often in measurement systems it is necessary to measure the effects of a contaminant such as glint in order to evaluate its possible effects on a tracking radar. In analyzing the effect of glint, for example, it is necessary first to use a location reference (such as those discussed in the previous section) to establish the actual position of the target; and second, to reduce the effects of other contaminants such as system noise, multipath, and cross polarization coupling. In analyzing the measurements it is normally important to correlate the measured parameters to the target parameters such as target type, aspect angle, velocity, direction, *et cetera*.

One technique which is useful for analyzing the characteristics of a monopulse return was proposed by Sherman[15] in which he postulates that the angular return from an extended target is complex in nature, and that the imaginary component of the target return is a function of the target extent. A point-source type of target will have no imaginary angular component ($\beta_i$ or $\Theta_i$), while for a complex target producing glint the measured $\beta_i$ or $\Theta_i$ will be a measure of the glint in that axis of the radar. Multipath will also produce an imaginary angular component as shown by Howard and Sherman[15] in Figure 14-25.

## 14.5 EXAMPLES OF ANGULAR MEASUREMENTS

Angular measurements can be grouped into two general classes, static range measurements and dynamic. For static range measurements the target or model of the target is situated on a turntable or a pylon, or is suspended from a gimbaled suport structure. In all these cases the aspect angle of the target can be precisely controlled, and the support can be rotated in a continuous fashion to enable complete characterization of the target.

For dynamic measurements, the target vehicle is observed during a portion of its trajectory. The target may be cooperative in the sense that its flight path is controlled for the purposes of the tests, and

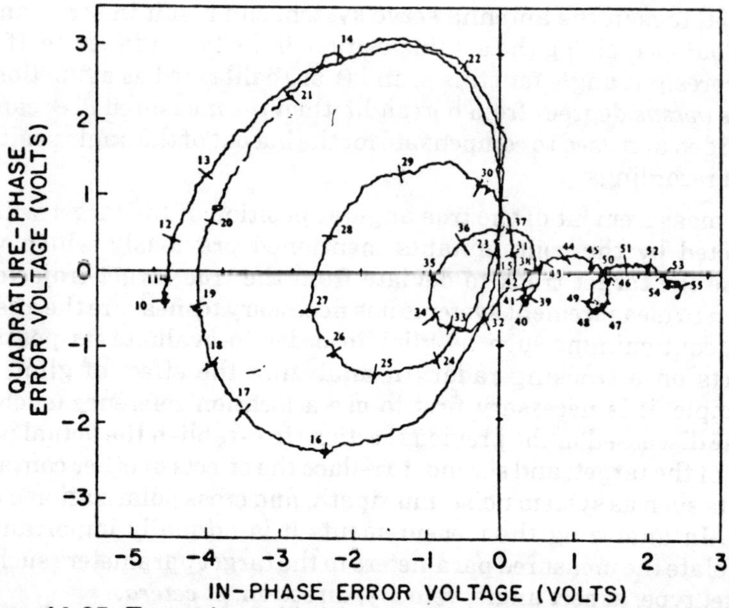


**Figure 14-25.** Target in multipath environment (reprinted from Howard and Sherman, ref. [16]; ©1974 IEEE).

may contain instrumentation which will supply data to help analyze the measured data. The target may also be a target of opportunity, in which case the only data obtained on its flight are these recorded by the sensor.

Included in this section are a few selected examples of the different types of measurements.

### 14.5.1 Static Multipath Measurements

Measurements were performed by Howard and Sherman[16] using a 5.83 GHz radar with the antenna boresight elevation angle pointed up 12.2 mrad. A target was situated on a tower 2550 feet from the radar, and the multipath reflective surface was an asphalt parking lot between the radar and the tower. The target was moved up the tower in one foot increments, which resulted in the measurement of in-phase and and quadrature phase error components as shown in Figure 14-25. The in-phase component of the error voltage is that which would be tracked by a radar, showing that potentially large tracking errors can be caused by multipath.

The Air Force Radar Target Scatter Facility (RATSCAT) has the capability of measuring glint from aircraft mounted on a rotating pylon. An example of a glint measured from a target drone [17] is shown in Figure 14-26.

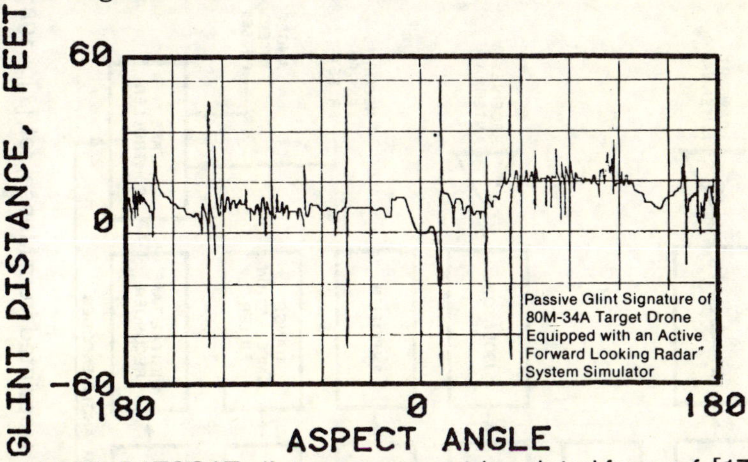


**Figure 14-26.** RATSCAT glint measurement (reprinted from ref. [17]).

### 14.5.2 Doppler Target Imaging Measurements

A coherent Doppler system can be configured to provide both fine range resolution and cross range resolution. An example of such a system described by Mensa[7] is shown in Figure 14-27 along with the image of a missile as shown in Figure 14-28.

### 14.5.3 Dynamic Measurement of Glint

An example of a glint measurement error is shown by Ewell[19] in Figure 14-29 for a C-45 aircraft. The probability distribution function of the measurement error for the aircraft is shown in Figure 14-30 and the probability density is shown in Figure 14-31. A major portion of the glint scintillations from the aircraft have frequency components of less than three Hertz, as were shown in Figure 14-17. This is a typical servo bandwidth for a tracking system. Thus the typical tracking radar would respond to a major portion of the glint scintillation thus resulting in erratic angle tracking. One effective method for reducing the glint tracking errors is to employ pulse-to-pulse frequency agility, in which the transmit frequency is rapidly scanned over a frequency range sufficient to decorrelate the predominant scatterers. While the glint will still be present in the same magnitude, the spectral energy distribution of the glint will be shifted to a higher frequency, and as a result the servo bandwidth will filter out most of the glint returns. A small amount of wander about the aircraft will still be experienced due to changes in the apparent centroid of the target returns.

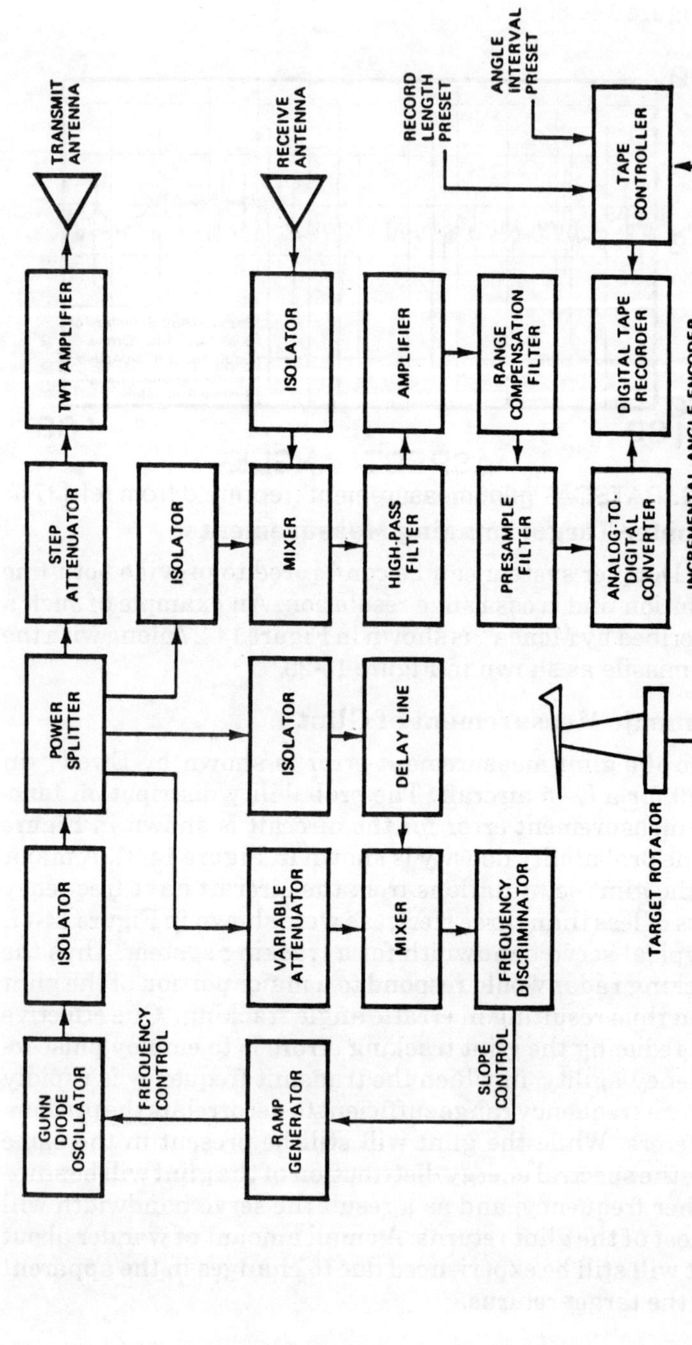


**Figure 14-27.** Block diagram of Doppler measurement system (reprinted from Mensa, ref. [7]; ©1981 Artech House, Inc.).

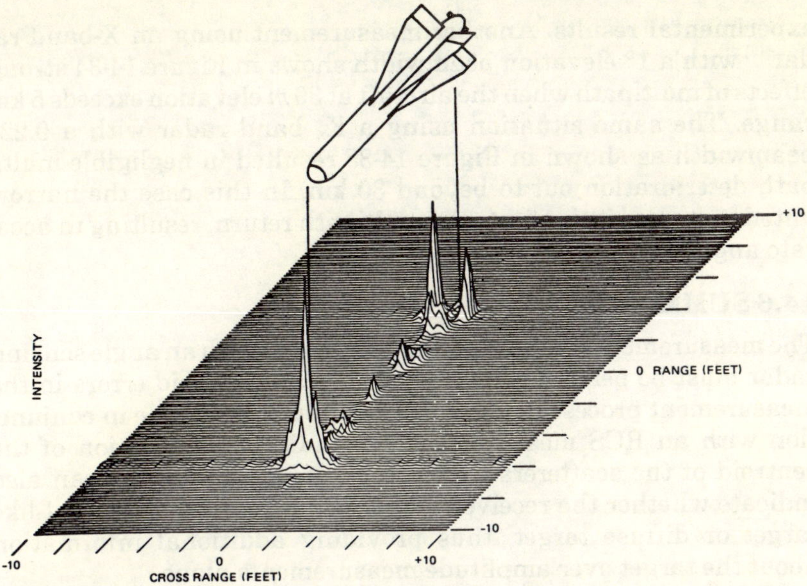


**Figure 14-28.** Image of missile body (reprinted from Mensa, ref. [7]; © 1981 Artech House, Inc.).

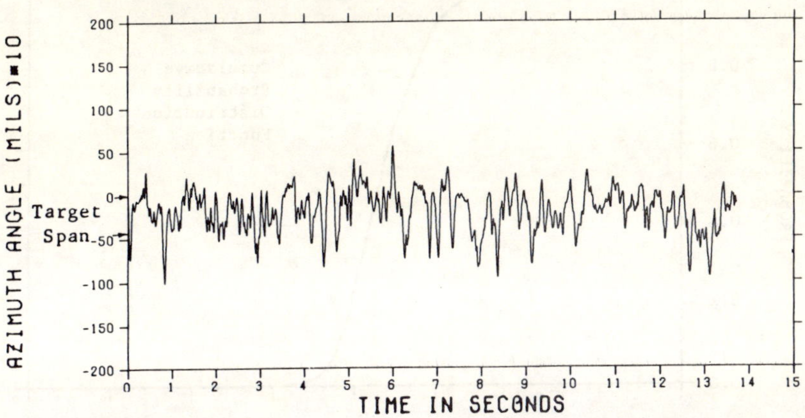


**Figure 14-29.** Angular errors for C-45 aircraft (reprinted from Ewell, ref. [19]; ©1981 Georgia Institute of Technology).

### 14.5.4 Dynamic Multipath Measurements

Multipath produced by tracking a low flying aircraft over Long Island Sound[18] is shown in Figure 14-32 along with the computed multipath effect shown in Figure 14-33. A comparison between the two figures shows good agreement between the computed and the experimental results. Another measurement using an X-band radar[19] with a 1° elevation beamwidth shows in Figure 14-34 strong effects of multipath when the aircraft at 30 *m* elevation exceeds 5 km range. The same situation using a $K_a$ band radar with a 0.22° beamwidth as shown in Figure 14-35 resulted in negligible multipath deterioration out to beyond 30 km. In this case the narrow antenna beamwidth rejects the multipath return, resulting in accurate angle tracking.

## 14.6 SUMMARY

The measurement of angle (or angular errors) with an angle sensing radar must be performed with extreme care to avoid errors in the measurement process itself. The measurement of angle in conjunction with an RCS measurement can indicate the location of the centroid of the scatterers within the antenna beam and can also indicate whether the received signal is due to a discrete point-like target or diffuse target, thus providing additional information about the target over amplitude measurements alone.

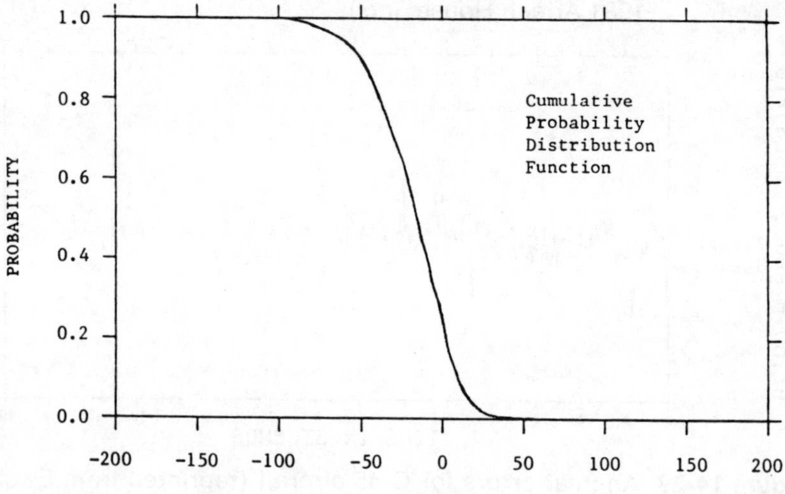


**Figure 14-30.** Probability distribution of angle error for C-45 aircraft (reprinted from Ewell, ref. [19]; © 1981 Georgia Institute of Technology).

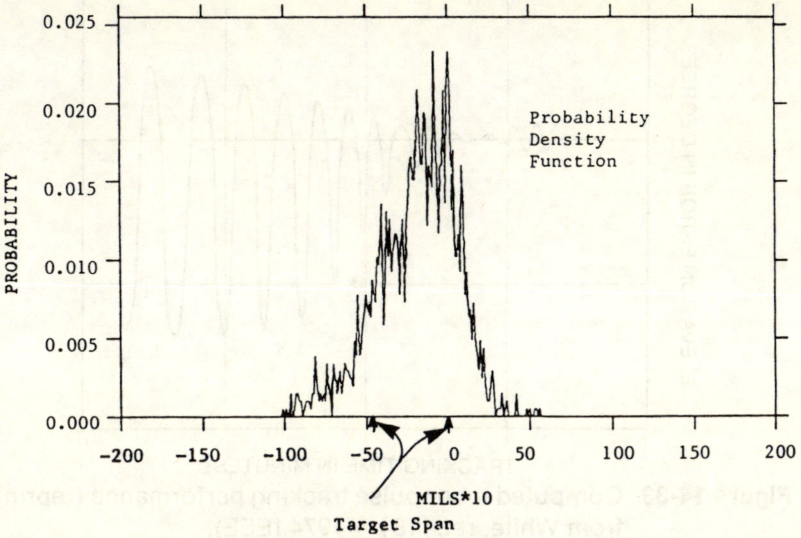


**Figure 14-31.** Probability density function of angle error for C-45 aircraft (reprinted from Ewell, ref. [19];© 1981 Georgia Institute of Technology).

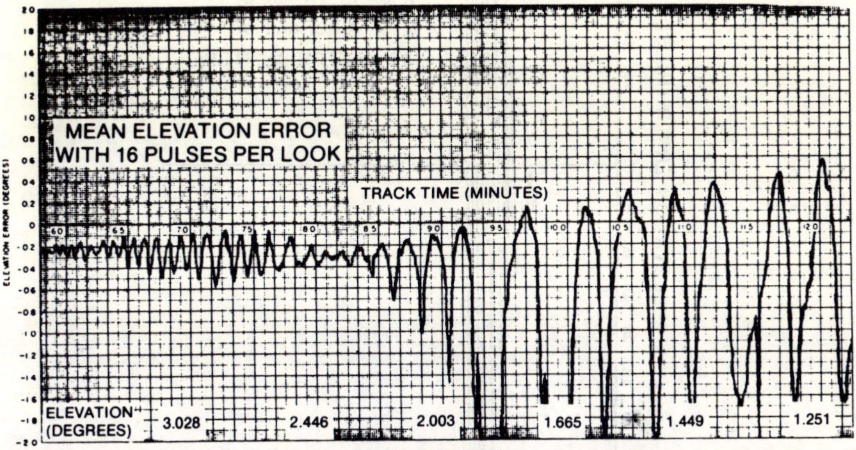


**Figure 14-32.** Experimental monopulse tracking performance (reprinted from White, ref. [18]; © 1974 IEEE).

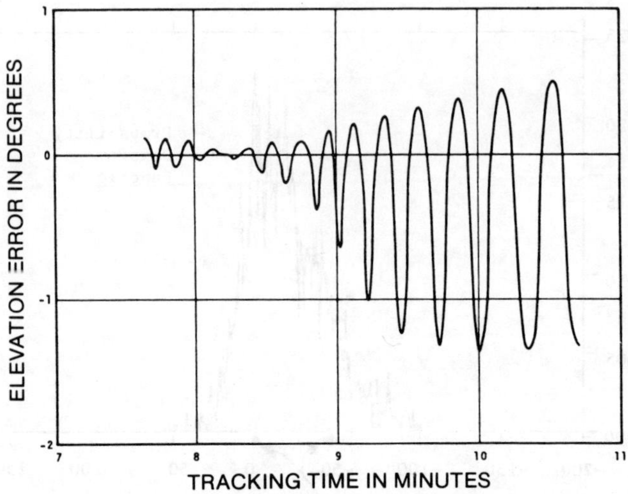


**Figure 14-33.** Computed monopulse tracking performance (reprinted from White, ref. [18]; © 1974 IEEE).

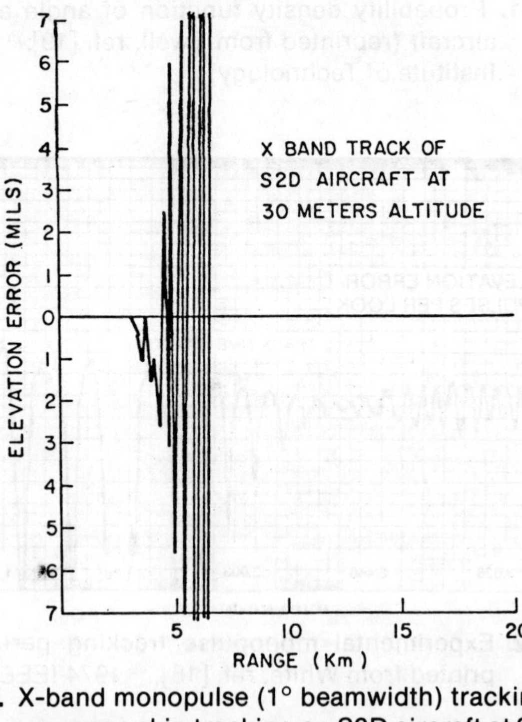


**Figure 14-34.** X-band monopulse (1° beamwidth) tracking error versus range skin tracking on S2D aircraft at 33 *m* altitude (reprinted from Cross, *et al.*, ref. [19];© 1976 *Microwave Journal*).

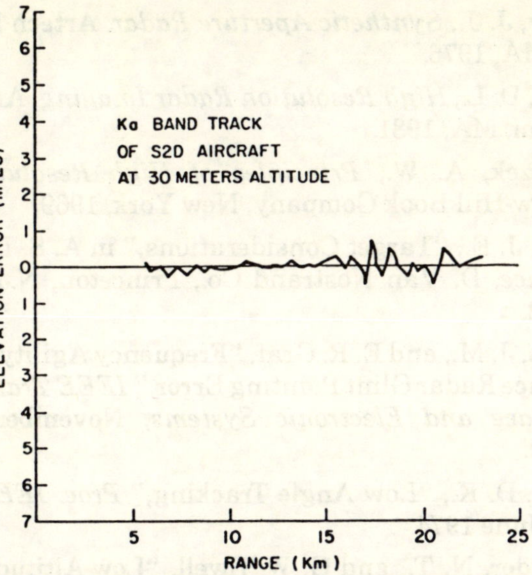


**Figure 14-35.** $K_a$ band monopulse (0.22° beamwidth) tracking error *versus* range skin tracking on S2D aircraft at 33 *m* altitude (reprinted from Cross, *et al.*, ref. [20];© 1976 *Microwave Journal*).

## REFERENCES

1. Dunn, J. H., and D. D. Howard, "Radar Target Amplitude, Angle, and Doppler Scintillation from Analysis of the Echo Signal Propagating in Space," *IEEE Transactions on Microwave Theory and Techniques*, Vol. MTT-16, No. 9, Sept. 1968, pp. 715-728.
2. Howard, D. D., "Radar Target Angular Scintillation in Tracking and Guidance Systems Based Upon Echo Signal Phase Front Distortion," *Proceedings of NEC*, Vol. 15, 1959, pp. 840-849.
3. Barton, D. K., *Radar Systems Analysis*, Artech House, Dedham, MA, 1976.
4. Rhodes, D. R., *Introduction to Monopulse*, McGraw-Hill Book Company, New York, 1975.
5. Moore, Richard K., *et al.*, "Microwave Remote Sensors," Chapter 9 of *Manual of Remote Sensing*, American Society of Photogrammetry, Falls Church, Virginia, 1975, pp. 399-537.

6. Kovaly, J. J., *Synthetic Aperture Radar*, Artech House, Dedham, MA, 1976.

7. Mensa, D. L., *High Resolution Radar Imaging*, Artech House, Dedham, MA, 1981.

8. Richaczek, A. W., *Principles of High Resolution Radar*, McGraw-Hill Book Company, New York, 1969.

9. Meade, J. E., "Target Considerations," in A. S. Locke, *et al., Guidance*, D. Van Nostrand Co., Princeton, N.J., 1955, pp. 420-442.

10. Loomis, J. M., and E. R. Graf, "Frequency Agility Processing to Reduce Radar Glint Pointing Error," *IEEE Transactions of Aerospace and Electronic Systems,* November 1974, pp. 811-820.

11. Barton, D. K., "Low Angle Tracking," *Proc. IEEE,* Vol. 62, No. 6, June 1974.

12. Alexander, N. T., and G. W. Ewell, "Low-Altitude Tracking Experiments," Final Report on Contract DAAK40-77-C-0192, Georgia Institute of Technology, Atlanta, Georgia, September 1978.

13. Leney, T. F., "Effects of Pre-Hybrid and Post-Hybrid Errors on Monopulse Radars," Masters Thesis, State University of New York at Buffalo.

14. Ewell, G. W., "Monopulse Performance When Receiving Nonuniform Waves," *IEEE Transactions on Aerospace and Electronic Systems*," May 1971.

15. Sherman, S. M., "Complex Indicated Angles in Monopulse Radar," Ph.D. Dissertation, University of Pennsylvania, Philadelphia, PA, December 1965.

16. Howard, D. D., and S. M. Sherman, "Experimental Results of the Complex Indicated Angle Technique for Multipath Correctin," *IEEE Transactions of Aerospace and Electronic Systems*, November 1974, pp. 779-787.

17. "The Air Force Radar Target Scatter Facility (RATSCAT) Facilities and Capabilities," U.S. Air Force Capabilities Brochure, 1978.

18. White, W. D., "Low Angle Radar Tracking in the Presence of Multipath," *IEEE Transactions on Aerospace and Electronics Systems,* November 1974, pp. 835-852.

19. Ewell, G. W., "Radar Angle Tracking," Section 6.1 of *Principles of Modern Radar*, Georgia Tech Short Course, November 1981.

20. Cross, D., *et al.*, "TRAXKX: A Dual Frequency Tracking Radar," *Microwave Journal,* September 1976, pp. 39-41.

# CHAPTER 15

# RADIOMETRIC MEASUREMENTS

**Robert D. Hayes**
**RDH, Inc.**



## 15.1 RADIOMETER SENSORS

It is known that every physical body not at absolute zero emits electromagnetic radiation at all frequencies, by virtue of the thermal energy the body possesses. This radiation is called thermal radiation, even though one may be speaking of the radio or optical portion of the emission spectrum. A blackbody is a body that absorbs all electromagnetic energy that falls on it. No such physical object exists, but many bodies, or arrangements of them, approximate a blackbody closely, at least over a small range of frequencies.

About 1900, Planck[1] presented a radiation law relating the brightness, $\beta$, of a blackbody at a temperature $T$ to frequency $f$ (*see* Sproull[1]). The brightness is the amount of power arriving at a unit area, over a unit solid angle, and in a given frequency interval; this can be stated as

$$\beta = \frac{2hf^3}{c^2} \frac{1}{e^{hf/kT} - 1} \; [watts\ m^{-2}\ rad^{-2}\ Hz^{-1}] \tag{15-1}$$

where

$h$ = Planck's constant (= 6.55 x $10^{-34}$ joule-sec)

$f$ = source frequency, hertz

$c$ = velocity of light (= 3 x $10^{8}$ meters/sec)

$k$ = Boltzmann's constant (= 1.38 x $10^{-23}$ joules/°K)

$T$ = source temperature (° K)

Planck's law holds strictly for blackbodies. Assuming that the sun radiates as a thermal source (blackbody) at thousands of degrees Kelvin, and that the observed frequency is in the region of a few hundred gigahertz (GHz) or less, then $hf << kT$. The brightness, with this approximation, reduces to the Rayleigh-Jeans law of the form

$$\beta = \frac{2f^2kT}{c^2} \tag{15-2}$$

Most bodies do not absorb perfectly, so they cannot be called blackbodies; instead, they are called gray bodies. Suppose a gray body absorbs only a fraction, $\sigma(f)$, of the incident radiation. (We note that $\sigma$ may be a function of frequency, as well as of the physical configuration of the material). Let $B(f)$ be the monocohromatic brightness of the gray body. A law due to Kirchhoff states that (*see* French[2]):

$$B(f) = \beta(f)\sigma(f) \tag{15-3}$$

Therefore, the monochromatic brightness of a gray body is always less than that of a blackbody.

Whatever energy a body does not absorb, it reflects. As $B(f)$ tends toward zero, the body becomes less emissive and more reflective. A perfect reflector emits no thermal radiation at all, but just as for blackbodies, there are no perfect reflectors either. The brightness temperature of a (gray) body is a frequently used concept. This is the temperature at which a true blackbody would have the same monochromatic brightness as the observed gray body at the frequency of observation.

For a source of random polarization, such as a thermal source, the power received by an antenna of a single polarization is given by

$$P = \frac{1}{2} \iint A \cdot \beta \cdot d\Omega \cdot df \; [watts] \tag{15-4}$$

where

$A$ = effective antenna area in specified direction ($m^2$)

$d\Omega$ = element of solid angle from the antenna

$df$ = frequency interval

The effective antenna area, $A$, is related to the antenna gain, $G$, specified with respect to an isotropic antenna by

$$A = \frac{C^2 G}{4\pi f^2} \tag{15-5}$$

Now, by substitution

$$P = \frac{1}{2} \iint \frac{C^2 G}{4\pi f^2} \frac{2f^2}{C^2} \cdot kT \cdot d\Omega \cdot df$$
$$= \frac{1}{4\pi} \iint Gkt \cdot d\Omega \cdot df \tag{15-6}$$

If the object being observed is large, relative to the antenna beamwidth, then the brightness of the object is independent of the beam position. This is the case with an extended source, where the temperature is constant in the region of the solid angle, $d\Omega$, subtended by the antenna. If the antenna has a beamwidth less than the subtended angle of the object and has low sidelobes, the radiation received from the other sources is negligible and the integrated gain function, $G$, over the solid angle is $4\pi$; the expression for the received power then becomes

$$P = k \int T \, df \tag{15-7}$$

This equation is usually presented in the form

$$P = kT_a \int df \tag{15-8}$$

where $T_a$ is the effective blackbody temperature of the extended source as seen by the antenna over the frequency region $\int df$. The power received by a radiometer, then, is linearly related to the temperature of the source at which the antenna is pointed.

With these preliminaries, one may now make a few general comments about radiometers. A radiometer is merely a device for noting the reception of radiant energy. Usually, the radiometer is a quantitative instrument as well, measuring in some way the amount of incident energy. For example, a thermometer with a bulb coated with lampblack could be used as a radiometer in the infrared region. In the radio wave region of the electromagnetic spectrum, a radio receiver is the most satisfactory radiometer, although such devices as bolometers have their uses in the millimeter-wave range.

In 1928, Nyquist[3] observed that electrons moving at random in a resistor would produce a voltage across that resistor. When this resistor is placed as a matched load termination on a generator, then the noise power available is given by

$$P = kT\Delta f\,[watts] \tag{15-9}$$

or

$$P = kTB\,[watts] \tag{15-10}$$

This is the same power observed from an antenna when the antenna is enclosed in a box which is radiating as a blackbody. The concept of power transfer in an impedance-matched system is satisfied and the conservation of energy is fulfilled.

The noise figure of a linear two port network which is matched at both input and output is defined as

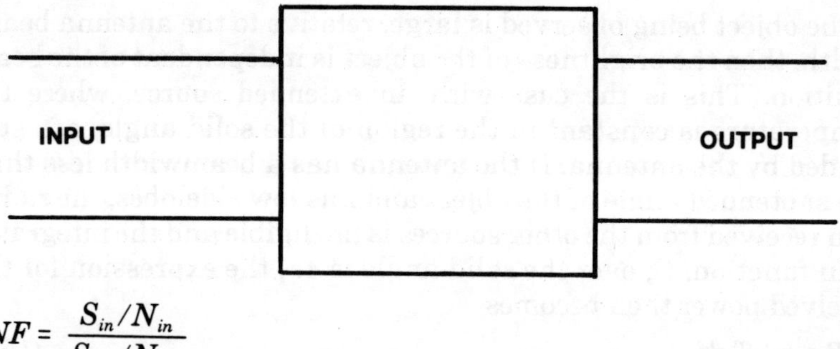


$$NF = \frac{S_{in}/N_{in}}{S_{out}/N_{out}} = \frac{N_{out}}{N_{in}}\frac{S_{in}}{S_{out}} = \frac{N_{out}}{kT_oB}\frac{1}{G} \tag{15-11}$$

where

$G$ = network gain

$T_o$ = room temperature

Now the output noise, $N_{out}$, can be written as the input noise plus the noise generated inside the network, or

$$N_{out} = kT_oBG + N$$
$$= (kT_o B + kT_e B)G \qquad (15\text{-}12)$$

where $T_e$ is effective temperature introduced by the network.

Thus,

$$NF = \frac{kT_oBG + kT_eBG}{kT_oBG} \qquad (15\text{-}13)$$

or

$$NF = 1 + \frac{T_e}{T_o} \qquad (15\text{-}14)$$

In the familiar form of equivalent temperature

$$T_e = T_o (NF - 1) \qquad (15\text{-}15)$$

When several networks are cascaded, then the total noise power at the output can be written as

$$P = kB (T_1G_1G_2G_3 \ldots + T_2G_2G_3 \ldots + T_3G_3 \ldots + \ldots) \qquad (15\text{-}16)$$

and the overall effective temperature becomes

$$T_e = T_1 + \frac{T_2}{G_1} + \frac{T_3}{G_1G_2} + \ldots \qquad (15\text{-}17)$$

or in terms of noise figure

$$NF_e = NF_1 + \frac{NF_2 - 1}{G_1} + \frac{NF_3 - 1}{G_1G_2} + \ldots \qquad (15\text{-}18)$$

It is important to list the effective temperature and noise figure of some passive elements. The noise temperature of an attenuation which has a loss value of $L$ is given by

$$T_{att} = T_o (L - 1) \qquad (15\text{-}19)$$

and the noise figure is

$$NF_{att} = L \qquad (15\text{-}20)$$

From the two concepts of cascaded networks and attenuators, it is a simple matter to determine the effective temperature and noise figure of an amplifier connected to an antenna by way of a lossy transmission line at temperature $T_o$, where $L = 1/G$.

$$\textit{System temperature} = T_{ant} + T_e \qquad (15\text{-}21)$$
$$= T_{ant} + LT_r + T_o (L - 1) \qquad (15\text{-}22)$$

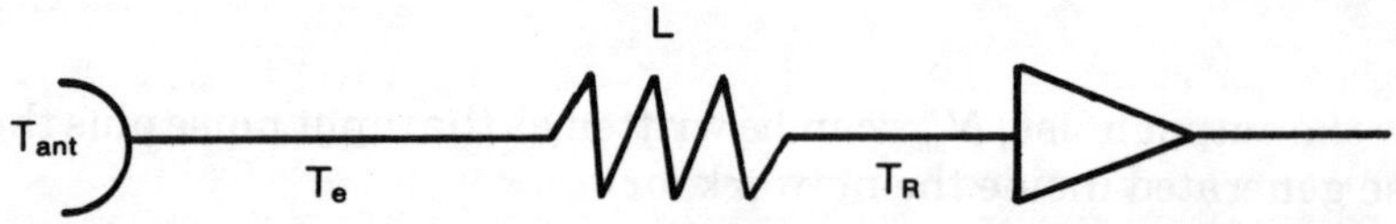


and from the noise figure of cascade networks, the overall noise figure is given by

$$NF_{system} = NF_1 + \frac{NF_2 - 1}{G}$$

$$= L + L(NF_R - 1)$$

$$= LNF_R \tag{15-23}$$

Noise in a radiometer is thus the limiting factor in determining the minimum detectable temperature change, and thus, discrimination between surfaces.

## 15.2 RADIOMETRIC MEASUREMENTS

### 15.2.1 Emissivity

Any object at temperature $T$ has a state of thermal energy and emits electromagnetically as previously discussed. Under the concept of Kirchhoff's law, this spectral emission is equivalent to spectral absorption. In a model of a surface, the energy incident upon the body will be partially absorbed, partially reflected, and partially transmitted. With the requirements of conservation of energy, and taking advantage of such a theorem in network synthesis,[4] it has been shown that for a gray body

$$\rho + \epsilon + \tau = 1 \tag{15-24}$$

where

$\rho$ = reflectivity

$\epsilon$ = emissivity

$\tau$ = transmissivity

Consider the material and physical conditions shown below:

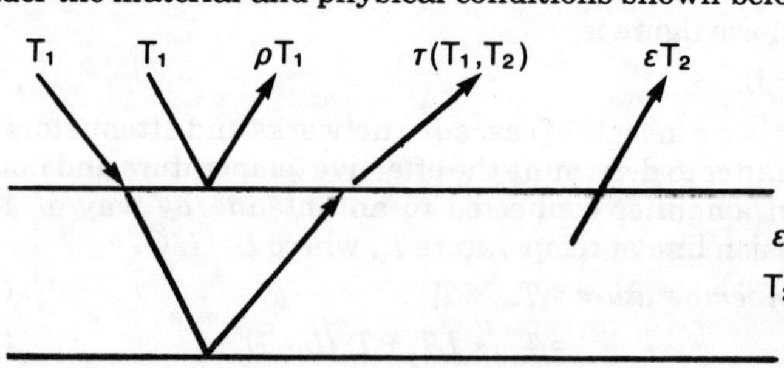

The total effective temperature received by the radiometer is a function of the material properties and the material temperature, as well as the temperature reflected by the material. For a homogeneous body several skin-depths thick, the effects of transmissivity are neglected, and to a first order approximation only the emissivity and reflectivity will contribute to the measured effective temperature. The network equation reduces to

$$\rho + \epsilon = 1 \qquad (15\text{-}25)$$

The quantities of $\rho$, $\epsilon$ and $\tau$ are molecular properties of the material, and usually are functions of frequency. In addition to the molecular properties, the material shape, surface roughness, and bistatic viewing angle will affect the total power reflected from the sources and thus, the equivalent temperature observed by the radiometer.

### 15.2.2 Passive Cross Section

The passive detection of metallic ground based targets is dependent on the radiometer temperature contrast between the background and the apparent target temperature. Consequently, as the antenna beam passes from the background to the target, a temperature differential is observed which is a function of the ratio of target to beam intercept area. This relationship can be expressed as:

$$\Delta T_A = C \Delta T_t \; \frac{A_t}{A_b} \qquad (15\text{-}26)$$

where

$\Delta T_A$ = the observed temperature change

$\Delta T_t$ = the target-to-background temperature contrast

$A_t$ = the target area projected normal to the line of sight

$A_b$ = the 3 dB beam area normal to the line-of-sight at the target range, and

$C$ = an antenna constant of proportionality for small beam fill targets ($C < 1$).

Determination from measured data of the values for $\Delta T_t$, $A_t$, and $C$ to the desired degree of accuracy is difficult to achieve. The expression $\Delta T_t A_t$ has been selected as the measure of target cross section, rather than the usual $\Delta T_t A_t C$, partially because it follows naturally from the usual measurement techniques and patially because it is of greater utility in systems analysis. The value of $C$ is dependent upon the efficiency of the antenna, the geometry of the target, and the target beam fill ratio; consequently, measurements containing $C$ in the cross sectional values specify the target cross section as measured within the constraints of the measurement antenna.

It is essential to consider the effect of sidelobes on passive measurements to reduce the signal-to-noise ratio of the measured contrast. The effect of the poor sidelobes manifests itself as the reduced antenna proportionality constant, $C$, previously discussed. However, using a calibration radiometer and known size flat-plate reflectors, the antenna proportionality constant can be eliminated from the measurement. This fact, however, requires that the contributions from other sources through the sidelobes remain the same for both the calibration measurements and target measurements. This appears to be a reasonable assumption since the target and calibration reflectors can be located in the same general background areas, and all should be measured from the same depression angles within a short time span. When a calibrated radiometer is not available, the temperature differential can be determined by comparison with the minimum detectable temperature ($T_{min}$) measured for the system. When a high (>10) noise figure ($NF_R$) exists, the effects of observed antenna temperature ($T_A$) variations on the output noise level can be seen from the expansion for $T_{min}$ as given by Taylor for a Dicke[5] type radiometer,

$$\Delta T_{min} = \frac{1}{\pi^{1/2}} \frac{(NF_R - 1)T_o + T_A}{(\tau_{rc} B_R)^{1/2}} \tag{15-27}$$

to be on the order of 2% ($T_A$ is typically 240° K to 310° K) where

$NF_R$ = the receiver noise figure

$B_R$ = the receiver bandwidth

$\tau_{rc}$ = the post-detection filter time constant, and

$T_o$ = 290° K

The assumption of constant sidelobe contributions as the beam scans the target usually is made when high gain and mechanically stable antennas or standard gain horns are employed. Consequently, the sidelobe contributions have known patterns and proportionality constants. Using the receiver output noise level to establish the scale factor, the value of the observed temperature change, $\Delta T_A$, between the background and target can be calculated, and the value of $C$ from measurements can be evaluated. From a form of Equation (15-26), the cross section can be calculated to be

$$\Delta T_t A_t = \frac{\Delta T_A A_b}{C} \tag{15-28}$$

This technique for measuring target contrast is not as desirable as using a calibrated receiver since the accuracy of the measurement is based on the continued constancy of the system noise figure. For the case where $T_A \cong T_o$ (as is a typical measurement case), the signal-to-noise ratio of the measured contrast can be stated as

$$\frac{S}{N} = \frac{\Delta T_A}{\Delta T_{min}} = (C\Delta T)\,\frac{A_t}{A_b}\cdot\frac{(B_R\tau_{rc})}{K_R F_R T_o} \qquad (15\text{-}29)$$

where $K_R$ is the receiver constant dependent upon the receiver configuration. From this expression, it is easy to see that the reduction of the antenna proportionality constant results in a lower signal-to-noise ratio. From this expression, the expected signal-to-noise ratio for the system under clear skies and low humidity can be determined, depending on the percentage beam fill of the target.

To extend the concept of what temperature will be observed by the antenna, it is required to know the antenna far-field pattern. In a well designed antenna, the distribution of power in the far field will generally occur as follows:

a. 60% to 65% of power within $\Theta_{3\,dB}$

b. 35% to 40% of power within remainder of main beam

c. 1% to 3% of power within sidelobes

The effective temperature seen by a radiometer is a combination of the temperatures entering through the various portions of the antenna beam. As a typical measurement configuration, it is reasonable to assume: half of the total sidelobe power is derived from the ground, half is derived from the ambient air mass, and the main beam of the antenna is directed at the ground or the target area of interest. The radiometrically measured temperature for a beam-filled extended surface is given by

$$T_{measured} = E\epsilon T_{ground} + E(1-\epsilon)T_{sky} + (1-E)\,\frac{T_{air} + T_{ground}}{2} \qquad (15\text{-}30)$$

where

$E$ = antenna efficiency

$\epsilon$ = emissivity of ground

$T$ = temperature as subscripted

If the target area contains two materials, such as a large metal object and ground, then the partial beam-fill conditions exist and the observed temperature is given by

$$T_{measured} = E \frac{A_{metal}}{A_{beam}} (1 - \epsilon_m) T_{sky} + E\left(1 - \frac{A_{metal}}{A_{beam}}\right) \epsilon_g T_{ground} + (1 - E)\left(\frac{T_{air} + T_{ground}}{2}\right) \quad (15\text{-}31)$$

where

$A$ = area as subscripted

$\epsilon_m$ = emissivity of metal

$\epsilon_g$ = emissivity of ground

$E$ = antenna efficiency

$T$ = temperature as subscripted

### 15.2.3 Field Measurement Calibration

A flat plate reference target should be adjusted in tilt angle so that the zenith of the sky is reflected toward the radiometer, as shown by Figure 15-1.

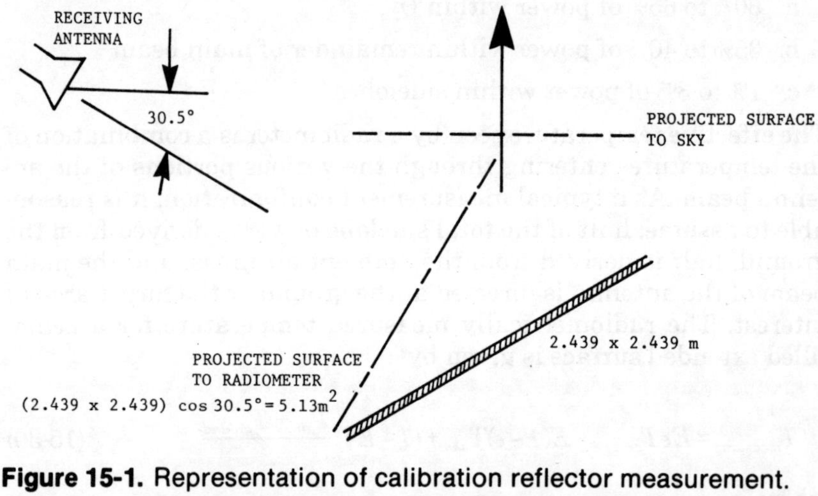


**Figure 15-1.** Representation of calibration reflector measurement.

Data on atmospheric attenuation when looking at the zenith can be used to estimate the expected temperature contrast between a ground area only and a ground area containing a tactical target. The effective sky temperature is then adjusted by measuring the zenith sky temperature with a calibrated radiometer. The following discussion is related to expected sky temperatures, the target-to-ground contrast temperatures, and those values which might be obtained during measurements. For example, for field measurements:

$$T_{sky} = T_{space}t + T_{atmos}(1 - t) \tag{15-32}$$

$$50 = 3t + 298(1 - t)$$

$$t = (298 - 50)/(298 - 3) = 0.84067 \tag{15-33}$$

$$\alpha = 0.7537 \text{ dB/air mass}$$

where:

$T_{sky}$ = sky temperature

$T_{space}$ = apparent temperature of space (3°K)

$t$ = atmospheric transmission

$\alpha$ = attenuation constant.

Using this value of transmission, an effective temperature at ground level can be determined by integrating over the effective sky illuminating the target area. The atmospheric losses, the sky temperature as a function of angle, and the integrated sky temperature at the target area, when looking at a given depression angle, can be tabulated. Beam area is a function of range, and thus, from any experiment, is a function of depression angle. The general temperature equation for comparing target and ground only is given by

$$T = \frac{A_{target}}{A_{ground}}(T_{ground} - T_{target})E \tag{15-34}$$

where $E$ is the antenna energy efficiency factor ($E = 0.60$ for a horn) within the half-power beamwidth.

For example, if the sky temperature was measured to be 50°K and the ground temperature was 298°K, then from Equation (15-34):

$$T = [(5.13)/(43.2)](298 - 50)(0.6) = 17.7°\text{K} \tag{15-35}$$

The actual measured temperature could range ±2°K. Considering the errors associated with orienting an 8 foot by 8 foot plate and a typical voltmeter indicator, this would be good agreement between measured and calculated temperature differences.

## 15.3 EXAMPLE PROBLEMS

The following examples illustrate typical radiometric temperature calculations.

### 15.3.1 Example One, Radiometric Detection of Vehicle (16 GHz)

**Temperature of target (reflectivity of 1) can be approximated to be**

$$4° \ \frac{140}{180} + 6° \ \frac{20}{180} + 300° \ \frac{20}{180} = 37°K \tag{15-36}$$

**The effective temperature-area as seen by a radiometer in Figure 15-2 when looking at the ground is given by**

$$(T_{gnd})\ [A\,(3dB\,)]\ (efficiency) \tag{15-37}$$

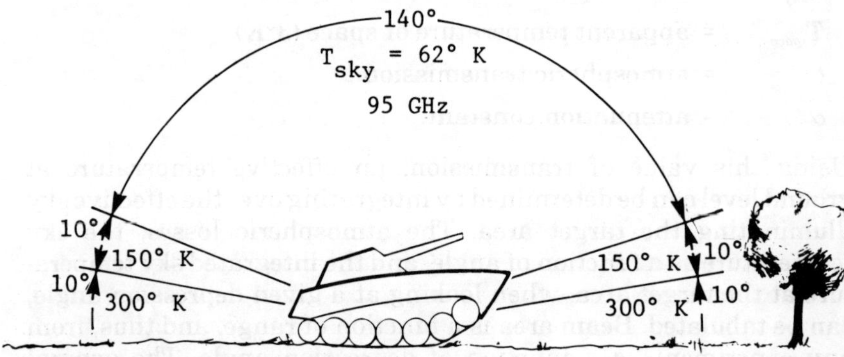


**Figure 15-2.** Example one scenario.

**The effective temperature-area seen by the radiometer when a target having a cross-sectional area of 10 $m^2$ is placed in the footprint then becomes**

$$(T_{gnd})\ (A_{3dB}\ -\ A_{target})\ (efficiency) + (T_{target})(A_{target})(efficiency)$$

**and the contrast between ground only and ground-with-target becomes simply**

$$T = \frac{(T_{ground} - T_{target})\ (A_{target})}{A_{3\,dB}}\ (efficiency) \tag{15-38}$$

**Typical contrast temperatures as a function of range are listed below:**

| Range (*m*) | Δ*T* (°K) |
|---|---|
| 500 | .21 |
| 400 | .33 |
| 300 | .59 |
| 200 | 1.33 |
| 100 | 5.34 |

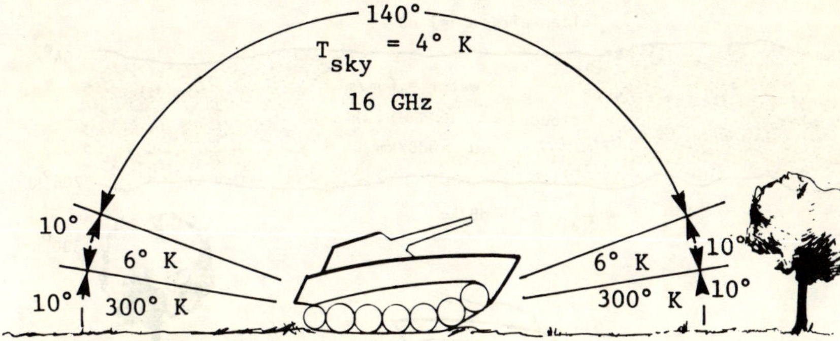


**Figure 15-3.** Example two scenario.

### 15.3.2 Example Two, Radiometric Detection of Vehicle (95 GHz)

Consider now the use of a 95 GHz passive radiometer in the final stage of tracking, as shown in Figure 15-3. The target temperature can be approximated in this case to be

$$60° \frac{140}{180} + 150° \frac{20}{180} + 300° \frac{20}{180} = 98°K \qquad (15\text{-}39)$$

For the same size missile antenna ($D = 0.12\, m$) as used at 16 GHz, the $\Theta_{3dB}$ at 95 GHz becomes 1.87°. With the same care in antenna design and with the same efficiency, *et cetera*, as before, the beam area and temperature contrast are

| Range ($m$) | Beam Area ($m^2$) | $\Delta T$(°K) |
|---|---|---|
| 500 | 208.0 | 5.34 |
| 400 | 133.0 | 8.35 |
| 300 | 75.0 | 14.80 |
| 200 | 33.0 | 33.70 |
| 100 | 8.3 | 134.00 |

which produces adequate temperature contrast, even at 500 meters range, for passive tracking.

### 15.3.3 Example Three, Radiometric Detection in Final Homing Phase

Let us consider in more detail a 95 GHz passive radiometer tracking in the final phase of homing, under an overcast cloudy sky, as shown in Figure 15-4.

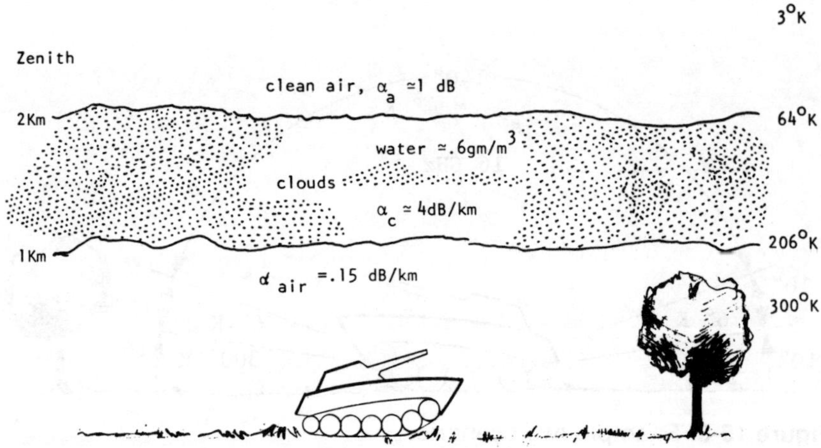


**Figure 15-4.** Example three scenario.

The target temperature illuminated by the sky becomes

$$T_{sky} = 206e^{-\alpha L}\left(\frac{160}{180}\right) + 300\ \frac{20}{180} + 300\,(1 - e^{-\alpha L})\ \frac{160}{180}$$

$$= 177° + 33° + 9°$$

$$= 219°K \tag{15-40}$$

Including path loss, the target-plus-clutter temperature contrast to clutter alone becomes

$$\Delta T = \frac{target\ area}{beam\ area}\ (T_{gnd} - T_{target})\,(e^{-\alpha L})\,(efficiency) \tag{15-41}$$

The beam area and temperature contrast are

| Range (*m*) | Beam Area ($m^2$) | ΔT (°K) |
|---|---|---|
| 500 | 208.0 | 2.1 |
| 400 | 133.0 | 3.2 |
| 300 | 75.0 | 5.7 |
| 200 | 33.0 | 13.0 |
| 100 | 8.3 | 43.0 |

### 15.3.4 Example Four, MMW Communication Systems

To introduce the material which is to follow, the communication range equation will be employed along with propagation parameters and typical component characteristics to demonstrate how systems operating at millimeter wavelengths might be expected to perform. The power received at a point in free space is given by

$$P_r = \frac{G_T P_T}{4\pi R^2} A_e \tag{15-42}$$

where

$A_e$ = effective area of the receiver antenna

$G_T$ = gain of the transmitter antenna

$P_T$ = power output of the transmitter

$R$ = range between transmitter and receiver

The gain of a directional antenna can be expressed in terms of effective aperture size and wavelength by

$$G = \frac{4\pi A_e}{\lambda^2} \tag{15-43}$$

or in terms of the beamwidths, a commonly accepted expression (accounting for efficiency) is

$$G = \frac{27{,}000}{\Theta_{az}\,\Theta_{el}} \tag{15-44}$$

where

$\Theta_{az}$ = half-power azimuth beamwidth in degrees

$\Theta_{el}$ = half-power elevation beamwidth in degrees

The half-power beamwidth of a well designed circular aperture (parabola or horn/lens) with sidelobes about 23 dB below the mainlobe can be expressed by

$$\Theta = 70\,\lambda/D \tag{15-45}$$

where

$D$ = diameter of antenna

$\lambda$ = carrier wavelength

By substitution of the various equations, the received power can be expressed as

$$P_r = \left[\frac{27{,}000}{4\pi\cdot 70^2}\right]^2 \left[\frac{D_R D_T}{\lambda^2}\right]^2 \left[\frac{\lambda}{R}\right]^2 P_t \tag{15-46}$$

Noise is generated in a receiver and will compete with the signal received. The noise is given by

$$N = kTB\cdot NF \tag{15-47}$$

where

$k$ = Boltzmann's constant (= $1.38 \times 10^{-23}$ watts/°K Hz)

$T$ = device temperature, degrees Kelvin

$B$ = receiver bandwidth, Hz

$NF$ = RF to IF noise figure

By substitution and clearing of constants, the signal (received power)-to-noise ratio in free space reduces to

$$S/N = 4.64 \left[\frac{D_R D_T}{\lambda R}\right]^2 \frac{P_T}{B\,NF} \tag{15-48}$$

where

$P_T$ is in milliwatts

$\lambda$, $D_R$, $D_T$ are in centimeters

$R$ is in kilometers

$B$ is in megahertz

When the losses resulting from transmission through trees, rain, and grass are considered, then

$$S/N = 4.64 \left[\frac{D_R D_T}{\lambda R}\right]^2 \frac{P_T}{B \cdot NF\,(L_C R_C)\,(L_W R_W)(L_T R_T)} \tag{15-49}$$

where

$R_c$ = path range containing gas absorption, usually equal to $R$

$R_W$ = path range containing water particles

$R_T$ = path range containing trees

Now consider our example where

Transmit Power = 10 mW CW

Receiver Noise Figure = 10

Receiver Bandwidth = 6 MHz (clear channel TV)

The signal-to-noise equation now reduces to

$$S/N = 174 \left(\frac{1}{\lambda R}\right)^2 \frac{1}{(L_C R_C)\,(L_W R_W)\,(L_T R_T)} \tag{15-50}$$

Values of atmospheric,[6,7] rain,[8] and tree[9,10] attenuation obtained from the literature are tabulated in Table 15-1. Graphical presentations of Equation (15-50) are shown in Figures 15-5 and 15-6.

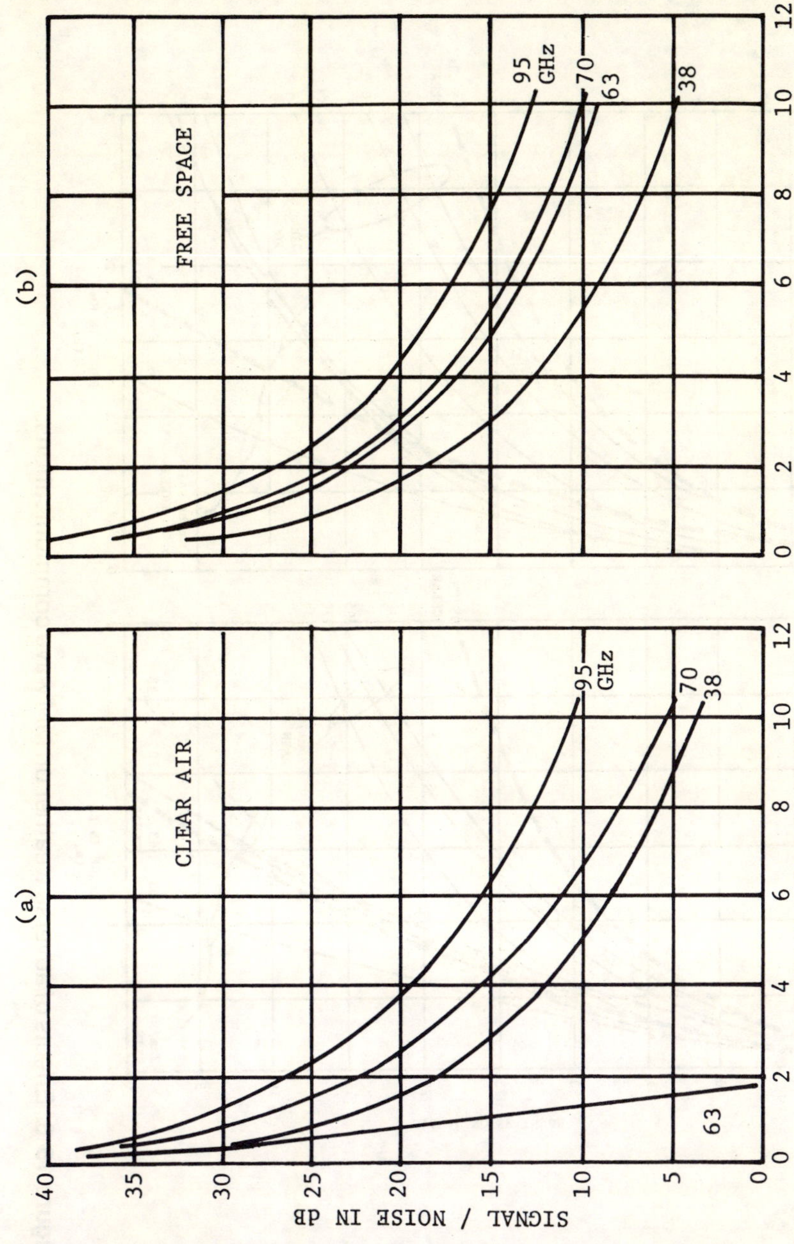


**Figure 15-5.** Signal-to-noise for MM wave communications.

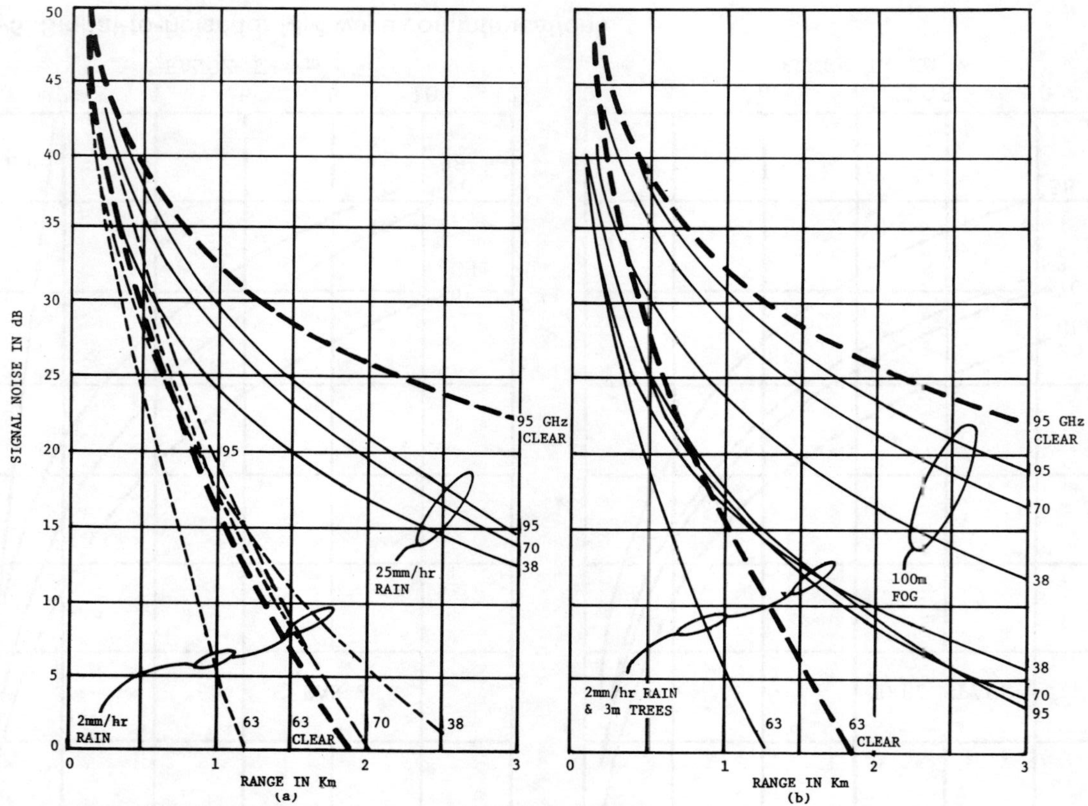


**Figure 15-6.** Effects of adverse weather on MM wave communications.

**TABLE 15-1**
**Propagation Losses**

| FREQUENCY | 38 GHz | 63 GHz | 70 GHz | 95 GHz |
|---|---|---|---|---|
| | | Losses in dB/km | | |
| $L_C$ Molecular $O_2$ and $H_2O$ | 0.068 | 13.0 | 0.45 | 0.18 |
| $L_W$ 2 mm/hr rain | 0.62 | 1.38 | 1.6 | 2.6 |
| $L_W$ 25 mm/hr rain | 6.3 | 10.5 | 11.3 | 13.0 |
| $L_W$ 30 m visibility fog | 2.5 | 5.5 | 7.0 | 12.0 |
| $L_W$ 100 m visibility fog | 0.2 | 0.5 | 0.54 | 1.1 |
| | | Losses in dB/meter | | |
| $L_T$ wet summer trees | 6.3 | 8.4 | 9.0 | 12.0 |
| $L_T$ dry summer trees | 3.5 | 3.9 | 3.95 | 4.1 |
| $L_T$ winter trees | 2.4 | 3.1 | 3.4 | 3.8 |

## 15.4 SUMMARY

Millimeter waves offer a region of the electromagnetic spectrum which to date has neither been fully exploited for passive radiometric detection of objects, nor for line-of-sight wideband communication systems. In this chapter a brief description of radiometric concepts and examples of the ability to passively detect the presence of vehicles, even under adverse weather conditions, is presented. System performance is improved when highly efficient antennas (low sidelobes) are employed along with wideband low-noise receivers.

Wideband communication systems capable of supporting television transmission can be implemented with good signal-to-noise ratios for short haul truck lines (4-5 km) during moderate rains, heavy fog, or through wet trees.

## REFERENCES

1. Sproull, R. L., *Modern Physics*, John Wiley & Sons, New York, 1956.
2. French, A. P., *Principles of Modern Physics*, John Wiley & Sons, New York, 1958.
3. Nyquist, H., "Thermal Agitation of Electric Charge in Conductors," *Physical Review*, Vol. 32, 1928, p. 110.
4. Gillemin, E. A., *Synthesis of Passive Networks*, John Wiley & Sons, New York, 1957.
5. Dicke, R. H., *Review of Scientific Instruments*, Vol. 17, 1946.

6. Hayes, R. D., "Total Atmospheric Attenuation of Millimeter Wavelengths," Doctoral Thesis, Georgia Institute of Technology, Atlanta, Georgia, 1964.

7. Lukes, G. D., "Penetrability of Haze, Fog, Clouds, and Precipitation..." Center for Naval Analysis Study, Report No. 61, May 1968.

8. Currie, N. C., F. B. Dyer, and R. D. Hayes, "Analysis of Radar Rain Return at Frequencies of 9.375, 35, 70, and 95 GHz," Technical Report No. 2 on Contract DAAA 25-73-C-0256, Georgia Institute of Technology, Atlanta, Georgia, February 1975.

9. Currie, N. C., *et al.*, "Millimeter Foliage Penetration Measurements," *IEEE AP-S International Symposium Digest*, October 1976.

10. Hayes, R. D., class notes: "Principles of Radar," School of Electrical Engineering, Georgia Institute of Technology, Atlanta, Georgia.

## ADDITIONAL REFERENCES

1. Evans, G., and C. W. McLeish, *RF Radiometer Handbook*, Artech House, Dedham, Massachusetts, 1977.

2. *Proc. IRE*, "Radio Astronomy Issue," January 1958.

3. Skolnik, M. I., ed., "Passive Detection," *Radar Handbook*, Chapter 39, McGraw-Hill Book Company, New York, 1970.

4. Janza, F. J., ed., *Manual of Remote Sensing*, Vol. I, American Society of Photogrammetry.

5. Button, K. J., ed., *6th International Conference on Infrared and Millimeter Waves*, MTT, December 1981.

6. Hayes, R. D., N. C. Currie, and J. A. Scheer, "Reflectivity and Emissivity of Snow and Ground at MM Waves," *Record of the 1980 IEEE International Radar Conference*, April 1980.

# INDEX